# 天然气开发理论与实践

（第九辑）

贾爱林　何东博　郭建林　编著

石油工业出版社

## 内 容 提 要

天然气开发是油气田开发的重要组成部分。本书分为综合篇、方法篇、地质应用篇和气藏应用篇,汇总了国内一批天然气开发领域专家的最新研究成果与心得,可以为天然气开发提供理论参考和方法借鉴。

本书可供从事天然气开发的科研人员使用,也可以作为高等院校相关专业师生的参考用书。

### 图书在版编目(CIP)数据

天然气开发理论与实践. 第九辑 / 贾爱林, 何东博, 郭建林编著. — 北京: 石油工业出版社, 2021.5
ISBN 978-7-5183-4384-3

Ⅰ.①天… Ⅱ.①贾… ②何… ③郭… Ⅲ.①采气-文集 Ⅳ.①TE37-53

中国版本图书馆 CIP 数据核字(2020)第 228599 号

---

出版发行:石油工业出版社
　　　(北京安定门外安华里 2 区 1 号　100011)
　　网　　址:www.petropub.com
　　编辑部:(010)64523708
　　图书营销中心:(010)64523633
经　　销:全国新华书店
印　　刷:北京中石油彩色印刷有限责任公司

2021 年 5 月第 1 版　2021 年 5 月第 1 次印刷
787×1092 毫米　开本:1/16　印张:27
字数:670 千字

定价:230.00 元
(如出现印装质量问题,我社图书营销中心负责调换)

版权所有,翻印必究

# 前　　言

中国天然气开发虽然历史悠久,但规模化工业开发的时间并不长,特别是在相当长的时间内,仅为地区性产业。进入21世纪以来,天然气勘探开发取得了快速发展,探明储量连续快速增加,产量跨过千亿立方米大关,进入世界产气大国的行列。

随着天然气开发的不断深入,常规天然气生产格局基本形成,即以鄂尔多斯盆地、四川盆地、塔里木盆地为核心的三大基地和以南海及柴达木盆地为核心的基地。五个地区的产量占全国总产量的85%以上。

在常规天然气继续作为开发主体的同时,近几年非常规天然气的开发也取得了一定的进展。在常规天然气领域,中国虽然资源基础较为雄厚,但开发对象都比较复杂,主要的气藏类型为低渗透—致密砂岩气藏、高压—凝析气藏、碳酸盐岩气藏、疏松砂岩气藏、火山岩气藏和高含硫气藏。过去10多年,面对日益复杂的开发对象,气田开发工作者坚持"发现一类,攻关一类,形成一套配套技术"的思路,成功开发了各类气藏并形成了系列配套技术与核心专项技术。在非常规天然气开发领域,主要集中在煤层气与页岩气的攻关,虽然目前成本与效益仍困扰着开发步伐,但开发技术思路与手段已日趋清晰,并形成了一定的规模产量。

总结过去的成果,我们认为中国天然气工业在过去10多年的快速发展主要得益于以下三个方面:一是坚持资源战略,储量持续增长;二是技术不断配套完善,使复杂气藏开发成为可能;三是坚持创新驱动,挑战技术极限。

展望未来,天然气工业方兴未艾,天然气产量将继续保持增长势头,但随着开发阶段的深入,天然气工业将由快速上产转变为上产与稳产并重的开发阶段,面对这一开发局面的变化,过去以有效开发主体技术为核心的攻关,将向气田稳产技术、提高采收率技术及不同类型气田的开发方式与开发规律等方向进行转变,天然气开发技术必将进入更加丰富与成熟的阶段。

《天然气开发理论与实践》文集立足于中国天然气开发的最新成果与前瞻技术,收集了国内具有代表性的论文,按照综合篇、方法篇、地质应用篇与气藏应用篇四个类型进行汇编出版。文集的连续出版,希望在对中国天然气开发理论与技术总结的同时,也对广大的科研、生产工作者有所启迪,共同促进中国天然气事业的不断发展。

# 目 录

## 综 合 篇

天然气产业一体化发展模式研究与实践 ················ 马新华　胡　勇　何润民（3）

抑制我国天然气对外依存度过快增长的对策与建议 ········ 陆家亮　唐红君　孙玉平（14）

中国天然气发展形势研判与对策建议 ································· 赵文智（26）

## 方 法 篇

气藏开发全生命周期不同储量的意义及计算方法 ········· 位云生　贾爱林　徐艳梅 等（37）

鄂尔多斯盆地东缘致密砂岩气藏动态储量计算方法研究
　　················································ 王泽龙　唐海发　杨佳奇 等（46）

Unified Approach to Optimize Fracture Design of Horizontal Well Intercepted by Primary and
　　Secondary Fracture Networks ··············· Junlei Wang　Yunsheng Wei　Wanjing Luo（54）

考虑裂缝变导流能力的致密气井现代产量递减分析 ··· 孙贺东　欧阳伟平　张　晃 等（85）

基于蒙特卡洛随机模拟的油气储量不确定性评价 ········· 尹　涛　杨屹铭　靳锁宝 等（99）

毛细管力曲线转换方法探讨 ································ 刘兆龙　张永忠　黄伟岗 等（109）

致密砂岩气藏井网密度优化与采收率评价新方法········ 高树生　刘华勋　叶礼友 等（116）

气藏水侵与开发动态的实验综合分析方法············· 徐　轩　梅青燕　陈颖莉 等（126）

致密砂岩气藏井网加密优化 ································ 胡　勇　梅青燕　王继平 等（141）

## 地质应用篇

川中合川气田须二段致密砂岩储层"甜点"研究 ······ 张满郎　谷江锐　孔凡志 等（153）

致密砂岩气藏储渗单元研究方法与应用——以鄂尔多斯盆地二叠系下石盒子组为例
　　················································ 郭建林　贾成业　闫海军 等（168）

高磨地区灯四段岩溶古地貌分布特征及其对气藏开发的指导意义
　　················································ 闫海军　彭　先　夏钦禹 等（182）

四川盆地九龙山气田珍珠冲组砂砾岩储层评价及有利区优选
　　················································ 张满郎　孔凡志　谷江锐 等（198）

Study on the Effects of Fracture on Permeability with Pore-Fracture Network Model
　　···································· Chunyan Jiao　Yong Hu　Xuan Xu et al（213）

塔里木盆地克拉 2 气田储层综合定量评价 ………………… 徐艳梅　刘兆龙　张永忠 等（223）

鄂尔多斯盆地东部奥陶系古岩溶型碳酸盐岩致密储层特征、形成机理与天然气富集
…………………………………………………………… 王国亭　程立华　孟德伟 等（231）

塔里木盆地库车坳陷深层大气田气水分布与开发对策
…………………………………………………………… 贾爱林　唐海发　韩永新 等（246）

Control of Fault Related Folds on Fracture Development in Kuqa Depression, Tarim Basin
………………………………… Yongzhong Zhang　Jianwei Feng　Baohua Chang et al （260）

靖边气田低效储量评价与可动用性分析 ……………… 贾爱林　付宁海　程立华 等（275）

苏里格致密砂岩气田水平井开发地质目标优选 ……… 刘群明　唐海发　冀　光 等（284）

苏里格大型致密砂岩气田储层结构与水平井提高采收率对策
…………………………………………………………… 唐海发　吕志凯　刘群明 等（292）

苏里格致密砂岩气田潜力储层特征及可动用性评价 …… 王国亭　贾爱林　闫海军 等（302）

# 气藏应用篇

Evaluation of Dynamic in Ultra-deep Naturally Fractured Tight Sandstone Gas Reservoirs
……………………………………………… Ruilian Luo　Jichen Yu　Yujin Wan et al （317）

Optimization of Managed Drawdown for A Well with Stress-Sensitive Conductivity Fractures:
Workflow and Case Study ……………… Yunsheng Wei　Junlei Wang　Ailin Jia et al （327）

Production Behavior Evaluation on Multilayer Commingled Stress-Sensitive Carbonate
Gas Reservoir ……………………………… Jianlin Guo　Fankun Meng　Ailin Jia et al （349）

苏里格致密砂岩气藏大井组混合井网立体开发技术 … 张　吉　范倩倩　王　艳 等（370）

国内外大型碳酸盐岩气藏开发规律研究 ………………… 孙玉平　陆家亮　刘　海 等（379）

黄骅坳陷千米桥潜山凝析气藏开发经验与启示 ………… 初广震　韩永新　周宗良 等（386）

苏 6 区块气藏剩余储量评价及提高采收率对策 ………… 董　硕　郭建林　郭　智 等（394）

Distribution Characteristics of the Mudstone Interlayer and Their Effects on Water Invasion
in Kela 2 Gas Field ……………………… Yongzhong Zhang　Yong Sun　Zhaolong Liu et al （405）

基于数值试井法的神木气田多层压裂气井产能评价 …… 刘姣姣　刘志军　刘　倩 等（417）

# 综合篇

# 天然气产业一体化发展模式研究与实践

## 马新华[1] 胡 勇[1] 何润民[2]

(1. 中国石油西南油气田公司；2. 中国石油西南油气田公司天然气经济研究所)

**摘要**：中国天然气发展已经进入"黄金时代"，对于保障国家能源安全和实现美丽中国目标具有重要战略作用。天然气产业链涉及上游勘探开发、中游运输储存、下游销售利用，协调一致是产业发展的内生需求。基于此，构建基于产业链上中下游协同、技术创新支撑与管理创新保障的天然气产业一体化发展模式，以资源、管网、市场高度集中的川渝地区为例分析该模式的实际应用情况，表明该模式在推动产业协调发展、实现安全保供、优化区域能源结构和实现经济社会和谐稳定发展方面成效显著。从而提出天然气产业发展战略思考：发挥央企职能，在坚持市场化改革基本方向前提下，资源与市场重合区域实行一体化运营是最佳选择，有助于降低全产业链供应成本，同时对加快页岩气发展也有重要意义。

**关键词**：天然气产业；一体化；动因；发展模式；实践；战略思考

天然气作为优质清洁高效的绿色低碳能源，增加供应、扩大利用、提高天然气利用水平是加快转变经济发展方式、促进节能减排和绿色发展最现实的选择。随着天然气在一次能源中的消费比重不断增长，中国天然气发展已经进入"黄金时代"。《能源发展"十三五"规划》明确提出，2020年天然气在一次能源中的消费比重力争达到10%；《能源生产和消费革命战略（2016—2020）》要求，到2030年，一次能源消费结构中天然气占比将达到15%左右；伴随天然气消费量快速增长，天然气对外依存度不断增加，2018年对外依存度达到45.3%，跃升为全球第二大天然气进口国[1]。国家能源安全和行业发展需求，催生了中国天然气产业发展不断优化和提升的强大需要。

作为世界上最早开发利用天然气的地区，四川盆地是中国现代天然气工业的摇篮，长期引领全国天然气产业发展，远早于全国形成了成熟完整的天然气产业链，具备了天然气资源、管网及市场在物理实体上的一体性。经过60余年的发展探索、总结完善，创新形成了上游勘探开发、中游输送储存、下游销售利用之间"牵一发而动全身"的一体化运营模式，并经实践证明成效显著、意义重大。本文以此为基础对天然气产业一体化发展模式创新与实证进行研究，提出天然气产业发展的战略思考，在深化油气体制改革的当下，为中国天然气产业发展提供一条可供复制与推广的新路子和一个可供参考的成熟范式与模板。

# 1 天然气产业一体化发展内涵与动因

## 1.1 天然气产业一体化发展内涵与现状

天然气产业属于采掘行业和基础产业，由其产业链、相关企业及生产经营管理核心业务活动构成。根据"木桶理论"原理,产业链的协调发展并不是由某一子链决定,只有各环节都实现了协调发展,才能推动整条产业链的产生、完善与成熟。天然气作为一种特殊的能源,其生产和消费几乎同时进行,使得产运销具有了物理实体上的系统性[2],决定了天然气产业上、中、下游各环节必须紧密衔接、协调一致[3],也是天然气合同以"照付不议"形式存在的根本依据[4]。根据产业组织、产业管理和产业布局相关理论,天然气产业作为一个技术密集型的高风险、高投入产业,规模经济和范围经济效应十分明显,产业关联度也非常高,关联效应突出[5]。因此,天然气产业一体化发展是适应天然气产业发展特征与发展环境的独特的一种产业经济发展方式,遵循产业经济相关理论,以提高天然气在一次能源中的比重为主线,是在充分适应产业特征和发展环境基础上,通过上游勘探与开发、中游输送与储存、下游销售与利用的互动和优化,推动天然气全产业链一体化整体发展,最终实现产业价值最大化、资源配置最优化、产业协调可持续发展和社会福利最大化目标。

基于产业经济视角的一体化是指企业根据资源流动的方向,利用自身在技术、产品、劳动力、市场等方面优势,不断向深度广度发展,形成统一的经济组织的战略。在全球能源需求从高碳向低碳发展成为全社会共识的前提下,国际大石油公司积极构造上下游一体化的天然气产业链,在全球范围内加快天然气业务发展:埃克森美孚积极推进天然气业务的纵向一体化,在扩大资源基础、特别是页岩气等非常规油气资源开发的同时,积极推进天然气下游产业链的发展;能源巨头英国石油公司(BP)、壳牌集团(Shell)等全球性运作的石油公司,着眼于全球范围内的效益实现,分区域独立开展一体化运营,Shell 在生产全业务链分别独立使用优化软件(PIMS, Process Information Management System, DPO, Distribution Planning Optimization 等),其中炼化业务领域的 PIMS 模型成熟度在业内处于领先地位;BP 在物流优化中利用 IMOS (Inventory Management and Operations Schedualing)模型与 DPO 模型互动应用提升了一体化运营整体水平。国内而言,中国石油化工集团公司以炼化生产 PIMS 优化软件为主,兼顾了海外原油选择数据库和下游 DPO 优化软件的开发和应用以实现上下游的衔接;中国石油天然气股份有限公司使用的炼化物料优化与排产系统(Advanced Planning and Scheduling, APS)和基于 CDM 和 DPO 的成品油一次物流系统,为实现上中下游业务协调提供了重要支撑[6]。

## 1.2 天然气产业一体化发展的动因

### 1.2.1 天然气产业的自然属性与生产经营的内在需求

天然气产业属于资源采掘业,生产经营的核心是根据市场需求不断探索地下天然气资源,把投入资本转化为储量,采用先进的开采工艺技术,将气藏中的可采储量开采出来,成为可利用的商品气,并通过管道输送到用户。天然气是优质高效的清洁能源,其发展有着不同于其他化石能源的特殊性和区别于一般能源产业的特殊性,主要体现在勘探生产环节的高风险性、输配环节的自然垄断性、利用环节的不均衡性、物理实体上的系统性以及技术与装备的专有性等方面。作为一种带有自然垄断属性的不可再生的稀缺资源,天然气还具有不易储存的内生属性,这一属性从根本上制约了天然气资源的运输方式,必须严重依赖于管道运输[7]。然而,天然气管道铺设规模大、风险高、用途单一、投资数额巨大、建设周期长、技术要求高,并且严重受制于天然气资源的生成客观规律和分布位置的自然性、资源分布的广度、气藏分布的隐蔽性等客观条件制约。因此,天然气产业的显著特点是上游的勘探开发、中游管输和下游的天然气销售利用之间形成紧密联结的产业链体系,产业链一体化经营才能从根本上保证天然气产业的健康可持续与高质量发展。

### 1.2.2 天然气资源安全保供的必然要求

2018年9月5日,国务院印发《关于促进天然气协调稳定发展的若干意见》(以下简称《意见》)中,"产供储销,协调发展;有序施策,保障民生"两条基本原则核心均为保障天然气安全供应。中国天然气供应的三大来源为:国内自产、管道天然气进口和LNG进口。虽然中国"西气东输、北气南下、海气登陆、就近供应"的天然气供应四大格局已经初步形成,但供需缺口依然巨大,供应安全不容乐观。由于资源短缺,供气安全问题历来是天然气发展史上的重大和焦点问题,为政府、社会和企业各界高度关注。

天然气资源供应安全保障是一项系统工程,需要相应的组织体系、制度创新、管理创新和市场开拓,其核心内容是要素资源在资源、输配、市场之间的循环流转及其应用。天然气产、运、销、用联动性本质上要求各个环节必须紧密配合协调一致,任何一个环节出问题都可能造成经济和社会损失巨大,任何一个环节又都不可能单独承担起天然气供应安全的责任。因此,天然气产输销一体化管理是保障供气安全和应对突发事件的最佳模式。一方面,凭借一体化管理优势,在天然气资源的组织、管网的调度以及对市场的及时反馈方面具有强大高效的协调能力,能够形成强有力的天然气供应安全保障机制;另一方面,遇到停气检维修、事故中断供气等特殊情况时,高效的上中下游协调能够保证应急条件下快速处理问题,历次冬季保供和应对气荒的实践也反复证明,一体化管理在保障供气安全方面发挥了难以替代的重要作用。

### 1.2.3 实现信息对称降低交易费用的有效途径

天然气产业上游主要存在行政垄断,其他企业进入将受到技术壁垒、法律法规和矿权冲突等的限制,但现有资料无法论证行政垄断将使得在市场内的天然气企业失去提升效益、降低成本的外在压力和内在动力;天然气产业中游存在着庞大的基础设施如天然气管网、储气库、净化厂等因其巨大的建设成本而形成的自然垄断;天然气产业下游存在着城市燃气等利用上的高度自然垄断。历史和实践表明,一体化绝不是整体的无序扩张,而是主营业务的做大做强,产业链存在的三种垄断虽然因信息不对称会造成一定程度的非经济性,但是总体而言是利大于弊的。天然气产业的一体化是囊括天然气上中下游的庞大而又复杂的工程,一方面这是以一种昂贵的具体的行动来进行的信号传递,同时也可以利用统一的体制机制更有效地来进行信息甄别,减少市场出现逆向选择的问题,从而提升市场效率,降低不必要的市场摩擦。最为重要的是,降低供气成本是实现天然气产业快速发展的重大命题,天然气产业上中下游一体化管理能够缩短各环节之间的交易进程,显著降低中间交易与结算管理成本;且高效协调上中下游生产运行,产输销各环节无缝衔接,无论是"以销定产"还是"以产定销",都能快速反应,降低生产运行成本,实现整体效益最大化。

## 2 天然气产业一体化发展模式创新

天然气产业一体化发展是一个复杂系统,受政策变化、产业的发展与转移、技术与社会进步、区域发展环境、天然气资源条件、市场竞争等多重因素的综合影响,天然气产业一体化发展必然存在一个随着外部环境变化不断发展演变的过程,这一过程的演进与天然气工业发展的阶段性、天然气资源供给、国民经济的周期变化与天然气用户结构息息相关。以产业组织、产业关联、产业布局、交易费用等一系列产业经济发展相关理论为基础和指导，天然气产业一体化发展模式创新，必须以尊重产业发展的系统性和协调性、适应产业特殊的地理根植性、坚持企地协调与政府引导为前提，促进天然气产业链勘探开发、输送储存、销售利用各环节及其间技术、管理等多要素协同作用提升产业整体功能价值。基于此，立足产业经济发展相关理论以及天然气产业一体化发展动因，集成创新构建天然气产业一体化发展模式架构[8]，见图1。

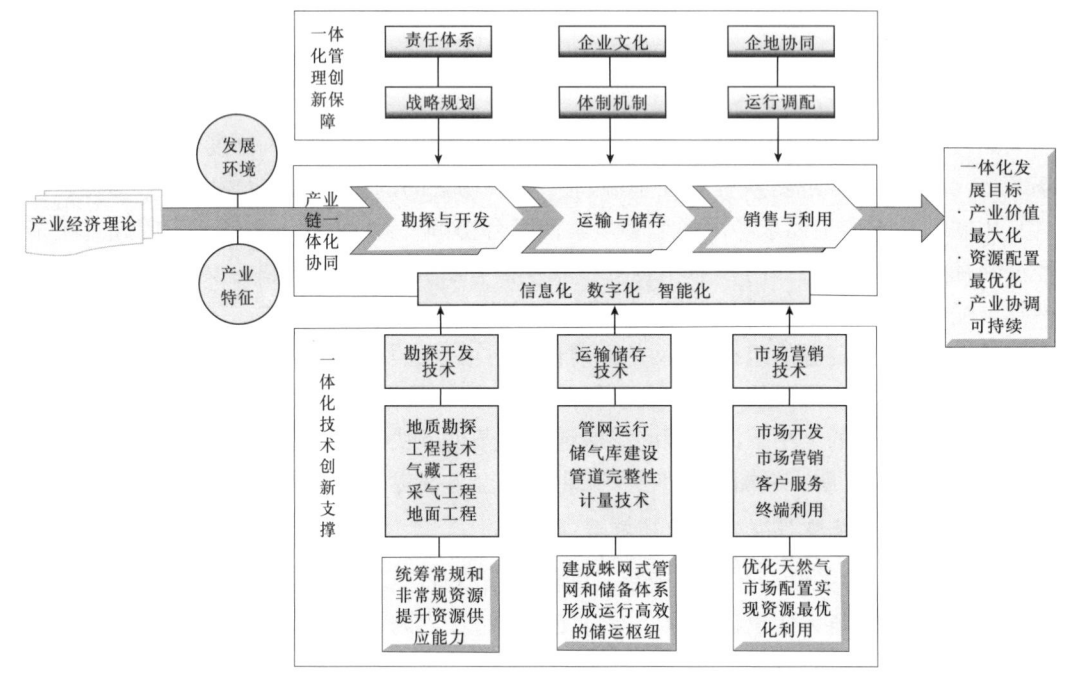

图 1 天然气产业一体化发展模式结构

天然气产业链一体化协同、一体化技术创新支撑与一体化管理保障是天然气产业一体化发展模式的三大重要组成部分，其中，产业链一体化协同是模式的核心与主线，是模式结构中最主要的组成部分；为保证核心主线有条不紊的推进和实现，必须要依靠技术创新和管理创新两大辅助功能的发挥。三大部分紧密联系、牢不可分，完整地反映了天然气产业一体化发展的深刻内涵与广泛外延，共同促进天然气产业一体化发展模式整体价值和目标实现。

## 2.1 天然气产业链一体化协同

天然气产业链一体化协同是将天然气产业链视为一个复杂的自适应系统，以产业链上勘探开发、输送储存、销售利用等产输销主要环节为系统的组成要素，进行多环节间多要素的组合与优化，协同实现产业链整体价值最大化和功能最优化。因此，充分考虑上游气源供气能力、中游管道输气以及储气库注采气能力、下游用户用气量波动等因素的限制和影响，统筹天然气生产、资源配置、管网布局、市场营销、竞争应对，合理安排产能、管网投资以及分配天然气流量，将天然气合理的输配给用户，确保产运销各环节无缝衔接、协调发展以实现最大的经济和社会效益，是产业链一体化协同需要解决的根本问题。

在天然气产业一体化发展模式框架下,为了更好地描述产业链一体化协同这条主线功能价值的发挥,采用数学工具进行产业链一体化协同实现过程建模,寻求解决问题的理性方案。基于供应链管理和系统优化理论、产业经济与技术经济相关理论、需求预测与市场营销相关理论,将天然气产业一体化协同系统抽象为一个由上游天然气生产气田、中游外部气源下载节点、调峰储备库、输气管道交叉,以及下游用气市场等多节点集 $V$ 和其构成的边集 $E$ 组成的有向图 $G=(V,E)$;以最大化利润和社会福利综合绩效为目标,分别考虑销售收入、生产成本、运输成本、储气库注气成本、采气成本、固定投资成本等,建立天然气产业链一体化协同的数学模型[9]:

$$\max z \sum_{t \in T} \sum_{j \in D} a_{jt} d_{jt} p_{jt} - \sum_{i \in S} c_{it}^s \sum_{t \in T} \sum_{j \in J} q_{ijt} - \sum_{i \in S} \sum_{J \cup C \cup J / |j| j} = \sum_{J \cup C \cup D} c_{ij} q_{ijt} - \sum_{t \in T} \sum_{i \in C} f_i^M m_{it} - \sum_{t \in T} \sum_{i \in C} f_i^R r_{it} - \sum_{i} \sum_{j} \sum_{\phi \in \phi} \frac{F_{ij\phi} |Y|}{|T|} \sum_{1 \in T} s_{ij\phi}^t \tag{1}$$

其中,$S$ 表示天然气上游供应点集合,$D$ 表示下游天然气需求点集合,$J$ 表示天然气中游管道连接或交叉点集合,$C$ 表示储气库集合,$T$ 表示时间段集合,$Y$ 表示每年计划周期集合,$P$ 表示投资管道可选集合,$\phi$ 表示投资管道类型集合。该公式需要满足以下几个约束。

注气约束:表示储气库注气量与管径流量之间的守恒关系,即储气库注入天然气的量等于与之连接的管道流向其的总气量。

$$m_{jt} = \sum_{i \in J} q_{ijt}, \forall j \in C, \forall t \in T \tag{2}$$

采气约束:表示储气库采气量与管径流量之间的守恒关系,即储气库采出天然气的量等于其流向与之连接的管道的总气量。

$$r_{jt} = \sum_{i \in J} q_{jit}, \forall j \in C, \forall t \in T \tag{3}$$

生产能力约束:表示气源供应量不能超过其能力约束,即气源流向与之连接的管道的总气量不能超出产能。

$$\sum_{j \in t} q_{ijt} \leq Q_{it}^s, \forall i \in S, \forall t \in T \tag{4}$$

需求满足情况:表示各需求点需求被满足的情况,即与需求点连接的管道流向该需求点的天然气总量等于当期需求与被满足率的乘积。

$$\sum_{i \in t} q_{ijt} = \alpha_{jt} d_{jt}, \forall j \in D, \forall t \in T \tag{5}$$

流量守恒约束:表示在各个管网交叉节点,天然气进出流量必须守恒,即与节点连接的管道流向该点的天然气总量等于该点流向与之连接管道天然气总量。

$$\sum_{i \in S \cup J \cup C} q_{ijt} = \sum_{i \in S \cup J \cup C} q_{jit}, \forall j \in J, \forall t \in T \tag{6}$$

累计净注气量约束:表示任何时间点储气库累计注气量与采气量之差加上初始气量要小于储备库储备能力。

$$Q_i^0 + \sum_{t=0}^{k}(m_{it} - r_{it}) \leq Q_i^C, \forall i \in C, \forall k \in T \tag{7}$$

注采气量约束:表示任何时间点储气库累计注气量与初始气量之和大于累计采气量。

$$Q_i^0 + \sum_{t=0}^{k}(m_{it} - r_{it}) \geq 0, \forall i \in C, \forall k \in T \tag{8}$$

期末安全库存约束:计划期末储气库累计注气量与初始气量之和大于安全库存水平。

$$Q_i^0 + \sum_{t=0}^{|T|}(m_{it} - r_{it}) \geq Q_i^{|T|}, \forall i \in C \tag{9}$$

满足率下限约束:表示每个时期各需求点的需求满足率必须达到最低保供要求。

$$\alpha_{jt} \geq 1_j, \forall j \in D, \forall t \in T \tag{10}$$

满足率范围:

$$\alpha_{jt} \in [0,1], \forall j \in D, \forall t \in T \tag{11}$$

采气能力约束及范围:

$$0 \leq m_{it} \leq M_i, \forall i \in C, \forall t \in T \tag{12}$$

注气能力约束及范围:

$$0 \leq r_{it} \leq R_i, \forall i \in C, \forall t \in T \tag{13}$$

管道输送能力约束:节点$(i,j)$之间的天然气输送量不能超过现有管道与被选中建设管道在当期的状态条件下的总输送能力。

$$q_{ijt} \leq \frac{Q_{ij}}{Y|Y|} + s_{ij\varphi}^1 \cdot \frac{Q_{ij\varphi}^p}{|Y|} + s_{ji\varphi}^t \cdot \frac{Q_{ji\varphi}^p}{|Y|}, \forall i,j \in J$$
$$i \neq j, \forall \varphi \in \phi, t \in T \tag{14}$$

管径方案状态转化:投资决策的决策量与状态量之间的状态转移,即当期某管道投资状态是上期状态与决策量 $x_{ij\varphi}$ 共同决定的。

$$s_{ij\varphi}^t = s_{ij\varphi}^{t-1} + x_{ij\varphi t}, \forall i,j \in J, i \neq j, \forall \varphi \in \phi, t \in T \tag{15}$$

管径方案约束:整个计划期投资两节点间投资方案只能选择一种。当计划期内同一对节点间允许重复投资多条管道时,该约束可以松弛掉。

$$\sum_{t \in T} \sum_{\varphi \in \Phi} \chi_{ij\varphi t} \leq 1, \forall i,j \in J, i \neq j \tag{16}$$

管径方案选择定义式:

$$x_{ij\varphi t}, s_{ij\varphi}^t \in \{0,1\}, \forall i,j \in J, i \neq j, \forall \varphi \in \phi, t \in T \tag{17}$$

利用 Benders 分解算法,基于天然气管网投资、生产供需调度及管网流量分配动态数据,对模型进行求解,可快速得到上游气源和下游市场动态变化条件下满足产业链一体化协同发展需求的生产方案、储气库调峰方案、需求满足方案,从而实现对模式主线的优化。

天然气事关民生福祉和社会稳定,使得产业发展不仅仅受到经济指标的影响,更成为关于全民社会福利的重要资源分配利用的重大决策事项。上述数学模型最大化目标函数中全面考虑了经济效益和社会福利综合绩效体系的天然气上中下游共同利益,能够为做好供需衔接、明确保供次序、及时协调解决影响平稳供气的矛盾和问题从而优化实现天然气全产业链一体化发展目标提供了可操作的实现程序。

## 2.2 天然气产业一体化技术与管理创新

天然气产业一体化技术创新是在创新驱动发展的战略指引下,强调科技创新驱动的巨大价值,通过天然气勘探开发技术系列、天然气输送储存技术系列、天然气市场销售技术系列的创新和天然气信息化技术系列的发展与技术进步,实现科技创新驱动产业发展的第一生产力作用,为天然气产业链一体化协同的实现提供重要动力。基于产业链的视角,天然气勘探开发技术系列需要综合考虑地质勘探、工程技术、气藏工程、采气工程、地面工程(包括集输与净化)等技术子系列,天然气输送储存技术系列则涵盖管网运行、储气库建设、管道完整性、计量技术等,天然气市场销售技术系列则包括了市场开发、市场应用、客户服务、终端利用等方面,天然气信息化技术系列是充分利用大数据、网络平台、信息技术手段等统筹全产业链的一体化技术服务[10]。

天然气产业一体化管理创新是在深化改革创新促高质量发展的宏观背景下,通过战略规划、体制机制、生产运行、责任体系、企业文化、企地协同等,为天然气产业链一体化协同提供重要的组织和制度保障。战略规划一体化布局从战略、规划、计划等层面将天然气资源、管网与市场紧密联系在一起,形成统筹协调发展的总框架;体制机制一体化构建通过上市、未上市业务一体化管理、业务链管理一体化、专业管理一体化和区域管理一体化,形成产业链融合深度发展格局;生产运行一体化调配按照"按效排产、以效定销、产销联动、安全保供"进行资源配置、管网管控、市场保供;责任体系一体化主要落实可持续的天然气勘探开发责任、以安全环保为主线的生产运营责任、以人为本的员工发展责任、服务国计民生的社会责任的全面履行;企业文化一体化建立要充分构建天然气精神文化、天然气制度文化、天然气行为文化、天然气物质文化;企地协同一体化推动是充分利用一体化优势,构建和谐企地关系,助推区域经济健康稳定发展。

# 3 天然气产业一体化发展实践——以川渝地区为例

天然气产业一体化发展模式是在川渝地区60余年的艰苦探索和不断完善中丰富发展起来的,既是产业经济发展相关理论在天然气产业发展中的理论应用,又经历了川渝地区60余年天然气产业发展实践检验,因此,它既是来源于实践的创新理论,又是指导实践的创新发展产物。过去60余载,川渝地区推进天然气产业一体化协同发展,就取得了一系列重要成效[11],取得成效的主要实践包括:

## 3.1 充分利用管网和资源优势,推动上中下游协调发展

### 3.1.1 充分利用管网优势,完善产销价值提升通道

将天然气资源转化为经济效益,管网在天然气产销规划与整体布局优化中起到关键性作

用。川渝地区立足天然气资源、供气管网优势,建设大型环形输气管网和与之配套的支线管道和气田集配气管网,已建成的"三横、三纵、三环、一库"的骨干管网,综合输配能力达到每年 $300×10^8 m^3$ 以上,建成了全国甚至世界上独特的蛛网式环形管网系统,有效协调上下游的原省功效,有利于造就完整发达的区域天然气一体化产业链。蛛网式管网系统的核心价值在于通过环状化、网状化及压力分布式布局,提高管网覆盖密度和市场适应能力,进而促进市场高效发展;对产业发展的作用是确保安全稳定供气,高效培育市场,形成完整产业链并协调发展、联通区域内外市场,成就产业一体化特色。

### 3.1.2 充分发挥资源优势,协同政府实现市场发展

利用产运销一体化中天然气勘探开发和市场销售紧密结合的优势,与地方政府共同规划、开发、布局天然气市场,把天然气市场发展和促进地方社会经济发展结合起来,引导地方政府出台有利于天然气业务发展的政策,进行科学的资源配置和市场供需平衡。同时,以转变发展方式、调整产业结构、优化资源配置、提升经济增长质量为目的,引导政府规划、优化产业布局,从而提升天然气在区域能源结构中的地位和作用,发展了管网周边潜在的产业集群,积极拓展了新的天然气利用方式,如 LNG 的利用项目等。

## 3.2 供储销联动实现资源调度,有效应对市场变动

20 世纪末以来,天然气市场供需先后经历了供大于求、供需不断紧张、供需宽松和供需紧张的阶段性波浪式发展行情。在天然气产业一体化发展模式下,针对不同的天然气市场供需形势,采用适应的资源调度方式。

### 3.2.1 当天然气供大于求时,采用"以销定产"

销量是运作机制的主要输入条件,不受输配业务环节制约,而其他决策均受制于销量。在这种方式下,天然气生产组织采取"以销定产、以销促产"和"按效排产"的策略,严格按照气田开发效益排序,优先安排高效气田生产,降低天然气单位操作成本和安全成本。天然气销售则采取"以效定销"和灵活的天然气价格,在用气淡季实行优惠价格或对大工业用户实行按用气量递减的阶梯价格,努力提高天然气销售量。

### 3.2.2 当天然气供不应求时,采用"以产定销"

产量是输配业务各环节生产决策的主要依据,是运作机制的主要输入条件,不受输配业务环节制约,而其他决策均受制于产量。在该方式下,天然气生产组织和销售采取"以产定销、有保有压"的策略,严格执行国家《天然气利用政策》的用气保障顺序,确保市场供需平衡和稳步发展;通过输配气管网与天然气生产和销售无缝对接对市场需求的有力保障和有效管控,确保供应安全。近年来,即使在用气最紧张的时期和季节,川渝地区再未出现城镇居民生活用气"有气无力"现象和 CNG 出租车加气排大队的情形。

## 3.3 强化安全保供,提升应急调峰能力

### 3.3.1 加大上游勘探开发,保障天然气资源可持续供应

实施上游勘探开发一体化,把彼此分散独立的勘探与开发紧密结合为一个有机的整体,相互协调,相互配合,高效完成油气资源向油气产量的转化。在勘探取得突破,对含油气区有一个整体认识的基础上,将高产富集区块优先投入开发,同时,在重点区块突破的同时,在开发中

继续深化新层系和新区块的勘探工作,为勘探的扩边连片探明提供指向,表现出"勘探中有开发,开发中有勘探",全面挖掘生产潜力,实现快速上产。

### 3.3.2 多元化储气系统建设,提高战略储备与调峰能力

我国的天然气资源分布不均匀,供给侧与需求侧局部失衡,近年来在部分地区出现了"气荒"现象,其中民用天然气消费极不均衡,因此,多元化储气系统建设对于确保安全平稳供气意义重大。川渝地区充分利用地下储气库储备与调峰、气田储备与调峰、管道储备与调峰、建设事故备用气源储气站储备与调峰等多种储气方式并举,通过储气库采气,解决了季节调峰,保障民用和工业用户平稳用气。例如,2016年末,中亚天然气供应国因极端天气和装置故障等原因减少供应量,相国寺储气库立即启动应急采气方案,确保了北方地区的天然气供应需求;2017年11月,罗家寨高含硫净化厂故障全停、页岩气停气碰口期间,储气库立即提升采气量,保障了市场供应。

## 3.4 优化区域能源结构,促经济社会和谐稳定发展

### 3.4.1 提升天然气消费比重,推进供给侧结构性改革

川渝地区推进能源供给侧结构性改革,能源转型速度加快,天然气占清洁能源消费比重持续提升(图2)。川渝地区天然气的消费利用,不但提高了工业产品质量和人民生活质量,更重要的是,减轻了企业和社会能源消费对煤炭和成品油的依赖,优化了能源消费结构。多年来,川渝地区能源消费结构中天然气所占份额一直是全国最高,而煤炭则比全国平均水平约低5个百分点以上。2017年,天然气在全国一次能源消费结构中的比例为7.2%,而川渝地区的天然气在一次能源消费结构中的比例高达15%,超过了挪威、芬兰、希腊、瑞士、新加坡等经济发达国家和亚太地区平均水平(10.6%),接近日本(16.6%)和韩国(14.9%);川渝地区城镇气化率达到87%,名列全国前茅。

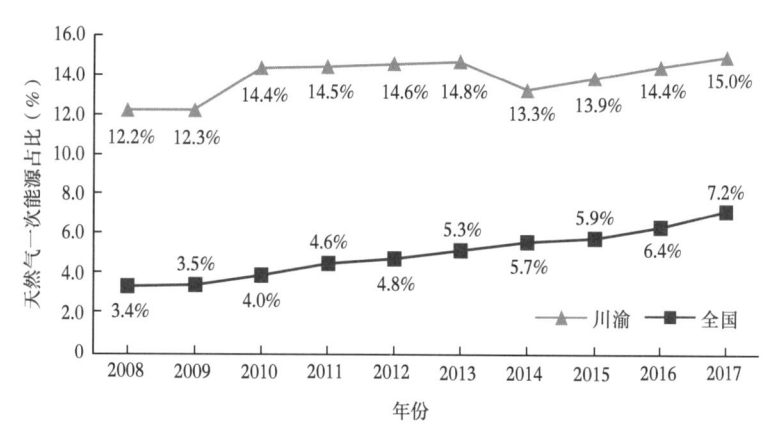

图2 川渝地区天然气一次能源消费比重走势图

### 3.4.2 助力生态文明建设,保障区域经济社会发展

与煤炭相比,天然气燃烧利用的最大优势之一是在于其环境效益。川渝地区天然气产业的发展,对节能减排、生态环境改善和生态文明建设起到重要作用,成为绿色经济增长新引擎。

2008—2017年生产的天然气可替代煤 $2.06×10^8$ t 标煤，综合减排 $CO_2$ $3.67×10^8$ t，给四川盆地乃至全国带来了更多的"天然气蓝"。同时，天然气产业一体化发展还推动形成和发展了以天然气为原料或燃料的产业集群，对区域社会经济发展做出了重要贡献。2008—2017年，四川省和重庆市的 GDP 分别保持了 7.7%~15.1% 和 9.3%~17.1% 较高年度增速，天然气的贡献功不可没。实证分析认为，2008—2017年期间中国石油在四川盆地共生产天然气 $1546×10^8$ $m^3$，可带动地区 GDP 增长 1.22 万亿元[12]。通过天然气消费利用带动相关产业的发展，每年新增了数万个就业机会，促进了社会的和谐稳定[13]。2008—2017年期间中国石油为四川盆地年均创造了 55.5 万个就业岗位，对四川盆地就业贡献率为 1.49%，即四川盆地每 1 万名员工中，有天然气产业的职工 149 个（表1）。

表1 2008—2017年天然气生产对地区社会经济发展贡献

| 年份 | 2008 | 2009 | 2010 | 2011 | 2012 | 2013 | 2014 | 2015 | 2016 | 2017 |
|---|---|---|---|---|---|---|---|---|---|---|
| 天然气产量($10^8$ $m^3$) | 148.33 | 150.32 | 153.62 | 142.06 | 131.52 | 127.79 | 137.3 | 154.8 | 190.1 | 210.2 |
| 对地区 GDP 的贡献值(亿元) | 994 | 1007 | 1030 | 952 | 1132 | 1100 | 1182 | 1333 | 1637 | 1810 |
| 对地区 GDP 的贡献率(%) | 1.90 | 2.20 | 3.33 | 4.63 | 5.61 | 5.63 | 5.69 | 5.27 | 5.38 | 6.03 |
| 对地区财税的贡献率(%) | 4.56 | 7.06 | 10.38 | 10.35 | 10.45 | 10.69 | 11.27 | 11.79 | 11.65 | 11.97 |
| 对地区就业的贡献率(%) | 1.40 | 1.44 | 1.49 | 1.30 | 1.54 | 1.55 | 1.54 | 1.54 | 1.54 | 1.55 |

## 4 天然气产业发展的战略思考

天然气产业高质量发展是适应能源革命要求、构建安全高效现代能源体系、实现美丽中国目标的内生需要[14]，也是中国深入推进可持续发展战略的重要选择。天然气产业一体化发展是一项考虑产业链上中下游多环节多要素耦合问题的复杂系统工程，自勘探开发、储集管输到销售利用各个环节的一体化发展都具有积极重要的意义和作用[15]。作为一项具有低碳性、民生性和价值性的战略工程，需要用辩证思维综合看待，既要放眼全球市场化改革浪潮、贯彻和践行中国油气改革精神，又要充分考虑区域实际情况、多路并举[16]、差异化发展，进行综合的战略性考量。

（1）中央企业是保障国家能源安全的可靠力量。立足中国当前天然气工业结构，无论自产还是进口，中国石油、中国石化、中国海油三家中央企业，在保障国家能源安全中扮演非常重要的作用；同时，要实现天然气产业长期稳健发展，必须处理好企地关系，兼顾企地利益，为天然气产业稳健发展创造良好的外部环境。

（2）坚持天然气产业市场化改革基本方向。天然气产业市场化改革是国家的大政方针，也是行业发展的必然趋势，必须严格执行。但在产输销分离格局下，如何实现全产业链的协调发展，则是需要考虑的新问题。2017年冬季保供实践证明，在产输销分离格局下，保持全产业链的协调发展仍然非常重要，对于产地与消费地完全重合的局部地区，保持一体化运营与管理格局仍是最合理、最经济有效的运作模式。

（3）资源与市场重合区域实行一体化运营是最佳选择。在全国总体实施产输销分离改革的背景下，针对部分资源与市场重合区域，由于存在与地方协调发展的特殊性，形成了唇齿相依的紧密关系,采用一体化模式更有利于产业链协调与地方经济发展。

(4)天然气产业一体化对加快页岩气发展具有重要意义。页岩气发展在中国天然气行业具有重大的战略意义,通过页岩气发展可以实现资源战略接替、保障国家能源安全。与常规气不同,页岩气开发具有需要专用的技术与设备、开采难度大、成本高和递减快等特殊性。从已有经验看,采取纵向一体化战略发展是页岩气非常现实的选择,通过实施一体化战略,与地方政府共同发展页岩气产业,实现多方共赢,更有利于页岩气产业的未来发展。

(5)一体化有助于降低全产业链供应成本。从最近几年的国家和地方相关政策看,降低用气企业成本是国家政策导向,采用天然气产业一体化运营模式,采取直供等方式,可以减少中间环节,不仅能够降低天然气上游开发与供应成本,对于天然气"黄金终端"的形成与发展也具有强大的推动作用,可以降低与用户的沟通与协调成本,从总体上降低全产业链交易成本。

60多年的实践证明,在四川盆地等生产区域与消费区域高度重叠的特殊区域,采取天然气产业一体化运营模式,对于和谐企地关系、推动上中下游产业链的协调发展、有效响应市场需求、降低交易成本、促进供需平衡、推进区域国民经济社会发展都发挥了重要作用。因此,在坚持市场化改革的原则下,在外部环境没有根本改变的情况下,在川渝地区等产输销高度重合的特殊区域继续实施天然气产业一体化运营模式,是合理的也是必要的。

## 参 考 文 献

[1] 刘朝全,姜学峰. 2018年国内外油气行业发展报告[M]. 北京:石油工业出版社,2019.
[2] 白兰君. 天然气经济学[M]. 北京:石油工业出版社,2001.
[3] 李士伦,汤奠,王希奠. 建立上、中、下游一体化天然气工业体系问题的思考[J]. 石油工业技术监督,2005,21(5):13-16.
[4] 李海涛. 浅谈天然气照付不议合同的有关条款[J]. 国际石油经济,2000,8(5):51-52.
[5] 郑红玲,刘肇民,刘柳. 产业关联乘数效应、反馈效应和溢出效应研究[J]. 价格理论与实践,2018(4):122-125.
[6] 西南油气田天然气上中下游一体化运营体制机制研究[R]. 北京:国家发改委综合运输研究所,2019.
[7] 李志刚,张吉军,苟建林. 天然气开发企业产运销一体化协调发展评价指标体系设计[J]. 生态经济,2012,248(1):113-116.
[8] 马新华. 大然气产业一体化发展模式[M]. 北京:石油工业出版社,2019.
[9] 川渝地区天然气产运储销协同优化模型与应用研究[R]. 成都:中国石油西南油气田公司天然气经济研究所,2018.
[10] 胡勇,姜子昂,何春蕾,等. 天然气产业科技创新体系研究与实践[M]. 北京:科学出版社,2016.
[11] 马新华,胡勇,王富平. 四川盆地天然气产业一体化发展创新与成效[J]. 天然气工业,2019(7):1-6.
[12] 姜子昂,周建,付斌,等. 天然气产业生态文明建设重大问题研究——以西南战略大气区建设为例[M]. 北京:科学出版社,2017.
[13] 何润民,熊伟,杨雅雯,等. 中国天然气市场发展分析与研究[J]. 天然气技术与经济,2018,12(6):21-24.
[14] 潘继平,杨丽丽,王陆新,等. 新形势下中国天然气资源发展战略思考[J]. 国际石油经济,2017,(25)6:12-18.
[15] 邹才能,赵群,陈建军,等. 中国天然气发展态势及战略预判[J]. 天然气工业,2018,(38)4:1-11.
[16] 郭焦峰,薛子文. 多路并进推动石油天然气体制革命[N]. 中国石油报,2015-04-21.

# 抑制我国天然气对外依存度过快增长的对策与建议

陆家亮  唐红君  孙玉平

（中国石油勘探开发研究院）

**摘要**：天然气对外依存度是一个国家的净进口气量占该国国内天然气消费总量的百分数，其高低不仅反映了该国与国际天然气市场之间的融合度，同时也反映了对国外天然气资源的依赖程度和安全供应的保障程度。进入21世纪以来，天然气在中国能源消费结构中的战略地位日渐突出，国产天然气供不应求，国内天然气产量与市场需求量的增速差不断扩大，天然气对外依存度快速攀升，2018年已达到44%。鉴于2018年中国的石油对外依存度已超过70%，天然气对外依存度的不断攀升无疑会给中国的能源安全带来更为严峻的挑战。为了保障中国的能源安全，系统分析了中国天然气对外依存度的变化趋势及其对中国社会经济发展有可能带来的风险，从一次能源消费总量、天然气消费大国生产消费状况以及天然气进口大国的净进口量与全球天然气贸易总量之比等3个方面探讨了中国天然气合理的对外依存度，进而提出了抑制中国天然气对外依存度过快增长的对策与建议：(1)多气并举，提高国内天然气供应能力，夯实国内天然气"压舱石"地位；(2)发展可再生能源，加快形成有效替代，减轻天然气供应压力；(3)坚持底线思维，提高天然气利用效率，抑制不合理需求。

**关键词**：中国；天然气；供需变化；对外依存度；抑制增长；对策与建议；多气并举；可再生能源；利用效率

进入21世纪以来，天然气作为低碳、高效、经济、安全的清洁能源，已成为促进经济增长、社会和环境可持续发展的重要主体能源之一，战略地位日益突出。以2004年底"西气东输"工程正式商业运行为标志，我国天然气实现了跨越式发展，天然气产量由当年的 $415\times10^8\,m^3$ 增长到2018年的 $1610\times10^8\,m^3$，年均增速为10.2%；天然气消费量从2004年的 $397\times10^8\,m^3$ 增长到2018年的 $2786\times10^8\,m^3$，年均增速达15.0%。由于国内天然气产量的增速低于消费量的增速且增速差不断扩大，为满足市场需求，中国从2006年开始进口LNG、2010年开始进口中亚管道气，进口渠道、规模不断增加，天然气对外依存度快速攀升，2018年已达44%[1-3]。天然气对外依存度过高，将给中国带来能源供应安全和社会经济利益的双重挑战，中国能源供应的长期安全性将面临严峻考验。因此，探讨中国合理的天然气对外依存度，探究抑制天然气对外依存度过快增长的措施与对策，对于确保天然气供应的长期性、稳定性和可靠性，以及保障国家能源安全都具有十分重要的意义。

## 1 中国天然气对外依存度的变化趋势

### 1.1 天然气对外依存度的演变历程

天然气对外依存度是指一个国家净进口气量占其国内天然气消费总量的比例。自2006年6月中国第一个LNG接收终端——广东深圳大鹏LNG接收站建成投产，当年从澳大利亚进口 $65\times10^4\,t$ 的LNG，由此，我国成为天然气净进口国，并且进口规模逐年增长，2009年进口LNG达到 $550\times10^4\,t$。特别是2010年开始引进中亚管道气以来，进口LNG和管道气合力推动中国天然气进口量的快速增长,2018年中国超越日本成为全球第一大天然气进口国,全年共计进口天然气 $1254×10^8m^3$,其中LNG为 $734×10^8m^3$、管道气为 $520×10^8m^{3[2]}$。从2008—2018年,期间除了2015年的天然气进口量增速仅为个位数之外,其他年份都保持了两位数以上的增长,而其中2010年的进口气量同比增速则高达127%(图1)。

图1 2006—2018年中国天然气进口构成及年均增幅统计图

伴随着进口气规模的快速增长,中国天然气对外依存度也不断攀升,特别是从2009年到2013年,天然气对外依存度增长最快。天然气对外依存度在2013年突破30%之后,2014、2015年基本稳定,从2016年起又快速上升,2018年已达44%(图2)。

图2 2000—2018年中国天然气产量、消费量及对外依存度统计图

## 1.2 天然气对外依存度较快增长的原因分析

对清洁能源的渴求推动了中国天然气消费量的快速增长。2008—2018年,中国天然气消费量从 $807×10^8m^3$ 增长到 $2786×10^8m^3$,年均增速13.2%。其中2008、2010、2011、2012、2013年的增长率分别达到16.1%、20.6%、22.1%、12.5%、14.2%[2,3],这几年天然气消费量的快速增

— 15 —

长,为中国第一个天然气黄金发展期奠定了基础。但在 2015、2016 年,受全球经济环境低迷、国际油价大跌的影响,国内天然气消费量增速已降至 10% 以内,2015 年的增长率仅为 4%,2016 年回升到 6.4%。2017 年以来中国经济形势稳中向好,GDP 增幅企稳反弹,在国家一系列政策的推动下,同时叠加煤炭、原油等替代能源价格从底部上行[4-7],2016 年国内天然气需求量增速触底反弹,2017 年天然气需求量迎来爆发,重返快速上升的通道,进入第二个天然气黄金发展期。2017、2018 年天然气消费量年均增速超过 17%,年绝对增长量超过 $300 \times 10^8 m^3$,2018 年天然气在中国一次能源消费结构中的占比达到 7.8%。

受天然气资源禀赋的影响,国内天然气产量的增长速度及幅度均有限。从资源总量看,中国已发现的天然气资源量总体不富裕,人均资源量较少,只相当于世界人均的七分之一。从资源勘探开发状况看,一方面,克拉 2、靖边、榆林、崖 13-1、普光、涩北等主力大气田已经进入产量递减阶段,老井总体综合产量递减率在 10% 左右,每年都需要有相当比例的新建产能用于弥补老井产量递减,上产能力有限;另一方面,资源劣质化加剧,新增天然气探明储量的 80% 以上都属于低渗透气藏和非常规气藏,规模效益开发难度非常大。2008—2018 年,国内天然气产量从 $803 \times 10^8 m^3$ 增长到 $1610 \times 10^8 m^3$,年均增速仅为 7.2%。其中,增速最高的为 2010 年的 12.2%,增速最低的为 2016 年的 1.7%。

综上所述,中国天然气的资源禀赋决定了国内天然气产量的增长速度赶不上天然气消费量的增长速度,国产气供应量有限而天然气消费量却快速增长,由此导致我国天然气对外依存度的较快增长。

## 1.3 天然气对外依存度发展趋势研判

基于天然气对外依存度的概念,分析、研究和预判中国天然气对外依存度的发展趋势,关键是需要确定国内天然气市场需求量和国产气的供应前景。

### 1.3.1 天然气市场需求量展望

目前中国仍处于工业化和城市化"双快速"发展阶段,经济将持续快速发展,能源需求量迅速增加。巨大的市场空间、不断完善的产业体系、节能减排政策的有效实施,都为中国天然气产业发展创造了难得的机遇和条件,未来天然气市场需求量仍将快速增长。国内外不同的研究机构分别对中国未来天然气需求量进行了预测[8-12](表1)。虽然受对市场发展规律认识程度和预测方法模型可靠性等的影响,不同机构预测的结果有所差异,但总体上中国天然气需求量都将有较大幅度的增长,2030 年需求量将介于 $(4800 \sim 6000) \times 10^8 m^3$,2035 年需求量将介于 $(6200 \sim 6370) \times 10^8 m^3$。

表 1 国内外不同机构对中国天然气需求量的预测结果表

| 预测年份 | 研究机构 | 预测结果($10^8 m^3$) ||||
|---|---|---|---|---|---|
| | | 2020 年 | 2025 年 | 2030 年 | 2035 年 |
| 2016 | 中国能源研究会 | 2900 | | 4800 | |
| 2017 | 中国石油规划总院 | 3200~3300 | 4300~4800 | 5300~6000 | |
| 2018 | 中国石油经济技术研究院 | 3200 | 4500 | 5500 | 6200 |
| 2018 | 国际能源署(IEA) | | 4640 | 5590 | 6370 |
| 2019 | 英国石油公司(BP) | 3100 | 4000 | 4850 | 5652 |

2016年，国家发布《能源生产和消费革命战略（2016—2030）》明确提出：2020、2030年一次能源消费总量分别控制在 $50\times10^8$ t 标准煤和 $60\times10^8$ t 标准煤以内，天然气消费量占比要分别达到10%和15%左右。据此测算，中国2020、2030年天然气消费量应分别达到 $3700\times10^8\ m^3$ 和 $6700\times10^8\ m^3$ 左右。

综合分析上述各个机构的预测结果，结合我国未来的国民经济发展规划、能源结构转型、大气污染治理及市场承受能力等因素，笔者分析研究后认为：2020、2025、2030、2035年我国的天然气需求量将分别为 $3200\times10^8\ m^3$、$(4200\sim4500)\times10^8\ m^3$、$(5200\sim5800)\times10^8\ m^3$、$(6200\sim7000)\times10^8\ m^3$。

### 1.3.2 国产气供应前景分析

中国天然气资源类型多，资源丰富，增储上产潜力大。笔者多年来一直从事国家和中国石油天然气集团有限公司的天然气发展规划研究，对中国天然气开发潜力进行了比较系统的分析。特别是2016年以来，利用笔者团队多年研究积累集成开发的"天然气产量预测系统"，结合不同的资源特点，详细预测了国内常规气（含致密气）、煤层气和页岩气的产量，并按时间顺序叠加得到了中国天然气产量的发展趋势[13]，2020年中国天然气产量将介于 $(1750\sim1850)\times10^8\ m^3$，2035年产量将介于 $(2800\sim3300)\times10^8\ m^3$（表2）。

在进行天然气供需平衡分析时，国内天然气供应量需在产量的基础上考虑商品率，即商品气量。统计分析结果表明，近10年国内天然气商品率约为90%[14]，据此测算出2020、2025、2030、2035年国产商品气量将分别为 $(1575\sim1665)\times10^8\ m^3$、$(1890\sim2205)\times10^8\ m^3$、$(2295\sim2700)\times10^8\ m^3$、$(2520\sim2970)\times10^8\ m^3$。

表2　中国天然气产量趋势预测结果表

| 气藏类型 | 2020年 | 2025年 | 2030年 | 2035年 |
| --- | --- | --- | --- | --- |
| 常规气（含致密气）（$10^8\ m^3$） | 1500~1550 | 1800~1900 | 2000~2200 | 2000~2200 |
| 煤层气（$10^8\ m^3$） | 60~80 | 100~150 | 150~200 | 200~300 |
| 页岩气（$10^8\ m^3$） | 190~220 | 200~400 | 400~600 | 600~800 |
| 合计 | 1750~1850 | 2100~2450 | 2550~3000 | 2800~3300 |

注：据本文参考文献[13]。

### 1.3.3 天然气对外依存度变化趋势

基于上述天然气供需趋势的预判结果，笔者认为，无论最终归属于低需—高供、低需—低供、高需—高供和高需—低供其中的哪一种供需情景，中国天然气对外依存度都将大幅度增加，预计最快将在2020年突破50%、2025年超过60%（图3）。

总而言之，一方面，为满足中国国民经济持续发展、改善能源消费结构、打赢蓝天保卫战、实现绿色发展和减排承诺等，未来中国天然气需求量将持续快速增长；另一方面，受天然气资源禀赋的影响，国产气增速有限。因此，在未来一段时间内，天然气对外依存度上升已是不可避免的趋势。天然气对外依存度不断升高无疑将给国家的能源安全带来更加严峻的挑战，并且对外依存度越高风险就越大。对此，我们应该有清醒的认识。

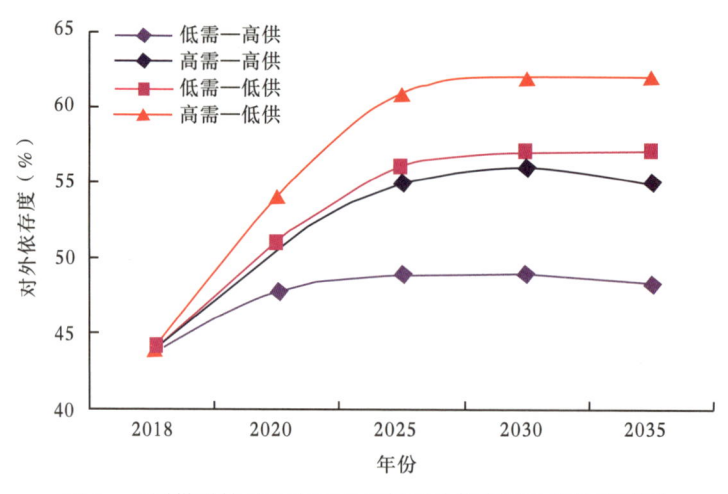

图3　不同供需情景下中国天然气对外依存度的变化趋势图

## 2　天然气对外依存度相关问题的探讨

### 2.1　合理的天然气对外依存度之争

一个国家天然气对外依存度的高低可以反映出该国对国外天然气的依赖程度。一般来说,一种商品的对外依存度越高,则表明该种商品对进口的依赖程度就越大,与世界的关系也就越密切,受世界市场价格波动以及供应安全等因素的影响也就越大。

关于什么是合理的天然气对外依存度的问题,常常会有不同的声音。有些人基于传统的石油对外依存度"50%"的国际警戒线,认为合理的依存度应控制在一定的范围内,比如在50%、60%或70%以内;也有一些人以目前日、韩等国天然气100%依靠进口为例,认为当今处于全球资源共享的时代,只要天然气资源买得到、用得起就可以,天然气对外依存度就是一个"伪命题",其高低对中国油气供应安全不会构成致命性打击。其实,上述两种观点都仅看到了表面上的对外依存度数据,明显忽略了背后正在崛起的中国国情的差异性和国际政治因素的复杂性。

伴随着中国天然气需求量的快速增长以及俄乌斗气殃及欧洲天然气"断供"事件的反复上演[15],特别是2017年冬季国内高峰供气期部分地区发生"供气紧张"以后,中国不断上升的天然气对外依存度引起了公众的担忧。

较之于石油,天然气对外依存度过高隐藏的风险则更为严重。与石油已经形成全球性统一市场和标杆价格体系的大宗商品不同,无论是管道气还是LNG,天然气尚未形成全球统一的市场交易和价格结算体系,天然气的供需还带有明显的区域性特点。加之天然气消费量与需求量的峰谷差明显,以北京市为例,峰谷差高达10倍以上,在当前中国天然气储备设施严重滞后和不足的情况下,天然气的对外依存度不仅仅是年均对外依存度,而更重要的则是高峰期对外依存度的问题。虽然目前中国天然气对外依存度尚低于石油,但考虑到用气的峰谷差,天然气的安全形势远远超越了石油的安全形势。如果进口气出现一定的波动,中亚气源、中亚管线等运输通道出现问题,天然气保供将会面临严峻的挑战,特别是冬季后果更不可想象。某进口气源国的"意外"减少供应量是造成 2017 年"供气紧张"的重要原因之一。由于中亚邻国与中国的气候相似，用气高峰期易重叠，容易导致供气量波动。对于这种情况，有关部门应该给予高度重视。

石油的高依存度已经给国际上"中国能源威胁论"话题提供了借口，天然气对外依存度的快速攀升无疑是火上浇油。在全球多级分化明显、地缘政治日趋紧张，以美国为首的西方势力对中国政策日趋强硬的背景下，天然气供应能力的巨大缺口对于中国的能源安全形成了极大的挑战，并且有可能带来经济、外交、军事等一系列连锁反应。在中国原油对外依存度已超过 70% 的情况下，必须防止天然气也步其后尘，成为新形势下影响中国能源安全的又一根"导火索"。

国外的天然气资源毕竟受他国掌控，风险是肯定存在的。随着中国大量进口天然气，其战略属性不断增强，对供应安全的要求也越来越高。作为一个发展中的天然气消费大国，在没有定价主导权的情况下，一半以上的需求量依靠进口是非常不安全的。过高的对外依存度，一旦国际天然气市场和进口环节有任何的风吹草动，都有可能引起国内市场的连锁反应。因此，笔者认为，将中国天然气对外依存度控制在一定水平线之下是非常必要的。

## 2.2 关于中国合理的天然气对外依存度的探讨

如上所述，对于什么是合理的天然气对外依存度众说纷纭，但是有两点认识却是一致的：(1) 合理的天然气对外依存度需要充分考虑其资源禀赋、消费规模、经济水平、地缘政治、军事外交能力等因素；(2) 天然气对外依存度是动态变化的，具有阶段性特征。例如美国，20 世纪 90 年代之前，国内天然气产量与消费量基本平衡，对外依存度都在 3% 以下；20 世纪 90 年代以后，国内产量基本稳定，而消费量却不断增长，天然气对外依存度不断升高，2005 年达到最高的 18%；2007 年以后，随着页岩气产量的迅猛增长，天然气对外依存度迅速降低，到 2017 年已摇身一变成为天然气净出口国（图 4）；2018 年天然气净出口量达到 $166×10^8 m^3$，天然气出口量增长势头迅猛，未来相当长一段时间内美国将从早期的争资源转向争市场，改变了世界天然气供需格局。

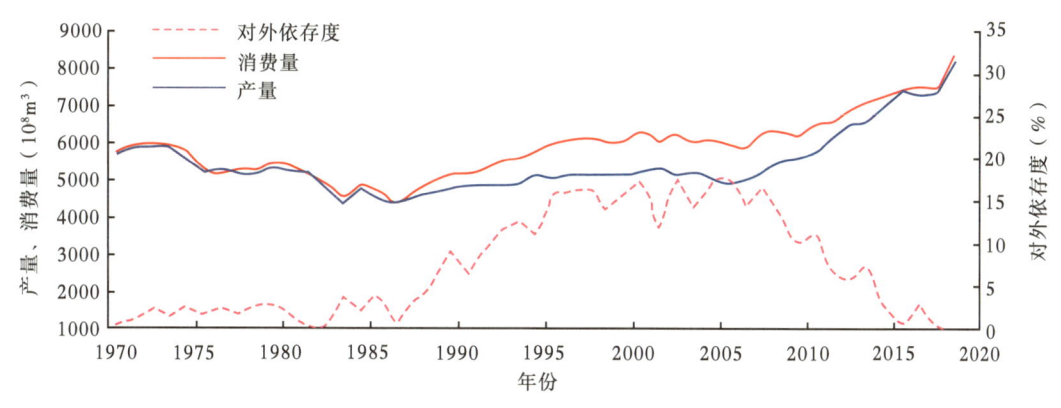

图 4 1970—2018 年美国天然气对外依存度变化图

因此，要探讨中国合理的天然气对外依存度，可以从一次能源消费总量、天然气消费大国生产消费状况、天然气进口大国的进口量（指净进口量，下同）与全球天然气贸易总量之比等方面来分析。

## 2.2.1 一次能源消费量与经济总量

从一次能源消费量与经济总量来看,中美两国是当今全球绝对的能源消费和经济大国。2017年全球一次能源消费量排名前十位的国家当中,中美合计占全球能源消费总量的39.7%,合计GDP占比为39.2%(表3)。因此,中美两国在世界能源消费中拥有绝对的比重和影响力,两国的能源消费情况不但能决定现阶段世界能源消费的形势,而且还将影响未来的走势。

表3 2017年全球能源消费量排名前10的国家单位GDP能耗情况统计表

| 国家 | 能源消费量<br>[$10^6$t(油当量)] | 能源消费量<br>占比 | GDP<br>(亿美元) | GDP<br>占比 | 单位GDP能耗<br>[$10^6$t(油当量)/亿美元] |
|---|---|---|---|---|---|
| 中国 | 3132 | 23.2% | 122377 | 15.2% | 0.026 |
| 美国 | 2235 | 16.5% | 193906 | 24.0% | 0.012 |
| 印度 | 754 | 5.6% | 25975 | 3.2% | 0.029 |
| 俄罗斯 | 698 | 5.2% | 15775 | 2.0% | 0.044 |
| 日本 | 456 | 3.4% | 48721 | 6.0% | 0.009 |
| 加拿大 | 349 | 2.6% | 16530 | 2.0% | 0.021 |
| 德国 | 335 | 2.5% | 36774 | 4.6% | 0.009 |
| 韩国 | 296 | 2.2% | 15308 | 1.9% | 0.019 |
| 巴西 | 294 | 2.2% | 20555 | 2.5% | 0.014 |
| 伊朗 | 275 | 2.0% | 4395 | 0.5% | 0.063 |
| 全球 | 13511(总计) | 100%(总计) | 806838(总计) | 100%(总计) | 0.017(平均) |

注:能源消费量数据来源于BP 2018年;GDP数据来源于2018年世界银行(现价美元)。

美国于20世纪40年代实现工业化,进入发达经济初期,到20世纪70年代已处于发达经济时代。当时美国一次能源消费量占全球的30%左右,石油在一次能源消费量中的占比在45%左右,对外依存度也在45%左右。能源的对外依赖性带来了经济的脆弱性,特别是在两次石油危机中暴露无遗。于是尼克松政府不得不提出了"能源独立"计划,经过长达半个世纪的不懈努力,通过加快非常规油气勘探开发产业化步伐和加快基础设施建设、完善天然气产业链等措施,美国已成为世界第一大石油天然气生产国,已逐渐摆脱对中东等外部石油天然气的高度依赖,即将实现能源独立的目标。这不仅对于美国能源、经济安全具有重大的意义,也为世界各国的能源发展提供了借鉴,尤其是对能源需求量急剧上升、油气对外依存度不断攀高的中国来说,更具有重要的现实意义。

当前中国正处于重工业化阶段,预计到2030年才能全面实现工业化[16]。发展成为工业化国家,既是实现"中国梦"的一个重要经济内涵,也是实现中华民族伟大复兴的必然要求。大量的能源消耗是发达国家工业化顺利完成的必要条件。这就意味着未来中国的经济发展还要在很大程度上依赖于能源,特别是油气。目前中国的油气对外依存度不断攀升,能源供应安全面临挑战。与此同时,随着中国经济总量的提升,"一带一路"建设逐步深入,人民币国际化趋势不断推进,美国已把中国视为其继续称霸全球的主要对手,扼制中国崛起、阻碍人民币走向全球的步伐成为其必然的战略选择,中美贸易摩擦将成为常态化。中美两国能源形势背向发展,将进一步拉大中美两国的综合国力差距,当前中国油气对外依存度已远超美国最高时期的水平。因此,必须对一次能源消费总量予以约束,提高能源效率,设置能源消费总量"天花板",特别是要千方百计地增加国内油气供应能力,将油气资源的自给率提高到能够保障中国经济相对独立的水平,有效抑制对外依存度的过快增长,将进口量控制在当国际油气市场波动时不至于引起国内经济大幅度波动的范围以内,特别是天然气对外依存度要充分考虑其区域性和峰谷差等特点,确保安全平稳供气。

### 2.2.2 天然气消费大国的生产消费状况

从天然气消费大国的生产消费状况来看,2018年天然气消费量超过$2000×10^8m^3$的有4个国家:美国($8171×10^8m^3$)、俄罗斯($4545×10^8m^3$)、中国($2786×10^8m^3$)和伊朗($2256×10^8m^3$)。其中俄罗斯和伊朗天然气资源基础雄厚,2018年剩余天然气可采储量排名全球前两位。俄罗斯是传统的天然气出口大国,也是长期以来欧洲天然气市场的主要供应者;伊朗天然气开发程度较低,拥有巨大的天然气出口潜力。美国曾经是天然气净进口国,从天然气对外依存度变化来看,进口气量占比较高的时期,对外依存度始终保持在10%~18%,而且美国从国外进口的天然气几乎都来自加拿大,政治和经济风险都相当低;借助于页岩气革命,2017年美国由原来的天然气进口国转变为出口国,在不久的将来,还有望成为全球最大的LNG出口国。

也就是说,在上述4个天然气消费大国当中,唯有中国是需要大量进口天然气的国家,目前的天然气对外依存度已经远高于当年美国的最高对外依存度,并且未来还将呈现出继续上升的趋势。鉴于全球的天然气资源充足,从理论上讲,中国可以从国外获得足够的资源,但不能忽视减少供应、中断供应以及地缘政治风险等。俄罗斯和美国都在天然气领域使用过地缘政治手段:俄罗斯多次威胁给乌克兰断气;美国则在近期的贸易摩擦当中,无论是对中国还是欧洲,均把大幅度增加美国LNG进口量作为谈判筹码。中国被美国视为最主要的竞争对手,也是其重点打压的对象,既然现在可以断供芯片,未来也有可能断供LNG。对美国LNG依赖程度越高,则有可能带来的地缘政治风险也越大。天然气对外依存度过高,将影响供应安全和经济安全。因此,中国应千方百计地抑制当前天然气对外依存度过快上涨的趋势,将安全平稳供气的主动权牢牢掌握在自己的手里。

### 2.2.3 天然气进口量与全球天然气贸易总量之比

从历史上天然气进口大国的进口量与当时的全球天然气贸易总量之比来看,中国天然气进口量不宜超过全球天然气贸易总量的20%。历史上天然气年进口量超过$1000×10^8m^3$的国家有3个:美国、日本和中国。其中美国在页岩气革命成功之前,为满足国内的需求量也曾大量进口天然气,1995年进口量与全球天然气贸易总量之比达到峰值,为20.1%,之后逐渐降低;日本天然气资源极其匮乏,长期依靠进口LNG来满足需求(主要用来发电),1991年进口量与全球天然气贸易总量之比达到最高的16.5%,之后逐渐降低,目前在10%左右(图5)。中国自2006年开始进口天然气以来,随着进口量的不断增加,进口量与全球天然气贸易总量之比也呈上升趋势,2018年达到了9.8%(图6)。

为了让"中国能源威胁论"不攻自破,同时在中国天然气进口量大幅度上升的同时,不至于过多挤占其他国家的天然气进口空间,类比于美国,考虑将中国天然气进口量控制在不超过全球天然气贸易总量的 20% 为宜。根据 2010 年以来全球天然气贸易量增长趋势[17],预测 2035 年全球天然气贸易总量约为 $1.6×10^{12} m^3$,由此测算中国天然气进口量应控制在 $3200×10^8 m^3$ 以内,若按中国 2035 年天然气最高需求量方案 $7000×10^8 m^3$ 测算,对外依存度则应控制在 46% 左右。

图 5 美国和日本 1990—2018 年天然气净进口量及其全球贸易量之比变化图

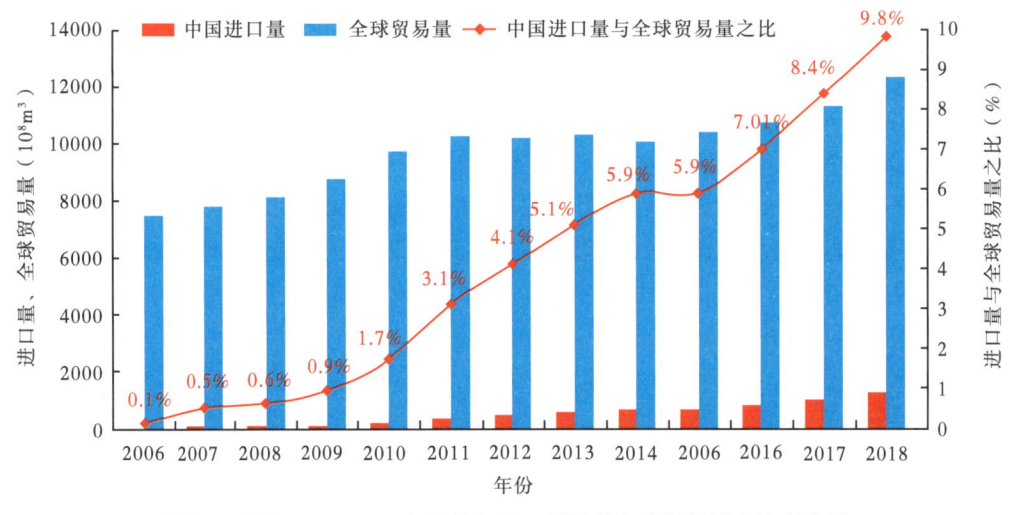

图 6 中国 2006—2018 年天然气进口量及其全球贸易量之比变化图

综上所述,中国既是经济大国也是天然气消费大国,天然气对外依存度过高将影响供应安全和经济安全,需要树立底线思维。因此,中国天然气供应绝不能再步原油的后尘,应千方百计抑制当前对外依存度过快上涨的趋势,将安全平稳供气的主动权牢牢掌握在自己的手里。着眼于长远,应将我国的天然气外依存度控制在 50% 以内。

## 3 抑制中国天然气对外依存度过快增长的对策与建议

为抑制天然气对外依存度的过快增长,需要在供需侧两端同时发力,做到开源节流。在供应侧,一方面实施多气并举,做大国内天然气供应能力;另一方面做好可替代能源发展这篇文章,增加可再生能源的供应量。在需求侧,积极做好厉行节约,提高天然气利用效率,压缩不合理的需求。

### 3.1 多气并举,提高国内天然气供应能力,夯实国内天然气"压舱石"地位

加大天然气上游勘探开发力度是天然气行业供给侧结构性改革的主要发力点。目前中国大多数盆地的天然气勘探程度都比较低,常规气资源探明率不到20%、页岩气和煤层气资源探明率不到2%,天然气勘探开发潜力非常大。要突出陆上深层、海洋深水和非常规天然气三大领域,加大勘探开发投入,加强重大理论攻关和关键技术研发,完善资源开发政策,围绕国产气供应情景的高方案目标,落实勘探开发工作部署。常规气(含致密气)立足于四川、鄂尔多斯、塔里木三大盆地和海域,加快新区增储建产步伐和提高老区储量动用率、采收率,进一步扩大生产规模。全面推进川渝地区海相页岩气规模开发,重点突破3500m以深海相页岩气效益开发技术,实现规模效益发展。煤层气在立足鄂尔多斯盆地和沁水盆地已开发老区规模效益开发的同时,加快蜀南地区、准噶尔盆地、二连盆地等的煤层气勘探评价和开发建设。

此外,破解制约煤制气、生物气、天然气水合物等资源开发利用的政策困境和技术瓶颈,加强试点示范项目建设,积极推进规模效益发展,拓宽天然气供应渠道,力争2035年国产气供应量再增长$(500\sim800)\times10^8m^3$,从而进一步降低对外依存度。在环保和成本可承受的条件下,加快大唐克旗煤制气、伊犁庆华煤制气、内蒙古汇能煤制气和伊犁新天煤制气等项目建设,同时加快开展煤炭地下气化攻关,力争2035年产量达到$(200\sim300)\times10^8m^3$。生物气在改善农村人居环境、推动生态循环农业建设等方面都具有重要的意义,加大中国粮食主产区(秸秆资源)、东北和西南林区生物质资源的利用,据不完全统计,以中国农村每年产生秸秆等农业废弃物约$9\times10^8t$、畜禽粪便等垃圾约$30\times10^8t$测算,考虑一定的开发利用率,预计2035年生物气产量潜力可达$(300\sim500)\times10^8m^3$。中国天然气水合物资源丰富,2017年在南海海域天然气水合物开发试采已经取得了成功,应进一步加大天然气水合物开发评价和技术攻关,一旦开发技术成熟配套,国内天然气总产量还有较大的上升空间。

### 3.2 发展可再生能源,加快形成有效替代,减轻天然气供应压力

发展可再生能源应是未来解决中国天然气对外依存度过快增长问题的重要途径。壳牌是全球大型能源公司中可再生能源业务最多的公司,可再生能源已成为其第五大核心业务,所研发的CIS薄膜电池在太阳能零散电力利用领域占约17%的市场份额,硅基太阳能业务年发电超过80MW,风能年发电量已超过500MW,并在全球建有多个风力发电厂[18]。

如果太阳能、风能、地热等非化石能源发展起来,将形成对天然气等化石能源的有效替代,降低对天然气等化石能源的需求量,可以在一定程度上抑制天然气对外依存度的过快增长,有助于安全平稳供气、保障国家的能源安全。建议中国加大对相关产业的支持力度,通过自主开发和合作开发等多种模式,进一步扩大可再生能源的供应规模。

## 3.3 坚持底线思维,提高天然气利用效率,抑制不合理需求

明确底线思维,有限的资源满足不了无限的消费需求。优化天然气的消费方向,通过技术和管理创新推动能源利用效率不断提升,抑制不合理消费,有效抑制天然气对外依存度的过快增长。建议加强全社会用气管理,优化配置天然气资源;优化负荷管理,促使天然气需求量在不同时序上合理分布,实现移峰填谷;优化能效管理,改变用户的用气行为,采用先进的节能技术提高终端用气效率;改变原来单纯增加能源供应量的传统观念,转向由供给和需求双向制约提供保障的观念;价格是市场经济中重要的调节供需的杠杆,应充分发挥其杠杆作用,抑制快速增长的天然气需求量。

## 4 结论

(1)中国已进入实现中华民族伟大复兴的关键阶段,国民经济持续稳定发展,改善能源消费结构、打赢蓝天保卫战、实现绿色发展和减排承诺以及人民过上美好生活的愿望将持续推动天然气需求量的快速增长,近期中国天然气对外依存度继续上升已是不可避免的趋势。

(2)天然气的市场区域性和用气峰谷差等特点决定了其对外依存度过高所隐藏的风险,较之于石油,天然气的这种风险则更为严重。为了将安全平稳供气的主动权牢牢掌握在自己的手里,彻底避免"气荒"再现,抑制我国天然气对外依存度过快增长是非常必要的。

(3)供需两侧同时发力,通过采取多气并举做大国内天然气供应能力、大力发展可再生能源实现有效替代、提高天然气利用效率压缩过度需求等措施,从长远来看将中国天然气对外依存度控制在50%、甚至更低是可行的。

### 参 考 文 献

[1] 陆家亮,赵素平. 新常态下中国天然气勘探开发战略思考[J]. 天然气工业,2015,35(11):1-8.
[2] 中国石油经济技术研究院. 能源数据统计[R]. 北京:中国石油经济技术研究院,2019.
[3] 刘朝全,姜学峰. 2018年国内外油气行业发展报告[M]. 北京:石油工业出版社,2019.
[4] 国务院. 打赢蓝天保卫战三年行动计划[EB/OL]. (2018-07-03). http://www.gov.cn/zhengce/content/2018-07/03/content_5303158.htm.
[5] 国家发改委,国家能源局,财政部,等. 北方地区冬季清洁取暖规划(2017—2021年)[EB/OL]. (2017-12-20). http://www.ndrc.gov.cn/zcfb/zcfbtz/201712/t20171220_871052.html.
[6] 环境保护部,国家发改委,工业和信息化部,等. 京津冀及周边地区2017—2018年秋冬季大气污染综合治理攻坚行动方案[EB/OL]. (2017-08-21). https://baike.so.com/doc/26802424-28116994.html.
[7] 国家发改委. 关于降低非居民用天然气门站价格并进一步推进价格市场化改革的通知[EB/OL]. (2015-11-18). http://www.ndrc.gov.cn/gzdt/201511/t20151118_758904.html.
[8] 中国能源研究会. 中国能源展望2030[M]. 北京:经济管理出版社,2016.
[9] 中国石油规划总院. 中国油气供需展望[R]. 北京:中国石油规划总院,2018.
[10] 中国石油经济技术研究院. 2050年世界与中国能源展望(2018版)[R]. 北京:中国石油经济技术研究院,2018.
[11] British Petroleum. BP world energy outlook (2019)[R]. London:BP,2019.
[12] International Energy Agency. World energy outlook[R]. Paris:IEA,2018.
[13] 陆家亮,赵素平,孙玉平,等. 中国天然气产量峰值研究及建议[J]. 天然气工业,2018,38(1):1-9.

[14] 霍瑶, 杨依超, 韩永新, 等. 提高我国天然气商品率的主要技术与措施[J]. 天然气工业, 2016, 36(9): 141-145.

[15] 林雪丹, 任彦. 谈判破裂 俄乌"斗气"风波再升级[J]. 中国经济周刊, 2014, 24(16): 78-79.

[16] 中国社会社科院. 工业化蓝皮书: 中国工业化进程报告(1995—2015)[M]. 北京: 社会科学文献出版社, 2017.

[17] British Petroleum Company PLC. BP statistical review of world energy 2019[R]. London: BP, 2019.

[18] 陈建军, 王南, 唐红君, 等. 持续低油价对中国油气工业体系的影响分析及对策[J]. 天然气工业, 2016, 36(3): 1-6.

# 中国天然气发展形势研判与对策建议

赵文智

（中国石油勘探开发研究院）

**摘要**：天然气是世界范围从化石能源向绿色清洁能源过渡的桥梁，中国也不例外，在未来能源结构优化和绿色发展中占有重要地位。然而，中国的天然气需求却不可以想用多少就用多少，这由三方面因素决定：中国天然气资源总体比较丰富，但资源品质相对较差，且埋藏深度较大，天然气高峰产量不会很高，预计年产在 $(2500\sim2600)\times10^8m^3$ 左右；中国经济发展对天然气需求增长既快又总量很大，高峰需求量在 $(5500\sim6500)\times10^8m^3$ 左右甚至更高，天然气进口量很快会超过国内自产气量；天然气是与民生息息相关的必需品，如出现供应短缺，将影响社会安定，对外依存度不宜超过 50%。本文基于对中国天然气资源潜力、供需关系与供气安全因素研究等，预测了中国天然气未来发展形势，提出了相关对策建议。

**关键词**：天然气；产量；消费量；对外依存度；安全形势

自 2000 年"西气东输"工程正式批准启动以来，中国天然气消费量以每年 15% 的速度快速发展。尤其是 2017 年，在蓝天计划、煤改气等环保政策和经济快速发展的驱动下，中国天然气需求量同比增加 15%，占全球天然气需求增量的三分之一。与此同时，中国天然气产量难以满足天然气消费量快速增长的需要，天然气进口量逐年增加，预计 2019 年将会成为全球最大的天然气进口国[1-2]，2018 年对外依存度已经超过 43%，预计未来还将快速增长，如不采取措施引导，天然气对外依存度将很快超过石油，成为制约国家安全的重大因素之一。针对中国天然气资源潜力、市场发展与未来供需关系，本文分析了中国天然气未来供需趋势，对合理制定中国天然气对外依存度安全上限与如何尽早应对风险，提出了相关建议。

## 1 中国天然气发展现状

中国天然气工业经过几十年的快速发展，目前已进入发展调整期。近几年天然气产量增速有所放缓，形成以常规气为主体、非常规气产量快速增长的发展态势。目前，中国天然气市场表现出三个"快速增长"的特点：一是天然气消费量快速增长；二是天然气进口量快速增长；三是天然气对外依存度快速增长。

中国天然气产量自新中国成立之初就一直持续增长，由建国初期的年产仅 $0.07\times10^8m^3$，增长到 2017 年的 $1487\times10^8m^3$[3]（图 1），大体上经历了四个发展阶段：第一个阶段是 1949 年到 1971 年，天然气工业处于缓慢发展期，产量由年产不足 $0.1\times10^8m^3$ 增长到 1971 年的 $37\times10^8m^3$，历时 22 年，实现了中国天然气产量从无到有的发展，这期间产量增量主要来源于小型气田发现和油田伴生气开发；第二个阶段是 1972 年到 2003 年，天然气工业进入规模增长期，历时 31 年，产量由 1972 年 $50\times10^8m^3$ 增长至 2003 年的 $341\times10^8m^3$，年均增速达 6.4%，该阶段气产量增量主要来自中型气田发现与新投入开发油田伴生气开发；第三个阶段从 2004 年至

2014年,为天然气工业快速发展期,历时十年,产量从 410×10⁸m³ 猛增到 1307×10⁸m³,年均增速 12.3%,在鄂尔多斯盆地长庆、塔里木盆地库车、四川盆地川渝气区和青海等地发现一系列大气田,成为产量增长的主要贡献者;第四个阶段是 2015 年至今,天然气工业进入发展调整期,天然气产量年均增速降到 5% 以下,大气田发现数量减少,资源品质也有变差趋势。

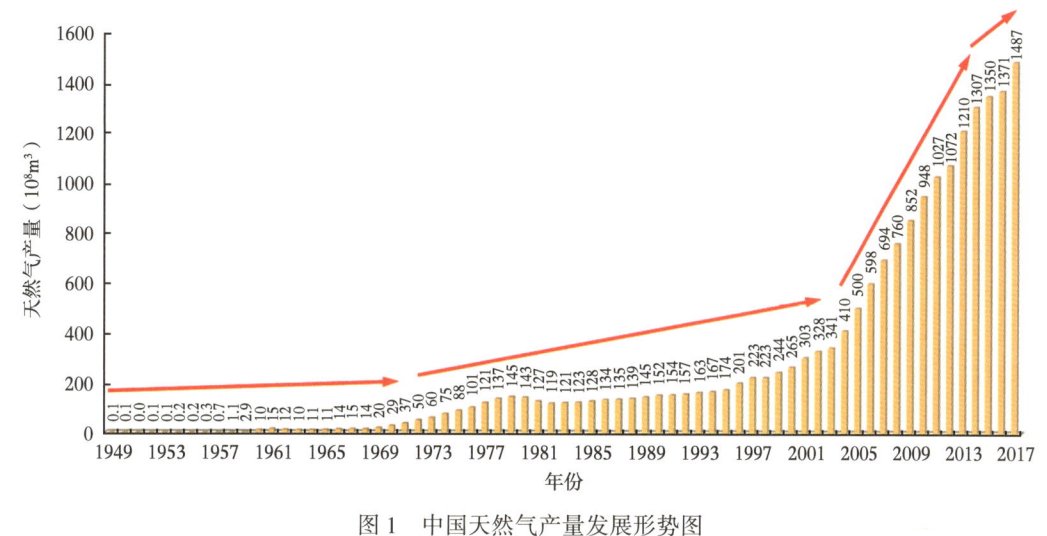

图 1 中国天然气产量发展形势图

随着天然气勘探领域、成藏理论、开发技术取得新突破,中国非常规天然气开发利用地位越来越重要,产量占比逐年增加,改变了中国天然气产量构成[4-10]。非常规天然气产量比例由 2000 年的 3% 快速增加至 2017 年的 33%(图 2)。其中,致密气、页岩气、煤层气产量分别达到 350×10⁸m³、90×10⁸m³ 和 49×10⁸m³。致密气是目前非常规气产量贡献的主体,页岩气具有较大的增长潜力,煤层气也有较好的增长预期。假以时日,随着工程技术的进一步创新发展,非常规气产量将很快占据中国天然气总产量的半壁江山。

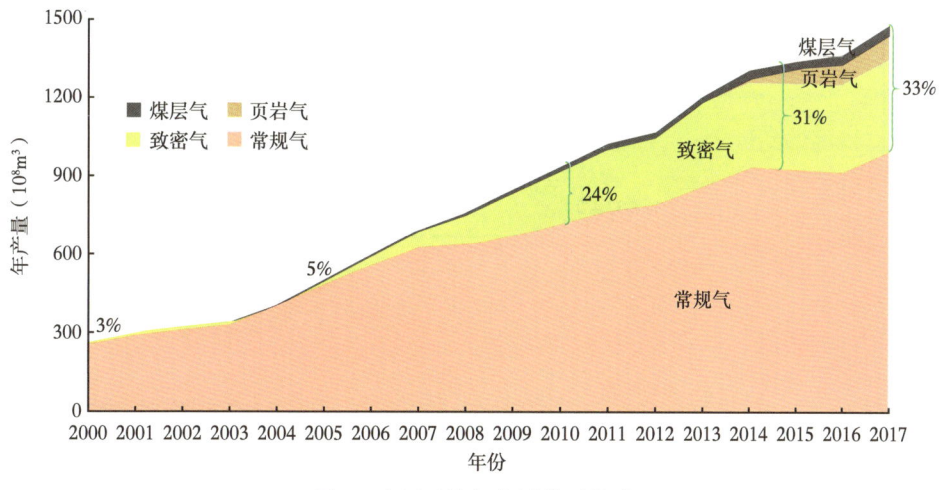

图 2 中国天然气产量类型构成

受蓝天计划、城镇化等国家重点发展规划的驱动,中国天然气消费量增长强劲。2000年国务院批准启动"西气东输"工程,天然气市场消费量由 $245×10^8m^3$ 快速增长至2017年的 $2373×10^8m^3$,年均增速达15%(图3)。其中,2004年西气东输工程全部完成建设并实现全线商业运营,2012年西气东输二线全部建成投产,成为推动中国天然气市场发展最为迅猛的10年。2013年受国际天然气价格、国际地缘政治和经济形势及中国经济发展等多方面因素的影响,天然气消费量增速有所放缓,天然气市场进入发展的调整期[11]。2017年是一个特殊的年份,受一些国家政策调整影响,天然气消费量同比增加 $315×10^8m^3$,用气增量为往年同期增量的2.5倍,特别是当年冬季,中国液化天然气交易量和价格都创历史新高[12],标杆价格最高达到10064元/t,是世界天然气市场同期价格的近10倍。

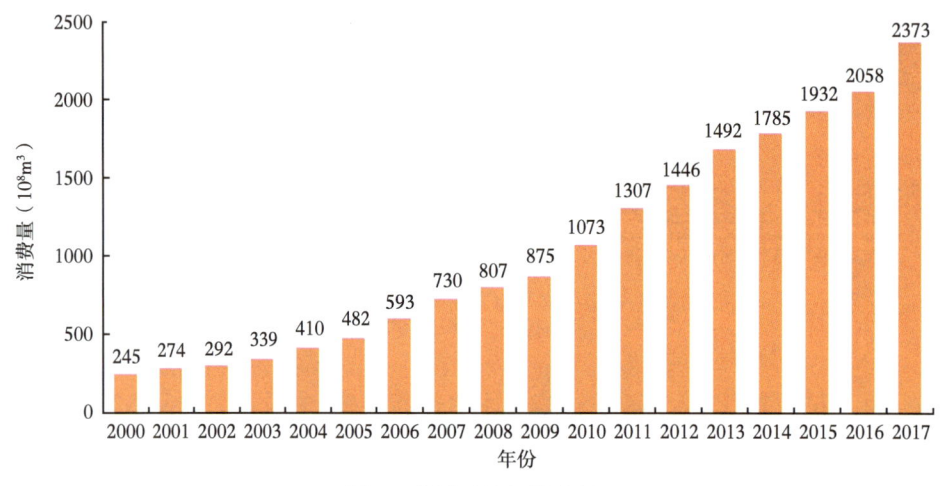

图3 中国天然气消费量

为了满足日益增长的天然气消费需求,中国从2006年开始进口天然气,总体进口气量呈现出快速增长态势。尤其是最近两年,国内天然气对外依存度攀升较快,2017年达到39%,2018年超过43%,未来这一趋势还将持续上升(图4)。另一方面,中国天然气在一次能耗中占比较低,

图4 中国天然气对外依存度(据国家发改委数据,2018)

— 28 —

2017年天然气在中国一次能源消费中的比例为6.9%,远低于全球23.9%的平均水平。

## 2 中国天然气供需形势

从天然气资源潜力、各主要气区资源探明程度看,中国天然气工业尚处于快速发展阶段,年增探明储量处于高峰增长期。Weng旋回模型预测显示,2030年以前中国天然气探明地质储量增量规模年均大致可保持在$(6000～8500)×10^8m^3$,表明中国天然气储量高峰增长还将持续相当长时间(图5)。从资源总量与资源构成来看,中国常规气与非常规气的可采资源总量大致相当,但非常规气资源探明率很低,未来储量增长的潜力更大。预计在2030年前后,投入建产的天然气储量中,非常规气的比例将超过常规气。如果以2015年中国天然气产量($1350×10^8m^3$)作为基数,中国天然气产量具备实现倍增的可能性,时间节点大致在2030年前后(图6)。其中,常规气在稳产基础上保持微幅度上产,非常规气则在现有基础上实现有规模的上产,产量占比将超过一半。应该指出,伴随着中国天然气产量的倍增发展,天然气开发的

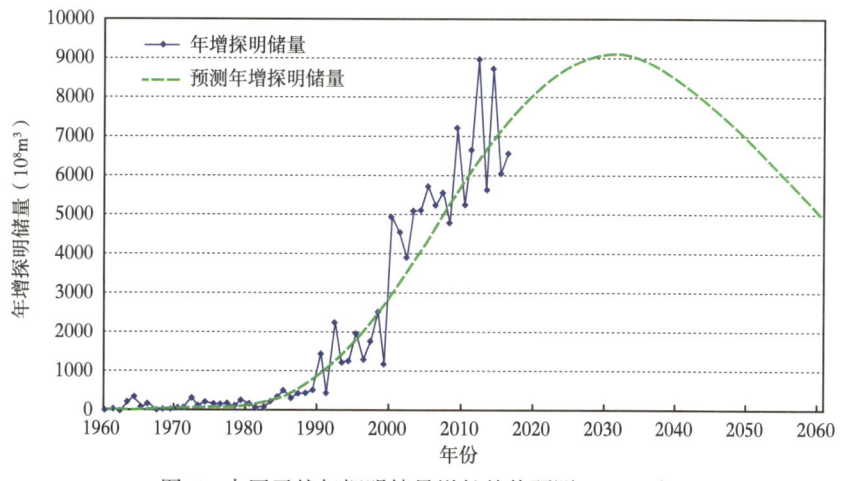

图5 中国天然气探明储量增长趋势预测(Weng法)

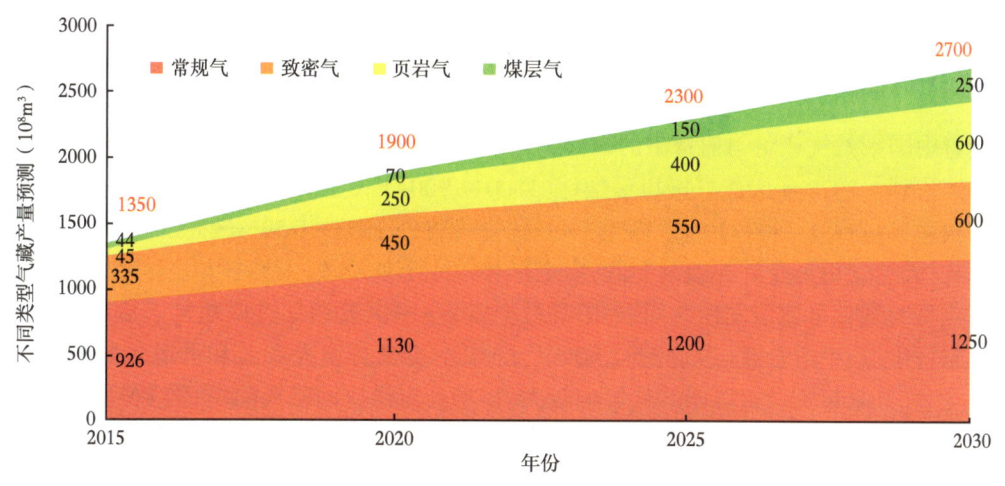

图6 中国常规与非常规气产量预测

施工工作量将成倍增加,此外因投入建产的资源多数为非常规资源,品质相对较差,天然气开发效益能否保证是值得关注的问题,这两方面是制约中国天然气产量能否实现倍增发展的主要受限点。

随着国民经济发展对天然气需求的不断增长,特别是国家绿色低碳发展战略的实施,中国天然气消费量将持续快速增长。2018 年 1—8 月份天然气消费量为 $1804×10^8m^3$,比去年同期增长 18.2%。预测 2020 年中国天然气消费量将达到 $(3500～3800)×10^8m^3$ 左右,同期天然气产量预计只有 $1900×10^8m^3$;2030 年天然气消费量将达到 $(6000～6500)×10^8m^3$ 左右,产量估计很难超过 $(2600～2700)×10^8m^3$(图 7)。

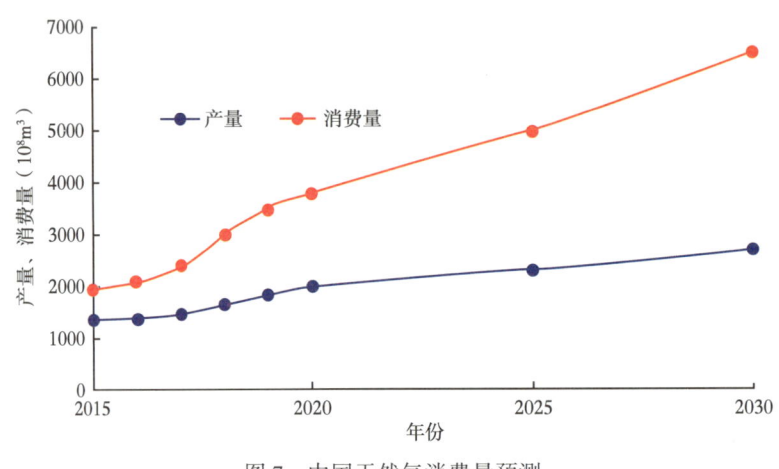

图 7 中国天然气消费量预测

根据国家统计局和国务院发展研究中心发布的统计数据,中国用于发电、工业燃料、城市燃气和交通运输四个领域的用气量增长强劲,成为推动中国天然气需求快速增长的龙头(表 1)[13]。2015 年—2020 年,工业领域天然气消费量为中国天然气消费量的主体。随着发电用气量的增加,预计在 2022 年发电用气量将会超过工业用气量,并且这一趋势将会继续持续到 2030 年。未来,用电量持续增加和相关政策的发布都会进一步推动天然气需求增长。

## 3 天然气安全形势与地位

随着中国天然气产量与消费量的缺口不断扩大,天然气对外依存度快速攀升,天然气安全形势不容乐观。2020 年天然气进口量预计 $(1800～1900)×10^8m^3$,对外依存度接近 49%。2030 年天然气进口量预计 $3500×10^8m^3$ 以上,对外依存度将处于 55%～59% 区间。与中国石油安全形势相比,天然气更易受外部因素制约,安全形势挑战更加严峻;天然气与民生密切相关,如果民用气一旦断气就可能引发社会问题,影响社会安定;中国天然气进口地过于集中,其中管道气主要来自中亚、中缅管线(占比超过 90%),LNG 主要来自澳大利亚和卡塔尔(占比 71%),受买方竞争、开采投资及资源国政局影响较大,市场稳定供应面临风险;虽然 LNG 占进口气比例逐年增加,但运输通道安全保障也存在较大不确定性。同时,中国天然气储备形势也不容乐观。中国共投产储气库 25 座,2017 年工作气量 $173×10^8m^3$,占年消费量比例仅为 7.3%。而美国共有储气库 419 座,2016 年总工作气量 $1364×10^8m^3$,占年消费比例 17.5%。相比之下,中国盆地地质条件复杂,储气库建设不论从规模还是储气能力增长都具有很大挑战,不是想搞多大就能搞多大,也不是想搞多快就能搞多快。未来相当长一段时间,中国在储气库选址和储气规模建设上都难以满足调峰需要,预计2030年中国储气库储气能力可达到$450×10^8m^3$左右,届时储气量占消费量的比例约8%,中国储库调峰短板将长期存在,特别是随着消费量规模增长,储气调峰能力建设面临的压力将越来越大。

表1 中国分行业天然气消费量预测(单位:$10^8m^3$)

| 行业 | 2015年 | 2016年 | 2020年 | 2025年 | 2030年 |
| --- | --- | --- | --- | --- | --- |
| 工业 | 671 | 734 | 1150 | 1500 | 1800 |
| 城市 | 411 | 434 | 750 | 1100 | 1350 |
| 发电 | 353 | 414 | 1100 | 1650 | 2000 |
| 交通 | 238 | 255 | 500 | 750 | 950 |
| 化工 | 259 | 241 | 300 | 350 | 400 |

基于以上分析,本文建议要提早关注中国天然气对外依存度上限,并及时从政策上加以引导。这里所说的对外依存度上限有两层含义:一是由哪些因素介入确定依存度的上限;二是依存度上限数值该怎么确定。关于确定依存度上限的因素,本文重点考虑四方面因素:一是天然气的自产量,这是一个国家最安全的保证;二是天然气的储备量,这是一个国家应对供气突发事件的重要保证条件;三是天然气利用领域及用量大小,以及其在该领域的天然气用气量如果出现战时状态是否能够被其他途径替代(煤化工和地下煤制气等);四是天然气用量在考虑被替代时,不与石油用量的增加重叠,因为后者在中国也存在严重的供应安全问题。

基于上述考虑,本文对中国天然气对外依存度上限确定提出如下计算等式,即:

$$依存度上限 = (自生产气量+最大储备量)/消费总量 \tag{1}$$

或

$$依存度上限 = [需求总量-(自生产量+不可替代量)]/需求总量 \tag{2}$$

这里所说的不可替代量,是指在考虑出现战时状态情况下,综合各领域用气平衡,所不能被替代又必须确保供应的天然气数量,主要包括城市用气、化工用气和交通用气等。需要说明的是,化工用气和交通用气如遇战时状态,改用石油或煤化工替代,以减少用量。但考虑到中国石油供应安全形势也很严峻,改用石油替代并不能从根本上解决国家安全面临的问题,所以不列入可替代用量范畴,纳入必保用气量考虑。而工业用气和发电用气在出现气源输入被切断情况时,完全可以改用煤基液化产品利用和加大煤发电来替补,这部分用气量纳入可替代用量考虑,不包括在不可替代量中。

基于以上分析,在对中国天然气自产量、储备量建设趋势、各领域天然气用量规模预测及天然气总需求量分析基础上,提出中国天然气对外依存度上限为50%,这个数值是在国内自产气量、储备气量与需求气量三元平衡基础上得到的,笔者认为这是保证中国天然气发展安全的关键数值,既不能按60%考虑,更不能按70%考虑,这是天然气利用与民生息息相关性决定的。

## 4 措施和建议

为保中国天然气供应安全的长久平安,提出以下措施和建议:

(1)设立中国天然气对外依存度上限,天花板为最高不突破50%。国家应该按照国内天然气产量增长规模与趋势、储气库与储气量建设规模与节奏,用天然气对外依存度天花板规定约束天然气的总体用量,并以此为基础,做好相关领域用气规划。中国是一个正在崛起的新兴大国,未来10至20年是发展的关键期,也是西方大国打压的对象。国家应该统筹设定天然气在国内一次能源消费结构的比例,本文认为以15%为宜,不能参照世界平均值(23.9%)来规划中国的天然气用量规模,需要量力而行,适可而止。

(2)进一步加强勘探开发技术创新力度,不断夯实中国天然气自生产能力和规模。通过加强新区新领域勘探,积极扩大天然气资源发现,做大产量增长的基础。同时,加快关键工程技术进步的步伐,最大限度释放常规低品位天然气资源和非常规天然气储量的潜在上产能力,重点攻关深层或超深层天然气勘探理论技术、重点富气盆地勘探新理论与新技术、天然气藏立体开发理论技术、海域深水天然气勘探开发理论技术、非常规"人工气藏"开发理论技术、深层页岩气勘探开发理论技术、中低阶煤层气勘探开发理论技术及复杂地质条件储气库建库技术等,为充分挖掘低品位天然气资源在国家天然气上产和稳产中的主力军作用提供可靠的科技支撑。

(3)建立全国互联互通的天然气储运体系,进一步增强调峰保供能力。根据国际经验,储气设施工作气量应占天然气消费量的15%以上,而中国在2017年储气库工作气量却不足天然气消费量的10%。因此,应加快推进国内储气库设施建设,降低外部因素制约。同时,通过加强管网与储气设施的互联互通,保障高峰供气能力。

(4)开辟多元化天然气进口通道,保障稳定的市场供应。中国进口气来源国过于单一,从土库曼斯坦和澳大利亚两个国家进口天然气比例高达61%。目前投运的LNG接收站年接收能力接近$800×10^8 m^3$,未来LNG进口量将远超过现有LNG接收能力。面对这一系列问题,一方面应加快中俄天然气管道东线的建设与西线的论证进度,另一方面应积极建设沿海LNG接收站,扩大LNG进口能力。

(5)加大天然气生产与调峰政策扶持力度,激发发展活力。加强未动用储量和尾矿利用的政策支持,例如:借鉴国家西部大开发优惠政策,将所得税率由25%降为15%,资源税减免或降低到2%~3%。实行差别化税费政策,加大对深层、深水天然气资源开发利用的财税优惠力度,延续并完善页岩气和煤层气开发补贴政策,给予致密气不低于$0.2$元$/m^3$的财政补贴。合理切割天然气产、运、销产业链中的利润分配关系,国家相关政策应向上游有所倾斜,提高天然气勘探开发的积极性。推动储气调峰设施相关扶持政策,在调峰气价、气源等方面给予储气调峰主体相关优惠政策。

### 参考文献

[1] IEA. Market Report Series:Gas 2018[R]. 2018.
[2] IGU. Global Gas Report 2018[R]. 2018.

[3] 国土资源部. 全国油气矿产资源储量通报(2016)[R]. 2017.
[4] 赵文智,董大忠,李建忠,等. 中国页岩气资源潜力及其在天然气未来发展中的地位[J]. 中国工程科学, 2012,14(7):46-52.
[5] 赵文智,王红军,钱凯. 中国煤成气理论发展及其在天然气工业发展中的地位[J]. 石油勘探与开发, 2009,36(3):280-289.
[6] 赵文智,胡素云,李建忠,等. 中国陆上油气勘探领域变化与启示——过去十余年的亲历与感悟[J]. 中国石油勘探,2013,18(4):1-10.
[7] 赵文智,王红军,卞从胜,等. 中国低孔渗储层天然气资源大型化成藏特征与分布规律[J]. 中国工程科学,2012,14(6):31-39.
[8] 郭焦锋,高世楫,赵文智. 中国页岩气已具备大规模商业开发条件[N]. 中国经济时报,2015-04-20(005).
[9] 赵文智,李建忠,杨涛,等. 中国南方海相页岩气成藏差异性比较与意义[J]. 石油勘探与开发,2016,43(4):499-510.
[10] 贾爱林. 中国天然气开发技术进展及展望[J]. 天然气工业,2018,38(4):77-86.
[11] 杨建红. 中国天然气市场可持续发展分析[J]. 天然气工业,2018,38(4):145-152.
[12] 杨光,刘小丽. 2017年中国天然气发展形势与政策分析及2018年展望[J]. 中国能源,2018,40(1):15-18.
[13] 国家能源局石油天然气司,国务院发展研究中心资源与环境政策研究中心.
[14] 国土资源部油气资源战略研究中心. 中国天然气发展报告(2018)[M]. 北京:石油工业出版社,2018.

# 方法篇

# 气藏开发全生命周期不同储量的意义及计算方法

位云生[1]　贾爱林[1]　徐艳梅[1]　方建龙[2]

(1. 中国石油勘探开发研究院;2. 中国石油勘探与生产分公司)

**摘要**:储量评价贯穿气藏开发始终,系统梳理气藏不同储量计算方法,对认识和开发气藏具有重要意义。前期评价阶段,采用容积法或体积法计算探明地质储量;开发方案实施阶段,落实可动用储量,指导井位部署;规模开发后,采用物质平衡与现代递减方法计算动态储量和可采储量,指导技术政策优化;开发中后期,采用数值模拟方法落实剩余储量,指导挖潜部署。对比来看,储集空间结构、流体赋存状态和气藏边界条件是优选储量计算方法的重要依据。

**关键词**:探明地质储量;可动用储量;动态储量;可采储量;剩余储量;储量评价

储量一直是国内外油气矿产资源勘探开发中的一个重要参数[1]。一个油气田从发现、探明到开发直到气田废弃的不同勘探开发阶段,人们对地下油气田地质条件和开发规律的认识不断变化,所获取的参与油气储量计算的各项参数精度也在不断提高。勘探提交控制储量或有重大发现后,开发前期评价开始介入,标志着油气田正式进入开发阶段。通过前期评价对气田基本静态参数在开发尺度下进行初步评价认识,提交探明地质储量,编制试采方案;通过试采评价,利用动态特征验证静态参数,更全面地认识或核实气井生产能力及储量可动用性,落实可动用储量,编制开发方案;在大规模产能建设和开发过程中,采出程度达到10%或井网基本完善后,需要进行储量复算;进入开发动态跟踪和调整阶段,需要进行动态(地质)储量和(经济)可采储量评价,评价现有经济技术条件下的气藏采收率,到开发中后期,评价气藏剩余储量,为措施挖潜提供依据。纵观气藏整个开发过程,储量评价贯穿始终。本文立足不同开发阶段对储量评价的需求,梳理不同类型气藏不同储量的计算方法,深化对气藏的全面认识,提高气藏采出程度和开发水平。

## 1 气藏储量内涵

国外气藏储量通常指剩余经济可采储量,包括P1(proved)、P2(probable)和P3(possible),具有时间属性,反映现状,计算时地面标准条件是15℃、0.101MPa,矿权改变可能改变储量大小和级别,具有经济属性,可以进行交易。

国内气藏储量通常指气藏地质储量,是指在地层原始状态下,气藏中天然气的总储藏量,没有时间属性,反映原始表象,计算时地面标准条件是20℃、0.101MPa,不因权属不同改变储量大小,不具有经济属性,获得地质储量是天然气勘探的最终目标,也是气田开发的前提。在气田开发阶段,气藏储量一般是指探明地质储量,在不同开发阶段,还涉及可动用储量、动态(地质)储量、(经济)可采储量、剩余储量等[2]。以常规气藏为例,用图解法说明开发阶段不同储量的关系和范围(图1),不同阶段储量的意义和计算方法不同。动态储量,也称为动态地质储量,即通过气藏的生产动态数据确定出的气藏地质储量[3]。动态储量与探明地质储量的比值,为气藏储量的可动用程度。可采储量数值上等于最终累计产气量,可采储量或最终累计产气量与探明地质储量的比值,为气藏采收率,中间某一时间的采出气量与探明地质储量的比值,为气藏采出程度,采收率就是气藏废弃时的采出程度。剩余储量即为地层压力下降至某一值时气藏中剩余气量。

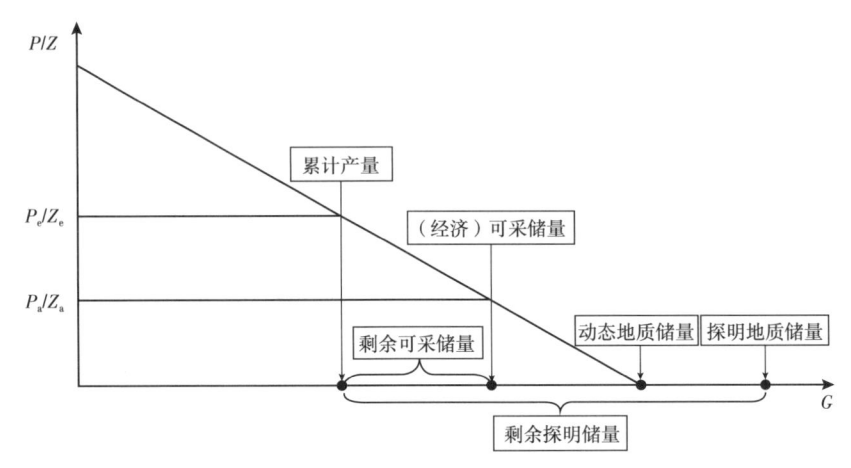

图 1 开发阶段储量分类

$P$ 表示地层压力;$Z$ 表示气体偏差因子;$G$ 表示气藏地质储量;下标 e、a 分别表示任意条件和废弃条件

## 2 前期评价阶段

在勘探提交控制储量或有重大发现后,开发开始介入,主要任务是认识气藏地质与开发特征,采用容积法或体积法评价探明地质储量,落实气藏开发储量基础,评价气田开发技术与经济可行性,优选成熟的气藏开发主体工艺技术,初步确定开发指标,编制开发概念设计。

### 2.1 前期评价模式

根据中国不同类型气藏地质特点和开发需求,形成四种开发前期评价模式和工作序列[4],提高评价成效,为探明地质储量计算提供参数论证依据(表1)。

表 1 不同前期评价模式的气田探明地质储量参数的评价方法

| 模式 | 突出特点 | 探明储量计算参数评价 ||||||
|---|---|---|---|---|---|---|---|
| | | 含气面积 | 有效储层厚度 | 孔隙度 | 含气饱和度 | 岩石密度 | 含气量 |
| 常规构造气藏评价模式 | 储层结构和气水关系简单 | 二维或三维地震、少量评价井 | | 岩心分析、测井解释、试气资料分析 | | — | — |
| 大型优质气田试采评价模式 | 规模大,评价要求高 | 三维地震、评价井、试采数据 | | | | — | — |
| 滚动评价模式 | 断块边界确定难度大 | 高精度三维地震、关键位置一定密度的评价井 | | | | — | — |
| 非常规气田多阶段评价模式 | 分级分区评价 | 三维地震、少量评价井 | | | | 称重法 | 保压取心或折算法 |

## 2.2 探明储量计算

目前投入开发和评价的天然气藏类型较多,大体分为常规气藏和非常规气两大类。天然气在地下的赋存状态不同,有气态和固态两种,气态又分为游离气和吸附气。针对不同类型天然气藏特点和天然气赋存状态,探明储量计算方法也不同。

常规气和致密气储层孔隙以毫米和微米级无机孔为主,地层中天然气以游离气态存在,吸附气量极小,可以忽略不计,故常规气藏和致密气探明地质储量通常采用容积法计算[5-8](表2)。其中对于凝析气藏而言,先利用容积法计算凝析气总地质储量,然后再分别计算干气和凝析油地质储量[5,9]。

页岩储层孔隙以纳米有机孔为主,地层中天然气主要以游离气和吸附气两种状态存在,故页岩气探明地质储量分两部分,游离气地质储量 $G_f$ 采用容积法计算,吸附气地质储量 $G_a$ 采用体积法计算[10-12](表2)。煤层气是储存在煤层中的天然气,地下以吸附在煤基质颗粒表面的吸附气为主,故煤层气探明地质储量采用体积法计算[13](表2)。

天然气水合物是由甲烷等气体分子和水分子在高压低温环境下($3\sim20$MPa,$0\sim10$℃)形成的固态物质,外观像冰且点火即可燃烧,故又称可燃冰[14]。目前已发现的天然气水合物具有三种基本笼型晶体结构[14,15],探明地质储量通常采用类体积法计算(表2)。

表2 不同类型天然气藏探明储量计算方法

| 气藏类型 | 地下赋存状态 | 探明储量计算方法 | 计算公式 | 符号说明 |
|---|---|---|---|---|
| 常规气和致密气 | 气态 | 容积法 | $G = 0.01Ah\phi S_{gi}/B_{gi}$ | $G$—探明地质储量,$A$—含气面积,$h$—有效厚度,$\phi$—有效孔隙度,$S_{gi}$—原始含气饱和度,$B_{gi}$—原始气体体积系数,$\rho_s$—岩石密度,$C_a$—吸附气含气量,$\alpha$—沉积层孔隙中含天然气水合物体积比例,$B$—天然气水合物分解甲烷的膨胀系数 |
| 页岩气和煤层气 | 游离气态 | 容积法 | $G_f = 0.01Ah S_{gi}/B_{gi}$ | |
| | 吸附气态 | 体积法 | $G_a = 0.01Ah\rho_s C_a$ | |
| 天然气水合物 | 固态 | 类体积法 | $G = 0.01Ah\phi\alpha B$ | |

## 2.3 可采储量标定

可采储量又分为技术可采储量和经济可采储量,技术可采储量指在现行的技术条件下和政府法规下,预期从已发现的气藏中,最终可以采出的天然气总量,也称最终可采储量,数值上等于最终累计产量 EUR(Estimated Ultimate Recovery);经济可采储量至当前已实施的或肯定要实施的技术条件下,按当前的经济条件(价格、成本、新增投资等)估算的、可经济开采的天然气总量[16]。国际上所讲的可采储量通常是指经济可采储量,而国内所讲的可采储量通常指技术可采储量。在提交探明储量阶段,国内一般类比同类型气藏的技术采收率和经济采收率,从而标定气藏的技术可采储量与经济可采储量。

# 3 开发方案编制与实施阶段

开发方案是指导气田开发的重要技术文件,是产能建设、生产运行管理、市场开发、长输管道立项的依据。该阶段的主要任务是探明地质储量分类与评价、可动用储量评价、产能评价、开发方式、开发指标与风险论证、组织建产及探明地质储量复算等。

## 3.1 探明储量分类评价

不同类型气藏储量品质差异较大,不同储量品质的建产效果差异也较大。因此,应根据储层物性、储量丰度、气井产能、开发的难易程度和技术经济条件等对储量进行分类,可根据不同储量丰度区的气井产量的经济性制定分类标准,分级落实储量规模,作为分级评价气井开发指标的基础[17]。

开发效益较好的气藏,如龙王庙组气藏,储量可按照孔隙度大小,结合评价井动态特征,分为主体区(孔隙度>6%)储量、过渡区(孔隙度4%~6%)储量、低渗透区(孔隙度2%~4%)储量,对应的单井初期产量分为>100×10⁴m³/d、(30~100)×10⁴m³/d、<30×10⁴m³/d。边际效益气藏,如苏里格气田,则按照开发的经济性分为效益储量(内部收益率>8%)、低效储量(内部收益率6%~8%)、难动用储量(内部收益率<6%)。

## 3.2 可动用储量评价

评价不同类型储量的可动用性,确定可动用储量规模,作为气藏工程方案设计的储量基础。对常规气藏而言,确定储层物性下限是储层评价和可动用储量评价的基础[18],方法较多,不同方法的原理和适用对象有所差异,同一类型气藏,适用的不同方法确定的物性下限值也不尽相同,一般根据不同方法评价结果,综合确定储层物性下限(表3)。

表3 确定储层物性下限的方法及适用条件

| 序号 | 方法 | 原理 | 适用气藏和条件 |
|---|---|---|---|
| 1 | 孔隙度—渗透率交会法 | 利用储层孔隙度—渗透率关系图确定孔隙度下限值,该方法精度不够,仅供参考 | 所有气藏,岩心常规实验数据 |
| 2 | 孔隙结构法 | 利用压汞数据得到渗透率累计贡献值99.9%时的孔喉大小,再根据孔喉中值与常规物性参数的关系,截取物性下限值 | 所有气藏,压汞数据 |
| 3 | 最小含气喉道半径法 | 根据驱动力和毛细管阻力平衡关系,可计算不同气藏高度气体进入岩石孔隙的最小喉道半径,计算最大驱替压力,再根据不同气藏高度下孔隙度与含水饱和度及驱替压力关系,确定孔隙度下限 | 有水构造气藏,压汞数据、岩心常规实验数据 |
| 4 | 钻井液侵入法 | 根据不同层位岩心中氯化盐的含量判断储层物性,该方法精度不够,仅供参考 | 水基钻井液钻井,氯化盐测试数据 |
| 5 | 气—水相对渗透率法 | 根据不同物性岩心气水相对渗透率曲线,确定含水饱和度上限 | 有水气藏,气水两相相渗实验 |
| 6 | 试气法 | 试气成果与孔隙度—电阻率交会图结合确定孔隙度下限 | 所有气藏,试气资料多时适用 |
| 7 | 水锁实验法 | 根据水锁实验得到岩心束缚水饱和度,再根据孔隙度与含水饱和度关系,确定孔隙度下限 | 所有气藏,水锁实验和常规实验 |
| 8 | 核磁共振实验法 | 利用核磁共振实验,建立束缚水饱和度与孔隙度的关系,根据束缚水饱和度值确定孔隙度下限 | 所有气藏,核磁共振实验 |

续表

| 序号 | 方法 | 原理 | 适用气藏和条件 |
|---|---|---|---|
| 9 | 压汞实验的"J函数"评价法 | 利用岩心实验数据建立"J函数",根据储层实际物性数据,得到储层毛细管压力,再根据不同气藏高度下孔隙度与含水饱和度及驱替压力的关系,确定孔隙度下限 | 所有气藏,压汞数据、岩心常规实验数据 |
| 10 | 邻区类比法 | 类比邻区构造、储层等特征,确定物性下限 | 所有气藏,同类型气藏对比 |

特别提出表3中压汞实验的"J函数"评价法,这种方法国外使用较为广泛,特别在数值模拟评价中应用最为普遍。原因是一般情况下实验所得的岩心样品毛细管压力曲线只能表征储层中的某一点的储渗特性,由于储层往往具有非均质性,为了表现整个储层段的毛细管压力特性则需要同时考虑储层段的孔隙度、渗透率和流体性质的变化,从而引入"J函数"的概念。实际应用时,首先将岩心样品的压汞实验数据进行"J函数"的处理分析,绘制一条代表整个储层的"J函数"曲线图[19]。然后将室内毛细管压力校正为气藏条件下的毛细管压力后,根据气水密度差计算对应的气藏高度,从而建立不同气藏高度条件下孔隙度与含水饱和度的关系变化曲线,确定储层流体的可动用性。

### 3.3 探明地质储量复算

在气田开发方案实施过程中,应做好跟踪分析和地层对比工作,不断加深对气藏的认识,如发现气藏地质情况有变化,研究后及时提出井位调整意见和补充录取资料要求,同时做好动态监测及分析。对于连通性较好的气藏,当采出程度达到10%时,应开展气藏探明地质储量复算;对非均质性较强或复杂断块气藏,井网基本完善后,应开展气藏探明地质储量复算,修正由于构造、主要储层参数及气藏温压系统等变化造成的储量评价误差,从而降低气藏开发风险。

## 4 开发方案跟踪与调整阶段

开发方案实施建设阶段结束后,评价的主要任务是开发动态跟踪与开发方案调整。跟踪分析动态储量、可采储量或最终累计产量、气藏储量采出程度、剩余储量、气藏采收率等关键指标,指导编制开发调整方案。

### 4.1 动态储量计算

根据影响动态储量计算的气藏类型和能量补给方式,可将气藏大体分为两类:连通性较好的整装气藏和连通性差的岩性或非常规气藏。前者又可分为定容气藏、封闭气藏和水驱气藏[20]。

#### 4.1.1 连通性较好的整装气藏

##### 4.1.1.1 定容气藏

顾名思义是指天然气开采过程中气藏容积一直不发生变化的气藏,这是一种理想的气藏模型。计算动态储量的常用方法是物质平衡法[21],对于定容气藏,物质平衡方程见式(1),是经典的压降法计算动态储量的公式,也是实际气藏物质平衡方法的基础。

$$\frac{P}{Z} = \frac{P_i}{Z_i}\left(1 - \frac{G_P}{G}\right) \tag{1}$$

其中,$P_i$、$P$ 分别表示原始地层压力和采出气量为 $G_p$ 时的地层压力,MPa;$Z_i$、$Z$ 分别表示原始条件下气体偏差因子和地层压力为 $P$ 时的气体偏差因子,无量纲;$G_p$ 表示地层压力下降到 $P$ 时的累计产气量,m³;$G$ 表示气藏地质储量,m³。

#### 4.1.1.2 封闭气藏

指无相连水体、没有外界补给的气藏,即在开采过程中,与水体或外界之间没有质量和能量交换。大多数岩性圈闭(透镜体等)气藏,基本上都属于封闭气藏。封闭气藏在开采过程中,随着孔隙压力降低,气藏的容积会有所减小,与定容气藏不同。通过地层压力校正,物质平衡法仍是封闭气藏储量计算的有效方法[22,23],修正后的方程见式(2)。

$$\frac{P}{Z}(1 - C_c\Delta P) = \frac{P_i}{Z_i}\left(1 - \frac{G_p}{G}\right) \tag{2}$$

其中,$C_c$ 表示气藏容积压缩系数,$C_c = \frac{C_p + S_{wc}C_w}{1 - S_{wc}}$,反映气藏容积随地层压力的变化程度,MPa$^{-1}$;$C_p$、$C_w$ 分别表示岩石(孔隙体积)压缩系数和地层水压缩系数,MPa$^{-1}$;$S_{wc}$ 表示束缚水饱和度,小数。

其他适用封闭气藏的动态储量计算方法很多,比较成熟、准确(表4)。连通性较好、构造型的封闭气藏可采用物质平衡法、弹性二相法、数值模拟等方法评价,计算较简便快速。

**表4 主要动态储量计算方法及适用条件**

| 方法类型 | 方法名称 | 气藏类型 | 适用条件 | 可靠程度 | 适用范围 气井 | 适用范围 气藏 |
|---|---|---|---|---|---|---|
| 传统方法 | 物质平衡法 | 各类气藏 | 采出程度大于20% 至少两个静压测试点 | 可靠 | √ | √ |
| 传统方法 | 弹性二相法 | 定容封闭气藏 | 拟稳态,定产生产 | 较可靠 |  | √ |
| 传统方法 | 不稳定试井法 | 定容封闭气藏 | 各开发阶段 | 较可靠 | √ |  |
| 传统方法 | 数值模拟法 | 各类气藏 | 地质模型、生产动态 | 可靠 | √ | √ |
| 现代方法 | Fetkovich | 定容封闭气藏 | 定压生产 | 较可靠 | √ |  |
| 现代方法 | Blasingame | 各类气藏 | 变产量、变压力 | 较可靠 | √ |  |
| 现代方法 | AG | 各类气藏 | 变产量、变压力 | 较可靠 | √ |  |
| 现代方法 | NPI | 各类气藏 | 变产量、变压力 | 较可靠 | √ |  |
| 现代方法 | FMB | 各类气藏 | 变产量、变压力 | 较可靠 | √ |  |
| 现代方法 | 解析法 | 各类气藏 | 解析拟合求解 | 较可靠 | √ |  |

#### 4.1.1.3 水驱气藏

水驱气藏,是指与周围水体(边水、底水)相连通,当气藏开采时,边底水会因气藏压力的下降而侵入气藏,从而驱替气藏中的天然气[24,25]。水驱气藏开采到一定程度就会产水,使得气藏的开发动态变得复杂,动态储量评价需要考虑水体侵入的影响。进一步校正地层压力后,仍可采用物质平衡法计算气藏动态储量,修正后的方程见式(3)。

$$\frac{P}{Z}(1 - C_e \Delta P - \omega) = \frac{P_i}{Z_i}\left(1 - \frac{G_p}{G}\right) \tag{3}$$

式中 $\omega$ ——气藏存水体积比,即气藏存水量占气藏容积的百分数,$\omega = W/GB_{gi}$,无量纲;

$W_e$、$W_p$、$W$ ——水侵量、产水量、存水量(气藏被水侵占据的孔隙体积),$W = W_e - W_p B_w$,$m^3$;

$B_{gi}$、$B_w$ ——原始条件下的气体体积系数和地层水体积系数,无量纲。

另外,数值模拟方法、Blasingame、AG、NPI、FMB 及解析方法均可用于水驱气藏的动态储量计算[25],核心是校正地层压力。

#### 4.1.2 连通性较差的岩性或非常规气藏

由于气藏连通性较差,通过气藏整体压力测试或产量递减均无法实现对全气藏压力的掌握,因此,以井为单元,首先评价单井动态储量,然后通过单井动态储量采用面积加权或类比推算到全气藏,从而计算全气藏动态储量。

对于非常规气而言,由于储层极其致密,关键参数难以准确测量或获取,动态储量评价时常联合概率法使用,给出动态储量的范围[26]。

### 4.2 可采储量计算

气藏投入规模开发后,可采储量可通过多种方法进行计算。技术可采储量与动态储量的差别,就是废弃条件(废弃条件和废弃压力),两者差值等于气藏达到废弃条件时没有采出的那部分动态储量[27]。在采用气藏工程方法和数值模拟方法计算动态储量时,考虑废弃压力和废弃产量,计算得到的储量就是技术可采储量,即 EUR。而经济可采储量的计算,需要采用现金流量法和经济极限法进行详细计算,由于所需的经济参数较多,且参数变化趋势预测难度大,因此,经济可采储量在国内应用较少。

### 4.3 剩余储量计算

剩余储量是气田开发中后期评价的重要内容。由于储层条件的非均质性和布井的不均匀性,剩余储量多呈分散状分布,因此剩余储量评价包括剩余储量分布和剩余储量大小。剩余储量评价是气田开发动态监测中至关重要的一项内容,它是指导开发调整方案的重要依据,也是气田提高采收率的关键因素。

剩余储量评价可采用气藏工程方法和数值模拟方法,而气藏工程方法仅能计算剩余储量大小[28]。目前主要采用数值模拟方法准确直观地反映剩余气分布,计算剩余储量大小。具体研究步骤是首先进行气藏静动态精细描述,包括精细沉积微相划分、井点测井精细解释、储层净毛比校正、有效砂体空间展布及连通性分析、单井泄气范围评价等,然后采用确定性和随机建模方法,建立气井静态和生产动态资料综合约束的精细地质模型,最后根据单井生产历史和区块生产动态,进行数值模拟动态曲线,校正动用范围和动用程度,精细刻画剩余储量三维空间分布及剩余储量大小,为调整井部署和措施挖潜提供依据和基础。

## 5 结论

不同类型气藏不同开发阶段关注的储量不同,对应的计算方法也不同。通过本文的系统梳理和对比研究,得出以下认识:

(1)前期评价阶段,探明地质储量评价是关键。常规气藏采用容积法计算;非常规气主要分游离气和吸附气两部分,分别采用容积法和体积法计算,天然气水合物以固态赋存于储层中,初步以地下固态甲烷体积和地下到标况下的膨胀系数进行计算。

(2)开发编制与实施阶段,可动用储量评价是核心。确定储量是否可动用,主要是论证储层动用的物性下限,论证方法很多,"J 函数"评价法是目前国内外常用的一种方法。

(3)开发方案跟踪与调整阶段,动态储量、可采储量和剩余储量评价是重点。气藏开发到中后期,开发规律较为明确,开发动态资料较完善,各计算方法均可适用,动态储量、可采储量及剩余储量评价较为靠实。动态储量和可采储量对于封闭气藏和水驱气藏而言,均可采用传统物质平衡方法、现代产量递减方法进行计算,但应注意地层压力的校正。剩余储量的评价对于开发中后期的动态调整非常关键,采用地质建模和动态数值模拟手段,落实剩余储量分布和大小,指导挖潜部署。

## 参 考 文 献

[1] 贾爱林,闫海军,郭建林,等. 全球不同类型大型气藏的开发特征及经验[J]. 天然气工业,2014,34(10):33-46.

[2] 叶庆全,袁敏. 油气田开发常用名词解释[M]. 北京:石油工业出版社,2002.

[3] 李熙喆,刘晓华,苏云河,等. 中国大型气田井均动态储量与初始无阻流量定量关系的建立与应用[J]. 石油勘探与开发,2018,45(6):1020-1025.

[4] 贾爱林. 中国天然气开发技术进展及展望[J]. 天然气工业,2018,38(4):77-86.

[5] 陈元千. 油气藏工程实践[M]. 北京:石油工业出版社,2011.

[6] 孙贺东,王宏宇,朱松柏,等. 基于幂函数形式物质平衡方法的高压、超高压气藏储量评价[J]. 天然气工业,2019,39(3):56-64.

[7] 赵宽志,张丽娟,郑多明,等. 塔里木盆地缝洞型碳酸盐岩油气藏储量计算方法[J]. 石油勘探与开发,2015,42(2):251-256.

[8] 丁显峰,刘志斌,潘大志. 异常高压气藏地质储量和累积有效压缩系数计算新方法[J]. 石油学报,2010,31(4):626-628.

[9] 郭平,欧志鹏. 考虑水溶气的凝析气藏物质平衡方程[J]. 天然气工业,2013,33(1):70-74.

[10] 张烈辉,陈果,赵玉龙,等. 改进的页岩气藏物质平衡方程及储量计算方法[J]. 天然气工业,2013,33(12):66-70.

[11] 李海涛,王科,张庆,等. 基于修正容积法计算页岩气井改造区原始天然气地质储量[J]. 天然气工业,2017,37(11):61-69.

[12] 周尚文,王红岩,薛华庆,等. 页岩过剩吸附量与绝对吸附量的差异及页岩气储量计算新方法[J]. 天然气工业,2016,36(11):12-20.

[13] 王星锦,王伟. 煤层气储量计算方法[J]. 天然气工业,1998,18(4):24-27.

[14] 戴金星,倪云燕,黄士鹏,等. 中国天然气水合物气的成因类型[J]. 石油勘探与开发,2017,44(6):837-848.

[15] Wei Zhang, Jinqiang Liang, Jiangong Wei, et al. Origin of natural gases and associated gas hydrates in the Shenhu area, northern South China Sea: Results from the China gas hydrate drilling expeditions [J]. Journal of Asian Earth Sciences, 2019, 183.

[16] SY/T 6098—2010. 天然气可采储量计算方法[S]. 中华人民共和国石油天然气行业标准, 国家能源局, 2010.

[17] 郭智, 贾爱林, 冀光, 等. 致密砂岩气田储量分类及井网加密调整方法——以苏里格气田为例[J]. 石油学报, 2017, 38(11): 1299-1309.

[18] 贾爱林, 付宁海, 程立华, 等. 靖边气田低效储量评价与可动用性分析[J]. 石油学报, 2012, S2(12): 160-165.

[19] Tarek Ahmed. Reservoir engineering handbook[M]. Elsevier, 2006.

[20] 李传亮. 油藏工程原理[M]. 北京: 石油工业出版社, 2005.

[21] 张立侠, 郭春秋, 蒋豪, 等. 物质平衡—拟压力近似条件法确定气藏储量[J]. 石油学报, 2019, 40(3): 337-349.

[22] 王会强, 彭先, 李爽, 等. 裂缝系统气藏动态储量计算新方法——以四川盆地蜀南地区茅口组气藏为例[J]. 天然气工业, 2013, 33(3): 43-46.

[23] 李新峰. 一种提高独立封闭有水气藏储量计算精度的新方法[J]. 天然气工业, 2004, 24(3): 94-97.

[24] 黄全华, 方涛. 低渗透产水气藏单井控制储量的计算及产水对储量的影响[J]. 天然气工业, 2013, 33(3): 33-36.

[25] 唐圣来, 罗东红, 闫正和, 等. 中国南海东部强边底水驱气藏储量计算新方法[J]. 天然气工业, 2013, 33(6): 44-47.

[26] 陈劲松, 韩洪宝, 年静波, 等. 概率法在页岩气未开发最终可采量评估中的应用——以北美某成熟页岩气区块为例[J]. 天然气工业, 2018, 38(7): 52-58.

[27] 曹毅民, 丁蓉, 赵启阳, 等. 煤层气可采储量计算方法的评价与应用[J]. 天然气工业, 2018, S1(12): 50-56.

[28] 陈元千, 唐玮. 油气田剩余可采储量、剩余可采储采比和剩余可采程度的年度评价方法[J]. 石油学报, 2016, 37(6): 796-801.

# 鄂尔多斯盆地东缘致密砂岩气藏动态储量计算方法研究

王泽龙[1,2,3]　唐海发[3]　杨佳奇[4]　吕志凯[3]　成　伟[5]　刘群明[3]

（1. 中国科学院大学；2. 中国科学院渗流流体力学研究所；3. 中国石油勘探开发研究院；
4. 中建环能科技股份有限公司；5. 湖北能源集团）

**摘要**：鄂尔多斯盆地是中国最大的天然气生产基地，近年来，大宁—吉县区块成为了鄂尔多斯盆地东缘煤系地层致密砂岩气藏勘探开发的重大突破，初步形成了年产 $5×10^8m^3$ 产能规模，生产效果良好。但气田还处于开发评价的早期阶段，地质认识不深入，气井生产规律不明确。动态储量的确定对于明确气田开发规模和稳产时间有重要作用。

本文针对以大宁—吉县气田为代表的鄂尔多斯盆地东缘致密砂岩气藏针对该区块气藏开发时间短，地质认识不清，致密气井生产初期递减迅速，后期低产量长期生产的特征，对比不同动态储量计算方法的特点与适用性，建立了一套综合评价该类型气藏动态储量的方法体系，并以大宁—吉县区块 10 口气井为例，运用该套方法体系计算了单井动态储量与泄流面积。

**关键词**：致密砂岩气藏；气藏动态储量；鄂尔多斯盆地东缘

随着苏里格、子州—米脂、神木等致密气田的勘探开发，鄂尔多斯盆地已经成为中国天然气产量最高的盆地。鄂尔多斯盆地东缘大宁—吉县区块，构造上隶属于盆地东部的晋西挠褶带。西部为延长区块，北部为石楼西合作区块，西北为长庆清涧勘查区块，勘探面积约 $6000km^2$。与盆地中北部的伊陕斜坡区的苏里格、子洲—米脂、神木等气田相比，同样具有致密砂岩气成藏的有利条件，无论在烃源岩、沉积体系及储层类型等具有类似的成藏条件，主要含气层系和目的层基本一致。东缘主体处于盆地生烃强度大于 $20×10^8m^3/km^2$ 的生烃中心，主要目的层上古生界发育辫状河水道、三角洲分流河道和沿岸沙坝等多种砂体类型，砂岩厚度可达近 100m，孔隙度为 4%～10%，为典型的致密砂岩储层。多种类型的砂体与煤系烃源岩构成源内和近源成藏组合，构造形态在桃园断层等主干断层以西为平缓的单斜构造，埋藏深度 1500～2000m，埋藏深度适中，具有良好的成藏条件。

中国石油煤层气公司 2013 取得该区块的矿权，2015 年开始致密气的生产。目前气田仍处于早期评价阶段，动态储量的确定对于明确气田开发规模和稳产时间有重要作用。前人就气藏动态储量的确定做了很多研究，提出了多种计算方法，这些方法大多需要提供准确的气藏物性参数，或要求气井流动达到拟稳态，达到较高的采出程度。但由于致密气藏，特别是鄂尔多斯盆地东缘开发时间相对较短，气藏参数掌握不准确的这类低渗透率储层，还没有形成有效的计算方法。

气藏动态储量是指气藏联通孔隙内，在现有开采技术水平条件下设想地层压力降为零时能有效流动的气体,折算到标准条件的体积量之和[2]。动态储量是可靠的地质储量,能直接反映气井的最终累计产量,实用性强。其计算方法是根据生产动态数据,如井底流压、产气量、产液量等,折算出气藏中可以流动的天然气储量,即为气藏动态储量。随着气井生产时间的推移,动态资料逐渐增多,计算结果也更加准确。

# 1 致密气井动态储量计算方法

20世纪40年代以来,中外学者对于油气藏动态储量做了大量的研究,提出了多种计算方法。考虑到致密气井生产的适用性,常用的计算方法包括产量累计法,流动物质平衡方法、采气曲线法,以及产量不稳定分析法等。

## 1.1 产量累计法

根据气田生产经验,可认为累计产气量与生产时间符合以下关系:

$$G_p = a - \frac{b}{t} \tag{1}$$

当生产时间 $t$ 接近无穷大时,此时 $G_p$—$t$ 的关系趋近于它的水平渐近线,$a$ 值即为气藏动态储量(图1)[1]。

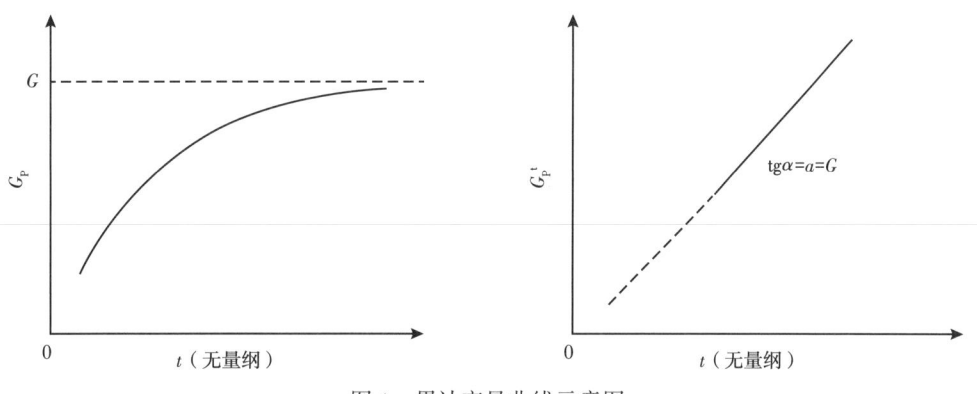

图1 累计产量曲线示意图

这种方法简便易行,不需要获得地层压力等地质属性,但要求产量发生正常持续递减时才可应用,一般采出程度要求达到40%~50%。

## 1.2 流动物质平衡方法

气藏的物质平衡方程为

$$G = \frac{G_p B_g - (W_e - W_p B_w)}{B_g - B_{gi}} \tag{2}$$

以视地层压力表示的压降方程为

$$\frac{P}{Z} = \frac{P_i}{G_p} \left[ \frac{G - G_p}{G - (W_e - W_p B_w) \frac{P_i T_{sc}}{P_{sc} Z_i T}} \right] \tag{3}$$

当没有连通的边底水时,可是为定容气藏,其物质平衡方程可简化为

$$GB_{gi} = (G-G_p)B_g \tag{4}$$

定容气藏的压降方程为

$$\frac{P}{Z} = \frac{P_i}{Z_i}\left(1 - \frac{C}{G}\right) \tag{5}$$

$$\frac{P}{Z} = \frac{P_i}{G_p} \tag{6}$$

令 $a = \dfrac{P_i}{Z'_i}, b = \dfrac{a}{G'}$,则

$$\frac{P}{Z} = a - bG_p \tag{7}$$

即 $\dfrac{P}{Z}$ 和 $G_p$ 呈线性关系,因此可根据不同阶段视地层压力求算累计采气量。在已知初始视地层压力 $\dfrac{P_i}{Z_i}$ 和投产后任意时期的视地层压力 $\dfrac{P}{Z}$ 以及当时的累计采气量 $G_p$ 时,可解得气藏储量为

$$G = \frac{G_p \dfrac{P_i}{Z_i}}{\dfrac{P_i}{Z_i} - \dfrac{P}{Z}} \tag{8}$$

对于非均质性气藏,压降线一般出现三段:初始段、直线段和上翘段(图2),这是由于初期产量大,采气速度高,但低渗透区补给速度不足,形成初始段陡降;后期采气速度低,低渗透区补给相对较高,形成末段上翘。在计算动态储量时,应采取中期直线段,通过原始地层压力点作直线的平行线与横轴相交所得的储量[3]。

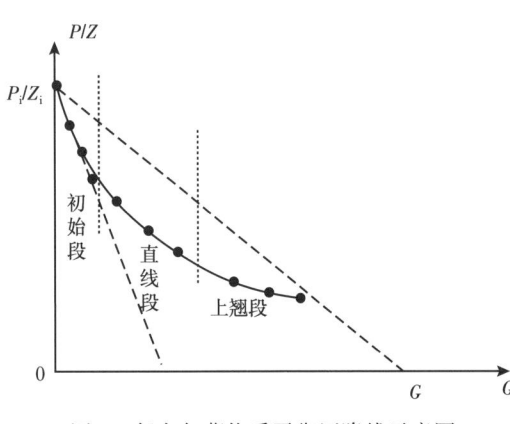

图2 复杂气藏物质平衡压降线示意图

## 1.3 采气曲线法

在一些特殊情况下,开井生产长时间不能关井,但气井具有稳定试井和开采资料,此时用采气曲线法试凑法进行气井生产史拟合来估算气井控制储量是比较有效的方法之一。该方法主要是借助物质平衡方程和二项式采气方程,建立的一种分析方法,可计算一口探井或气藏试采早期所反映出的气井控制储量。

假设稳定试井产能方程:

$$P_e^2 - P_{wf}^2 = Aq + Bq^2 + C \tag{9}$$

结合气藏物质平衡方程,可推导出定容气藏的储量计算方程:

$$\left[P_i\left(1 - \frac{G_p}{G}\right)\right]^2 - P_{wf}^2 = Aq + Bq^2 + C \tag{10}$$

该方法的基本思路是假设在某一控制储量条件下,联立求解气藏物质平衡方程式和气井二项式采气方程,进而计算气井采气生产曲线,与实际的采气生产曲线相匹配,最后确定有关参数[4-6]。

### 1.4 压力—产量递减法

对生产处于递减期的定容封闭气藏,在衰竭开发方式下,视地层压力和气藏产量均不断衰减,根据物质平衡原理,具有如下关系:

$$\frac{P}{Z} = a + bQ_g \tag{11}$$

其中:

$$\alpha = \frac{P_i}{Z_i}\left(1 - \frac{Q_{gi}}{GD_a}\right) \tag{12}$$

$$b = \frac{P_i}{Z_i GD_a} \tag{13}$$

其中,$D_a$ 为递减率。任取两时刻的视地层压力和产量用下式计算气藏的原始储量:

$$G = \frac{E\left[(Q_{gi} - Q_{g2})\frac{P_1}{Z_1} - (Q_{gi} - Q_{g1})\frac{P_2}{Z_2}\right]}{D_a\left(\frac{P_1}{Z_1} - \frac{P_2}{Z_2}\right)} \tag{14}$$

### 1.5 产量不稳定分析法

产量不稳定分析法又称高级递减分析法,是拟合相关气井生产动态曲线特征图版来进行动态储量的估算,是以气井生产动态(产量、流动压力等)为基础,引入不稳定试井分析的基本思想,对传统的产量递减法进行了改进,建立了气井生产曲线特征图版,把气井早期的不稳定流动段和后期的边界流动段结合起来,通过特征图版拟合计算储层物性、表皮系数、井控半径和动态储量。主要分析特征图版包括 Arps 图版、Fetkovich 图版、Blasingame 图版、Agwarl-Gardner 图版、NPI(Normalized Pressure Integrate)图版、Transient 图版、流动物质平衡(flowing materialbalance)等[7-10]。

## 2 气井动态储量计算方法

鄂尔多斯盆地东缘致密气藏是一个定容气藏,自 2014 年中国石油煤层气公司开发大宁—吉县区块致密气藏以来,气藏动态分析工作经过不断摸索与创新,建立了"储量是基础,压力是根本,产能是核心"的动态分析思路。针对该区块气藏开发时间短,地质认识不明确,致密气藏初期递减迅速,后期低产量长期生产的特征,建立了一套多种方法综合评价动态储量的方法体系。

分析以上动态储量计算方法,产量累计法为经验方法,对于有较长生产时间的气井估算较为准确,采出程度较高一般要求达到40%以上。对于大宁—吉县区块,生产时间普遍在三年以内,故产量累计法估算的井控动态储量可靠性较差;流动物质平衡法要求压力波达到边界,地层压力下降后应用结果较为可靠,对于大宁—吉县区块的老井,此方法计算得到的动态储量比较可靠;采气曲线法同样用到了物质平衡方程,而且有修正后的气井产能方程辅助,因而计算结果较为准确;压力—产量递减法,对于衰竭式开采的定容气藏结果较为可靠,因而可适用于鄂尔多斯盆地东缘致密气藏;Blasingame图版和Agarwal—Gardner图版都属于产量不稳定分析法,都是利用单井的生产历史数据与典型图版进行拟合,计算出井控动态储量[11-13],因而较好地适应了气井工作制度的频繁改变,对于地层压力测试点的依赖程度也很低,获得的结果最为准确。

在针对大宁—吉县区块致密气藏动态储量的计算过程中,我们首先应用可靠性较高的产量不稳定分析法,利用RTA软件优先使用Blasingame图版和Agarwal—Gardner图版框定气井的动态储量范围,由于拟合图版人为主观性比较强,容易产生较大误差,再利用流动物质平衡法、采气曲线法和压力—产量递减法反复校正,最后利用经验法(产量累计法)校验结果的可靠性。

在实际操作过程中,Blasingame图版中压力导数和无量纲化的产气量有明显的交点,可作为参考点来拟合实际的生产数据,而Agarwal—Gardner图版因没有明显的参考点,存在较大的人为误差(图3),故在实际图版法的应用中,应先拟合Blasingame图版,之后在Agarwal—Gardner图版中校验。

气井控制动态累加与区块整体计算的动态储量具有良好的对应关系,下面以10口致密气井为例,进行了单井控制动态储量的计算,结果见表1。

表1 大宁—吉县区块10口气井动态储量计算结果汇总　　　　单位:$10^8 m^3$

| 井号 | 产量累计法 | 流动物质平衡法 | 采气曲线法 | 压力—产量递减法 | Blasingame图版 | Agarwal—Gardner图版 |
| --- | --- | --- | --- | --- | --- | --- |
| 大吉1X-X | 4987 | 5011 | 5011 | 5324 | 5334 | 5313 |
| 大吉2X-X | 1706 | 2011 | 2011 | 1839 | 1855 | 1828 |
| 大吉3X-X | 95 | 103 | 103 | 75 | 88 | 88 |
| 大吉4X-X | 906 | 1193 | 1193 | 1061 | 1169 | 1027 |
| 大吉5X-X | 1621 | 1601 | 1601 | 1677 | 1741 | 1732 |
| 大吉6X-X | 2613 | 2631 | 2631 | 2933 | 2936 | 2937 |
| 大吉7X-X | 11614 | 11807 | 11807 | 11588 | 11384 | 11338 |
| 大吉8X-X | 7391 | 7538 | 7538 | 7537 | 7630 | 7728 |
| 大吉9X-X | 5029 | 5143 | 5143 | 5199 | 5145 | 5231 |
| 大吉10X-X | 12783 | 12993 | 12993 | 13278 | 12973 | 13214 |

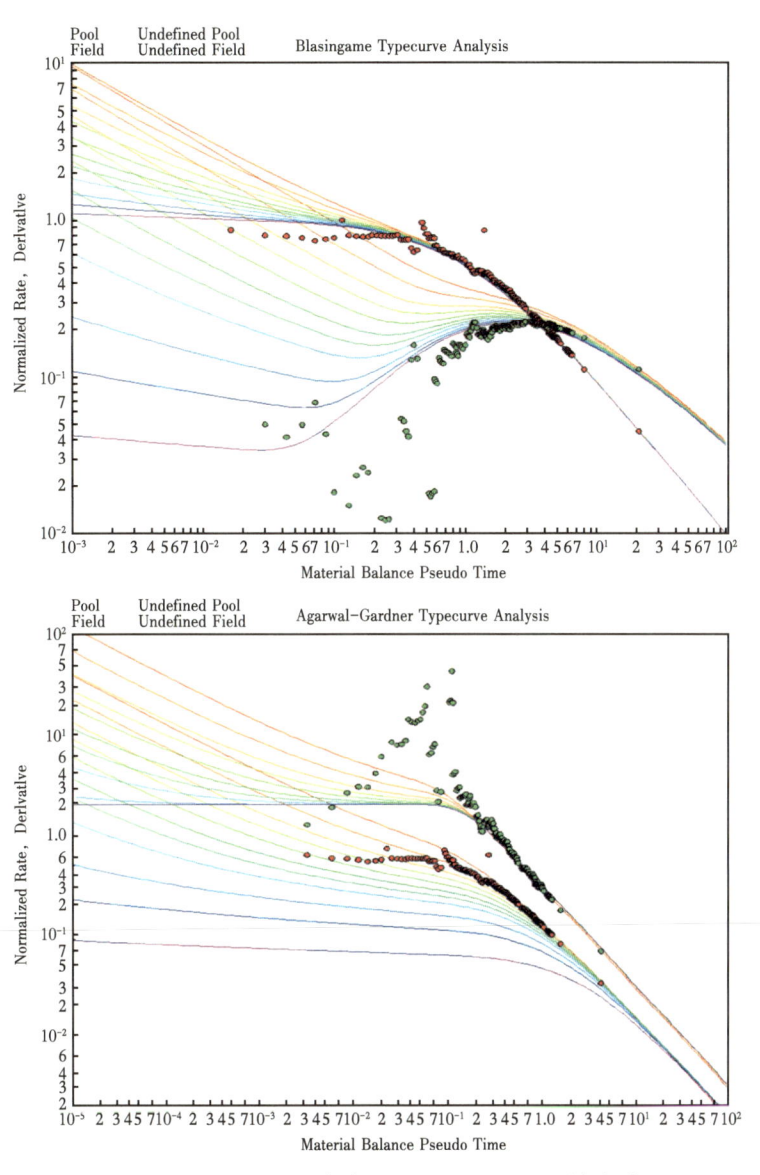

图 3  Blasingame 图版与 Agarwal—Gardner 图版拟合

综合以上方法,相互验证,所得的平均值即为比较可靠的单井控制动态储量,根据气藏基本物性参数,可计算出气井泄流半径(表2),并绘制气井泄流面积分布图(图4),用于核实气田整体的动态储量,也可以用于调整井网井距,为开打加密井,提高气田采收率提供依据。

表 2  大宁—吉县区块 10 口气井泄流半径统计表       单位:m

| 井号 | 泄气半径 | 井号 | 泄气半径 |
| --- | --- | --- | --- |
| 大吉 1X-X | 400.37 | 大吉 6X-X | 368.65 |
| 大吉 2X-X | 345.83 | 大吉 7X-X | 1080.23 |

续表

| 井号 | 泄气半径 | 井号 | 泄气半径 |
|---|---|---|---|
| 大吉 3X-X | 117.05 | 大吉 8X-X | 944.25 |
| 大吉 4X-X | 463.36 | 大吉 9X-X | 455.85 |
| 大吉 5X-X | 319.37 | 大吉 10X-X | 936.59 |

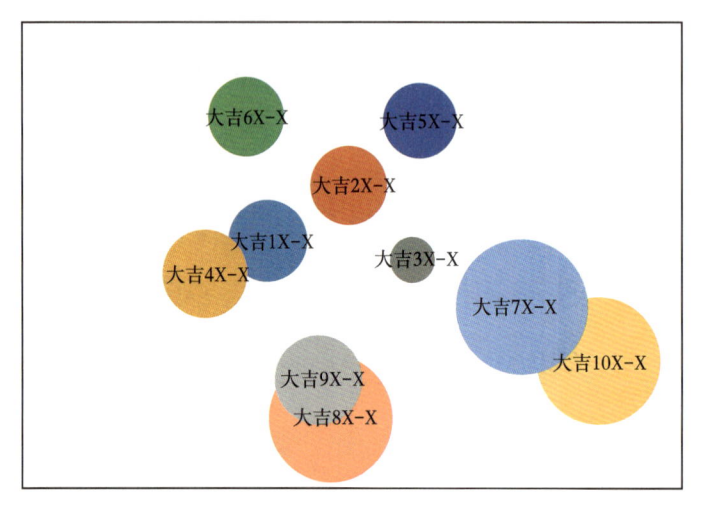

图 4　大宁—吉县区块 10 口气井泄流面积分布示意图

## 3　结论与认识

（1）鄂尔多斯盆地东缘致密气资源丰富,中国石油煤层气公司开发的大宁—吉县致密气田取得了较好的开发效果。针对该致密砂岩气藏地质特征,气井衰竭式开发形式,综合产量不稳定分析法、流动物质平衡法、采气曲线法、压力—产量递减法及产量累计法特点,多种方法相互校对,形成了一套有效评价鄂尔多斯盆地东缘致密气藏动态储量的方法。

（2）结合表 1 动态储量计算结果可以看出,采气曲线法和流动物质平衡法计算结果与产量不稳定分析法计算的储量接近,较为可靠;产量累计法计算出的储量总体偏低。产量累计法为经验统计方法,都要求气井或气藏要达到一定的采出程度,产量进入持续递减期时计算结果才较可靠,而大宁—吉县区块致密气井生产时间较短,开关井操作频繁,部分井产量不稳定,没有进入持续递减期。

（3）产量不稳定分析方法利用气井产量与压力测量数据,拟合渗流模型典型图版,理论上计算的动态储量更加接近真实情况。实际工作中,大多借助软件进行图版拟合,人为误差可能性大,故需要多种模型相互校正,综合应用采气曲线法、流动物质平衡法以及经验法校对,修正动态储量计算结果。

（4）气田单井控制动态储量累加与区块整体计算的动态储量具有良好的对应关系,对于密集井网开采的致密气藏,单井累加动态储量即为气藏整体动态储量。

## 参考文献

[1] 陈霖,熊钰,张雅玲,等. 低渗气藏动态储量计算方法评价[J]. 重庆科技学院学报,2013,15(5):31-34.

[2] 王京舰,王一妃,李彦军,等. 鄂尔多斯盆地子洲低渗透气藏动态储量评价方法优选[J]. 石油天然气学院学报,2012,34(11):114-117.

[3] 李士伦,等. 天然气工程[M]. 北京:石油工业出版社,2000.

[4] 李旭. 苏里格气田东区动态储量计算方法[J]. 石油化工应用. 2016,35(3):98-101.

[5] 李传亮. 油藏工程原理[M]. 北京:石油工业出版社,2005.

[6] 程时清,李菊花,李相方,等. 用物质平衡二项式产能方程计算气井动态储量[J]. 新疆石油地质,2005,4(2):181-182.

[7] 刘晓华. 气藏动态储量计算中的几个关键参数探讨[J]. 天然气工业,2009,29(9):71-74.

[8] Fetkovich M J, Vienot M E, Bradley M D, et al. "Decline curve analysis using type curves case histories," SPEFE(December 1987)637, Trans., AIME, 283.

[9] 王卫红,沈平平. 非均质复杂低渗气藏动态储量的确定[J]. 天然气工业,2004,24(7):80-82.

[10] 梅志宏. 低渗致密气藏单井动态储量计算方法研究[D]. 成都:西南石油大学硕士论文,2013.

[11] 何丽萍. 长庆气田动储量评价方法研究[D]. 西安:西安石油大学硕士论文,2010.

[12] Havlena D, Odeh A S: "The material balance as an equation of a straight line," JPT (August 1963)896; Trans., AIME, 228.

[13] Blasingame A, Johnston J L, Lee W J, et al. Type-curve analysis using the pressure integral method[C]. SPE California Regional Meeting, 5-7 April 1989,Bakersfield, California. SPE, 1989;SPE 18799.

# Unified Approach to Optimize Fracture Design of Horizontal Well Intercepted by Primary and Secondary Fracture Networks

Junlei Wang[1]  Yunsheng Wei[2]  Wanjing Luo[2]

[1. PetroChina Research Institute of Petroleum Exploration and Development (RIPED);
2. China University of Geosciences (Beijing)]

**Abstract**: The classical optimization design based on single-fracture assumption is widely applied in performance optimization for hydraulically fractured wells. The objective of this paper is to extend the optimal design to complex fracture network for achieving the maximum productivity index (PI). In this work, we established a pseudo-steady-state (PSS) productivity model of fractured horizontal well, which has the flexibility of accounting for the complexity of fracture-network dimensions. A semi-analytical solution was then presented in the generalized matrix format through coupling reservoir and fracture flowing systems. Subsequently, several published literatures on PSS productivity calculation of single fracture were used to verify this model, and a 3D transient numerical simulation of orthogonal fracture network was employed to make further verification. It is shown that results from our solutions agree very well with those benchmarked results. On the basis of the model, we provide detailed analysis on the productivity enhancement of fracture network and optimization workflow by using Unified Fracture Design (UFD). The results show that (1) the productivity index is determined by fracture conductivity and complexity (network size, spacing and configuration), and it is a function with regard to fracture complexity and conductivity when the influence of proppant volume is not considered; (2) in the constraint of a given amount of proppant known as UFD, the maximum PI would be achieved when the best balance between network complexity and conductivity was obtained; (3) it is more advantageous to minimize fracture complexity by creating relatively-simple-geometry fractures with smaller network size and larger fracture spacing in the condition of small and mediate proppant number; (4) it should be the design goal to generate complex network by creating relatively-complex-geometry fractures with larger network size and smaller fracture spacing in the condition of large proppant number; (5) increasing fracture complexity could reduce the optimal requirement of fracture conductivity. The proposed approach can provide guidance for hydraulic network-fracturing design for an optimal completion.

**Key words**: Fracture network, PSS productivity index, Unified Fracture Design, Proppant number, Optimum fracture dimensions

## 1 Introduction

Due to the large-scale energy demand and fast-growing investment in the worldwide development of unconventional resource, horizontal drilling and hydraulic fracturing serve as two reliable technologies in practice of recovering oil and gas from unconventional reservoirs. The key success is to generate very complex man-made fracture network that contacts large reservoir volume with appropriate conductivity, which is effectively propped by a large amount of proppants. Therefore, determination of optimum fracture design [azimuth, number, spacing, length, width of (partially) propped fracture etc.] is of great significance in maximizing productivity of fractured horizontal well.

Numerous papers have been published on the subject of optimizing the performance of multi-fractured horizontal well with planar fracture geometry. Beginning with the pioneering work of Prats[1], the theory of "Unified Fracture Design" was introduced by Economides et al[2] as a standard criterion for optimizing fracture design, relating the proppant volume to the maximum productivity index in PSS condition. In UFD, the optimum design was obtained by achieving the best compromise between the ability of the reservoir making fluid flow into fracture and the ability of the fracture conducting the fluid into wellbore. On the basis of the assumption of single-fracture configuration, several methods have been presented to calculate productivity index (PI) of fractured well in reservoir with different shapes, such as the works given by Valko and Economides[3], Daal and Economides[4], Meyer and Jacot[5], Bhattacharya et al[6] and Lu and Chen[7]. Extensive researchers have further investigated the effect of slanted fracture[8], penny-shaped fracture[9], anisotropic reservoir with different aspect ratio[10], nonuniform multiple fractures[11], varying width of hydraulic fracture[12], and radial flow inside hydraulic fracture near the wellbore[13] on the PI and optimum fracture dimensions. It is worth noting that these methods are only effective in the condition that all fractures have identical properties and are parallelly located in the equally-spaced configuration, so that the well model multiplies the single-fracture result by the number of fractures in the well. In reality, the complex fracture network is more common in the fractured reservoirs. The fracture designs based on single-fracture model are not appropriate for complex fractured reservoirs, because the inherent heterogeneity of fracture completion can lead to significantly lower productivity index than homogeneous completion[14-16].

Most current studies on fracture network with complex fracture geometry were performed to quantify the relationship between the fracture dimension and transient productivity[17-20]. By contrast, the mechanism on fracture optimization of complex fracture network is not well investigated. Gu et al[14] established an orthogonally intersecting fracture network and the target of achieving 1-year critical conductivity was recommended to select the optimum amount of proppant. Since the response is in the transient condition in their model, the characterization of effective drainage volume/stimulated reservoir volume (SRV) cannot be taken into account. Luo et al[21] proposed a general PSS productivity index modeling of complex fractures by use of semi-analytical approach, but their model has no ability of accounting for fracture-fracture intersection in network. Besides, the focus of Luo's work was put on analyzing the productivity index in the PSS condition, rather than the optimization of fracture dimensions using UFD.

Considering the successful applications of UFD in planar fracture and extensive studies on productivity of complex fracture network, UFD has the potential of achieving the optimal values of fracture-network conductivity and dimensions under the constraint of given proppant number[2]. Here, the proppant number describes the weighted ratio of propped fracture volume to a rectangular reservoir volume, which is defined as follows:

$$N_{prop} = \frac{2K_f}{K_m}\frac{V_{prop}}{V_{res}} = \sum_{n=1}^{N_f} \frac{4x_{fn}^2}{x_e y_e}\left(\frac{K_f w_{fn}}{K_n x_{fn}}\right) \tag{1}$$

where $N_f$ is the fracture-panel number, $V_{prop}$ is the proppant volume. Therefore, the main purpose of this paper attempts to fill the gap between the productivity calculation and the UFD extension in complex fracture network. This paper is twofold: (1) calculating productivity index of fracture network by use of semi-analytical approach, (2) achieving optimal balance between network conductivity and complexity by use of UFD theory. Based on the simulation results, some summaries and conclusions were made.

## 2 Model development

### 2.1 Physical model

According to the description in Fig. 1 presented by Fisher et al[22], the fracture network propped effectively could be grouped into two kinds: primary fractures directly connected to the wellbore and secondary fractures intersecting primary fractures. Generally, primary fractures refer to those that massive proppants are concentrated within, and secondary fractures are partially propped with less proppant distribution. In this study, fracture network is divided into two fundamental patterns as shown in Fig. 2, including primary fracture with secondary fracture (PFSF) and orthogonal fracture network (OFN). There are $N_{pf}$ primary fractures and $N_{sf}$ secondary fractures ($N_{pf}+N_{sf}=N_f$) in each stage. Some basic assumptions underlying the mathematical model are list as follows:

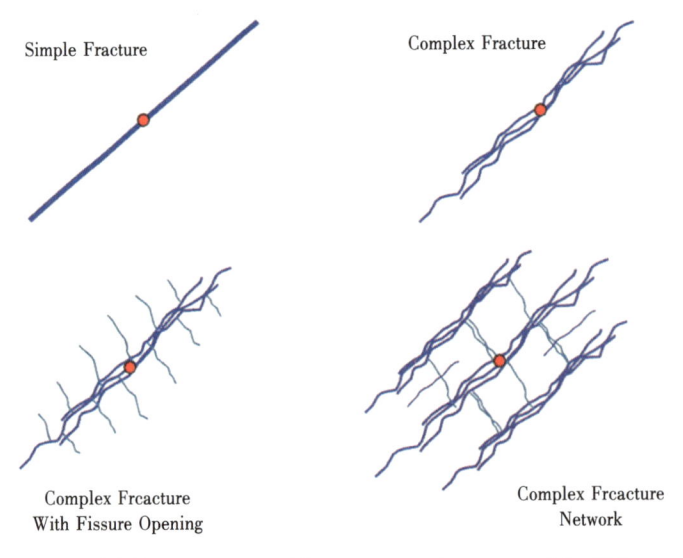

Fig. 1 Examples of different patterns of fracture networks presented by Fisher et al(2002)

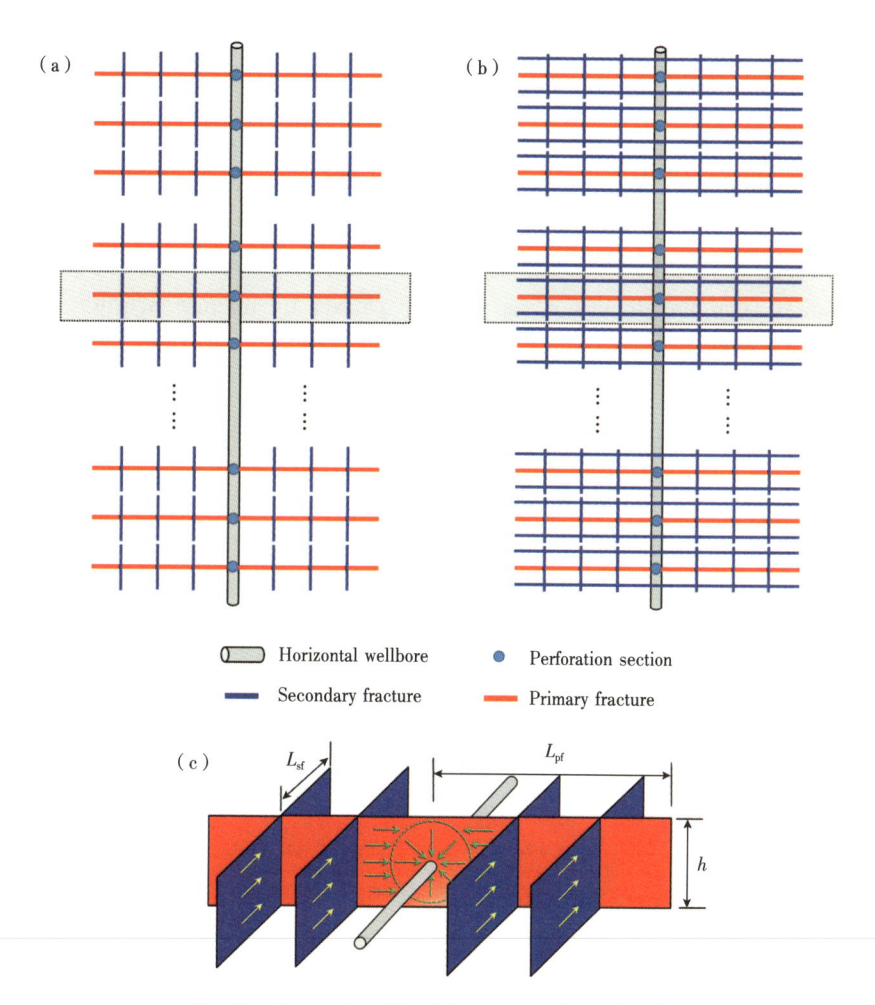

Fig. 2　Physical model of fractured horizontal well
(a) 2D aerial view of PFSF; (b) 2D aerial view of OFN; (c) 3D view of intersection between primary and secondary

(1) An isotropic, horizontal, slap drainage volume is bounded by a boxed reservoir, including overlying and underlying impermeable strata in vertical, and no-flow rectangular boundaries around reservoir in plane.

(2) The reservoir is of uniform thickness $h$, permeability $K_m$, and porosity $\phi_m$.

(3) The production process is assumed to be isothermal. Fluid flowing in both reservoir and fractures obeys Darcy's law, and the single-phase fluid is assumed slightly compressible with constant compressibility ctm and constant viscosity $\mu$.

(4) The flowing directly from reservoir to horizontal wellbore is negligible. The reservoir is produced only through a set of fully penetrating, finite-conductivity fracture network.

(5) Secondary fractures are intercepting primary fractures with arbitrary angle, and fractures have constant half-length $L_f$, width $w_f$, permeability $K_f$, and porosity $\phi_f$.

(6) The well is produced at constant production rate $Q$. Wellbore storage effect and fracture skin

effect are not considered in our model.

## 2.2 Mathematical model

In this section, the semi-analytical solutions for basic fracture panel will be presented. For the sake of simplicity, the models are established in the dimensionless form (Appendix A).

By taking Laplace and Fourier cosine transformation (detailed derivations are provided in the Wang and Jia work)[23], the general solution for fracture network could be achieved by using inversion transformation. Then the dimensionless pressure caused by fracture network in the reservoir could be further discretized into a set of fracture segments, which is rewritten as

$$p_{mD}(x_D,y_D) - p_{avgD} = \sum_{m=1}^{N_f} \sum_{i=1}^{M_m} q_{fDm,i} \int_{x_{Dm,i}-0.5\Delta x_{Dm}}^{x_{Dm,i}+0.5\Delta x_{Dm}} \Delta p_{mDm}(x_D,y_D,u_D) du_D \quad (2)$$

where $p_{avgD} = 2\pi t_D/(x_{eD} y_{eD})$, and $\delta p_{mDm,i}$ represents the dimensionless pressure caused by $i$-th uniform-flux segment of $m$-th fracture, and the detailed expression refers to Appendix B. As shown in Fig. 3, the functional relation between location of $i$-th segment on $m$-th fracture and Cartesian-coordinate system is given

$$\begin{cases} x_D = x_{ofDn} + x_{Dm,i}\cos\theta_m \\ y_D = y_{ofDn} + x_{Dm,i}\sin\theta_m \end{cases} \quad (3)$$

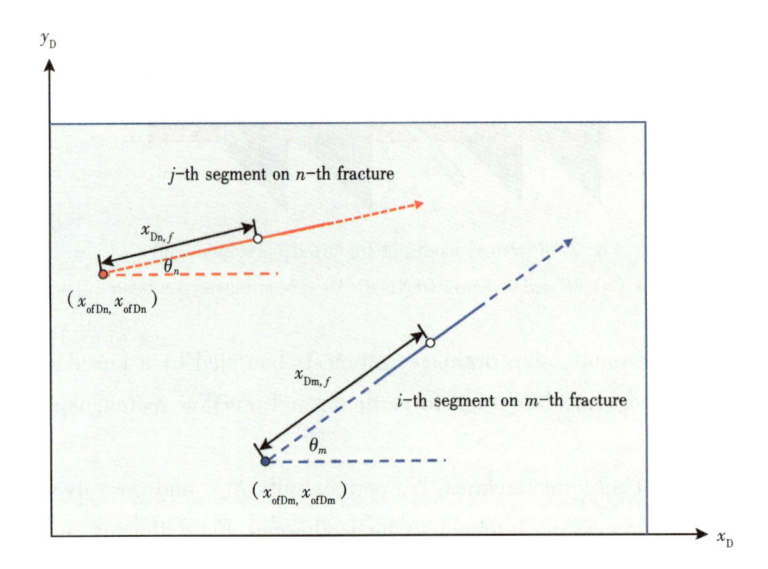

Fig. 3 Location of multiple fracture panels in a rectangular reservoir (2D aerial view)

It is noted that the main improvement on reservoir flowing model is to couple the reservoir flowing model with fracture-flowing model by constructing the generalized matrix format. Hence Eq. 3 can be rewritten in the form of matrix format, as following:

$$\begin{pmatrix} p_{\mathrm{mD1}}^{\mathrm{PSS}} \\ \vdots \\ p_{\mathrm{mDn}}^{\mathrm{PSS}} \\ \vdots \\ p_{\mathrm{mDN_f}}^{\mathrm{PSS}} \end{pmatrix} = \begin{pmatrix} \boldsymbol{p}_{\mathrm{mD1}} \\ \vdots \\ \boldsymbol{p}_{\mathrm{mDn}} \\ \vdots \\ \boldsymbol{p}_{\mathrm{mDN_f}} \end{pmatrix} - \begin{pmatrix} \boldsymbol{p}_{\mathrm{avgD}} \\ \vdots \\ \boldsymbol{p}_{\mathrm{avgD}} \\ \vdots \\ \boldsymbol{p}_{\mathrm{avgD}} \end{pmatrix} = \begin{pmatrix} \boldsymbol{A}_{1,1}^{R} & \cdots & \boldsymbol{A}_{1,m}^{R} & \cdots & \boldsymbol{A}_{1,N_f}^{R} \\ \vdots & & \vdots & & \vdots \\ \boldsymbol{A}_{n,1}^{R} & \cdots & \boldsymbol{A}_{n,m}^{R} & \cdots & \boldsymbol{A}_{n,N_f}^{R} \\ \vdots & & \vdots & & \vdots \\ \boldsymbol{A}_{N_f,1}^{R} & \cdots & \boldsymbol{A}_{N_f,m}^{R} & \cdots & \boldsymbol{A}_{N_f,N_f}^{R} \end{pmatrix} \cdot \begin{pmatrix} \boldsymbol{q}_{\mathrm{fD1}} \\ \vdots \\ \boldsymbol{q}_{\mathrm{fDn}} \\ \vdots \\ \boldsymbol{q}_{\mathrm{fDN_f}} \end{pmatrix} \quad (4)$$

where, $\boldsymbol{q}_{\mathrm{fDn}} = (q_{\mathrm{fDn},1}\cdots,q_{\mathrm{fDn},j}\cdots q_{\mathrm{fDn},M_n})^{\mathrm{T}}$, $\boldsymbol{p}_{\mathrm{mDn}}^{\mathrm{PSS}} = (p_{\mathrm{mDn},1}^{\mathrm{PSS}}\cdots p_{\mathrm{mDn},j}^{\mathrm{PSS}}\cdots p_{\mathrm{mDn},M_n}^{\mathrm{PSS}})^{\mathrm{T}}$, and the element of $q_{\mathrm{fDn},j}$ is the dimensionless influx of $j$-th segment of $n$-th fracture. $\boldsymbol{A}_{n,m}^{R}$ is the effect of n-th fracture on $m$-th fracture in the reservoir, which is written as

$$\boldsymbol{A}_{n,m}^{R} = \begin{Bmatrix} \Delta p_{\mathrm{mDm},1}^{\langle n,1 \rangle} & \cdots & \Delta p_{\mathrm{mDm},i}^{\langle n,1 \rangle} & \cdots & \Delta p_{\mathrm{mDm},M_m}^{\langle n,1 \rangle} \\ \vdots & & \vdots & & \vdots \\ \Delta p_{\mathrm{mDm},1}^{\langle n,j \rangle} & \cdots & \Delta p_{\mathrm{mDm},i}^{\langle n,j \rangle} & \cdots & \Delta p_{\mathrm{mDm},M_m}^{\langle n,j \rangle} \\ \vdots & & \vdots & & \vdots \\ \Delta p_{\mathrm{mDm},1}^{\langle n,M_n \rangle} & \cdots & \Delta p_{\mathrm{mDm},i}^{\langle n,M_n \rangle} & \cdots & \Delta p_{\mathrm{mDm},M_m}^{\langle n,M_n \rangle} \end{Bmatrix} \quad (5)$$

where $\Delta p_{\mathrm{mDm},i}^{\langle n,j \rangle}$ is the effect of $j$-th segment of $n$-th fracture on $i$-th segment of $m$-th fracture

On the basis of the previous work of author and co-author[24], fracture-network flowing could be divided into a set of planar fracture panels with multiple source terms as shown in Fig. 4. After using integral treatment, the final solution for n-th fracture could be generated in the following form:

$$p_{\mathrm{fDn}}^{\mathrm{PSS}}(x_{\mathrm{Dn}}) = [p_{\mathrm{fDn}}|_{x_{\mathrm{Dn}}=0} - p_{\mathrm{avgD}}] + \frac{2\pi}{C_{\mathrm{fDn}}} \left( \int_0^{x_{\mathrm{Dn}}} \mathrm{d}\xi \int_0^{\zeta} q_{\mathrm{fDn}}(\xi) \mathrm{d}\xi - \sum_{v=1}^{N_w} [q_{\mathrm{wfDn},v} G(x_{\mathrm{Dn}}, x_{\mathrm{wfDn},v})] \right) \quad (6)$$

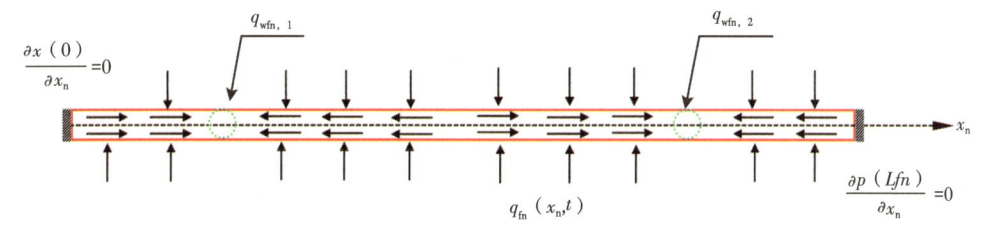

Fig. 4 Schematic of fracture flow model (2D aerial view)

Eq. (6) can be discretized into the matrix format, next:

$$\begin{pmatrix} p_{\mathrm{wfDn},w}^{\mathrm{PSS}} \\ \vdots \\ p_{\mathrm{wfDn},w}^{\mathrm{PSS}} \\ \vdots \\ p_{\mathrm{wfDn},w}^{\mathrm{PSS}} \end{pmatrix} = - \begin{pmatrix} p_{\mathrm{fDn},1}^{\mathrm{PSS}} \\ \vdots \\ p_{\mathrm{fDn},j}^{\mathrm{PSS}} \\ \vdots \\ p_{\mathrm{fDn},M_n}^{\mathrm{PSS}} \end{pmatrix} = \frac{2\pi}{C_{\mathrm{fDn}}} \begin{pmatrix} \alpha_{1,1}^{n,w} & \cdots & \alpha_{1,i}^{n,w} & \cdots & \alpha_{1,M_n}^{n,w} \\ \vdots & & \vdots & & \vdots \\ \alpha_{j,1}^{n,w} & \cdots & \alpha_{j,i}^{n,w} & \cdots & \alpha_{j,M_n}^{n,w} \\ \vdots & & \vdots & & \vdots \\ \alpha_{M_n,1}^{n,w} & \cdots & \alpha_{M_n,i}^{n,w} & \cdots & \alpha_{M_nM_m}^{n,w} \end{pmatrix} \cdot$$

$$\begin{pmatrix} q_{\text{fDn},1} \\ \vdots \\ q_{\text{fDn},j} \\ \vdots \\ q_{\text{fDn},M_n} \end{pmatrix} + \frac{2\pi}{C_{\text{fDn}}} \sum_{v=1}^{N_{n,w}} \begin{pmatrix} \beta_1^{n,w} & & & & \\ & \ddots & & & \\ & & B_j^{n,w} & & \\ & & & \ddots & \\ & & & & B_{M_n}^{n,w} \end{pmatrix} \cdot \begin{pmatrix} q_{\text{wfDn},v} \\ \vdots \\ q_{\text{wfDn},v} \\ \vdots \\ q_{\text{wfDn},v} \end{pmatrix} \quad (7)$$

where the elements are satisfied as

$$\begin{cases} \alpha_{j,i}^n = RS_{n,i}^{n,j} + RT_{n,j}^{n,w} \\ \beta_j^{n,w} = G(x_{\text{Dn},j} x_{\text{wfDn},v}) - G(x_{\text{wfDn},w}, x_{\text{wfDn},v}) \end{cases} \quad (8)$$

and detailed expressions of $RS_{n,i}^{n,j}$ and $RT_{n,i}^{m,w}$ are provided in Appendix C.

According to the fact that the pressure should be equivalent on the fracture face, continuity conditions are given as

$$p_{\text{mD}}^{\text{PSS}}(x_{\text{ofD}} + x_{\text{Dn},j}\cos\theta_n, y_{\text{ofD}} + x_{\text{Dn},j}\sin\theta_n) = p_{\text{fDn}}^{\text{PSS}}(x_{\text{Dn},j}) \quad (9)$$

and the flux relationship between fracture segment and individual fracture panel is satisfied as

$$\sum_{n=1}^{N_f} \left( \sum_{v=1}^{N_{n,w}} q_{\text{wfDn},v} \right) = \sum_{j=1}^{M_f} \left( \sum_{j=1}^{M_n} \widetilde{q}_{\text{fDn},j} \Delta x_{\text{Dn}} \right) \quad (10)$$

In addition, for each interconnection in fracture network, the inflow rate must equal to outflow rate to ensure the mass balance if the interconnection is not communicated with horizontal wellbore, which are written as

$$(\text{inflow})_i + (\text{outflow})_i = 0 \quad (11)$$

It is necessary to point out that this model uses the technique of varying spatial steps, where the length of fracture segment is uneven (Fig. 5). In general, the smallest spatial step is assigned to segments containing the source/sink, and then the step of adjacent segment increases until the maximum. Put another way, the segments close to the region of source/sink are refined. The technique can significantly improve the computational efficiency and accuracy. The dimensionless flux of fracture segments, dimensionless rate and pressure of sources could be achieved by solving Eq. 9 with the help of numerical algorithm. Thereafter, the dimensionless PI of $w$-th source of $n$-th fracture is obtained, which is expressed as

$$J_{\text{Dn},w}^{\text{PSS}} = \left( \frac{B\mu}{2\pi K_m h} \right) \left( \frac{q_{\text{wfn},w}}{p_{\text{avg}} - p_{\text{wfn},w}} \right) = \frac{q_{\text{wfDn},w}}{p_{\text{wfDn},w}^{\text{PSS}}} \quad (12)$$

and the total dimensionless PI of well is the sum of dimensionless PI of all sources connected to wellbore.

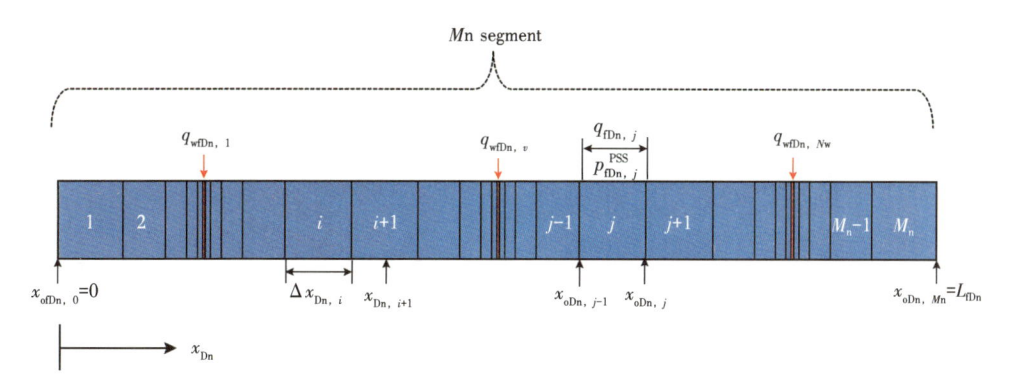

Fig. 5  Nth fracture divided into $M_n$ unequal segments.

## 3 Model verification

### 3.1 Validation with semi-analytical model

#### 3.1.1 Validation with Luo et al

First, we verify the case of multiple fractures which are directly connected to wellbore. Luo et al[21] presented a new PSS productivity-index solution for multiple fractures with non-planar angle. Since their results have exactly matched the classical results of Ozkan and Raghavan and Valko and Economides[3], they could be regarded as a benchmarked criterion. Our paper has the sufficient flexibility of accounting for the geometrical complexity of multi-fracture wings connected to wellbore. We selected two specific scenarios to make verification: (1) different non-planar angles, (2) different numbers of wings. Noted that $x_{eD} = y_{eD} = 10$, and wellbore is located at the center of rectangular reservoirs. Fig. 6 shows that our results are in excellent agreement with Luo et al[21] solutions, with an average relative error of less than 1%.

#### 3.1.2 Validation with Mao et al

The flow pattern in transverse fracture intercepting horizontal wellbore is comprised of linear-flow and radial-flow regimes. To account for the radial-flow regime, Mao et al[13] established a new model of a vertical hydraulic fracture in a horizontal well, where no flow contribution from the reservoir beyond the fracture tips is considered. Dimensionless normalized productivity index is presented as the ratio dimensionless productivity index of same penetration radio to an infinite conductivity, which is given by $\eta_D^{PSS} = J_{D,total}^{PSS} / J_{D,total}^{PSS}|_{C_{fD} = \infty}$. It is defined as fracture-flow efficiency. In our model, the skin of flow convergence is introduced to account for additional pressure drop caused by convergence flow around wellbore[25]:

$$S = \frac{1}{C_{fD}} \frac{2h_D}{L_{fD}} \left[ \ln\left(\frac{h_D}{2r_{wD}}\right) - \frac{\pi}{2} \right] \tag{13}$$

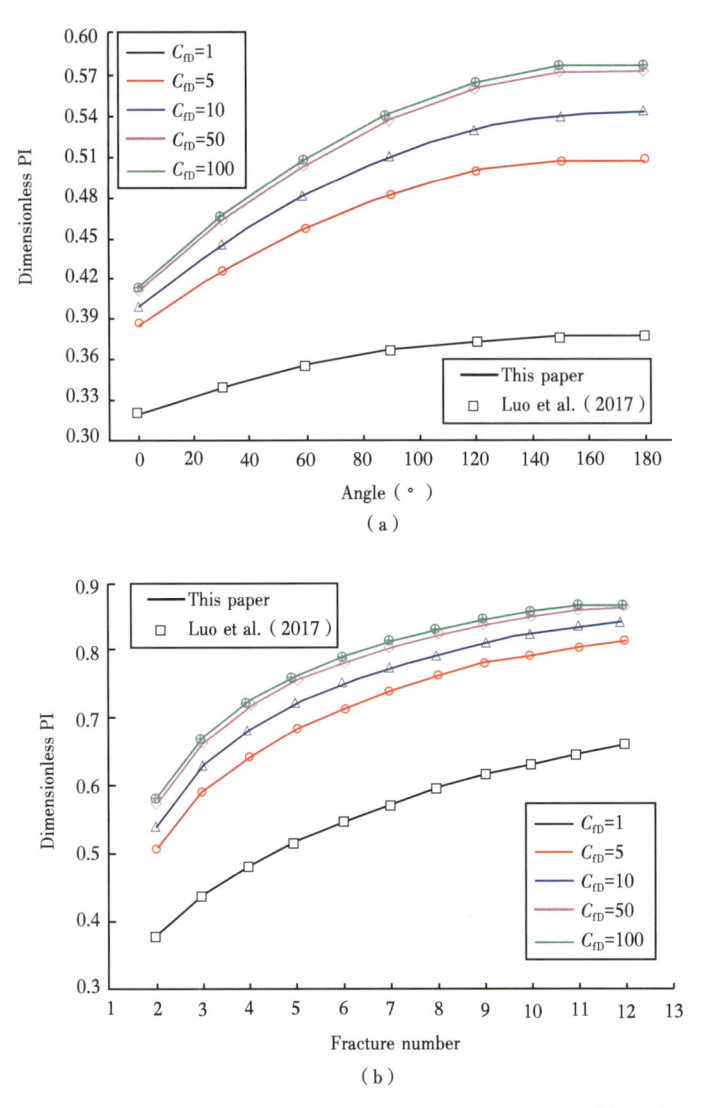

Fig. 6  Comparison of present results with solutions of Luo et al(2017) in dimensionless PI under (a) different angles and (b) different fracture-wing

To be consistent with the Mao et al[13] situation, we assume that $L_{fD} = x_{eD}$, $x_{eD}/y_{eD} = 1$, $r_{wD} = 0.0025$ and $h_D = 0.5$. Next, Eq. 13 multiplied by source term is added to the right hand side of Eq. 6. Fig. 7 shows the comparison results. It is shown that our results agree well with Mao results.

In addition, Bhattacharya et al [6] developed explicit empirical equations by quasi-polynomial ratio to compute optimum fracture dimensions and productivity index for given proppant number. Table 1 provides the comparison of optimum dimensionless conductivity and maximum dimensionless PI. Good agreement is also observed for different proppant numbers.

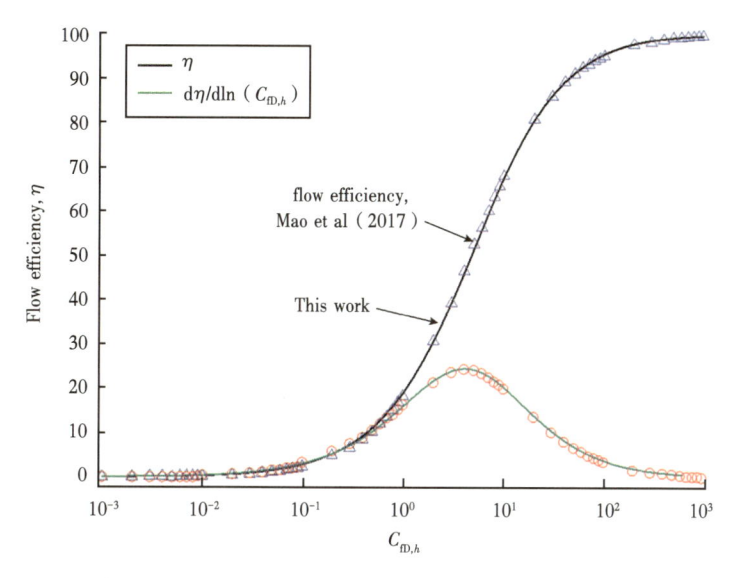

Fig.7　Comparison of present results with solutions of Mao et al (2017) under the condition of considering radial-flow regime

**Table 1　Comparison of the optimum & maximum values of this paper and Bhattacharya et al. (2012) for single fracture ($x_e/y_e=3$)**

| $N_{prop}$ | $C_{fD}$ This paper | Bhattacharya et al | Error(%) | $J_{D,max}$ This paper | Bhattacharya et al. | Error(%) |
|---|---|---|---|---|---|---|
| 1000 | 333.76 | 333.07 | 0.2065 | 5.6547 | 5.8750 | 3.7497 |
| 100 | 33.759 | 33.36 | 1.1900 | 4.8822 | 5.0211 | 2.7663 |
| 10 | 3.9885 | 3.8442 | 3.7545 | 2.2768 | 2.2132 | 2.8736 |
| 1 | 1.8517 | 1.8815 | 1.5851 | 0.7251 | 0.7522 | 3.6027 |
| 0.1 | 1.6120 | 1.6117 | 0.0043 | 0.3828 | 0.3664 | 4.4759 |
| 0.01 | 1.6117 | 1.6001 | 0.7236 | 0.2635 | 0.2798 | 5.8256 |
| 0.001 | 1.6117 | 1.6001 | 0.7311 | 0.2048 | 0.2188 | 6.3985 |

## 3.2　Validation with 3D numerical model

As we know, few works were conducted to calculate the PI of fracture network in PSS condition. Here, 3D numerical simulation was built to obtain transient PI. As shown in Fig. 8a, the grids surrounding the fractures are discretized logarithmically, ranging from 0.01m to 12.8m, to ensure that the grids near the fractures are refined enough to capture the flow behavior. The grids with a width of 0.01m, indicated by solid lines, are assigned different permeability to represent primary and secondary fractures with different conductivities. Basic parameters used in numerical model are list in Table 2.

## Table 2  Basic parameters used for numerical model validation

| Basic model parameter | value | Basic model parameter | value |
| --- | --- | --- | --- |
| Reservoir length(m) | 1149.5 | Spacing of SF(m) | 38.32 |
| Reservoir width(m) | 1149.5 | Fracture porosity(%) | 35 |
| Formation thickness(m) | 2.9 | Fracture width(m) | 0.0127 |
| Formation permeability(m$^2$) | 10$^{-16}$ | Production rate(m$^3$/s) | 1.157×10$^{-5}$ |
| Formation porosity(%) | 10 | Initial pressure(Pa) | 1×10$^8$ |
| Half-length of primary fracture(m) | 114.95 | Viscosity of fluid(Pa·s) | 0.001 |
| Half-length of secondary fracture(m) | 38.32 | Total compressibility(Pa$^{-1}$) | 4.35×10$^{-10}$ |

Our model calculated the value of the dimensionless PI, and compared with the numerical results. Fig. 8b shows the relationship between dimensionless transient pressure and dimensionless time $[t_D = K_m t/(\varphi_m \mu c_t L_{ref}^2)]$, and the function relation in PSS condition is satisfied as linear expression by

$$p_{wD} = 2\pi t_D/(x_{eD} y_{eD}) + 1/J_{D,total}^{PSS} \quad (14)$$

Fig. 8c shows that the transient PI based on estimated average reservoir pressure is quickly decreased to a constant when the response reaches PSS condition. Table 3 provides the PSS results of

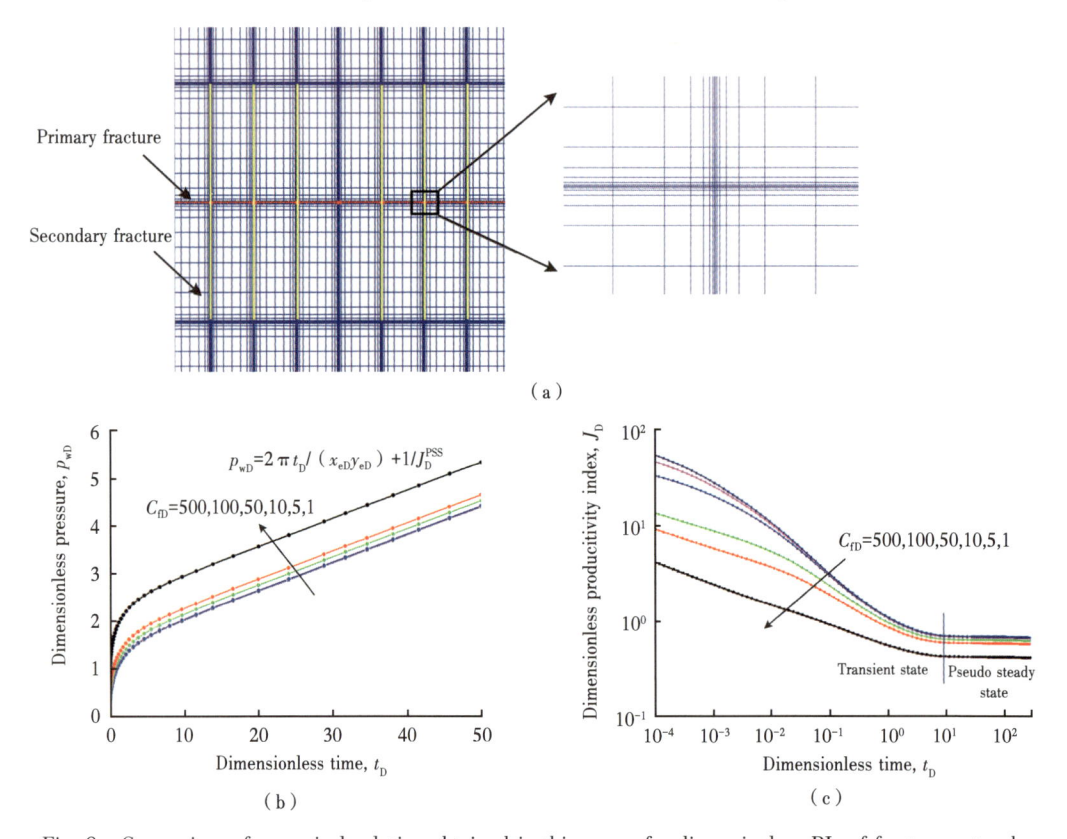

Fig. 8  Comparison of numerical solution obtained in this paper for dimensionless PIs of fracture network

dimensionless PI for two methods under different conductivities. It shows that the numerical results are in excellent agreement with our study in the PSS condition.

Table 3  Comparison of dimensionless PIs for fracture network

| $C_{fD}$ | Numerical | This paper | Error(%) |
|---|---|---|---|
| 1 | 0.3926 | 0.3952 | 0.6714 |
| 5 | 0.5518 | 0.5577 | 1.0697 |
| 10 | 0.5975 | 0.6026 | 0.8583 |
| 50 | 0.6429 | 0.6490 | 0.9522 |
| 100 | 0.6499 | 0.6557 | 0.8917 |
| 500 | 0.6598 | 0.6612 | 0.2119 |

# 4  Dimensionless PI of fracture network

The purpose of this section is to provide fundamental insight into the relation between PI and network properties. For simplicity of discussion, we put the focus on two fundamental fracture networks: primary fracture with secondary fracture (PFSF) and orthogonal fracture network (OFN). To highlight the difference between PFSF and OFN in network configuration, all fractures are of uniform conductivity and equally spaced. As a benchmarked line, we evaluated the productivity index of single fracture (SF) with various conductivities under different penetration ratios ($I_x$).

As shown in Fig. 9a, the PI monotonically increases with the increased conductivity under given penetration ratio, and the dimensionless PI of PFSF is larger than SF. The dimensionless PI reaches a plateau value ($J_{D,plateau}$) when dimensionless conductivity approximates to a large enough value, denoted as plateau dimensionless conductivity of $C_{fD,plateau}$. A larger value of $I_x$ corresponds to a higher value of $C_{fD,plateau}$. It is worth noting that the difference between SF and PFSF in the dimensionless PI would be more pronounced when dimensionless conductivity is higher than a critical value (denoted as $C_{fD,critical} \approx 0.3$). Fig. 9b presents the effect of stimulation ratio of PFSF to SF on the PI enhancement. When $C_{fD,critical} \approx 0.3$, the stimulation ratio in the case of $I_x \leqslant 1$ ranges from 0.8% to 1.5%. When $C_{fD} > C_{fD}$,critical, the stimulation ratio also increases with the increase in $C_{fD}$, which depends on $I_x$. Taking $C_{fD} = 10^3$ for example, the stimulation ratio is only 4.5% at $I_x = 0.01$, but it approximately reaches 200% at $I_x = 1.0$. It indicates that it is of more significance to create fractures with high penetration ratio and larger conductivity for enhancing productivity index.

To investigate the mechanism of critical dimensionless conductivity, the distribution of inflow fluid along primary fracture is presented in Fig. 10. In PFSF, the primary fracture contains one source (wellbore) and four sinks (the intersections of primary and secondary fractures). Therefore, PFSF is regarded as a primary fracture with several additional sinks. The well production is supplied by the fluid withdrawal caused by primary-fracture segments and additional sources. Additional source indicates that the reservoir fluid is withdrawn into secondary fracture, and then flows into pri-

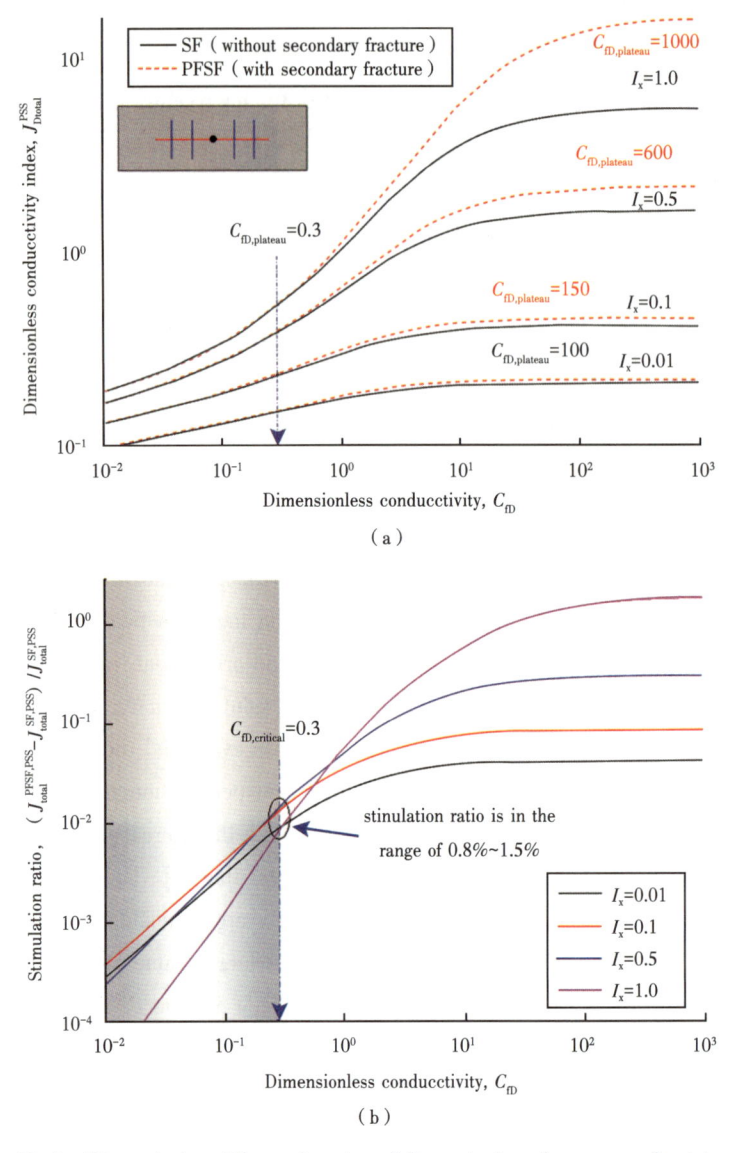

Fig.9  Dimensionless PI as a function of dimensionless fracture conductivity
and penetration ratio ($x_e/y_e = 3$, $L_{pf}/L_{sf} = 3$)

mary fracture. Fig. 10a shows the relationship between dimensionless conductivity and dimensionless rate of sinks embedded in fracture. The y-axis value indicates the ratio of sink-injection rate to well-production rate. When $C_{fD} = 0.3$, the percentage of sink-injection rate is only 5% in the contribution of well-production rate. The main contributor to well-production rate is the withdrawal caused by primary fracture. Therefore, the curves of SF and PFSF in Fig. 9a are overlapped when $C_{fD} < 0.3$. With the increasing of $C_{fD}$, the absolute value of sink/source rate increases until it reaches a constant when $C_{fD} = 10^3$. Fig. 10b reflects the flux density along primary fracture at $I_x = 1$. In the condition of smaller conductivity (e. g., $C_{fD} = 0.1 < C_{fD,critical}$), the fluid distribution along primary fracture are concentrated around the region closer to wellbore. As a result, the secondary-fracture sinks connected to the far-from region could not be effectively conducted throughout primary fracture into wellbore. In the condition of larger conductivity (e. g., $C_{fD} = 1000 > C_{fD,critical}$), all fracture segments in the region of primary fracture would take effect, and the sinks could be conducted throughout primary fracture into wellbore very well.

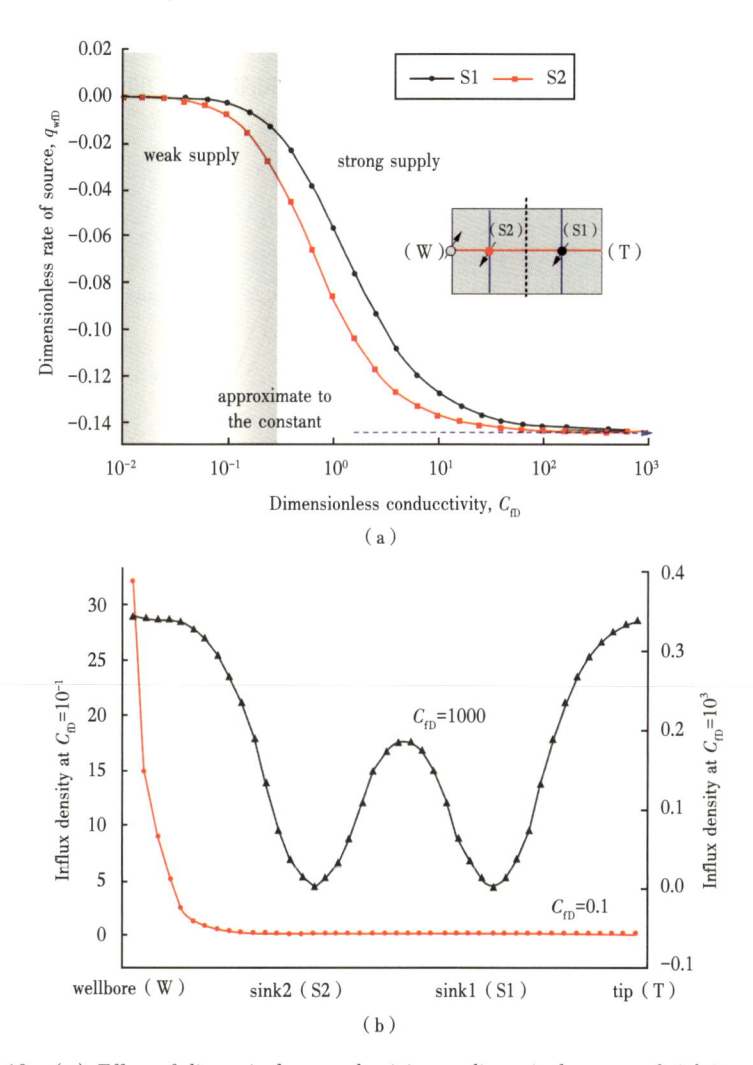

Fig.10 (a) Effect of dimensionless conductivity on dimensionless rate of sink/source;
(b) Dimensionless rate distribution along primary fracture for different values of $C_{fD}$ when $I_x = 1.0$

# 5 Optimum design for fracture network

Based on the definition of Eq. 1, the proppant number is converted into the following functional relationship:

$$N_{\text{prop}} = \frac{2K_f}{K_m}\frac{V_{\text{prop}}}{V_{\text{res}}} = \left(\frac{x_e}{y_e}N_x + \frac{I_y}{I_x}N_y\right)C_{fD}I_x^2 \quad (15)$$

where, $N_{\text{prop}}$ is the proppant number per stage, $x_e$ is well spacing, $y_e$ is stage spacing, $N_x$ is the number of fracture in parallel direction of primary fracture, $N_y$ is the number of secondary fracture perpendicular to primary fracture, $I_x$ is the penetration ratio of primary fracture to well spacing, $I_y$ is the penetration ratio of secondary fracture to stage spacing. Here, $x_e/y_e = 3$, and $I_x = I_y$. The effect of network properties on the dimensionless PI is presented in Fig. 11 ~ Fig. 12 for a set of given proppant numbers. In addition, the optimum results for fracture dimensions are provided in the tables in Appendix D.

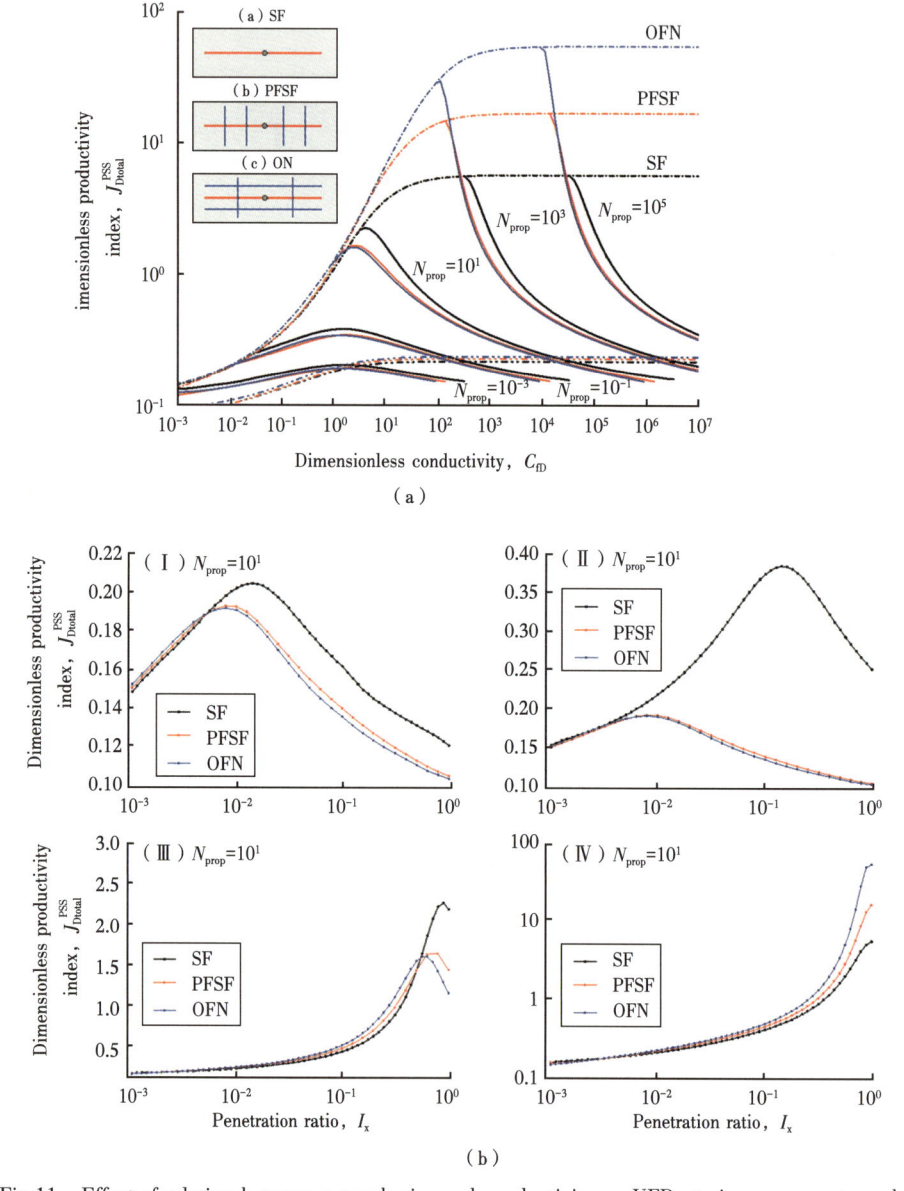

Fig.11 Effect of relation between network size and conductivity on UFD at given proppant number

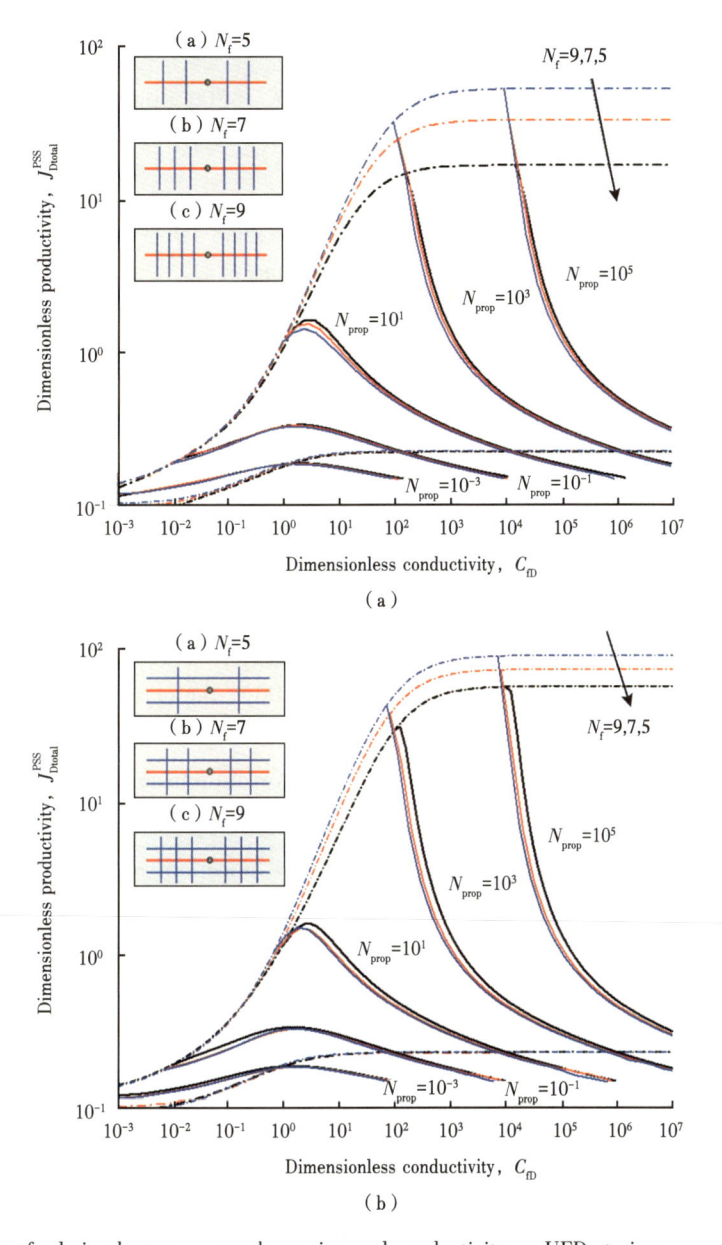

Fig.12 Effect of relation between network spacing and conductivity on UFD at given proppant number.

## 5.1 Relation between network size and conductivity

The network size is regarded as the drainage area covered by fracture network, so the fracture length is essentially of same significance as network size. Fig. 11 shows the dimensionless PI as the function of dimensionless conductivity and penetration ratio in each fracture-network pattern. For a given value of $N_{prop}$, there exists a maximum dimensionless PI, $J_{D,max}$. Associated with $J_{D,max}$ are the optimal $C_{fD}$ value, $C_{fD,opt}$, and the optimal $I_x$ value, $I_{x,opt}$. As expected, the PI approximately reaches the maximum at $C_{fD} = 1.6$ when $N_{prop} \leq 0.1$, independent of network pattern. As $N_{prop}$ increases, the maximum PI corresponds to a larger optimum values of $C_{fD,opt}$ and $I_{x,opt}$. In the case of SF, the maximum PI would reach the plateau value of $J_{D,plateau}$ when $C_{fD} > 300$ (infinite conductivity) and $I_x = 1$ (fully penetrating the drainage area) in the condition of $N_{prop} \geq 10^3$. What would happen in the case of PFSF and OFN?

In the condition of mediate ($1 \leq N_{prop} < 10$) and small proppant number ($N_{prop} \leq 0.1$), the value of $J_{D,max}$ for SF is larger than PFSF, and the optimum values of $C_{fD,opt}$ and $I_{x,opt}$ are also larger. Taking $N_{prop} = 10$ for example, the optimal value of $C_{fD,opt}$ in the case of PFSF is about 2.7911, and the optimal value of $I_{x,opt}$ is 0.7125. As comparison, the optimal value of $C_{fD,opt}$ in the case of SF is 3.9885, and the optimum value of $I_{x,op}$ is that 0.9142. The optimum values in the case of SF are larger than PFSF, which indicates that allocating the additional proppant to primary fracture could contribute to a higher PI. Put another way, the effect of additional length provided by secondary fractures in PFSF is slightly weaker than the additional fracture-length of SF. If the complexity of fracture network was increased to OFN, the maximum value of $J_{D,max}$ in the case of OFN would be smaller than PFSF (Table D-1). The lower PI in the case of OFN is related to the smaller fracture length (or network size). Here, $I_{x,opt(SF)} > I_{x,opt(PFSF)} > I_{x,opt(OFN)}$, while $J_{D,max(SF)} > J_{D,max(PFSF)} > J_{D,max(OFN)}$.

In the condition of large proppant number ($N_{prop} \geq 100$), the maximum PI is achieved when the fractures fully penetrate the drainage area ($I_{x,opt} = 1$), regardless of the proppant number (Table D-1). In this case, the stimulation ratio of PFSF to SF is more remarkable as the $N_{prop}$ increases. For example, the stimulation ratio is 49.9% when $N_{prop} = 10^2$, and it increases to 192.9% when $N_{prop} = 10^5$. The reason is that fracture conductivity plays a more important role in enhancing PI when the network approximately reaches the limited level ($I_x = 1$) in the condition of larger value of $N_{prop}$. The larger the network size, the more effectively the PI could be enhanced. Larger network would provide much room to allocate the proppant for improving network conductivity. As a result, the maximum PI of OFN is the largest in the condition of larger $N_{prop}$.

## 5.2 Relation between network spacing and conductivity

Fracture number consequently determines the network spacing. The larger the number of fracture was, the smaller network spacing would be. In the subsection, we change the secondary fracture number to investigate the effect of network spacing on dimensionless PI and optimum fracture design, as presented in Fig. 12.

In the condition of mediate ($1 \leq N_{prop} \leq 10$) and small proppant number ($N_{prop} \leq 0.1$), the maximum PI is slightly higher when fracture spacing is larger, where the largest value of $J_{D,max}$ is in the case of $N_f = 5$. In the presence of large-spacing network, the fracture conductivity requirements may be higher than that of small-spacing network (Table D-2~Table D-4). Taking $N_{prop} = 10$ for example, $J_{D,max(PFSF)}$ is approximately decreased by 5.8% and $C_{fD,opt(PFSF)}$ is decreased by 15.0% by decreasing fracture spacing (increasing fracture number from $N_f = 5$ to $N_f = 9$). Additionally, as the increasing of fracture complexity from PFSF to OFN, the optimum value of $C_{fD,opt}$ is decreased. Taking the scenario of $N_{prop} = 10$ and $N_f = 9$ for example (Table D-4), the optimal value of $C_{fD,opt(PFSF)}$ in the case of PFSF is 2.3714, while $C_{fD,opt(OFN)} = 2.0744$ in the case of OFN. It is explained that increasing fracture complexity almost or insufficiently offsets the decrease of fracture-conductivity requirement.

In the condition of large proppant number ($N_{prop} \geq 100$), the maximum PI could be significantly enhanced by decreasing network spacing. However, the enhancement rate would gradually slow down. In the case of PFSF, the stimulation ratio of $N_f = 7$ to $N_f = 5$ is 98.7%, while the stimulation ratio of $N_f = 9$ to $N_f = 7$ is decreased to 59.8% when $N_{prop} = 10^5$. Meanwhile, increasing network complexity from PFSF to OFN has the ability of further enhancing productivity index. Taking $N_{prop} = 10^5$ for example, the stimulation ratio of PFSF to OFN is 224.8% in the condition of $N_f = 5$, and the stimulation ratio is decreased to 108.5% in the condition of $N_f = 7$. Finally it would be decreased to 60.1% in the condition of $N_f = 9$. In addition, the conductivity requirements are inversely proportional to fracture complexity.

## 5.3 UFD extension to nonorthogonal network

In many applications, the fracture network may be so complex that the conceptual OFN model is overly simplified. An example of the Changning Shale in China was used to illustrate the potential of our approach in modeling nonorthogonal network. A rich data set was available for this well to enable the modeling of the complex hydraulic-fracture propagation. In this example, the network geometry and conductivity distribution of one stage were comprehensively determined by combining the results of microseismic imaging and the UFM[26,27]. Fig. 13 shows the details of fracture geometry and the conductivity distribution of the network, highlighting the primary (propped) fractures and secondary

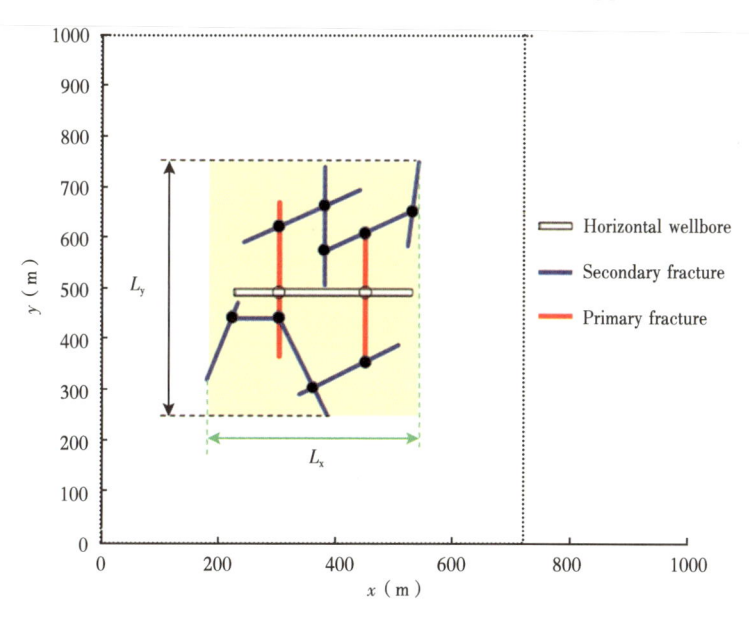

Fig.13 Schematic of non-orthogonal fracture network

(partially-propped) fractures. Fractures are interconnected with different inclined angles. Black circles indicate the fracture-fracture interconnections, and the hollow circles indicate the connected points of horizontal wellbore with the primary fractures.

The network size is defined as the rectangular drainage area ($L_x \times L_y$) covered by fracture network, assuming that $L_x/L_y = x_e/y_e$. According to Eq. 15, the proppant number of nonorthogonal fracture network is defined as,

$$N_{prop} = \frac{2K_f V_{prop}}{K_m x_e y_e} = \left( \sum_{n=1}^{N_{pf}} \frac{K_f w_{pfn}}{K_m} \frac{L_{pfn}}{x_e y_e} + \sum_{n=1}^{N_{sf}} \frac{K_f w_{sfn}}{K_m} \frac{L_{sfn}}{x_e y_e} \right) \quad (16)$$

where $N_{pf}$ is the number of primary fracture, $N_{sf}$ is the number of secondary fracture, $L_{pf}$ is the length of primary fracture and $L_{sf}$ is the length of secondary fracture.

Fig. 14 shows the maximum PI and optimum fracture dimensions of nonorthogonal network under

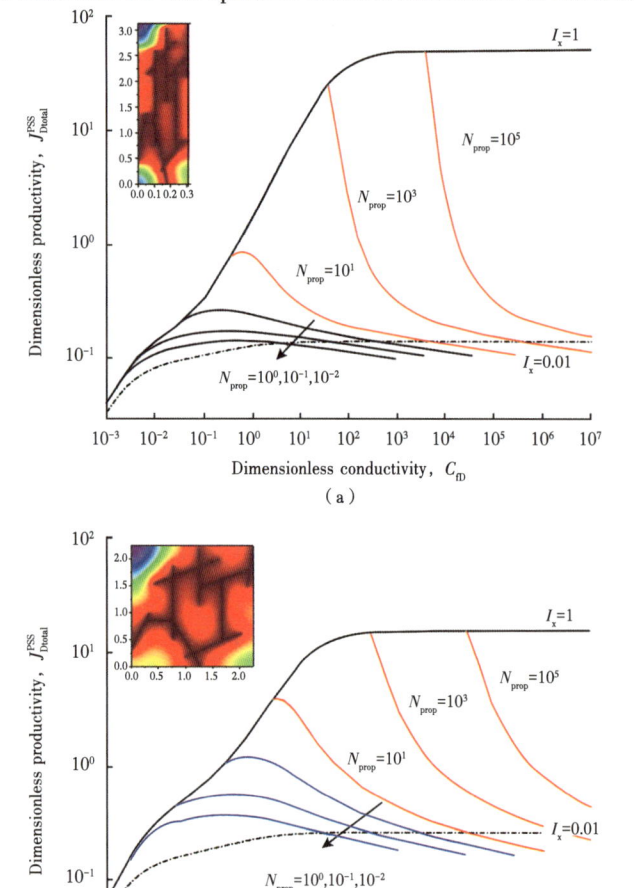

Fig.14 Optimum fracture dimensions of non-orthogonal network under (a) $x_e/y_e = 0.1$ and (b) $x_e/y_e = 1$

different aspect ratios of drainage area ($x_e/y_e$). The characterization of optimization curve is similar to the orthogonal network. Noted that the varying of the aspect ratio would result in the varying of intersection angles in network, so the difference of optimum UFD is more noticeable in Fig. 14a and Fig. 14b. In addition, the distribution of dimensionless difference between reservoir and average pressure ($p_{mD}-p_{avgD}$) is presented in Fig. 15. The results in Fig. 15 also confirm that increasing fracture conductivity and penetration ratio would contribute to a lower pressure drop and a higher PI.

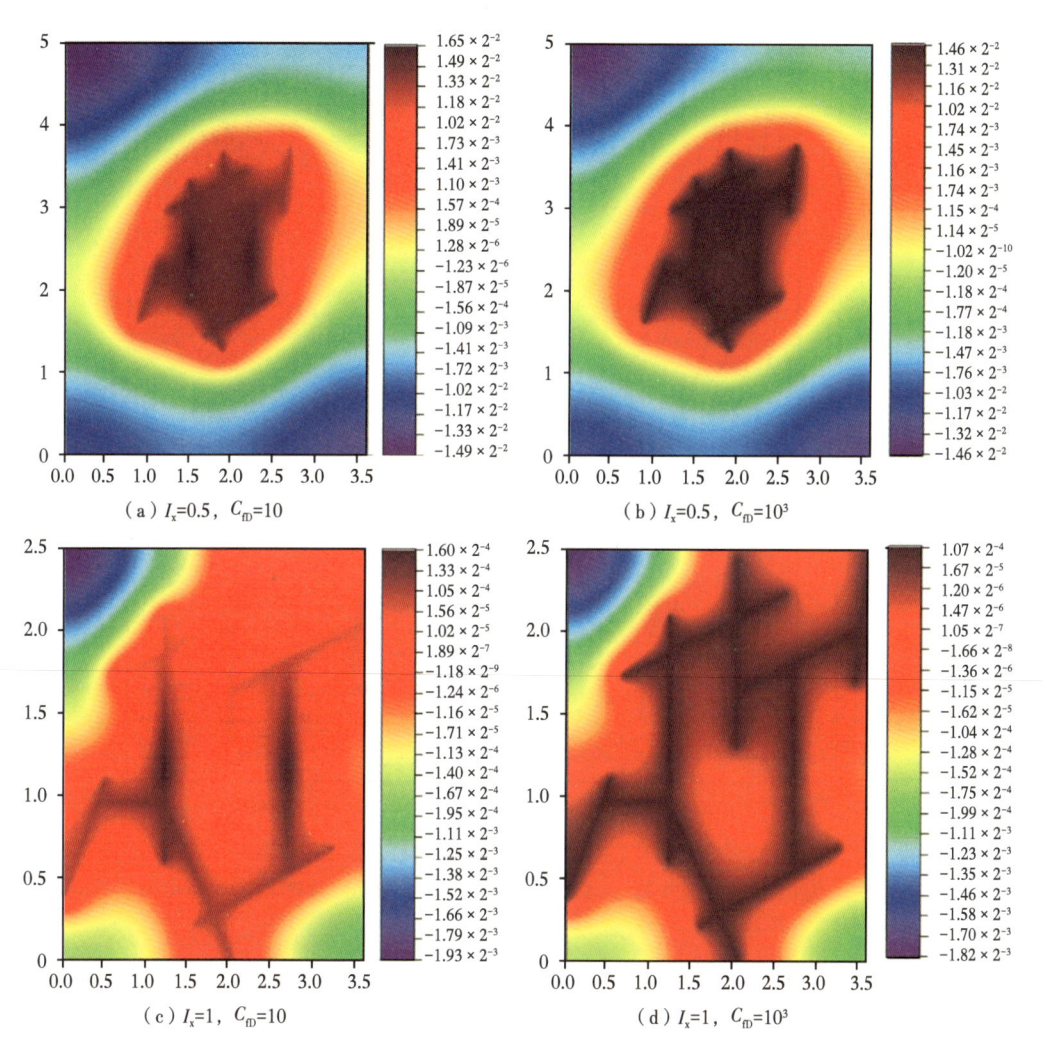

Fig.15 Pressure-difference field for different network conductivities and configurations

Using the new model, over 2000 cases were calculated to obtain the maximum PI of parameter combinations with different proppant numbers and aspect ratios. Eq. 16 can be rewritten as the following dimensionless form:

$$N_{prop} = I_x^2 C_{fD} \frac{y_e}{x_e} \Big( \sigma \sum_{n=1}^{n_{pf}} L_{pfD,n} + \sum_{n=1}^{n_{sf}} L_{sfD,n} \Big) \tag{17}$$

and

$$C_{fD} = \frac{K_f w_{pf}}{K_m(L_y/2)}, I_x = \frac{L_y}{y_e}, \sigma = \frac{C_{pfD}}{C_{sfD}} \quad (18)$$

Noted that the value in the bracket on the right hand side is a constant for a given fracture network. We take the optimum conductivity of secondary fracture corresponding the maximum PI as the objective function for each case. Type curves with $N_{prop}$, $x_e/y_e$ and optimum $C_{fDopt}$ are plotted in Fig. 16.

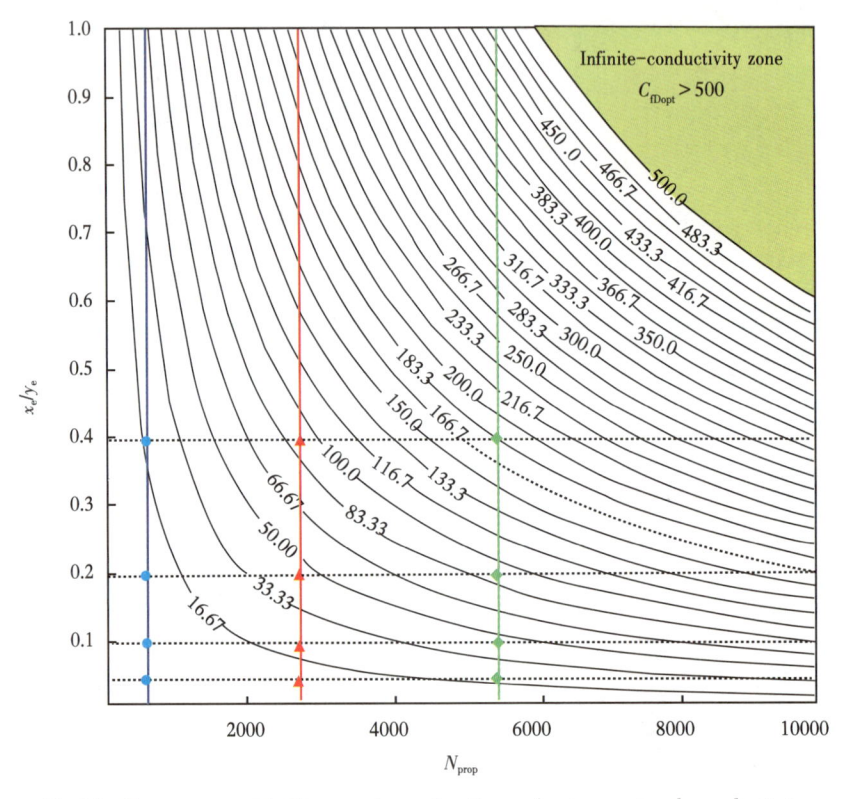

Fig.16　Type curves with $N_{prop}$, $x_e/y_e$ and optimum fracture-network conductivity

Case study is presented to illustrate the application of type curves to achieve the optimum fracture dimensions. The basic parameters for the reservoir and proppant properties based on the Changning Shale are list in Table 4. We set the values of the number of stage ($N_{stage}$) between 10 and 80, and the values of proppant volume ($V_{prop}$) between 45000kg and 450000kg. Using the parameters, we calculate the required values of $N_{prop}$ and $x_e/y_e$ in the model, which is given by

$$N_{prop} = \frac{M_{prop}}{\rho_p(1-\varphi_p)}, \left(\frac{x_e}{y_e}\right) = \frac{x_{ef}}{y_{ef}/N_{stage}} \quad (19)$$

Table 4  Reservoir and proppant parameters

| Basic model parameter | value | Basic model parameter | value |
|---|---|---|---|
| Porosity of proppant pack, $\phi_{prop}$ | 0.35 | Dimensions of reservoir Drainage area, $x_{ef} \times y_{ef}$ (m) | 1600×400 |
| Specific gravity of proppant, $\rho_{prop}$ (kg/m$^3$) | 26534 | Reservoir permeability, $k_m$ (mD) | 0.0001 |
| Reservoir thickness, $h$ (m) | 30 | Mass of proppant per well, Mprop (kg) | 45000; 225000; 450000 |
| Proppant permeability, $k_f$ (mD) | 200000 | | |

The final values of optimal conductivity are marked with blue circles for $V_{prop}$ = 45000kg, red triangulars for $V_{prop}$ = 225000kg and green rhombus for $V_{prop}$ = 450000kg in Fig. 16. In this example, two scenario are designed: In Scenario 1, a horizontal well with multistage transverse planar fractures is designed. In Scenario 2, a horizontal well with multistage nonorthogonal fracture networks is designed. As seen from Table 5, the maximum PI of fractured horizontal well depends on the proppant volume and the number of stage. The treatment of more proppant volume and more fractured stages could dramatically enhance the productivity of the well. However, the maximum PI of Scenario 1 may be higher than that of Scenario 2 when the proppant volume is not sufficient and the fractured stages is less.

Table 5  Optimum fracture dimensions for case study

| Fracture pattern | $V_{prop}$=45000kg ($N_{prop}$=548.4) | | | $V_{prop}$=225000kg, ($N_{prop}$=2742.0) | | | $V_{prop}$=450000kg, ($N_{prop}$=5484) | | |
|---|---|---|---|---|---|---|---|---|---|
| | $I_{x,opt}$ | $C_{fD,opt}$ | $J_{Dmax}$ | $I_{x,opt}$ | $C_{fD,opt}$ | $J_{Dmax}$ | $I_{x,opt}$ | $C_{fD,opt}$ | $J_{Dmax}$ |
| Planar fracture (Stage=40) | 1.000 | 54.801 | 574.92 | 1.000 | 273.98 | 741.54 | 1.000 | 547.96 | 769.64 |
| Non-orthogonal (Stage=10) | 0.897 | 18.952 | 191.57 | 0.952 | 92.65 | 235.57 | 1.000 | 179.58 | 242.56 |
| (Stage=20) | 0.918 | 15.79 | 457.20 | 0.964 | 45.91 | 807.09 | 1.000 | 86.34 | 893.17 |
| (Stage=40) | 0.938 | 9.86 | 712.84 | 0.988 | 22.43 | 2073.8 | 1.000 | 46.81 | 2730.7 |
| (Stage=80) | 0.965 | 5.46 | 833.07 | 1.000 | 16.87 | 3497.7 | 1.000 | 18.26 | 5703.6 |

# 6  Further consideration

The productivity index is determined by fracture complexity and conductivity. Without the influence of proppant volume, PI could be enhanced by increasing network complexity and fracture conductivity. The enhancement ratio of OFN to PFSF would be more remarkable in the condition of high fracture conductivity and large network size. Under the influence of given proppant number, these two key components are constrained to search the best compromise for maximizing productivity index.

However, when applying the findings of this work to the optimization of complex fracture network in the field, there are many other practical considerations that must be taken into account, such as varying conductivity, stress-sensitive conductivity and non-Darcy flow in the fracture. None of these issues is considered in this present work. Although these effects have not been addressed here, they can be readily incorporated into the model by using the dimension transformation in the fracture[28,29], this is

$$\xi_D = \xi_D(x_D) = \int_0^{x_D} \frac{d\zeta_D}{C_{fD}(\zeta_D)} \bigg/ \int_0^{l_D} \frac{d\zeta_D}{C_{fD}(\zeta_D)} \tag{20}$$

For extremely-low permeability reservoirs in which the transient condition lasts considerable time, based on the approach presented in this paper, the optimum UFD for transient condition can be modified by the following correction:[30]

$$C_{fD,opt}(t_D, N_{prop}) = \frac{1}{t_D} \int_0^{t_D} C_{fD,opt}^{transient}(\tau, N_{prop}) d\tau \tag{21}$$

For clarify of understanding, the study is limited to single-phase flow of liquids in homogeneous reservoir. To accurately obtain UFD, the two-phase (gas/water) flow can be incorporated by iteratively correcting the relative permeability to gas/water for each fracture segment[31,32], and the approach of incorporating discrete (natural) fracture model and multi-continuum medium model could be used to characterize different porosity types[24,33]. In addition, accurate characterization of the complexity for fracture network could be achieved by considering the geomechanics effect which determines the propagation of hydraulic fractures and the opening of natural fracture. In summary, it is important to investigate these fundamental mechanisms before the UFD can successfully satisfy the field constraints. To find the optimal amount of proppant and fractured stage per well an economic optimization is also needed. Those issues would be further addressed in the future works.

# 7  Conclusion

The focus of this paper is put on establishing a PSS productivity model and extending the application of UFD to maximize the productivity index of fracture network. According to the observations and discussions, some conclusions are further emphasized as follows:

(1) Based on a reliable solution for point source in a closed rectangular reservoir, a semi-analytical model for calculating PSS productivity of fracture network is developed to account for the details of network properties, especially for the complex geometry and finite conductivity.

(2) Network complexity and conductivity determines the reservoir area connected to fracture network and the conductive ability of making fluid flow throughout fracture network into horizontal wellbore.

(3) Without the influence of proppant number, the secondary fracture would play a significant role only if primary fracture was in the high level of conductivity and complexity (large network size, small fracture spacing, and generating OFN). The productivity index would reach a plateau value

when fracture network completely penetrated throughout the drainage area and was of infinite conductivity.

(4) Under the influence of proppant number, the maximum productivity index would be achieved only if optimum relation between fracture complexity and conductivity was searched. Fracture complexity should be minimized by generating small network size, large fracture spacing, and simple fracture geometry in mediate and small proppant number. Fracture complexity should maximized by setting large network size, small fracture spacing and complex fracture geometry in large proppant number. As a consequence, increasing fracture complexity could reduce fracture conductivity requirements.

## 8 Nomenclature

Field variables

$B$ = Volume factor
$c$ = Compressibility, $Pa^{-1}$
$h$ = Formation height, m
$K$ = Permeability, $m^2$
$L_{fn}$ = Length of $n$-th fracture, m
$N_f$ = Number of fracture
$N_p$ = Number of fracture segment
$N_v$ = Number of interconnection
$N_f$ = Number of hydraulic fracture
$p$ = Pressure, Pa
$p_g$ = Pressure of unit response, Pa
$Q$ = Production rate of horizontal well
$q_{wfn,v}$ = Production rate of $v$-th source in $n$-th fracture, $m^3/s$
$q_{fn}$ = Flux density along n-th fracture face, $m^2/s$
$r$ = Wellbore radius, m
$t$ = Time, s
$V$ = Volume, $m^3$
$x_f$ = Half-length of fracture, m
$x_{ofn}$ = Initial location of $n$-th fracture in the $x$-$y$ coordinate, m
$x_e$ = Length of drainage area, m
$x_{wfn,v}$ = $v$-th sink/source location within n-th fracture, m
$y_e$ = Width of drainage area, m
$y_{ofn}$ = Initial location of $n$-th fracture in the $x$-$y$ coordinate, m
$w_f$ = Hydraulic fracture width, m
$w_e$ = Equivalent hydraulic fracture width, m

$Z_g$ = Gas factor

$\mu$ = Fluid viscosity, Pa·s

$\varphi_f$ = Fracture porosity

$\Phi_m$ = Reservoir porosity

$\rho$ = Fluid density, g/m³

## Dimensionless variables

$C_{fD}$ = Dimensionless fracture conductivity

$J_D$ = Dimensionless productivity index

$t_D$ = Dimensionless time

$p_D$ = Dimensionless pressure

$\theta_m$ = Intersection angle between $m$-fracture and $x$-$y$ coordinate

## Subscripts

avg = Average pressure

D = Dimensionless

f = Fracture property

i = Initial condition

m = Reservoir property

prop = Proppant

res = Reservoir

w = Wellbore property

we = Effective wellbore radius

ref = Reference variable

## Reference

[1] Prats, M. Effect of vertical fractures on reservoir behavior-incompressible fluid case. SPE Journal, 1961, 105-118.

[2] Economides, M J, Oligney, et al. Unified fracture design bridging the gap between theory and practice. Orsa Press 2002.

[3] Valko, P P, Economides, et al. Heavy crude production from shallow formation: long horizontal wells versus horizontal fractures. SPE 50421 presented at the 1998 SPE International Conference on Horizontal Well Technology, Alberta, Canada, 1998, 1-4 November.

[4] Daal, J A and Economides, M J. Optimization of hydraulically fractured wells in irregularly shaped drainage areas. SPE-98047-MS presented at the 2006 SPE International Symposium and Exhibition, Lafayetee, 2006, 15-17 Feburary.

[5] Meyer, B R, Jacot, et al. Pseudosteady-state analysis of finite-conductivity vertical fractures. SPE 95941 presented at the 2005 SPE Annual Technical Conference and Exhibition, Dallas, Texas, 2005, 9-12 October.

[6] Bhattacharya, S Nikolaou, M, et al. Unified fracture design for very low permeability reservoirs. Journal of Natural Gas Science and Engineering, 2012, 9: 184-195.

[7] Lu, Y H, Chen, et al. Productivity-index optimization for hydraulically fractured vertical wells in a circular reservoir: a comparative study with analytical solutions. SPE Journal, 2016,12: 2208-2219.

[8] Sorek, N, Moreno, et al. Optimal hydraulic fracture angle in productivity maximized shale well design. SPE-170965-MS presented at the SPE Annual Technical Conference and Exhibition, Amsterdam, the Netherlands, 2014,27-29 October.

[9] Hagoort, J. The productivity of a well completed with a vertical penny-shaped fracture. SPE Journal, 2011,7: 401-410.

[10] Tovar, F D, Lee, et al. Horizontal hydraulic fracture design for optimal well productivity in anisotropic reservoirs with different aspect ratios. SPE-168688-MS presented at the Unconventional Resources Technology Conference, Denver, Colorado, USA,2013,12-14 August.

[11] Guk, V, Tuzovskiy, et al. Optimizing number of fractures in horizontal well. SPE-174772-MS presented at the SPE Annual Technical Conference and Exhibition, Houston, Texas, 2015,28-30 September.

[12] Paderin, G V. Modified approach to incorporating hydraulic fracture width profile in unified fracture design model. SPE-182034-MS presented at the SPE Russian Petroleum Technology Conference and Exhibition, Moscow, Russia, 2016,24-26 October.

[13] Mao, D M, Miller, et al. Influence of finite hydraulic-fracture conductivity on unconventional hydrocarbon recovery with horizontal wells. SPE Journal, 2017,1-18.

[14] Cipolla, C L, Warpinski, et al. The relationship between fracture complexity, reservoir properties, and fracture treatment design. SPE - 115769 - MS presented the SPE Annual Technical Conference and Exhibition, Colorado, USA, 2008,21-24 September.

[15] Yu, W, Luo, et al. Sensitivity analysis of hydraulic fracture geometry in shale gas reservoirs. Journal of Petroleum Science and Engineering, 2014,113:1-7.

[16] Gu, M, Kulkarni, et al. Optimum fracture conductivity for naturally fractured shale and tight reservoirs. Paper SPE- 171648 - MS presented at the SPE/CSUR Unconventional Resources Conference - Canada, Calgary, 2014,30 September-2 October.

[17] Zhou, W T, Banerjee, et al. Semianalytical production simulation of complex hydraulic-fracture networks. SPE 157367 presented at the SPE International Production and Operations Conference and Exhibition, Doha, Qatar, 2012,14-16 May.

[18] Jia, P, Cheng, et al. Transient behavior of complex fracture network. Journal of Petroleum Science and Engineering,2015,132: 1-17.

[19] Chen, Z M, Liao, et al. Performance of horizontal wells with fracture networks in shale gas formation. Journal of Petroleum Science and Engineering, 2015,133: 646-664.

[20] Cheng, L S, Jia, et al. Transient responses of multifractured systems with discrete secondary fractures in unconventional reservoirs. Journal of Natural Gas Science and Engineering,2017,41: 49-62.

[21] Luo, W J, Wang, et al. Productivity of multiple fractures in a closed rectangular reservoir. Journal of Petroleum Science and Engineering, 2017,157: 232-247.

[22] Fisher, M K, Wright, et al. Integrating fracture-mapping technologies to improve stimulations in the Barnett shale. SPE-77441-MS presented at SPE Annual Technical Conference and Exhibition, San Antonio, Texas, 2002,29 September-2 October.

[23] Wang, J L, Jia et al. A general productivity model for optimization of multiple fractures with heterogeneous properties. Journal of Natural Gas Science and Engineering, 2014,21: 608-624.

[24] Luo, W J, Liu,et al. Effects of discrete dynamic-conductivity fractures on the transient pressure of a vertical

[25] Mukherjee, H, Economides, et al. A Parametric Comparison of Horizontal and Vertical Well Performance. SPE Formation Evaluation. 1991, 7, 209-216.

[26] Zeng, S P, Zhang, et al. Model of multi-fracture stress shadow effect of optimization design for staged fracturing of horizontal wells. Natural Gas Industry, 2015, 35(3): 1-5.

[27] Bian, X B, Jiang, et al. A new post-fracturing evaluation method for shale gas wells based on fracturing curves. *Natural Gas Industry*, 2016, 36(2): 60-65.

[28] Luo, W J, Tang, et al. A semianalytical solution of a vertical fractured well with varying conductivity under non-Darcy-flow condition. SPE Journal, 2015, 20(5): 1028-1040.

[29] Luo, W J, Tang, et al. Mechanism of fluid flow along a dynamic conductivity fracture with pressure-dependent permeability under constant wellbore pressure. Journal of Petroleum Science and Engineering, 2018, 166: 465-475.

[30] Rueda, J I, Voronkov, et al. Optimum fracture design under transient and pseudosteady condition unsing constant fracture volume concept. SPE-94157-MS presented at the SPE Europec/EAGE Annual Conference, Madrid, Spain, 2005, 13-16 June.

[31] Yang, R Y, Huang, et al. A semianalytical approach to model two-phase flowback of shale-gas wells with complex-fracture-network geometries. SPE Journal, 2017, 12: 1808-1833.

[32] Chen, Z M, et al. A two-phase flow model for fractured horizontal well with complex fracture networks: transient analysis in flowback period. SPE-187486-MS presented the SPE Liquid-Rich Basins Conference-North America, Texas, USA, 13-14 September.

[33] Jia, P, Cheng, et al. A semi-analytical model for the flow behavior of naturally fractured formation with multi-scale fracture networks. Journal of Hydrology, 2016, 537: 208-220.

## Appendix A: Definitions for dimensionless variables

For the sake of simplicity, the dimensionless definitions are used in this paper.

$$p_{mD} = \frac{2\pi K_m h(p_i - p_m)}{q_{ref} \mu B}, p_{fD} = \frac{2\pi K_m h(p_i - p_f)}{q_{ref} \mu B}, t_D = \frac{K_m t}{\phi_m \mu c_{tm} L_{ref}^2} \quad (A-1)$$

For the fracture flow model, the $n$-th dimensionless fracture conductivity $C_{fD}$, dimensionless $n$-th fracture coordinate in the 1D coordinate $x_{Dn}$, and the dimensionless flow rate $q_{fDn}$ are defined as

$$C_{fDn} = \frac{K_{fn} w_{fn}}{K_m L_{ref}}, x_{Dn} = \frac{x_n}{L_{ref}} \in [0, L_{fDn}], q_{fDn} = \frac{q_{fn} L_{ref}}{q_{ref}} \quad (A-2)$$

Dimensionless rate of v-th source on n-th fracture is defined as

$$q_{wfDn,v} = q_{wfn,v} / q_{ref} \quad (A-3)$$

The relation between sources and flux distribution on $n$-th fracture is satisfied as

$$\sum_{v=1}^{N_w} q_{wfDn,v} = \int_{x_{ofDn}}^{x_{ofDn}+L_{fDn}} q_{fDn}(\xi) d\xi \quad (A-4)$$

and the dimensionless wellbore rate is the sum of all sources connected to wellbore. Other dimensionless spatial variables are given as

$$x_D = x/L_{ref}, y_D = y/L_{ref}, x_{eD} = x_e/L_{ref}, y_{eD} = y_e/L_{ref} \quad (A-5)$$

## Appendix B: Basic solution for uniform-flux segment

The aim of this section is to provide an expression of uniform-flux line solution as the accurate benchmark for calculating fracture-network productivity in bounded rectangular reservoir. It is different from the expression presented by Luo et al (2017).

The pressure governing equation for slightly compressibility fluid flowing in two-dimensional system is given as the following partial differential equation:

$$\left(\frac{\partial^2 p_m}{\partial x^2} + \frac{\partial^2 p_m}{\partial y^2}\right) + \frac{\mu B}{K_m h}\sum_{m=1}^{N_f}\int_0^{l_{fm}} q_{fm}(u,t)\delta(x - x_{ofm} - u\cos\theta_m)\delta(y - y_{ofm} - u\sin\theta_m)du = \frac{\phi_m \mu c_{tm}}{K_m}\frac{\partial p_m}{\partial t} \quad (B-1)$$

Where $q_{fm}(u,t)$ is the source term representing the flux distribution along $m$-th fracture per volume, and $\delta(\ )$ is the Dirac delta function. Under pseudosteady-state condition, the functional relation is satisfied as $\partial p_m/\partial t = \partial p_{avg}/\partial t$. According to the material balance equation in closed reservoirs, the derivative of average pressure is satisfied as $\dfrac{\partial p_{avg}}{\partial t} = -\dfrac{QB}{x_e y_e \phi_m}$, and $Q$ is the total rate of fluid withdrawal from the reservoirs. After using dimensionless definitions, Eq. B-1 is rewritten as

$$\nabla^2 p_{mD} + 2\pi\sum_{m=1}^{N_f}\int_0^{l_{fDm}} q_{fDm}(u_D, t_D)\delta(x_D - x_{ofDm} - u_D\cos\theta_m)\delta(y_D - y_{ofDm} - u_D\sin\theta_m)du_D = \frac{2\pi}{x_{eD} y_{eD}} \quad (B-2)$$

The outer reservoir boundaries are assumed as the no-flow condition, which are given as

$$\left.\frac{\partial p_{mD}}{\partial x_D}\right|_{x_D=0} = \left.\frac{\partial p_{mD}}{\partial x_D}\right|_{x_D=x_{eD}} = 0, \text{ and } \left.\frac{\partial p_{mD}}{\partial y_D}\right|_{y_D=0} = \left.\frac{\partial p_{mD}}{\partial y_D}\right|_{y_D=y_{eD}} = 0 \quad (B-3)$$

Taking Laplace and Fourier cosine transformation of Eq. B-2 ~ Eq. B-3 (detailed derivations are provided in the work of Wang and Jia[23]), the general solution could be obtained by using inversion transformation. Then it is further discretized into a set of fracture segments, which is written as

$$p_{mD}(x_D, y_D) = \sum_{m=1}^{N_f}\int_0^{l_{fDm}} q_{fDm}\delta p_{mDm}(x_D, y_D, u_D)du_D = \sum_{m=1}^{N_f}\sum_{i=1}^{M_m} q_{fDm,i}\underbrace{\int_{x_{Dm,i}-0.5\Delta x_{Dm}}^{x_{Dm,i}+0.5\Delta x_{Dm}}\delta p_{mDm}(x_D, y_D, u_D)du_D}_{\Delta p_{mDm,i}} \quad (B-4)$$

where $\delta p_{mDm,i}$ represents the pressure caused by $i$-th uniform-flux segment of $m$-th fracture, and the location is given in Fig. 3. The corresponding point-source solution is written as

$$\delta p_{mDm} = \frac{2\pi}{x_{eD} y_{eD}}t_D + \frac{2\pi}{x_{eD} y_{eD}}\delta p_{mDm}^{PSS} \quad (B-5)$$

and

$$\delta p_{mDm}^{PSS} = \frac{4\pi}{x_{eD}y_{eD}} \left\{ \begin{array}{l} + \sum_{v=1}^{\infty} \frac{\cos(\gamma_v y_D)\cos(\gamma_v y_{wDm})}{\gamma_v^2} + + \sum_{v=1}^{\infty} \frac{\cos(\beta_v x_D)\cos(\beta_v x_{wDm})}{\beta_v^2} \\ + \sum_{v=1}^{\infty} \cos(\beta_v x_D)\cos(\beta_v x_{wDm}) \sum_{v=1}^{\infty} \frac{\cos[\gamma_v(y_D + y_{wDm})] + \cos[\gamma_v(y_D - y_{wDm})]}{\gamma_v^2 + \beta_v^2} \end{array} \right\}$$

(B-6)

where $\gamma_v = v\pi/y_{eD}$, and $\beta_v = v\pi/x_{eD}$.

Substituting Eq. B-5 into Eq. B-4 results to the following expression in the PSS condition:

$$p_{mD}^{PSS}(x_D, y_D) = p_{mD} - \underbrace{\frac{2\pi t_D}{x_{eD}y_{eD}} \left( \sum_{m=1}^{N_f} \sum_{i=1}^{M_m} q_{fDm,i} \right)}_{p_{avgD}} = \sum_{m=1}^{N_f} \sum_{i=1}^{M_m} q_{fDm,i} \underbrace{\int_{x_{Dm,i}-0.5\Delta x_{Dm}}^{x_{Dm,i}+0.5\Delta x_{Dm}} \delta p_{mDm}^{PSS}(x_D, y_D, u_D) du_D}_{\Delta p_{mDm,i}}$$

(B-7)

After integrating Eq. B-6 from $x_{Dm,i} - 0.5\Delta x_{Dm,i}$ to $x_{Dm,i} + 0.5\Delta x_{Dm,i}$ with regard to $u_D$, we have the uniform-flux solution for fracture segment,

$$\Delta p_{mDm,i}^{PSS}(x_D, y_D) = \frac{4\pi}{x_{eD}y_{eD}} \sum_{v=1}^{\infty} \frac{\cos(\gamma_v y_D)\sin[\gamma_v(y_{wDm,i} \mp 0.5\Delta x_D \sin\theta_m)]}{\sin\theta_m \gamma_v^3} +$$

$$\sum_{v=1}^{\infty} \frac{\cos(\beta_v x_D)\sin[\beta_v(x_{wDm,i} \mp 0.5\Delta x_{Dm}\cos\theta_m)]}{\cos\theta_m \beta_v^3}$$

$$+ \frac{8\pi}{x_{eD}y_{eD}} \sum_{v=1}^{\infty} \sum_{v=1}^{\infty} \left\{ \begin{array}{l} \frac{\cos(\beta_v x_D)\cos(\gamma_v y_D)\cos[\beta_v x_{wDm,i} + \gamma_v y_{wDm,i}]}{\beta_v^2 + \gamma_v^2} \times \\ \left[ \frac{\sin[0.5\Delta x_D(\beta_v \cos\theta_m + \gamma_v \sin\theta_m)]}{\beta_v \cos\theta_m + \gamma_v \sin\theta_m} + \frac{\sin[0.5\Delta x_D(\beta_v \cos\theta_m - \gamma_v \sin\theta_m)]}{\beta_v \cos\theta_m - \gamma_v \sin\theta_m} \right] \end{array} \right\}$$

(B-8)

When calculating Eq. B-9, we need some relations to ensure the calculation accuracy, as follows:

$$\sum_{k=1}^{\infty} \frac{k\sin kx}{k^2 + a^2} = \frac{\pi}{2} \frac{\sinh a(\pi - x)}{\sinh a\pi}, \sum_{k=1}^{\infty} \frac{\cos kx}{k^2 + a^2} = \frac{\pi}{2a} \frac{\cosh[a\pi(1-x)]}{\sinh(a\pi)} - \frac{1}{2a^2},$$

$$\sum_{k=1}^{\infty} \frac{\cos kx}{k^2} = \frac{\pi^2}{6} - \frac{\pi x}{2} + \frac{x^2}{4}, \text{and } \sum_{k=1}^{\infty} \frac{\sin kx}{k^3} = \frac{\pi^2 x}{6} - \frac{\pi x^2}{4} + \frac{x^3}{13}[0 \leq x \leq 2\pi] \quad (B-9)$$

# Appendix C: Discretizing Eq. 6

On the location of $w$-th source, Eq. 6 can be rewritten as the following expression

$$p_{wFdn,w}^{PSS} - p_{fDn}^{PSS}(x_{Dn}) = \frac{2\pi}{C_{fDn}} \Big[ \sum_{v=1}^{N_w} q_{wfDn,v} \big[ G(x_{Dn}, x_{wfDn,v}) -$$

$$G(x_{wfDn,w}, x_{wfDn,v})] - \int_{x_{wfDn,w}}^{x_{Dn}} d\zeta \int_0^{\zeta} q_{fDn}(\xi) d\xi] \qquad (C-1)$$

Discretizing Eq. C-1 would obtain the expression with regard to $j$-segment of $n$-th fracture

$$p_{wfDn,w}^{PSS} - p_{fDn,j}^{PSS} = \frac{2\pi}{C_{fDn}} (U_{n,j}^{n,w} - R_{n,j}^{n,w}) \qquad (C-2)$$

where relevant expressions in Eq. C-1 are given as

$$U_{n,j}^{n,w} = \sum_{v=1}^{N_w} \tilde{q}_{wfDn,v} [G(x_{Dn}, x_{wfDn,v}) - G(x_{wfDn,w}, x_{wfDn,v})] \qquad (C-3)$$

and

$$R_{n,j}^{n,w} = \underbrace{\int_0^{x_{Dm,j}} d\zeta \int_0^{\zeta} \tilde{q}_{fDn}(\xi) d\xi}_{RS_{n,i}^{n,j}} - \underbrace{\int_0^{u_{wfDn,w}} d\zeta \int_0^{\zeta} \tilde{q}_{fDn}(\xi) d\xi}_{RT_{n,i}^{n,w}} \qquad (C-4)$$

where

$$RS_{n,i}^{n,j} = \begin{cases} x_{Dn,j} \Delta x_{Dn,j} - 0.5(x_{oDn,i}^2 - x_{oDn,i-1}^2), i < j \\ x_{Dn,j}(x_{Dn,j} - x_{oDn,j-1}) - 0.5(x_{Dn,j}^2 - x_{oDn,j-1}^2), i = j \\ 0, i > j \end{cases} \qquad (C-5)$$

and

$$RT_{n,i}^{n,w} = \begin{cases} (x_{wfDn,w} - x_{Dn,i}) \Delta x_{Dn,i}, i < 1 + \Xi \\ \left(x_{wfDn,w} - \sum_{i=1}^{\Xi} \Delta x_{Dn,i}\right)\left(0.5 x_{wfDn,w} + \sum_{i=1}^{\Xi} \Delta x_{Dn,i}\right), i = 1 + \Xi \\ 0, i > 1 + \Xi \end{cases} \qquad (C-6)$$

with $\Xi = M_n^w$ is the number of fracture segment in which $w$-th source of $n$-th fracture is located.

# Appendix D: Optimum fracture dimensions for fracture network

Please see Tables D-1 through D-4

**Table D-1  Optimum fracture dimensions for different fracture-network patterns**

| $N_{prop}$ | SF $C_{fD,opt}$ | $I_{x,opt}$ | $J_{D,max}$ | PFSF $C_{fD,opt}$ | $I_{x,opt}$ | $J_{D,max}$ | OFN $C_{fD,opt}$ | $I_{x,opt}$ | $J_{D,max}$ |
|---|---|---|---|---|---|---|---|---|---|
| $10^5$ | 33676 | 1.0000 | 5.77702 | 14239 | 1.0000 | 16.922 | 8931.9 | 1.0000 | 54.969 |
| $10^4$ | 3367.6 | 1.0000 | 5.7702 | 1423.9 | 1.0000 | 16.726 | 893.19 | 1.0000 | 51.111 |
| $10^3$ | 336.76 | 1.0000 | 5.6547 | 142.39 | 1.0000 | 14.684 | 114.31 | 0.9822 | 29.731 |
| $10^2$ | 33.676 | 1.0000 | 4.8822 | 14.239 | 1.0000 | 7.3206 | 14.629 | 0.7883 | 7.4218 |
| $10^1$ | 3.9885 | 0.9142 | 2.2768 | 2.7911 | 0.7125 | 1.6488 | 2.7582 | 0.5741 | 1.6212 |
| $10^0$ | 1.8517 | 0.4243 | 0.7251 | 1.8778 | 0.2758 | 0.5799 | 1.8069 | 0.2243 | 0.5876 |
| $10^{-1}$ | 1.6121 | 0.1438 | 0.3828 | 1.6089 | 0.0942 | 0.3485 | 1.5889 | 0.0756 | 0.3439 |
| $10^{-2}$ | 1.6117 | 0.0454 | 0.2635 | 1.6089 | 0.0298 | 0.2495 | 1.5889 | 0.0239 | 0.2468 |
| $10^{-3}$ | 1.6117 | 0.0144 | 0.2048 | 1.6089 | 0.0094 | 0.1937 | 1.5889 | 0.0075 | 0.1905 |

Table D-2  Optimum fracture dimensions for fracture network ($N_f=5$)

| $N_{prop}$ | PFSF $C_{fD,opt}$ | PFSF $I_{x,opt}$ | PFSF $J_{D,max}$ | OFN $C_{fD,opt}$ | OFN $I_{x,opt}$ | OFN $J_{D,max}$ |
|---|---|---|---|---|---|---|
| $10^5$ | 14239 | 1.0000 | 16.922 | 8931.9 | 1.0000 | 54.969 |
| $10^4$ | 1423.9 | 1.0000 | 16.726 | 893.19 | 1.0000 | 51.111 |
| $10^3$ | 142.39 | 1.0000 | 14.684 | 114.31 | 0.9822 | 33.731 |
| $10^2$ | 14.239 | 1.0000 | 7.3206 | 14.629 | 0.7883 | 7.4218 |
| $10^1$ | 2.7911 | 0.7125 | 1.6488 | 2.7582 | 0.5741 | 1.6212 |
| $10^0$ | 1.8778 | 0.2758 | 0.5799 | 1.8069 | 0.2243 | 0.5876 |
| $10^{-1}$ | 1.6089 | 0.0942 | 0.3485 | 1.5889 | 0.0756 | 0.3439 |
| $10^{-2}$ | 1.6089 | 0.0298 | 0.2495 | 1.5889 | 0.0239 | 0.2468 |
| $10^{-3}$ | 1.6089 | 0.0094 | 0.1937 | 1.5889 | 0.0075 | 0.1905 |

Table D-3  Optimum fracture dimensions for fracture network ($N_f=7$)

| $N_{prop}$ | PFSF $C_{fD,opt}$ | PFSF $I_{x,opt}$ | PFSF $J_{D,max}$ | OFN $C_{fD,opt}$ | OFN $I_{x,opt}$ | OFN $J_{D,max}$ |
|---|---|---|---|---|---|---|
| $10^5$ | 11141 | 1.0000 | 33.633 | 7817.2 | 1.0000 | 70.148 |
| $10^4$ | 1114.1 | 1.0000 | 32.366 | 781.72 | 1.0000 | 65.619 |
| $10^3$ | 111.41 | 1.0000 | 25.241 | 78.172 | 1.0000 | 37.911 |
| $10^2$ | 11.412 | 1.0000 | 7.7577 | 10.034 | 0.8556 | 7.8900 |
| $10^1$ | 2.6222 | 0.6509 | 1.6246 | 2.3225 | 0.5755 | 1.5413 |
| $10^0$ | 1.6889 | 0.2565 | 0.5712 | 1.6940 | 0.2131 | 0.5694 |
| $10^{-1}$ | 1.6089 | 0.0831 | 0.344 | 1.6193 | 0.0689 | 0.3336 |
| $10^{-2}$ | 1.6089 | 0.0263 | 0.2468 | 1.6193 | 0.0218 | 0.2409 |
| $10^{-3}$ | 1.6089 | 0.0083 | 0.1913 | 1.6193 | 0.0068 | 0.1899 |

Table D-4  Optimum fracture dimensions for fracture network ($N_f=9$)

| $N_{prop}$ | PFSF $C_{fD,opt}$ | PFSF $I_{x,opt}$ | PFSF $J_{D,max}$ | OFN $C_{fD,opt}$ | OFN $I_{x,opt}$ | OFN $J_{D,max}$ |
|---|---|---|---|---|---|---|
| $10^5$ | 9279.2 | 1.0000 | 53.76 | 6658.3 | 1.0000 | 86.078 |
| $10^4$ | 927.92 | 1.0000 | 50.393 | 665.83 | 1.0000 | 78.925 |
| $10^3$ | 92.792 | 1.0000 | 32.236 | 66.583 | 1.0000 | 41.671 |
| $10^2$ | 9.2792 | 1.0000 | 7.8442 | 9.4042 | 0.8719 | 8.0062 |
| $10^1$ | 2.3714 | 0.6162 | 1.5535 | 2.0744 | 0.5669 | 1.5188 |
| $10^0$ | 1.7356 | 0.2289 | 0.553 | 1.6378 | 0.2017 | 0.5542 |
| $10^{-1}$ | 1.5998 | 0.0754 | 0.3377 | 1.6008 | 0.0645 | 0.3411 |
| $10^{-2}$ | 1.5998 | 0.0238 | 0.2444 | 1.6008 | 0.0204 | 0.2456 |
| $10^{-3}$ | 1.5998 | 0.0075 | 0.1894 | 1.6008 | 0.0064 | 0.1875 |

# 考虑裂缝变导流能力的致密气井现代产量递减分析

孙贺东[1]　欧阳伟平[2,3]　张冕[2,3]　唐海发[1]　陈长骁[4]　马旭[5]　付中新[2,3]

(1. 中国石油勘探开发研究院;2. 中国石油川庆钻探长庆井下技术作业公司;
3. 低渗透油气田勘探开发国家工程实验室;4. 中石油煤层气有限责任公司忻州分公司;
5. 中国石油川庆钻探长庆指挥部)

**摘要**:针对水力压裂形成的裂缝导流能力会随空间以及时间发生变化这一特点,建立考虑裂缝空间、时间双重变导流和应力敏感效应的压裂直井渗流的新型数学模型,采用混合有限元方法对模型进行求解,获得新模型的 Blasingame 现代产量递减分析典型曲线。以此为基础,讨论了裂缝空间、时间双重变导流及应力敏感效应对 Blasingame 曲线形态的影响。研究表明:空间变导流效应主要表现为降低早期典型曲线值;时间变导流效应会使得产量及产量积分曲线"下掉",形成"S"形典型曲线;双重变导流则是两者效应的叠加;时间、空间变导流效应均不会延缓地层进入拟稳态阶段的时间。应力敏感会降低曲线值,但不会使得曲线"陡降";应力敏感会延缓地层进入拟稳态阶段的时间。忽略变导流效应和应力敏感效应,不会对井控动态储量的结果产生较大影响,但会给裂缝、储层参数解释带来较大误差。苏里格气田致密气压裂井的现代产量递减分析对比证实新模型可靠实用,且比常规模型更先进,可以用于致密气压裂井的动态分析。

**关键词**:变导流能力;现代产量递减分析;压裂;有限元;应力敏感;致密气;数值计算

现代产量递减分析方法[1]是近年来油气藏工程学科中进行单井、井组生产动态分析的新兴技术,它以不稳定渗流理论为基础,根据油气井的日常生产数据,采用 Blasingame[2,3] 递减典型曲线拟合分析压力和产量之间的关系,最终通过典型曲线拟合获取储层参数与井控储量。现代产量递减分析方法自 Blasingame 双对数图版拟合阶段开始得到广泛应用以来,许多学者将其扩展到不同类型井型[4-6]和储层[7-9]。水力压裂是目前提高致密气藏单井产量的最常用的方法,而压裂井现代产量递减分析除获取储层参数和井控储量以外,还可作为评价压裂效果、获取裂缝参数的技术手段。目前用于压裂井现代产量递减分析的渗流模型几乎均把裂缝假设成导流能力恒定[6-9],这将会给普遍存在变导流情况的压裂井生产数据分析带来较大误差。

裂缝变导流能力可分为空间变导流和时间变导流两类,空间变导流是由于压裂施工中支撑剂充填不均,裂缝为楔形缝,导流能力随裂缝延伸而变化;时间变导流是由于支撑剂破碎、嵌入地层、岩屑堵塞等因素导致裂缝导流能力随时间发生变化。裂缝导流能力变化大小主要与支撑剂铺置浓度、支撑剂物理性质、裂缝闭合应力以及岩石硬度有关,也与储层温度、流体性质以及盐水环境等因素有关。由于裂缝导流能力的影响因素多,理论上很难采用通用的数学公式加以表征,因此目前主要根据室内支撑剂导流能力测试来衡量裂缝变导流效应,再利用不同函数关系对测试数据进行回归分析。国内外学者做了大量支撑剂导流试验来研究裂缝变导流效应[10-12],其中空间变导流有线性、指数和对数关系 3 种变化形式,而时间变导流主要有对数关系和指数关系两种变化形式。不少学者根据裂缝变导流公式建立油气井不稳定渗流模型,然后应用到试井解释及生产动态分析[13-16]。然而,考虑裂缝空间与时间双重变导流的现代产量递减分析目前未查到相关的文献报道。

针对这一技术问题,本文根据目前常用的裂缝变导流关系式,建立一种同时考虑裂缝空间和时间变导流的致密气非稳态渗流数学模型。利用混合有限元方法对模型进行求解,获得 Blasingame 现代产量递减分析典型曲线,进而分析变导流因素对典型曲线的影响,最后通过现场实例应用论证模型的可靠性及实用性。

# 1 非稳态渗流数学模型

## 1.1 物理模型及假设条件

对均质有界储层中一口有限导流垂直压裂气井,物理模型假设条件为:(1)储层具有应力敏感效应,考虑渗透率应力敏感,忽略孔隙度应力敏感;(2)致密气藏为干气藏或者低含水饱和度气藏,水处于束缚水状态,不存在启动压力梯度效应,气体在储层中的流动为单相渗流,且满足达西定律;(3)水力压裂后形成一条有限导流裂缝,裂缝具有空间变导流和时间变导流特性,气体在裂缝中的流动为一维流动,考虑压裂液对储层的伤害,用裂缝表皮系数来衡量其对储层伤害的大小;(4)裂缝体积与井控体积相比非常小,裂缝渗透率远大于储层渗透率,裂缝压力降低造成裂缝内气体体积膨胀对整个流动的影响非常小,可忽略裂缝控制方程中拟压力对时间的导数项;(5)考虑气体压缩系数及黏度随压力变化而变化,气体压缩系数及偏差因子采用 DPR 方法[17]计算,气体黏度采用 Lee 方法[18];(6)不考虑井筒储集效应、温度变化等其他因素对流动的影响。

## 1.2 渗流数学模型

储层控制方程为[9]

$$\frac{\partial}{\partial x}\left[K_r(p)\frac{\partial \psi}{\partial x}\right]+\frac{\partial}{\partial y}\left[K_r(p)\frac{\partial \psi}{\partial y}\right]-\alpha_3 q_f = \alpha_1 \phi C_t(p)\mu(p)\frac{\partial \psi}{\partial t} \quad (1)$$

裂缝控制方程为

$$\frac{\partial}{\partial l}\left[K_f(l,t)\frac{\partial \psi}{\partial l}\right]+\alpha_3 q_f = 0 \quad (2)$$

初始条件为

$$\psi(x,y,0)=\psi_i \quad (3)$$

内边界条件为

已知井口产量 $\quad \dfrac{\partial \psi}{\partial l}\bigg|_{\Gamma_{in}} = \dfrac{\alpha_2 q_{sc}(t)T}{w_f K_f(0,t)h} \quad (4)$

已知井底压力 $\psi|_{\Gamma_{in}} = \psi_w(t)$ (5)

封闭外边界为

$$\left.\frac{\partial \psi}{\partial n}\right|_{\Gamma_{out}}$$ (6)

物质平衡方程为

$$\frac{\bar{p}}{Z} = \left(\frac{p}{Z}\right)_i \left(1 - \frac{G_p}{G}\right)$$ (7)

拟压力定义为

$$\psi(p) = 2\int_0^p \frac{p}{\mu Z}\mathrm{d}p$$ (8)

## 1.3 裂缝变导流公式

### 1.3.1 空间变导流

裂缝导流能力($K_f w_f$)随裂缝长度呈线性关系、指数关系与对数关系变化,计算公式分别为[14]

$$F_{cl} = F_{c0}(1 - al_D)$$ (9)

$$F_{cl} = F_{c0}\mathrm{e}^{-bl_D}$$ (10)

$$F_{cl} = F_{c0}[1 - c\ln(1 + l_D)]$$ (11)

### 1.3.2 时间变导流

裂缝导流能力随生产时间呈对数关系与指数关系变化,计算公式分别为[10,11]

$$F_{ct} = F_{ci}[1 - \beta\ln(1 + t)]$$ (12)

$$F_{ct} = F_{ci}\mathrm{e}^{-\gamma t} + F_{cr}$$ (13)

对目前常用的指数变化关系式(13)进行修改:

$$F_{ct} = F_{ci}(n\mathrm{e}^{-\frac{t}{8760C}} + 1 - \eta)$$ (14)

系数 $\eta$ 控制导流能力衰减幅度,系数 $C$ 控制导流能力的衰减速度,修改后系数的物理意义更明确,有助于生产数据解释及分析。

### 1.3.3 双重变导流

同时考虑空间变导流和时间变导流,即裂缝导流能力不仅随裂缝位置而变,同时还随着时间而变。以空间变导流和时间变导流均采用指数关系变化为例,推导出双重变导流计算公式:

$$F_c(l,t) = F_{ci0}\mathrm{e}^{-bl}(n\mathrm{e}^{-\frac{t}{8760C}} + 1 - \eta)$$ (15)

以此类推可以获得不同函数组合情况下的双重变导流公式。

## 1.4 储层渗透率公式

致密气藏可能会存在应力敏感、启动压力梯度及滑脱效应等非线性渗流特征。应力敏感效应主要表现为储层渗透率不再为常数,而是随有效应力的增加而减小。大量实验[20-22]表明致密气藏具有较强的应力敏感效应,对于分析长时间生产数据的产量递减分析来说必须予以考虑。气藏存在启动压力梯度主要是由于气水作用造成[23,24],这与油藏中的启动压力梯度有本质区别,因此气藏中存在一定程度的水是产生启动压力梯度的必要条件。此外,致密气藏可能还存在滑脱效应,滑脱效应会增加视渗透率,储层越致密,地层压力越低,滑脱效应越强,目前考虑滑脱效应的渗流模型主要应用于埋深较浅的煤层气中[25,26],而对于储层压力普遍较高的致密砂岩气藏,滑脱效应的影响还有待考察。

### 1.4.1 滑脱效应

滑脱效应的大小主要由滑脱因子与地层平均压力决定,可由 Klinkenberg 公式表征[27]:

$$K_g = K_\infty \left(1 + \frac{b}{p_m}\right) \quad (16)$$

为了考察致密砂岩气藏中滑脱效应对渗流影响的大小,采用苏里格气田岩心做了96次滑脱效应实验,结果见图1,经双对数线性回归得滑脱因子与绝对渗透率之间的关系式为

$$b = 0.0011 K_\infty^{-0.6061} \quad (17)$$

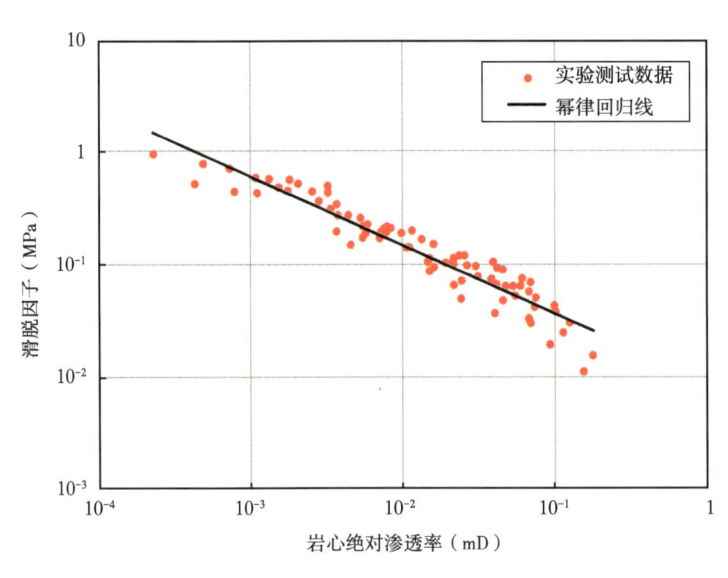

图 1  滑脱效应实验结果

假设致密气藏储层绝对渗透率为 0.1mD,原始地层压力 30MPa,生产后期的地层平均压力为 10MPa,根据式(16)与式(17)可计算得到滑脱效应对早、晚期储层视渗透率的影响分别为 0.12% 和 0.37%。即使储层绝对渗透率为 0.01mD,对早、晚期储层视渗透率的影响也仅有 0.50% 和 1.50%。计算该储层条件下滑脱效应对 Blasingame 曲线的影响如图2所示,图中曲线对比可知,考虑和不考虑滑脱效应情况下,两者曲线基本重合,说明其对产量递减曲线的影

响非常小,可以在致密气井产量递减分析过程中忽略滑脱效应的影响。

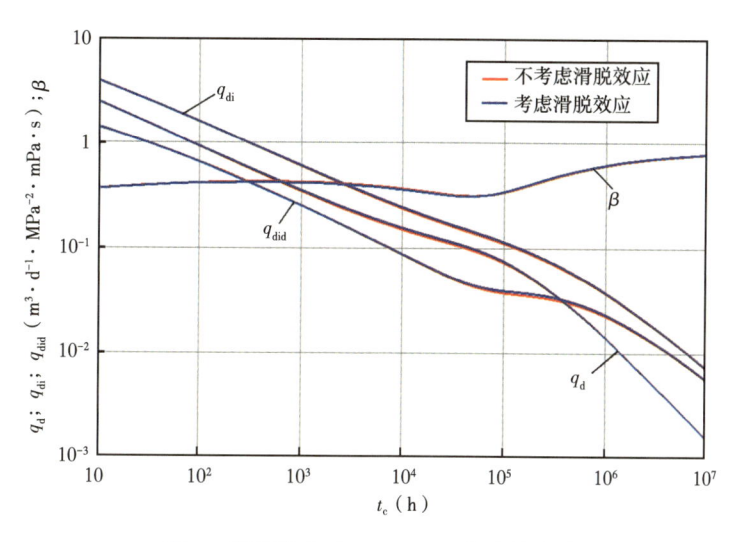

图 2 滑脱效应对 Blasingame 曲线的影响

### 1.4.2 应力敏感效应

储层应力敏感的评价方法具有很多种[28],目前多采用指数关系式和幂律关系式。鉴于幂律关系式回归实验测试数据的相关度更高,本文采用渗透率随有效应力呈幂律关系变化的公式[29]:

$$K_r(p) = K_{ri}\left(\frac{\sigma_e}{\sigma_{ei}}\right)^{-S_p} = K_{ri}\left(\frac{\sigma - p}{\sigma - p_i}\right)^{-S_p} \tag{18}$$

## 1.5 模型求解

采用混合有限元方法对模型进行求解[30,31]:

$$\iint_\Omega F_{eq} d\Omega = \iint_{\Omega_m} F_{eq} d\Omega_m + w_f \int_{\overline{\Omega_f}} - F_{eq} d\overline{\Omega_f} \tag{19}$$

将整个计算区域划分为两个部分,一个是二维流动的储层区域,一个是一维流动的裂缝区域。式中 $F_{eq}$ 代表流体流动方程, $\Omega$ 代表整个流动区域, $\Omega_m$ 代表储层流动区域, $\overline{\Omega_f}$ 代表裂缝流动区域。

利用 Galerkin 加权余量法[32]分别离散储层和裂缝的控制方程,得到储层区域二维有限元方程为:

$$\begin{aligned}
&AK_r(p)\left[b_ib_i + c_ic_i + \frac{\alpha_1\phi\mu(p)C_t(p)}{6\Delta tK_r}\right]\psi_i^{n+1} + AK_r(p)\left[b_ib_j + c_ic_j + \frac{\alpha_1\phi\mu(p)C_t(p)}{12\Delta tK_r}\right]\psi_j^{n+1} + \\
&AK_r(p)\left[b_ib_k + c_ic_k + \frac{\alpha_1\phi\mu(p)C_t(p)}{12\Delta tK_r}\right]\psi_i^{n+1} - \frac{K_r(p)L}{3}\frac{\partial\psi_i^{n+1}}{\partial n} - \frac{K_r(p)L}{6}\frac{\partial\psi_{j(k)}^{n+1}}{\partial n} \\
&= \frac{\alpha_1\phi\mu(p)C_t(p)}{6\Delta t}\psi_i^n + \frac{\alpha_1\phi\mu(p)C_t(p)}{12\Delta t}\psi_j^n + \frac{\alpha_1\phi\mu(p)C_t(p)}{12\Delta t}\psi_k^n
\end{aligned} \tag{20}$$

裂缝区域一维有限元方程为

$$\frac{F_c(l,t)}{L}\psi_i^{n+1} - \frac{F_c(l,t)}{L}\psi_j^{n+1} + F_c(l,t)\frac{\partial \psi_i}{\partial l} = 0 \tag{21}$$

由式(20)和式(21)建立储层区域的有限元刚度矩阵和裂缝区域的刚度矩阵,再将两者组合成系统刚度矩阵。根据窜流关系式,通过储层和裂缝单元叠加消除裂缝与储层交界处的边界项,即式(20)等号左端最后两项以及式(21)等号左端最后一项,具体组合方法可参照文献[30,31]。

裂缝导流能力、储层渗透率、气体黏度以及压缩系数均会随着时间而变,实际计算中每个时间步长根据各参数的具体计算方法来计算。此外,模型采用两种方法计算地层平均压力,在根据生产数据迭代获取初始地质储量 $G$ 值时采用联立物质平衡方程式(7)来计算地层平均压力,在全历史曲线拟合时采用网格单元压力加权平均的方法来计算地层平均压力。最后利用线性方程组求解器求解组合成的线性方程组,从而获得模型的解。

## 2 计算结果及对比分析

### 2.1 空间变导流下的典型曲线

假定储层渗透率为0.1mD,原始储层压力为30MPa,储层温度为100℃,有效厚度为8m,孔隙度为10%,圆形封闭半径为500m,天然气相对密度为0.7,裂缝半长为100m,单独考虑裂缝空间变导流,缝口导流能力为200mD·m。为了便于对比分析,可设置变导流系数 $a$、$b$、$c$ 的值使得缝端导流能力均为20mD·m。

通过计算得不同裂缝空间变导流关系下的 Blasingame 典型曲线(图3),对比 Blasingame 曲线计算结果可知,裂缝变导流能力对产量递减典型曲线影响主要在早期,空间变导流会降低早期典型曲线值,类似于裂缝表皮系数的影响,常规不考虑变导流的模型会造成解释得到的表

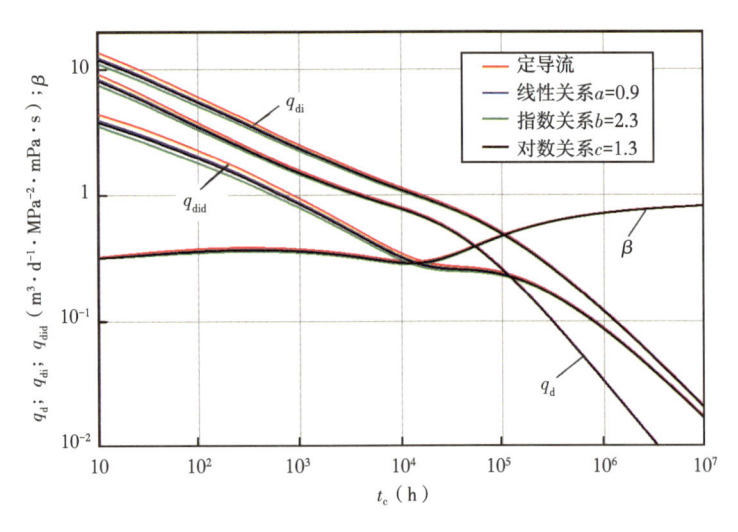

图3 空间变导流对 Blasingame 曲线的影响

皮系数偏大。不同变导流条件下晚期典型曲线几乎重合,说明变导流不影响压力的扩散速度,拟稳态阶段的出现时间不受其影响。另外,对比不同变导流关系曲线可以发现在缝端和缝口导流值相同的情况下,指数变化对典型曲线的降低幅度最大,线性关系最小,对数关系居中。

## 2.2 时间变导流下的典型曲线

采用相同的储层参数,计算分析单独考虑裂缝时间变导流的影响。假定裂缝初始导流能力为200mD·m,生产一年时间后衰减至导流能力值20mD·m,随后导流能力基本稳定。分别根据对数关系和指数关系变化计算时间变导流能力对 Blasingame 典型曲线的影响(图4)。分析计算结果可知,相比于标准定导流典型曲线,裂缝时间变导流对产量递减曲线的影响比较明显,其主要影响在中期,表现为产量曲线及产量积分曲线"下掉",形成"S形"。与空间变导流的影响类似,时间变导流对典型曲线晚期的影响非常小,时间变导流同样不会影响储层的压力扩散速度。此外,对比对数关系与指数关系条件下的典型曲线可知,对数关系下的典型曲线光滑度不如指数关系,另外对数关系对典型曲线的影响幅度要小于指数关系,这是因为对数关系整个过程的变化率比较平稳,而指数关系的变化率是由大变小。

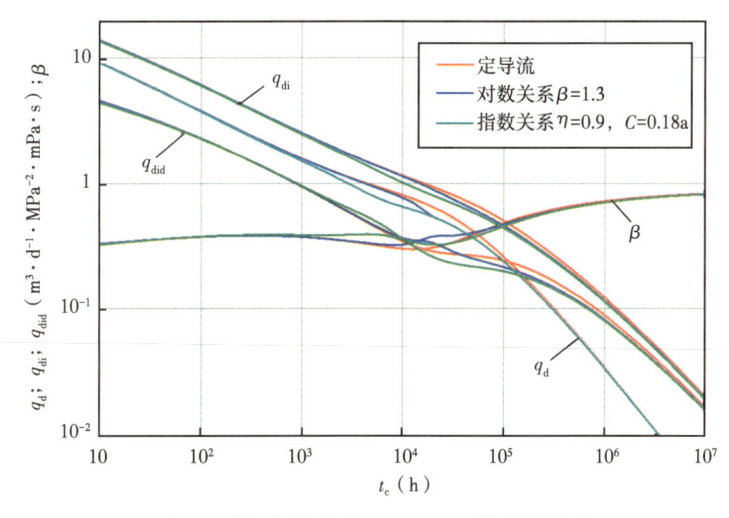

图 4 时间变导流对 Blasingame 曲线的影响

以对数变导流关系为例分析变导流系数对典型曲线的影响。图5为设定变导流时间系数 $C=0.1a$,不同变导流衰减系数 $\eta$ 条件下典型曲线对比。图6为设定变导流衰减系数 $\eta=0.9$,不同变导流时间系数 $C$ 条件下的典型曲线对比。由计算结果可知,裂缝导流衰减系数越大,产量曲线及产量积分曲线"下掉"的幅度越大,"S"形曲线越明显。裂缝变导流时间系数越大,即裂缝导流能力衰减速度越小,产量曲线及产量积分曲线"下掉"的开始时间越晚,其对典型曲线的影响也越小,当变导流时间系数很大时(图6中第4组曲线),其影响甚至可以忽略,或者很难辨别。由此可知时间变导流模型中裂缝衰减系数控制典型曲线"下掉"的幅度大小,而时间系数控制着典型曲线"下掉"的时间早晚。这一特点对于时间变导流裂缝井的产量递减曲线拟合具有重要指导意义。

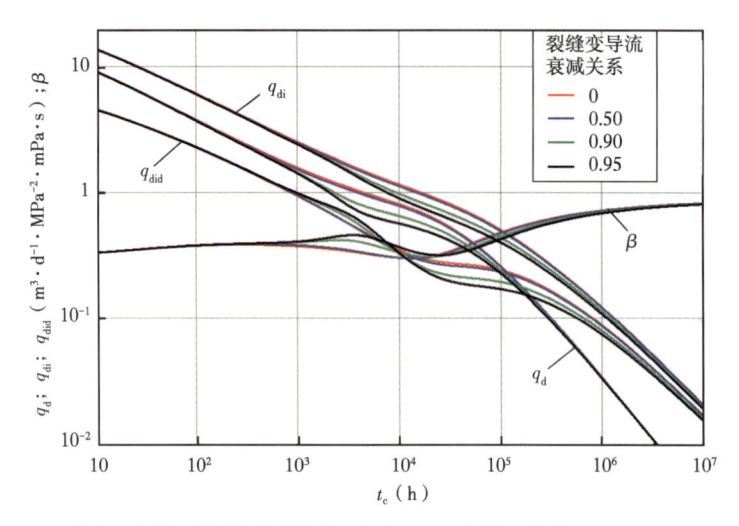

图 5 裂缝导流衰减系数对 Blasingame 曲线的影响（$C=0.1a$）

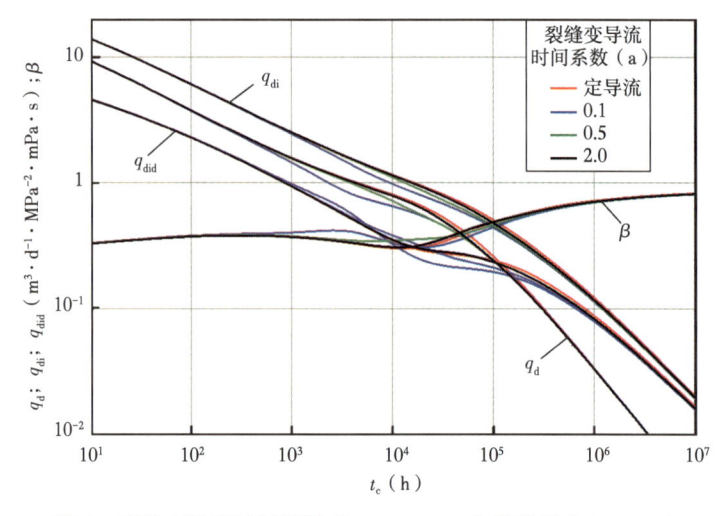

图 6 裂缝变导流时间系数对 Blasingame 曲线的影响（$\eta=0.9$）

## 2.3 双重变导流下的典型曲线

图 7 为不同裂缝变导流条件下的 Blasingame 曲线对比，其中空间变导流和时间变导流均采用指数变化关系，变导流系数 $b$ 值为 $2$，$\eta$ 值为 $0.9$，$C$ 值为 $0.2a$。对比曲线可知，典型曲线早期，曲线 1 和曲线 3 重合，而曲线 2 和曲线 4 重合，这说明典型曲线早期主要受空间变导流的影响；典型曲线中期，曲线 2 和曲线 4 相差较大，而曲线 3 和曲线 4 差别不大，这说明典型曲线中期受时间变导流和空间变导流双重影响，但时间变导流起主导作用；典型曲线晚期基本不受变导流的影响，所有曲线基本重合。综合对空间变导流和时间变导流的单因素影响分析可以发现，同时考虑空间变导流和时间变导流条件下的 Blasingame 曲线是两者效应的叠加。

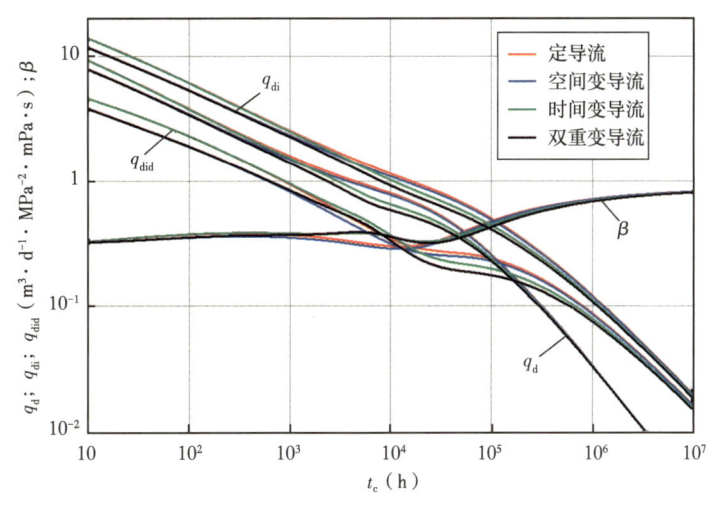

图 7 双重变导流对 Blasingame 曲线的影响

## 2.4 应力敏感与双重变导流共同作用下的典型曲线

致密气藏实际生产井中可能会存在裂缝双重变导流和应力敏感共同作用的情况,为此在模型中同时考虑裂缝双重变导流和地层应力敏感效应。假定上覆岩石压力为 60MPa,采用 2.3 节中相同的双重变导流系数,分别计算双重变导流作用下不同应力敏感系数的 Blasingame 典型曲线(图8)。对比可知,不同应力敏感系数条件下的曲线在早期阶段基本重合,但一段时间后曲线差异逐渐加大。与时间变导流的影响类似,应力敏感效应会造成中后期阶段产量递减典型曲线值减小,应力敏感系数越大曲线值减小的幅度越大。与时间变导流效应不同的是,应力敏感会使得地层渗透率减小,由此减缓压力扩散的速度,使得边界响应的时间推迟,图中 $\beta$ 曲线后期能够清晰地反映出该特征,应力敏感效应越强拟稳态开始的时间越晚,这种特点可用于区分应力敏感和时间变导流。

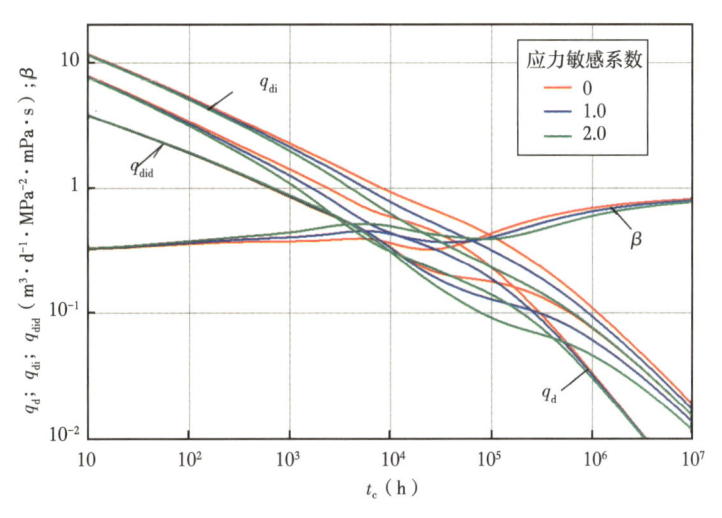

图 8 应力敏感与双重变导流共同作用下的 Blasingame 曲线

另外，晚期进入拟稳态后，变导流能力效应和应力敏感效应影响极小，即变导流能力效应和应力敏感效应不会对井控动态储量的拟合结果产生影响。

## 3 实例应用

采用本文所建立的致密气藏双重变导流渗流模型对苏里格气田某井的生产数据进行现代产量递减分析，并与常规模型的拟合结果进行对比来说明新模型的先进性与实用性。该井原始储层压力为32MPa，储层温度为107℃，孔隙度为13.7%，有效厚度为5.3m，含气饱和度为74%，天然气相对密度为0.6，生产时间约为2.5年，累计产量$848×10^4 m^3$。常规模型与新模型的产量递减典型曲线拟合效果如图9、图10所示，全历史日产量及累计产量曲线的拟合效果见图11，曲线拟合所获得的储层参数及裂缝参数见表1。从实测Blasingame产量曲线可知，该井产量曲线在生产时间不到1000小时后突然"下掉"，产量曲线具有明显的"S形"特征，非常符合时间变导流效应的特点。

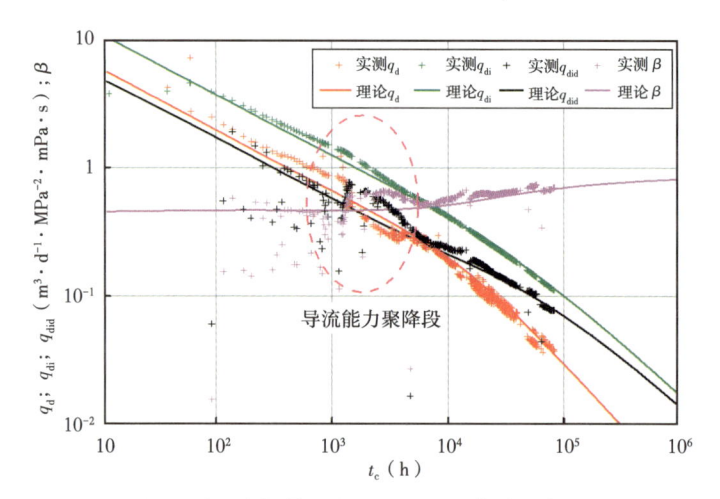

图9 采用常规模型的Blasingame曲线拟合图

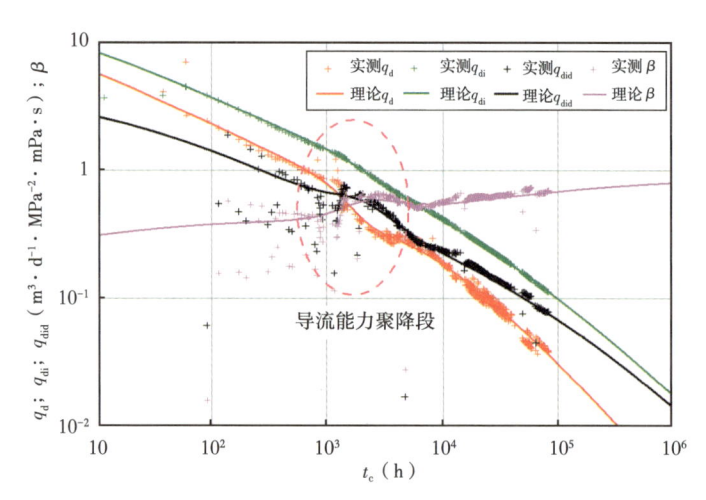

图10 采用本文模型的Blasingame曲线拟合图

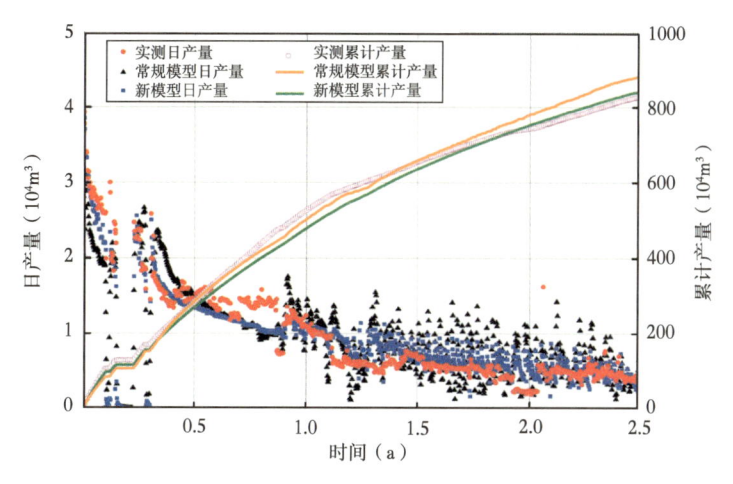

图 11　苏里格气田某井的全历史产量曲线拟合图

表 1　苏里格某井的产量递减分析结果

| 参数 | 储层渗透率（mD） | 裂缝半长（m） | 缝口初始导流（mD·m） | 裂缝表皮系数 |
|---|---|---|---|---|
| 常规模型 | 0.0235 | 165 | 860 | 0 |
| 本文模型 | 0.108 8 | 98 | 580 | 0.03 |

| 参数 | 空间变导流系数 | 时间变导流衰减系数 | 时间变导流时间系数(a) | 应力敏感系数 |
|---|---|---|---|---|
| 常规模型 | — | — | — | — |
| 本文模型 | 1.6 | 0.989 | 0.045 | 1.15 |

| 参数 | 井控半径（m） | 井控储量（$10^8 m^3$） | 当前地层平均压力（MPa） | 采出程度（%） |
|---|---|---|---|---|
| 常规模型 | 197 | 0.162 | 13.8 | 52.5 |
| 本文模型 | 203 | 0.172 | 14.8 | 49.4 |

从图中可以看出，考虑双重变导流效应和应力敏感效应可以对典型曲线以及全历史产量曲线进行比较理想的拟合，而常规模型无法对裂缝导流能力骤降段进行拟合，只能通过加大裂缝半长、裂缝导流能力，减小储层渗透率来大致拟合早期典型曲线的变化趋势，因此其解释的裂缝半长及裂缝导流能力都严重偏大。由此可见，新方法比常规模型方法更实用，更先进。此外，对比井控半径及井控储量可知，两者解释的结果基本一致，这进一步说明了变导流效应和应力敏感效应均不会对井控动态储量的结果产生较大影响。从解释得到的变导流系数可知，该井在早期生产过程中导流能力衰减幅度非常大，时间变导流系数达到了 0.989，这也是造成产量快速递减的主要原因。

# 4　结论

(1)建立了考虑裂缝空间、时间双重变导流和应力敏感效应的压裂直井渗流的新型数学模型;推导出一种新的指数型时间变导流表征公式,变导流系数物理意义更明确;采用混合有限元方法对模型进行求解,获得新模型的 Blasingame 典型曲线,可用于存在裂缝变导流致密气压裂井现代产量递减分析。

(2)裂缝变导流及应力敏感效应对 Blasingame 曲线的影响表现出不同的特点:空间变导流效应主要表现为降低早期典型曲线值;时间变导流效应会使得产量及产量积分曲线"下掉",形成"S 形"典型曲线;双重变导流则是两者效应的叠加;时间、空间变导流效应均不会延缓地层进入拟稳态阶段的时间。应力敏感会降低曲线值,但不会使得曲线"陡降";应力敏感会延缓地层进入拟稳态阶段的时间。忽略变导流效应和应力敏感效应,不会对井控动态储量的结果产生较大影响,但会给裂缝、储层参数解释带来较大误差。

(3)苏里格气田致密气压裂井的现代产量递减分析对比证实新模型可靠实用,且比常规模型更先进,可以用于致密气压裂井的动态分析。

## 符号注释

$A$——三角形网格面积,m²;$b$——气体滑脱因子,MPa;$b$,$c$——有限元单元系数;a,b,c——不同变化关系的回归系数;$C_t$——综合压缩系数,MPa⁻¹;$C$——裂缝变导流时间系数,a;$F_{cl}$——随裂缝长度变化的导流能力,mD·m;$F_{c0}$——缝口处裂缝的导流能力,mD·m;$F_{ct}$——随时间变化的导流能力,mD·m;$F_{ci}$——裂缝初始导流能力,mD·m;$F_{cr}$——裂缝残余导流能力,mD·m;$F_c$——随空间及时间而变的裂缝导流能力,mD·m;$F_{ci0}$——缝口处裂缝初始导流能力,mD·m;$G_p$——累计产量,10⁴m³;$G$——井控地质储量,10⁴m³;$h$——储层有效厚度,m;$i$,$j$,$k$——三角形网格结点序号;$K_f$——裂缝渗透率,mD;$K_g$——视渗透率,mD;$K_r$——储层渗透率,mD;$K_{ri}$——储层初始渗透率,mD;$K_\infty$——绝对渗透率,mD;$l$——裂缝控制方程的坐标轴,m;$l_D = l/x_f$——距缝口的无量纲距离;$L$——裂缝一维网格长度,m;$\bar{p}$——地层平均压力,MPa;$p_i$——地层初始压力,MPa;$p_m$——地层压力,MPa;$q_f$——单位时间单位体积储层进入裂缝的流体标况体积,1/d;$q_{sc}$——标准状况下气体产量,m³/d;$S_p$——渗透率应力敏感系数;$t$——生产时间,h;$t_c$——物质平衡时间,h;$T$——储层温度,K;$\Delta t$——时间步长,h;$w_f$——裂缝宽度,m;$x$,$y$——储层控制方程的坐标轴,m;$x_f$——裂缝半长,m;$Z$——气体偏差因子;$\alpha_1$——量纲换算系数,其值为 277.8;$\alpha_2$——量纲换算系数,取值 0.004;$\alpha_3$——常数,其值为 23.14$p_{sc}T/T_{sc}$;$\beta$,$\gamma$——不同变化关系的回归系数;$\eta$——裂缝变导流衰减系数;$\mu$——气体黏度,mPa·s;$\sigma$——上覆岩石应力,MPa;$\sigma_e$——储层有效应力,MPa;$\sigma_{ei}$——储层初始有效应力,MPa;$\Gamma_{in}$——内边界;$\Gamma_{out}$——外边界;$\phi$——有效孔隙度,%;$\psi$ 为拟压力,MPa²/(mPa·s);$\psi_i$——初始拟压力,MPa²/(mPa·s);$\psi_w$——井底拟压力,MPa²/(mPa·s);$\frac{\partial \psi}{\partial n}$——沿 $\Gamma$ 单位外法线方向 $n$ 的方向导数。

## 参 考 文 献

[1] 孙贺东. 油气井现代产量递减分析方法及应用[M]. 北京:石油工业出版社. 2013.
[2] Blasingame T A, Johnston J L, Lee W J. Type-curve analysis using the pressure integral method[R]. SPE 18799, 1993.

[3] Blasingame T A, Ilk D, Hosseinpour-Zonoozi N. Application of the B-derivative function to production analysis [R]. SPE 107967, 2007.

[4] Shih M Y, Blasingame T A. Decline curve analysis using type curves: horizontal wells[R]. SPE 29572, 1995.

[5] Marhaendrajana T, Schlumberger, Blasingame T A, et al. Decline-curve analysis using type curves-evaluation of well performance behavior in a multi-well reservoir system[R]. SPE 71517, 2011.

[6] Pratikno H, Rushing J A, Blasingame T A. Decline curve analysis using type curves-fractured wells[R]. SPE 84287, 2003.

[7] Clarkson C R, Jordan C L, Ilk D, et al. Production data analysis of fractured and horizontal CBM wells[R]. SPE 125929, 2009.

[8] Nobakht M, Clarkson C R, Kaviani D. New type curves for analyzing horizontal well with multiple fractures in shale gas reservoirs[J]. Journal of Natural Gas Science & Engineering, 2016, 10(1): 99-112.

[9] 魏明强, 段永刚, 方全堂, 等. 基于物质平衡修正的页岩气藏压裂水平井产量递减分析方法[J]. 石油学报, 2016, 37(4): 508-515.

[10] 俞绍诚. 陶粒支撑剂和兰州压裂砂长期裂缝导流能力的评价[J]. 石油钻采工艺, 1987, 9(5): 93-100.

[11] 温庆志, 张士诚, 王雷, 等. 支撑剂嵌入对裂缝长期导流能力的影响研究[J]. 天然气工业, 2005, 25(5): 65-68.

[12] 卢聪, 郭建春, 王文耀, 等. 支撑剂嵌入及对裂缝导流能力损害的实验[J]. 天然气工业, 2008, 28(2): 99-101.

[13] 李勇明, 罗剑, 郭建春, 等. 介质变形和长期导流裂缝性气藏压裂产能模拟研究[J]. 天然气工业, 2006, 26(9): 103-105.

[14] 牟珍宝, 樊太亮. 圆形封闭油藏变导流垂直裂缝井非稳态渗流数学模型[J]. 油气地质与采收率, 2006, 13(6): 66-69.

[15] 熊健, 马振昌, 马华, 等. 考虑变导流能力的垂直裂缝油井产能方程[J]. 复杂油气藏, 2013, 6(4): 52-54, 64.

[16] 高阳, 赵超, 董平川, 等. 致密气藏变导流能力裂缝压裂水平井不稳定渗流模型[J]. 大庆石油地质与开发, 2015, 34(6): 141-147.

[17] Dranchuk P M, Purvis R A, Robinson D B. Computer calculation of natural gas compressibility factors using the Standing and Katz correlation[R]. Edmonton: Annual Technical Meeting, 1973.

[18] Lee A, Gonzalez M, Eakin B. The viscosity of natural gases[J]. Journal of Petroleum Technology, 1966, 18(8): 997-1000.

[19] 孙贺东, 欧阳伟平, 张冕. 基于数值模型的气井现代产量递减分析及动态预测[J]. 石油学报, 2017, 38(10): 1194-1199.

[20] 杨朝蓬, 高树生, 郭立辉, 等. 致密砂岩气藏应力敏感性及其对产能的影响[J]. 钻采工艺, 2013, 36(2): 58-61.

[21] 肖文联, 李滔, 李闽, 等. 致密储集层应力敏感性评价[J]. 石油勘探与开发, 2016, 43(1): 107-114.

[22] 窦宏恩, 张虎俊, 姚尚林, 等. 致密储集层岩石应力敏感性测试与评价方法[J]. 石油勘探与开发, 2016, 43(6): 1022-1028.

[23] 贺伟, 冯曦, 钟孚勋. 低渗储层特殊渗流机理和低渗透气井动态特征探讨[J]. 天然气工业, 2002, 22(1): 91-94.

[24] 朱维耀, 宋洪庆, 何东博, 等. 含水低渗气藏低速非达西渗流数学模型及产能方程研究[J]. 天然气地球科学, 2008, 19(5): 685-689.

［25］肖晓春,潘一山. 滑脱效应影响的低渗煤层气运移实验研究［J］. 岩土工程学报,2009,31(10):1554-1558.

［26］肖晓春,潘一山. 考虑滑脱效应的水气耦合煤层气渗流数值模拟［J］. 煤炭学报,2006,31(6):711-715.

［27］Klinkenberg L J. The permeability of porous media to liquids and gases［J］. Socar Proceedings,1941,2(2):200-213.

［28］罗瑞兰,冯金德,唐明龙,等. 低渗储层应力敏感评价方法探讨［J］. 西南石油大学学报(自然科学版),2008,30(5):161-164.

［29］罗瑞兰,程林松,彭建春,等. 确定低渗岩心渗透率随有效覆压变化关系的新方法［J］. 中国石油大学学报:自然科学版,2007,31(2):87-90.

［30］Wan Y Z, Liu Y W, OuYang W P, et al. Numerical investigation of dual-porosity model with transient transfer function based on discrete-fracture model［J］. Applied Mathematics and Mechanics,2016,37(5):611-626.

［31］Wan Y Z, Liu Y W, Liu W C, et al. A numerical approach for pressure transient analysis of a vertical well with complex fractures［J］. Acta Mechanica Sinica,2016,32(4):640-648.

［32］张涤明,蔡崇喜,章克本,等. 计算流体力学［M］. 广州:中山大学出版社,1991.

# 基于蒙特卡洛随机模拟的油气储量不确定性评价

尹 涛[1] 杨屹铭[2] 靳锁宝[3] 江乾锋[4] 孟德伟[5]

(1. 中国石油西南油气田分公司天然气经济研究所；
2. 中国石油川庆钻探工程有限公司地质勘探开发研究院；
3. 中国石油长庆油田分公司气田开发事业部；
4. 中国石油长庆油田分公司勘探开发研究院；
5. 中国石油勘探开发研究院)

**摘要**：地质储量是油气田开发的重要指标和各个开发阶段方案设计与调整的核心基础，计算的准确与否会对油气田开发、管理、经济等目标产生重要影响。计算油气储量的常用方法有：容积法、压降法、类比法等，每种计算方法在参数选取过程中都存在静态或动态的不确定性，导致储量计算的误差。认识和评价参数不确定性，明确对油气藏储量及阶段剩余储量的影响程度十分重要。以储量计算应用中广泛且适用性较好的容积法为基础，应用蒙特卡洛方法模拟分析地质储量的不确定性，评价各变量参数的敏感性，与传统确定性方法对各物性参数取算术平均值不同，蒙特卡洛分析首先确定各参数变量的概率分布函数，进而通过足够次数的模拟计算给出基于不同风险的储量值，为投资决策提供参考。同时，模拟的储量成果和参数变量敏感性分析可指导三维地质建模参数调整，优选模型。

**关键词**：地质储量；容积法；蒙特卡洛模拟；不确定性评价；敏感性分析；模型优选

　　油气藏地质储量计算伴随着油气田勘探和开发的各个阶段，是制定科学、合理的开发方案，指导油气田勘探、开发工作，确定开发投资规模的重要依据。目前常用的计算地质储量方法有：容积法、压降法、类比法[1,2]等。其中容积法应用最为广泛，通过地质解剖、储层描述取得准确的静态数据，适用于各种圈闭类型、各种储集空间、各种驱动方式及油气藏的各个开发阶段[3,4]。压降法主要根据气藏的累计采气量与地层压力下降关系进行动态储量计算[5,6]。类比法则主要通过对比同类型、特征相近的油气藏，对钻探前的远景储量进行估算。上述三种常用方法外，一些新的储量计算方法也逐渐得到发展和应用[7,8]。但这些方法在计算过程的参数选取中都存在静态或动态的不确定性，影响储量的准确性。文中主要以适用性较好的容积法为基础，应用蒙特卡洛方法分析地质储量不确定性及各影响因素敏感性，获取地质储量的概率分布特征，给出基于不同风险基值，对应不同累计概率下的地质储量，明确参数变量对地质储量的敏感程度，为投资决策及气藏开发方案调整提供参考依据。相较于传统的确定性方法对每个物性参数取算术平均值得到唯一储量结果更有价值。同时模拟的储量成果和参数变量敏感性分析可以指导三维地质建模，调整参数，优选准确的属性模型，为数值模拟论证，开发指标预测提供可靠的模型基础，模拟验证过程采取逐小层地质储量验证，可以避免因层间属性差异而导致的误差，使储量计算有更高的准确度。

# 1 蒙特卡洛模拟

## 1.1 蒙特卡洛模拟原理

蒙特卡洛模拟(Monte Carlo simulation)[9-11]是一种以数理统计理论为指导的模拟技术。模拟的基本思想是将符合一定概率分布的大量随机数作为参数变量带入指定数学分析模型,得出目标变量的概率分布,进而了解多重参数对目标变量的综合影响以及目标变量的统计特性。

方法的数学表述为运用随机试验计算积分,所计算的积分为服从某种分布密度函数 $\psi(x)$ 的随机变量 $f(x)$ 的数学期望:

$$E[f(x)] = \int_{x_0}^{x_1} f(x)\psi(x)\mathrm{d}x \tag{1}$$

根据分布密度函数 $\psi(x)$ 抽取 $N$ 个子样 $(x_1,x_2,\cdots,x_N,)$,将对应 $N$ 个随机变量值 $f(x_1)$,$f(x_2)$,$\cdots$,$f(x_N)$ 的算术平均值作为积分估计值:

$$\overline{E}_N = \frac{1}{N}\sum_{i=1}^{N} f(x_i) \tag{2}$$

同理,对于多重变量函数,蒙特卡洛模拟即解决多重积分问题:

$$E = \int_{x_{10}}^{x_{1}1}\int_{x_{20}}^{x_{2}1}\int_{x_{30}}^{x_{3}1}\cdots\int_{x_{n0}}^{x_{n}1} f(x_1,x_2,x_3,\cdots,x_n)\psi_1(x_1)\psi_2(x_2)\psi_3(x_3)\cdots\psi_n(x_n)\mathrm{d}x_1\mathrm{d}x_2\mathrm{d}x_3\cdots\mathrm{d}x_n \tag{3}$$

对于目标函数 $y=f(x_1,x_2,x_3,\cdots,x_n)$,蒙特卡洛模拟根据各个变量的分布密度函数 $\psi_1(x_1)$,$\psi_2(x_2)$,$\psi_3(x_3)$,$\cdots$,$\psi_n(x_n)$ 抽取一组随机变量 $(x_{1i},x_{2i},x_{3i},\cdots,x_{ni})$,进而求取函数值 $y_i=f(x_{1i},x_{2i},x_{3i},\cdots,x_{ni})$,反复独立模拟多次($i=1,2,\cdots$),即可得到一组函数抽样数据 $(y_1,y_2,y_3,\cdots y_n)$,经过足够多次数的模拟,最终给出代表实际情况的函数 $y$ 的概率分布与数学特征。

式(1)—(3)中:$E$ 为函数期望;$x$ 为函数自变量;$f(x)$ 为目标函数;$x_0$,$x_1$,$x_{10}$、$x_{20}$,$\cdots$,$x_n0$、$x_11$、$x_21$,$\cdots$,$x_n1$ 为函数变量定义域;$\psi(x)$ 为 $x$ 的概率密度函数。

## 1.2 蒙特卡洛模拟的优点

与常规数值积分相比,蒙特卡洛积分的优点主要体现在多维度取点计算,在一重积分的情形下,蒙特卡洛方法的效率与常规数值积分相近甚至略低,但随着积分维度增加,常规数值积分的速度呈指数下降,蒙特卡洛方法的效率却基本不变,当积分维数达到4重甚至更高时,蒙特卡洛方法将远远优于常规数值积分方法。

## 1.3 蒙特卡洛模拟方法步骤

应用蒙特卡洛模拟的前提是要确定目标变量的数学模型及模型中各个变量的概率分布,进而按照给定的概率分布生成大量的随机数并代入模型,计算相应目标变量的随机结果,研究目标变量的统计学特征。具体步骤为:

(1)建立模拟目标的数学公式,即蒙特卡洛分析模型。
(2)确定蒙特卡洛分析模型的主要变量。
(3)依据实际数据,明确各变量的概率分布函数。常用概率分布主要有:正态分布、指数分布、二项式分布等(表1),通过模拟器Crystal Ball对各变量进行概率分布拟合确定。
(4)选择模拟器,确定模拟次数。
(5)依据变量分布函数随机抽样,录入分析模型,计算目标函数值,通过足够次数的模拟最终确定目标变量的概率分布及统计特征。

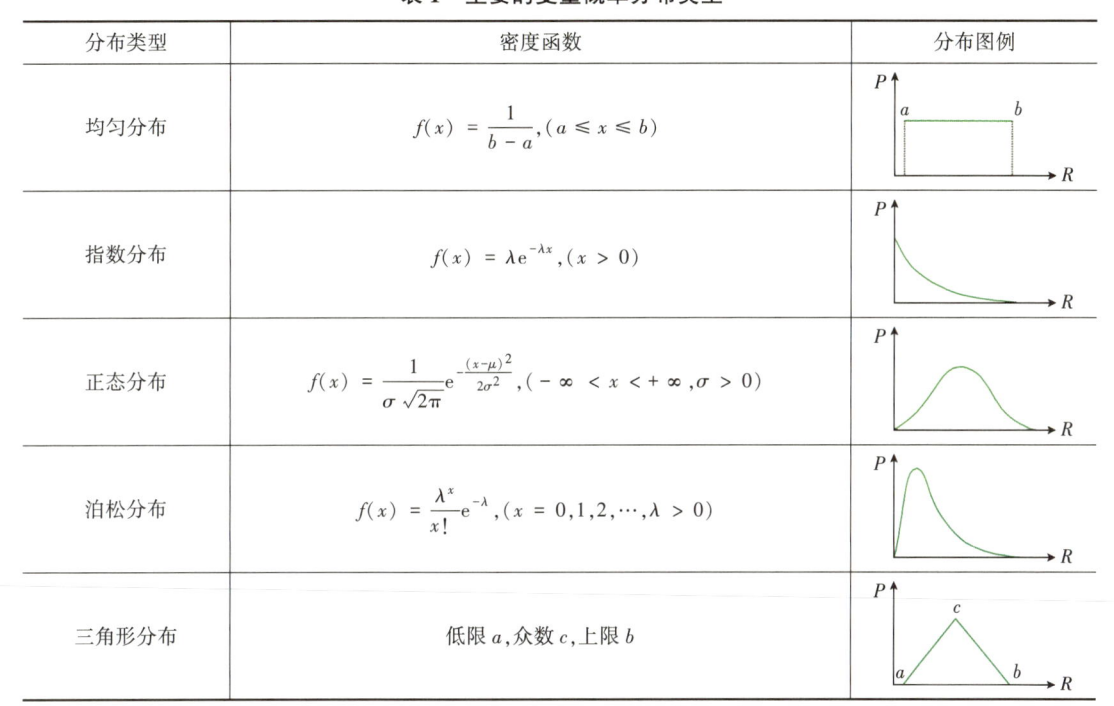

表1 主要的变量概率分布类型

| 分布类型 | 密度函数 | 分布图例 |
| --- | --- | --- |
| 均匀分布 | $f(x) = \dfrac{1}{b-a},(a \leq x \leq b)$ | |
| 指数分布 | $f(x) = \lambda e^{-\lambda x},(x > 0)$ | |
| 正态分布 | $f(x) = \dfrac{1}{\sigma\sqrt{2\pi}}e^{-\dfrac{(x-\mu)^2}{2\sigma^2}},(-\infty < x < +\infty, \sigma > 0)$ | |
| 泊松分布 | $f(x) = \dfrac{\lambda^x}{x!}e^{-\lambda},(x = 0,1,2,\cdots,\lambda > 0)$ | |
| 三角形分布 | 低限 $a$,众数 $c$,上限 $b$ | |

## 2 储量不确定性分析

### 2.1 储量模拟

将蒙特卡洛模拟应用于天然气地质储量不确定性分析,以容积法地质储量计算公式作为目标函数,容积法通过净毛比、孔隙度、含水饱和度逐级约束计算到含油气体积,经气体体积系数转换为地面条件下的气藏地质储量:

$$G = A \times h \times N_g \times \phi \times (1 - S_w)/B_g \quad (4)$$

式中 $G$——地质储量,$m^3$;
$A$——含气面积,$m^2$;
$h$——地层厚度,m;
$N_g$——净毛比,无因次;

$\phi$——孔隙度,小数;
$S_w$——含水饱和度,小数;
$B_g$——气体体积系数,无因次。

容积法计算包含 6 个参数,其中含气面积根据有效厚度等值线圈定,数值较为稳定,其余 5 项均会受到不同因素影响而发生变化。地层厚度随着层序延伸、构造起伏发生变化;净毛比、孔隙度、含水饱和度均来自井点测井,不同深度段具有不同的测点值;气体体积系数结合气藏不同深度的压力、温度条件及地面压力、温度条件计算得到不同的系列数据;蒙特卡洛模拟地质储量不确定性问题可归结为解决具有 5 重变量约束的 5 重积分问题,数学表述及模拟程序为(图 1):

$$G = \iiiint\limits_{h N_g \phi S_w B_g} G(h, N_g, \phi, S_w, B_g) \psi_1(h) \psi_2(N_g)(\phi) \psi_4(S_w) \psi_5(B_g) dh dN_g d\phi dS_w dB_g \tag{5}$$

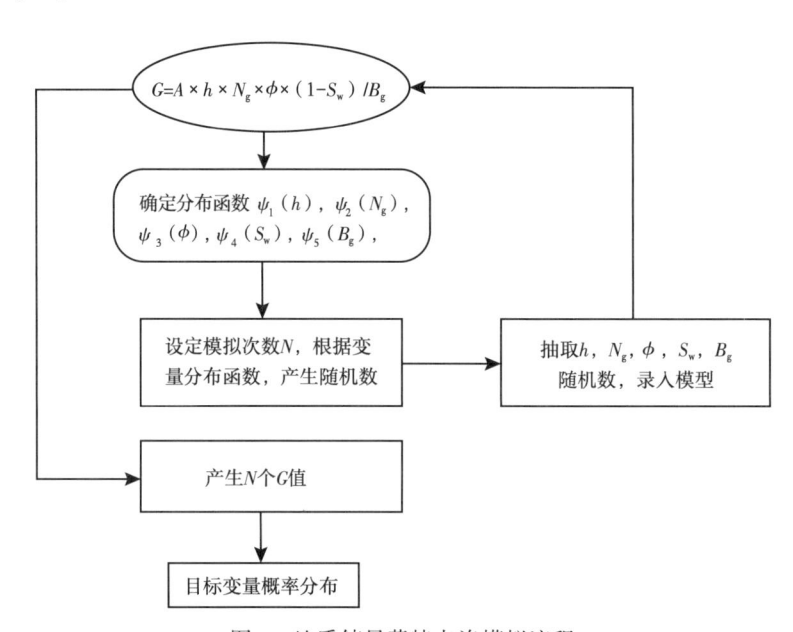

图 1 地质储量蒙特卡洛模拟流程

蒙特卡洛方法已广泛应用于工程物理及电力系统评估等方面,对大型项目、新产品等含有不确定因素的复杂决策系统模拟分析证实可靠[12-14]。选取苏里格气田 10 区块试验区应用蒙特卡洛模拟分析地质储量不确定性,试验区含气面积 22.98km²,垂向上分为 7 个小层,分别是 $H_8^{1-1}$、$H_8^{1-2}$、$H_8^{2-1}$、$H_8^{2-2}$、$S_1^1$、$S_1^2$、$S_1^3$。为避免层间属性差异引起整体储量计算的不合理,逐小层分别进行蒙特卡洛模拟,获得各小层地质储量的分布特征,并分析地质储量对各要素的敏感性,为地质模型建立及参数调整提供依据。

目前常用的蒙特卡洛模拟器有 Crystal Ball、Riskmaster 及 risk 三种,均以加载项方式挂靠在 Excel 环境下运行,分析比较方便。本次采用 Crystal Ball 模拟器,首先对各变量采集数据进行特征分析及概率分布拟合,优选出最合适的概率分布函数来量化参数变量的不确定性,设定足够多的模拟次数(10000 次),模拟器将根据变量概率分布的密度函数循环抽取随机数,录入目标函数,计算试验区地质储量,最终确定地质储量概率分布和不同累计概率下的地质储量(表2,图2)。

表2 $S_1^1$ 小层地质储量蒙特卡洛模拟成果

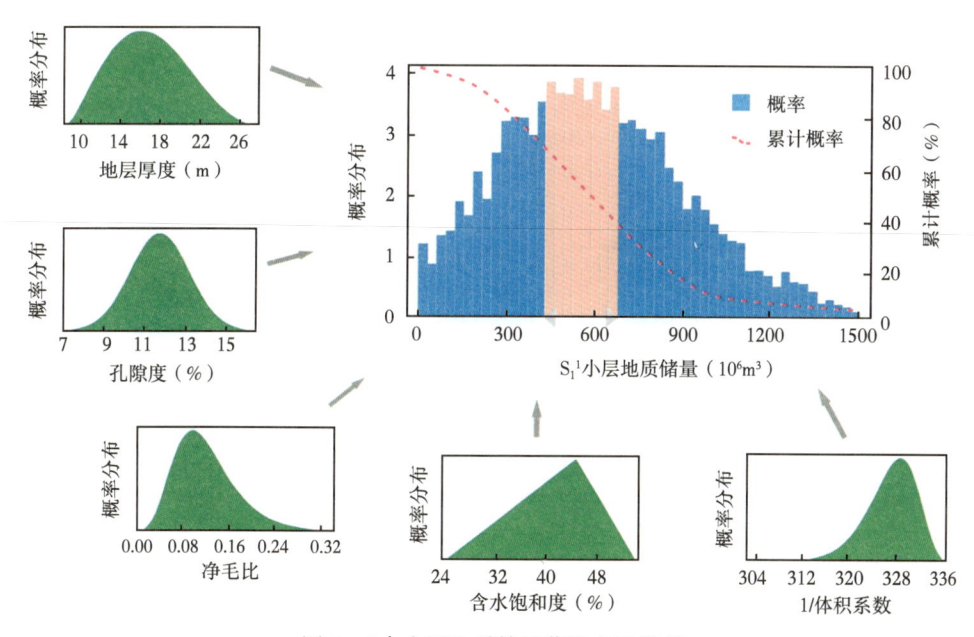

| 累计概率 | 地质储量($10^6 m^3$) |
| --- | --- |
| P100 | 0.19 |
| P90 | 205.93 |
| P80 | 320.97 |
| P70 | 415.01 |
| P60 | 498.89 |
| P50 | 580.9 |
| P40 | 664.38 |
| P30 | 758.05 |
| P20 | 867.56 |
| P10 | 1039.20 |
| P0 | 2019.23 |

图2 $S_1^1$ 小层地质储量蒙特卡洛模拟

图2和表2为试验区 $S_1^1$ 小层模拟的地质储量分布,累计概率代表地质储量小于对应模拟成果的可能性,如P90表示模型预测地质储量小于 $2.059×10^8 m^3$ 的概率为90%。模型分别计算了从0~100%各个概率区间的地质储量,相当于对不同的地质储量进行了风险量化,累计概率P90对应的储量为证实储量(proved reserves),累计概率P50对应的储量为概算储量或基

— 103 —

准储量(probable reserves),累计概率 P10 对应的储量为可能储量(possible reserves)[15],所模拟的储量系列值对气田开发方案优化和投资决策具有重要的指导意义。其中,模拟基准储量(P50)为 5.809×10$^8$m$^3$(表2),将作为研究区地质模型优选的储量依据。对其余 6 个小层,同样计算了地质储量分布和基准储量值(图3,表3)。

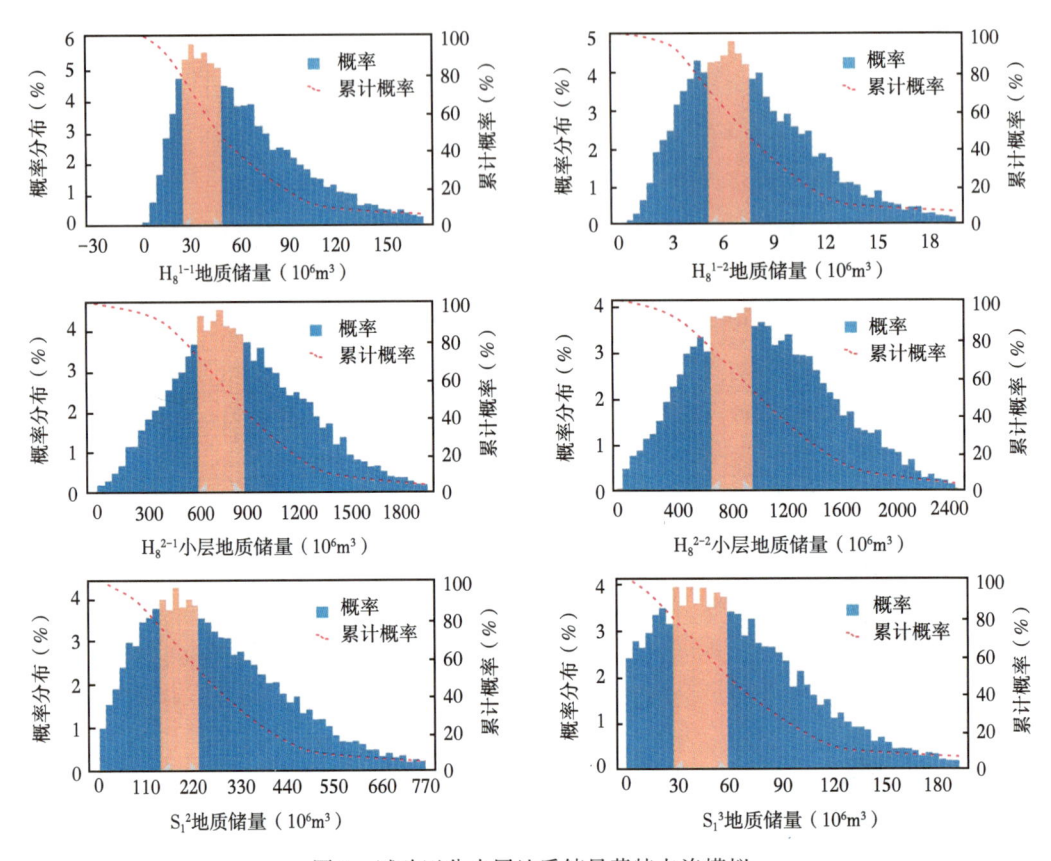

图 3　试验区分小层地质储量蒙特卡洛模拟

表 3　试验区分小层蒙特卡洛模拟基准储量值

| 小层号 | P50 基准储量(10$^6$m$^3$) |
| --- | --- |
| H$_8^{1-1}$ | 52.68 |
| H$_8^{1-2}$ | 7.20 |
| H$_8^{2-1}$ | 809.13 |
| H$_8^{2-2}$ | 968.07 |
| S$_1^1$ | 580.9 |
| S$_1^2$ | 246.27 |
| S$_1^3$ | 57.62 |

## 2.2 敏感性分析

蒙特卡洛多重变量模拟成果受到多个参数不同程度的影响,哪些变量具有决定性作用,哪些变量影响甚微需要开展敏感性分析。变量对目标函数影响预测值的敏感度包含两个方面:(1)变量针对目标函数的模型敏感度;(2)变量本身的不确定性(图4)。确定模型敏感度需要对变量参数与目标函数之间的相互关系进行代数分析,包括两者间全部的转换公式。

图4说明了两种情况下模型敏感度和变量不确定性对目标函数预测值的影响。图中变量与杠杆点之间的距离表示模型的敏感度,距离越大,模型敏感度越低。模型a中,变量参数有较高的不确定性,但因为模型敏感度较低使得变量对预测值不敏感。模型b中,变量的不确定性较低,但模型敏感度较高导致了变量对预测值的高度敏感。利用Crystal Ball模拟器的集成运算,通过对每个变量和预测值之间的相关性系数进行排序评价敏感度,相关系数对变量和预测值一起变化的程度进行了有效度量。如果一个变量和一个预测值有较高的相关系数,就意味着该变量对预测值有明显作用(包括不确定性和模型敏感度)。正相关系数代表随着变量的增加,预测值随之增加,负相关系数反之。敏感度分析图(图5)直观展示了对预测值影响最大及最小的变量,其正值越大,负值的绝对值越大,则目标函数对相应变量的不确定因素越敏感;反之,越不敏感。

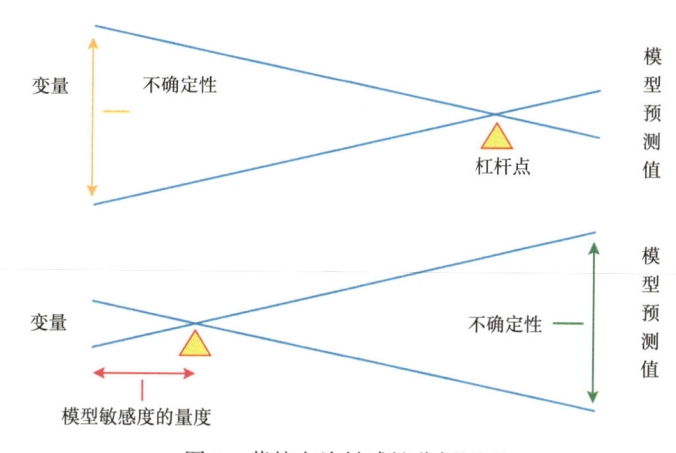

图4 蒙特卡洛敏感性分析原理

蒙特卡洛模拟容积法地质储量过程中主要受到地层厚度、净毛比、孔隙度、含水饱和度和气体体积系数的影响,根据模型公式可以看出各个变量与地质储量间的关系均为一次乘除的关系,因此各变量针对模型的敏感度是相同的,影响将主要来自变量本身的不确定性。分别对每个小层变量不确定性引起的地质储量敏感性进行了分析(图5)。

小层变量敏感性分析图反映不同小层间不同的参数变量对地质储量的影响程度有较大的差异。受储层非均质性强的影响,净毛比敏感程度普遍最高,体积系数相对变化小,对储量计算影响甚微。$H_8^{2-1}$—$S_1^2$小层作为主力储层,处于强水动力沉积环境,地层厚度变化较大,河道与心滩砂体相对发育,有效储层分布多且与非储层差异明显。孔隙度较好,含水饱和度较低且两者都具有较好的均质性,因此表现出地层厚度敏感性强而孔隙度和含水饱和度敏感性弱的特征。$H_8^{1-1}$、$H_8^{1-2}$及$S_1^3$小层沉积处于弱水动力环境,泥质含量高,储层物性差,有效储层发

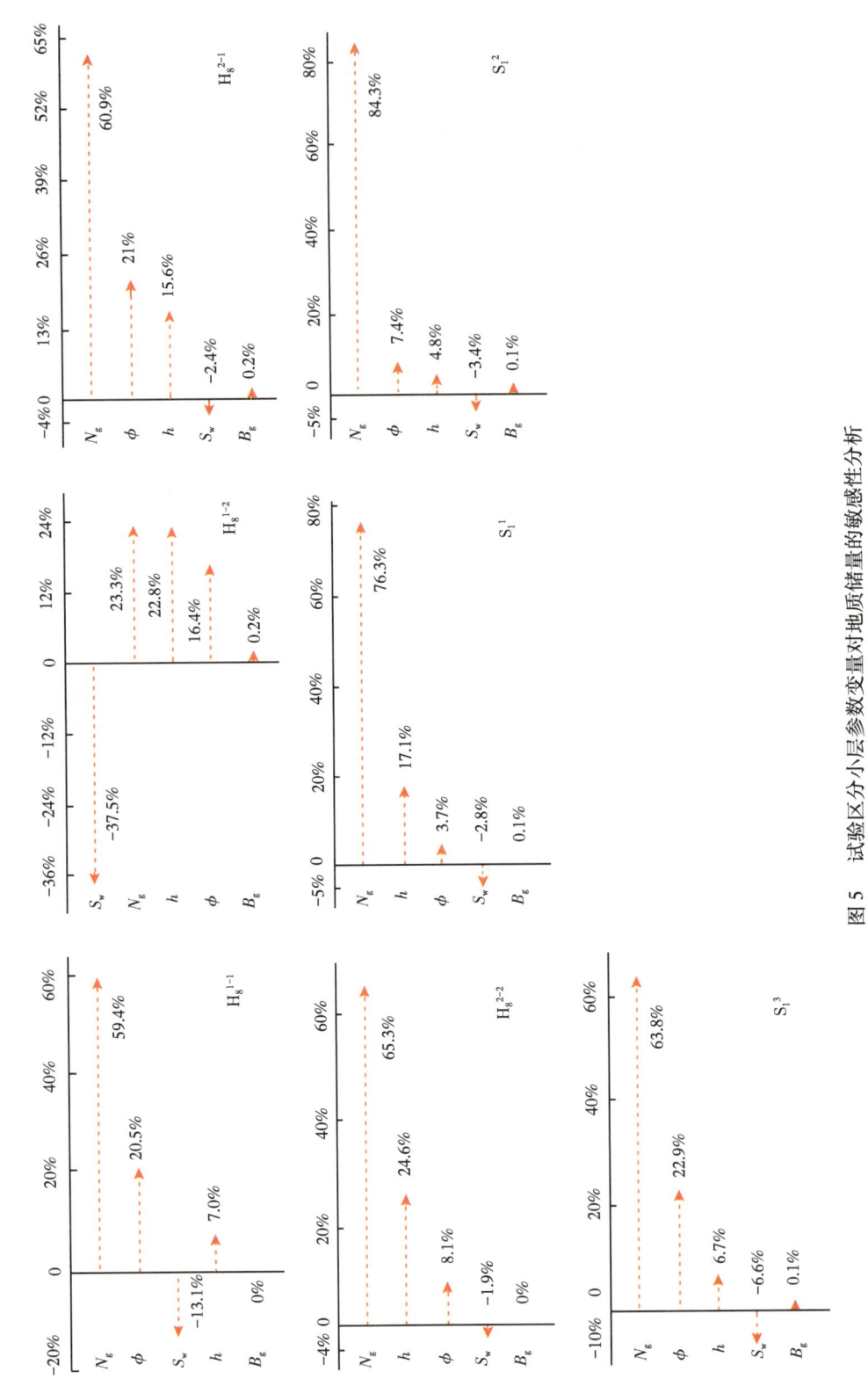

图 5 试验区分小层参数变量对地质储量的敏感性分析

$N_g$—净毛比；$h$—地层厚度；$\phi$—孔隙度；$S_w$—含水饱和度；$B_g$—体积系数

育较少,含水饱和度和孔隙度对储量计算具有较强的敏感性,特别对于物性最差的 $H_8^{1-2}$ 小层,含水饱和度超过净毛比成为最大的不确定性因素。敏感性分析有助于深入研究储量计算的不确定性,对影响储量成果的参数敏感性进行排序,找出对不同小层影响最大、最敏感及可忽略不计的变量因素,估算变量因素的变化引起的储量变动范围,实现缩短储量评价周期及提高评价准确性的目的。同时,分小层地质储量的敏感性分析可用来指导三维地质模型参数调整优选模型。

## 3 三维地质模型优选及实例分析

采用蒙特卡洛模拟成果优选地质模型,将模型计算的地质储量与蒙特卡洛模拟地质储量的概率分布作对比,在高概率分布区间约束下,分析三维模型储量与基准储量[15](P50)拟合程度,两者拟合良好则优选了准确的地质模型,反之依据敏感性分析成果调整各小层关键控制参数建立新的属性模型,循环拟合调整参数达到优选模型的目的。优选过程中分小层进行参数调整、验证,通过小层数据的严格控制使模型准确度更高。

针对苏里格10试验区,应用岩心及测井解释资料,结合野外露头地质知识库确定地层发育模式并进行地层精细划分与对比,统计分析储层物性参数特征,采用序贯高斯随机建模方法建立三维构造模型,沉积相模型和属性模型并计算地质储量[16-18]。将模型计算的地质储量与蒙特卡洛模拟储量进行对比,依据敏感性分析成果反复修正模型属性参数,参数调整过程中将确定性认识和随机模拟方法有机的结合,直至得到拟合良好的地质模型。最终优选的试验区地质模型储量与蒙特卡洛模拟基准储量相比,误差均在10%以内,其中主力产层 $H_8^{2-1}$、$H_8^{2-2}$ 及 $S_1^1$ 三个小层误差均小于5%(图6,表4)。

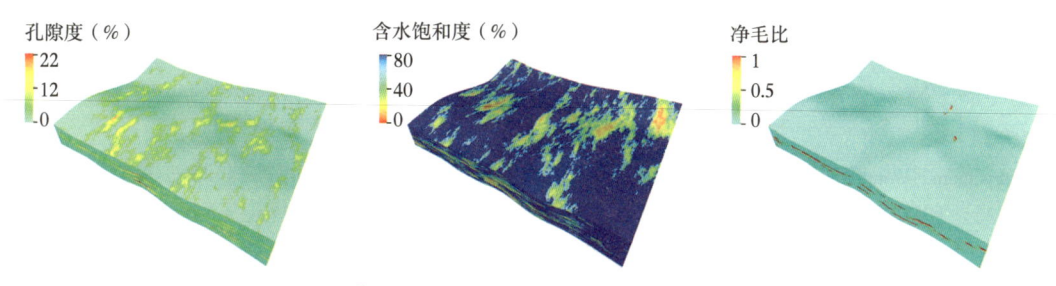

图6 苏里格10试验区优选建立的三维地质模型

表4 优选三维地质模型储量与蒙特卡洛模拟基准储量对比

| 小层号 | 优选三维模型 地质储量($10^6 m^3$) | 所处概率点 | 蒙特卡洛 P50 储量 ($10^6 m^3$) | 误差(%) |
|---|---|---|---|---|
| $H_8^{1-1}$ | 48.31 | P56 | 52.68 | 8.3 |
| $H_8^{1-2}$ | 6.66 | P56 | 7.20 | 8.1 |
| $H_8^{2-1}$ | 789.03 | P52 | 809.13 | 2.5 |
| $H_8^{2-2}$ | 931.93 | P52 | 968.07 | 3.7 |
| $S_1^1$ | 604.92 | P48 | 580.90 | -4.1 |
| $S_1^2$ | 223.61 | P56 | 246.27 | 9.2 |
| $S_1^3$ | 55.17 | P52 | 57.62 | 4.4 |

## 4 结论

（1）在容积法的基础上运用蒙特卡洛方法分析地质储量的不确定性，模拟计算油气地质储量的概率分布，提供不同累积概率下的地质储量（P0~P100），分别对应不同的风险基值，P90 对应的储量为证实储量，P50 对应的储量为基准储量，P10 对应的储量为可能储量，相较于传统的平均确定性方法，不同储量风险的量化有助于投资决策和开发方案制定。

（2）蒙特卡洛分析参数变量的敏感性，确定对各小层地质储量影响最大、敏感性最强的参数，分析结果可进一步认识和验证区域地质特征，对应不同的储层物性及非均质性表现出不同程度的影响，为储量评价提供参考依据。

（3）蒙特卡洛方法优选地质模型，将模型储量与蒙特卡洛模拟地质储量进行对比，依据敏感性分析成果，循环修正模型参数，调整过程中确定性认识和随机模拟有效结合，最终优选出与基准储量（P50）拟合较好的准确地质模型。

## 参 考 文 献

[1] 杨通佑, 范尚炯, 陈元千, 等. 石油及天然气储量计算方法[M]. 北京: 石油工业出版社, 1990.
[2] 贾成业, 贾爱林, 邓怀群, 等. 概率法在油气储量计算中的应用[J]. 天然气工业, 2009, 29(11): 83-85.
[3] 金强, 王伟锋, 信荃麟. 测井多井储层评价与石油储量计算[J]. 石油实验地质, 1994, 16(2): 152-156.
[4] 康志勇, 王永祥, 谢开宁, 等. 容积法储量计算方程合理性分析[J]. 特殊油气藏, 2012, 19(3): 31-34.
[5] 刘荣和. 气藏开发早期动态储量计算方法探讨[J]. 特殊油气藏, 2012, 19(5): 69-72.
[6] 高勤峰, 党玉琪, 李江涛, 等. 柴达木盆地涩北气田动态储量计算与评价[J]. 新疆石油地质, 2009, 30(4): 499-500.
[7] 余庆东. 地质储量计算方法研究[J]. 国外油田工程, 2007, 23(7): 39.
[8] 朱义东, 黄炳光, 章彤, 等. 异常高压气藏地质储量计算新方法[J]. 大庆石油地质与开发, 2005, 24(1): 10-12.
[9] G Fishman. Monte Carlo: Concepts, Algorithms, and Applications [M]. Academic Publishers, 2003.
[10] 詹姆斯 R 埃文斯, 戴维 L 奥尔森, 洪锡熙, 译. 模拟与风险分析[M]. 上海: 上海人民出版社, 2001.
[11] 陈立文. 项目投资风险分析理论与方法[M]. 北京: 机械工业出版社, 2004.
[12] 程强, 周怀春, 黄志锋. DRESOR 法对瞬态辐射传递问题的研究[J]. 工程热物理学报, 2006, 27(3): 472-474.
[13] 甄氼, 易红亮, 谈和平, 等. 散射性半透明介质内红外热辐射与相变耦合换热[J]. 红外与毫米波学报, 2011, 30(1): 42-47.
[14] 王景辰, 李孝全, 杨洋, 等. 基于交叉熵的蒙特卡洛法在发电系统充裕度评估中的应用[J]. 电力系统保护与控制, 2013, 41(20): 75-79.
[15] E C Capen. Probabilistic Reserves! Here at Last? SPE 73828 (2001) presented at the 1999 SPE Hydrocarbon Economics and Evaluation Symposium, Dallas, 20-23 March.
[16] 胡望水, 张宇焜, 牛世忠, 等. 相控储层地质建模研究[J]. 特殊油气藏, 2010, 17(5): 37-39.
[17] 贾爱林. 精细油藏描述与地质建模技术[M]. 北京: 石油工业出版社, 2010.
[18] 苗青, 周存俭, 罗日升, 等. 碳酸盐岩裂缝型油藏裂缝预测及建模技术[J]. 特殊油气藏, 2014, 21(2): 37-40.

# 毛细管力曲线转换方法探讨

刘兆龙　张永忠　黄伟岗　刘华林

(中国石油勘探开发研究院)

**摘要**：毛细管压力曲线是油气饱和度评价的重要资料,压汞曲线需要进行相应转换才能实际应用。本文利用采自克深气田的6块岩心样品的压汞与高速离心实验数据,运用常规方法及考虑黏土束缚水方法将压汞曲线转换为气驱水毛细管力曲线,并与高速离心法测得毛细管力曲线进行对比分析,结果表明,压汞转换后的曲线在高压段与高速离心法测得毛细管力曲线相差较大,通过参数拟合,得出了压汞毛细管力曲线与实际毛细管力曲与高速离心气驱水毛细管力曲线的转换系数。

**关键词**：克深气田；压汞毛细管力曲线；高速离心毛管力曲线；转换系数

## 1　前言

毛细管压力曲线在油气田勘探开发中发挥重要作用,是储层油气饱和度评价的重要资料[1]。测定岩石毛细管压力曲线的主要方法有半渗透隔板法、离心法和压汞法[2-4]。前两者的测量方法均属于气驱水模式,较接近于实际油藏的润湿条件,但因费用较高,在油气藏评价中应用相对较少。压汞法是在真空条件下将汞注入岩样,其测量速度快、成本相对较低,因此在油气田开发评价中得到广泛应用。压汞毛细管力曲线常常用来评价低渗透、致密气藏流体饱和度,由于实验室压汞曲线是岩样烘干抽真空高压注汞而获得,这与地下储层天然气充注实际情况差别较大,需要进行相应的换算得到地层条件下毛细管力曲线。在储层物性相对较好的情况下,将室内测试条件及气藏条件下汞、水岩石的润湿角及流体界面张力代入转换公式得到地层条件下的毛细管力曲线[5],但对于类似塔里木克深气田这种物性差的致密储层,运用上述方法换算所得毛细管力曲线与气驱水所得毛细管力曲线差别较大,应用于饱和度计算会带来偏差。

克深气田位于塔里木盆地库车坳陷克拉苏冲断带中东部,属典型的超深超高压裂缝性致密砂岩气藏,埋深6500~7000m,气藏压力普遍超过100MPa。储层为白垩系巴什基奇克组扇三角洲前缘—辫状河三角洲前缘水下分流河道砂岩,厚约300m,砂体相互叠置,砂地比高(>70%);夹层厚度一般小于2m,连续性差,未见明显隔层[6-10]。岩石类型主要为中细粒岩屑长石砂岩和长石岩屑砂岩。储集空间类型多样,主要有粒间孔(包括粒间溶孔及残余原生粒间孔)、粒内溶孔和裂缝。由于埋藏深,储层受强压实作用,孔喉细小、配位数低,排驱压力高,基质渗透率低,为低孔隙度、特低渗透率储层。157个压汞样品分析结果表明,克深地区孔喉半径更小,主要分布于0.01~0.1μm。孔隙度主要分布在2%~6%,基质渗透率主要分布在0.01~0.05mD,渗透率累计频率中值为仅为0.027mD。本文以克深气田为例,利用高压压汞资料与高速离心—核磁资料,分析了通过对高压压汞曲线转换地层条件下的毛细管力曲线与高速离心—核磁所得毛细管力曲线对比,分析了二者差异的主控因素,提出了高压压汞曲线的校正方法,对合理利用压汞曲线评价气水分布具有借鉴意义。

如何对压汞曲线进行校正,以应用于储层饱和度评价,对气田开发具有重要意义。

## 2 毛细管力曲线特征

### 2.1 岩心样品

选取克深2气田5口井6块岩心柱塞样品(样品来源于塔里木油田),样品均为洗盐后测试。首先进行了孔、渗测试,然后每个样品分为两段,分别进行了高速离心气驱水毛细管力曲线测试和抽真空高压压汞毛细管力曲线测试。基础数据如表1。高速离心测试样品采用189098mg/L的合成地层盐水加压饱和进行测试,最大离心力1000psi。高压压汞测试注入最大压力为55000psi。

表1 岩心基础数据

| 样品编号 | 样品来源 | 样品深度(m) | 空气渗透率(mD) | 孔隙度(%) |
| --- | --- | --- | --- | --- |
| 48A | Keshen 208 | 6610.66 | 0.220 | 3.1 |
| 9A | Keshen 2-1-5 | 6715.28 | 0.045 | 5.2 |
| 51A | Keshen 2-1-5 | 6730.77 | 0.059 | 5.4 |
| 3A | Keshen 2-2-3 | 6805.63 | 0.044 | 6.7 |
| 34A | Keshen 2-2-5 | 6776.75 | 0.039 | 7.4 |
| 22A | Keshen 2-2-8 | 6722.86 | 0.042 | 6.0 |

### 2.2 高速离心毛细管力曲线

应用高速离心气驱水法得到6块岩心的毛细管力曲线(图1)。可以看出:(1)随岩心渗透率升高,相同离心力下,岩心含水饱和度降低趋势明显,束缚水饱和度与岩石渗透率具有较好

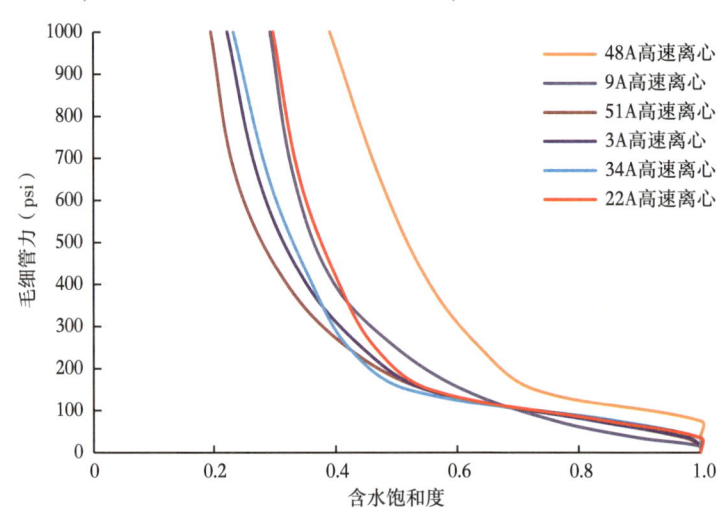

图1 高速离心法毛细管力曲线

相关性(图2);(2)不同渗透率样品曲线有交叉现象,这可能是由于微观孔喉非均质性造成,孔喉半径比、大小孔喉比例均影响毛细管力曲线形态。

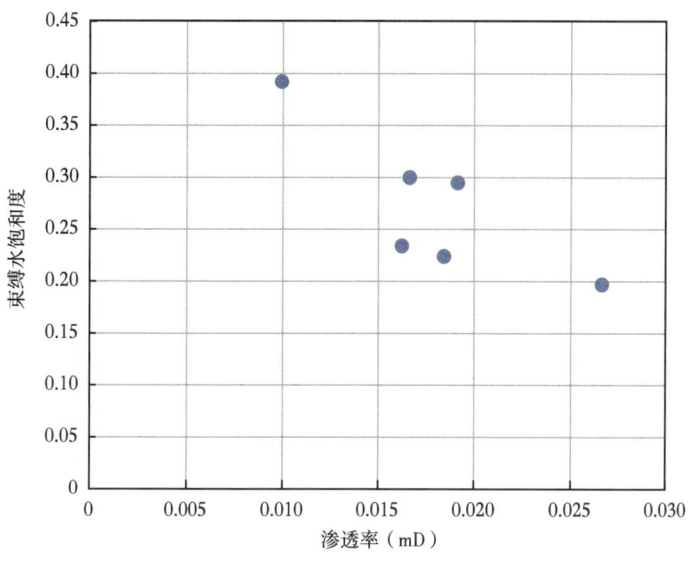

图 2 束缚水饱和度—渗透率相关图

## 2.3 高压压汞毛细管力曲线

由于克深气田储层渗透率低,6块岩心最高注汞压力55000psi(379.2MPa),测试结果如图3。可以看出,随着岩心渗透率增大,最大进汞饱和度有增大趋势。

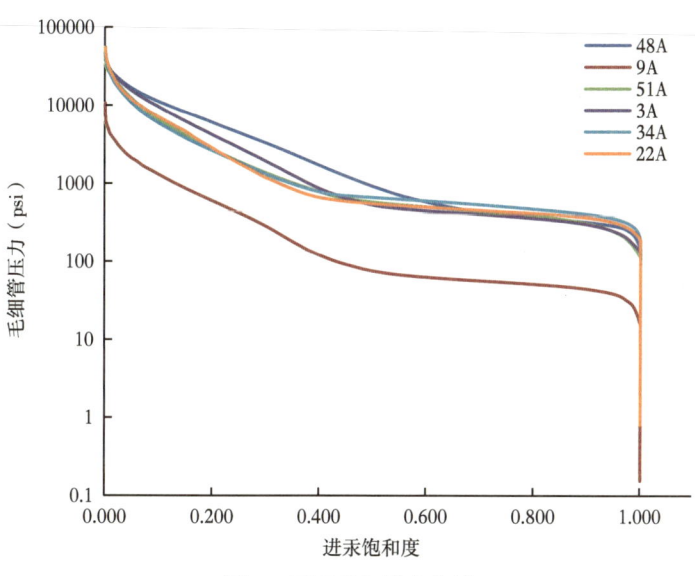

图 3 压汞毛细管力曲线

## 3 毛细管力曲线对比

### 3.1 压汞常规方法转换与高速离心毛细管力曲线对比

压汞法和离心法方法测定毛细管压力曲线时,所使用的流体体系不同,流体的表面张力和润湿角等均不同,因而使所测毛细管压力数值也不相同,需要进行相应的换算。通用换算公式如式(1):

$$Pc_{气水} = Pc_{气汞} \times (\sigma_{气水} \times \cos\theta_{气水})/(\sigma_{气汞} \times \cos\theta_{气汞}) \tag{1}$$

其中,$Pc_{气汞}$为压汞所测毛细管力;$Pc_{气水}$为换算后的水驱气毛细管力;$\sigma_{气水}$为气水界面张力,通常情况下一般取 72mN/m;$\sigma_{气汞}$为气汞界面张力,通常情况下一般取 480mN/m;$\theta_{气汞}$为气汞系统对岩石的润湿角,一般取 140°;$\theta_{气水}$为气水系统对岩石的润湿角,强湿润性岩石一般取 0°[11]。

将压汞法所测得的 $Pc_{气汞}$ 换算为离心法下的气水毛细管压力 $Pc_{气水}$,则 $Pc_{气水} = 0.196 Pc_{气汞}$。图 4 为 6 块样品经常规转换得到的气驱水的近似毛细管力曲线与离心法真实气驱水毛细管力曲线的对比。可以看出,常规转 $Pc_{气水} = 0.196 Pc_{气汞}$ 时,由压汞法经转换得到的毛细管力曲线主体部分明显低于实测离心法气驱水毛细管力曲线。润湿相饱和度为 60% 时,转换得到的毛细管力与实测离心法毛细管力差值为 10~70psi,说明常规方法的转换得到的 $Pc_{气水}$ 比实测偏小,使由压汞法换算得到的毛细管力数值较低。

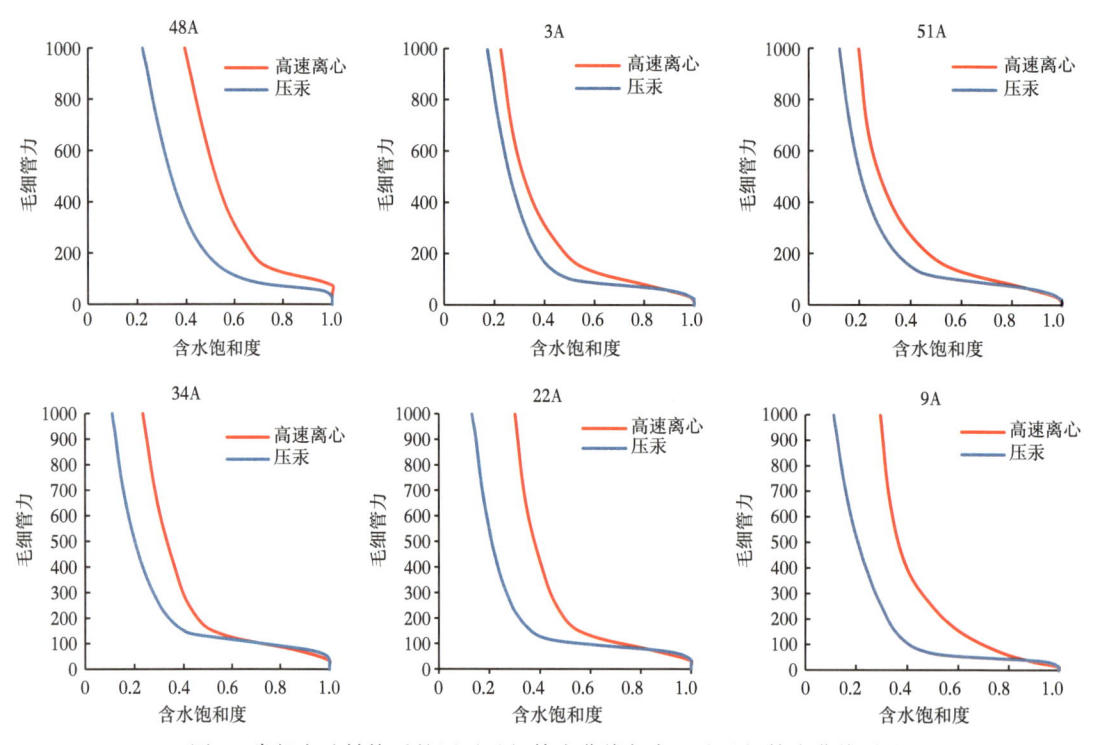

图 4 常规方法转换后的压汞毛细管力曲线与离心法毛细管力曲线对比

## 3.2 考虑黏土束缚水影响的压汞毛细管压力曲线校正

由压汞毛细管力曲线转换为气驱水毛细管力曲线时,除岩石的润湿角及流体界面张力影响外,黏土束缚水可导致毛细管力曲线的差异。黏土束缚水以静电力吸附于颗粒表面,这部分水在气驱水过程中难以被驱除,毛细管压力曲线反映的是除黏土束缚水部分之外的孔隙,而压汞法是岩样烘干后用汞驱替气,汞进入全部孔隙,因此转换时英考虑黏土束缚水影响。张冲等(2014)提出考虑黏土束缚水影响的校正方法(图5):假设黏土束缚水均匀分布在所有孔隙的表面,则在任意进汞压力条件下,实测进汞饱和度 $S_{Hg}$ 与校正后进汞饱和 $S^*$ 有如下关系(公式2):

$$S^* = S_{Hg}(1-S_{CBW}) \tag{2}$$

公式2中,$S_{CBW}$ 为黏土束缚水饱和度,把高速离心法测试中离心力1000psi时水饱和度近似作为黏土束缚水饱和度。同样,在任一给定的进汞饱和度条件下,若考虑黏土束缚水膜存在,充注孔喉半径应较烘干岩样小,进汞压力应更大,即

$$Pc^*/Pc = r/r^* = (1-S_{CBW})^{1/2} \tag{3}$$

其中,$Pc^*$ 为校正后的进汞压力;$Pc$ 为烘干岩样的进汞压力;$r$ 为烘干岩样喉道半径;$r^*$ 为有效喉道半径。

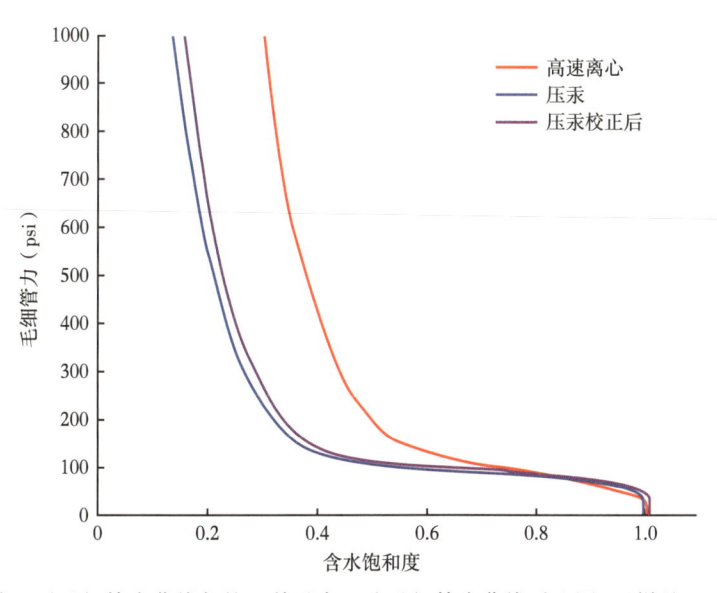

图5 校正后毛细管力曲线与校正前及离心法毛细管力曲线对比图(以样品22A为例)

校正后毛细管力曲线与校正前及离心法毛细管力曲线对比如图5,可以看出,进汞饱和度的校正系数随黏土束缚水相对体积的增加而减小,但校正量较小,与气驱水毛细管力曲线仍有较大偏差。可见黏土束缚水对于克深致密储层而言,黏土束缚水对毛细管力校正影响较小。压汞曲线转换曲线与离心法毛细管力曲线存在巨大差异,可能来自不同介质间界面张力和润湿角取值,由于压汞曲线为抽真空条件下,汞被高压挤进,其界面张力与岩石润湿角可能与通

常情况取值下存在差异[12,13]，但其真实情况下参数难以通过实验获得，同时由于储层微观非均质性与复杂性，真实的孔喉并非均匀的圆管，此外高压情况下的岩石应力敏感也会对曲线校正造成一定的影响[14]。基于上述情况，本文对压汞曲线参数与高速离心气驱水毛细管力曲线参数进行拟合。

## 3.3 压汞与气驱水毛细管力曲线拟合

考虑到压汞实验在抽真空条件和高压下，毛细管力变化控制因素较多且难以确定，因此在曲线拟合中对毛细管力进行拟合。通过对压汞和高速离心实验数据进行统计分析，在300psi（气驱水压力）以下压力段曲线差异不大，在300psi以上压力段曲线差异较大，经考虑黏土束缚水法进行校正，总体校正量较小，在300psi以上压力段曲线仍存在较大差异。分析其原因可能是注汞时随着压力增高其界面张力与润湿角可能存在变化，同时岩石应力敏感也可能对其具有一定影响。在此基础上提出了毛细管力曲线分段拟合转换方法，该方法原理是在同一饱和度条件下，将进汞压力数据乘以转换系数进而校正到与高速离心力一致的结果。在校正计算过程中发现，随着压力增大，毛细管力拟合系数（同一含水饱和度下进汞压力与离心力比值）有减小的趋势，在高于300psi，减小趋势变缓，通过计算确定高压段毛细管力曲线的最优拟合转换系数为3~4(图6)。

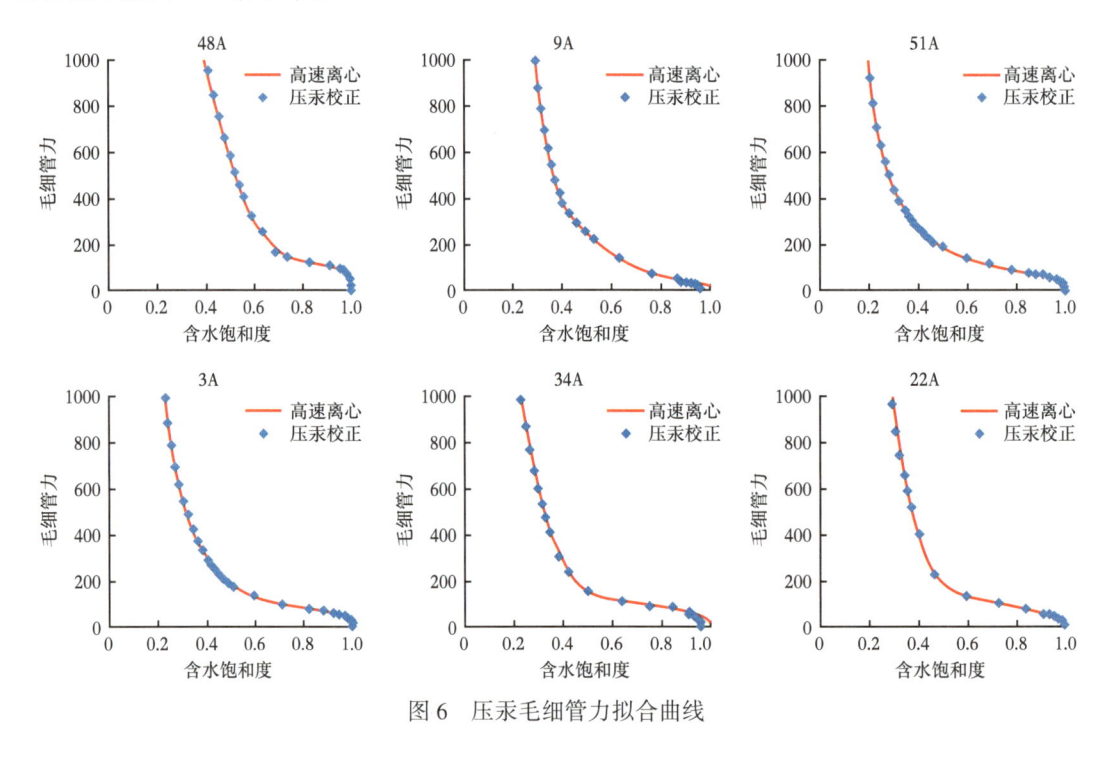

图6 压汞毛细管力拟合曲线

在利用毛细管力曲线进行气水关系识别、饱和度计算时还应校正到实际气藏条件下的毛细管力曲线，由于克深气藏温度超过120℃、压力普遍超过100MPa，气藏条件下的界面张力很难测得，可通过图版经换算后得到实际气藏条件下的毛细管力曲线，可应用于饱和度计算。

## 4 结论

高压压汞毛细管力曲线和高速离心—核磁所得毛细管力曲线相差较大,通过转换地层条件下和考虑黏土束缚水影响对压汞毛细管力曲线进行校正,校正后的曲线仍存在巨大差异。通过对克深气田6个样品的研究发现,在300psi以下转换地层条件下的压汞毛细管力曲线和高速离心毛细管力曲线差别不大,地压力下转换地层条件下的毛细管力曲线可以直接应用于饱和度的计算;在300psi以上通过拟合转换的方法,对压汞毛细管力曲线进行校正,最优拟合转换系数为3~4。该方法对于合理利用压汞曲线评价气水分布具有一定的推广和借鉴意义。

### 参 考 文 献

[1] 何更生,唐海.油层物理(第二版)[M].北京:石油工业出版社,2011.

[2] 朱林奇,张冲,石文睿,等.结合压汞实验与核磁共振测井预测束缚水饱和度方法研究[J].科学技术与工程,2016,16(15):22-29.

[3] 李传亮.半渗透隔板底水油藏油井见水时间预报公式[J].大庆石油地质与开发,2001,20(4):32-33.

[4] 李天降,李子丰,赵彦超,等.核磁共振与压汞法的孔隙结构一致性研究[J].天然气工业,2006,26(10):57-59.

[5] 张冲,张超谟,张占松,等.黏土束缚水对压汞毛管压力曲线的影响及校正[J].科技导报,2014,32(2):44-49.

[6] 张荣虎,王俊鹏,马玉杰,等.塔里木盆地库车坳陷深层沉积微相古地貌及其对天然气富集的控[J].天然气地球科学,2015,26(4):667-678.

[7] 张惠良,张荣虎,杨海军,等.超深层裂缝—孔隙型致密砂岩储集层表征与评价——以库车前陆盆地克拉苏构造带白垩系巴什基奇克组为例[J].石油勘探与开发,2014,41(2):158-166.

[8] 王俊鹏,张荣虎,赵继龙,等.超深层致密砂岩储层裂缝定量评价及预测研究——以塔里木盆地克深气田为例[J].天然气地球科学,2014,25(11):1735-1745.

[9] 赵靖舟,戴金星.库车前陆逆冲带天然气成藏期与成藏史[J].石油学报,2002,23(2):5-10.

[10] 贾承造.中国塔里木盆地构造特征与油气[M].北京:石油工业出版社,1997.

[11] Schowalter T T. Mechanics of Secondary Hydrocarbon Migration and Entrapment[J]. AAPG Bulletin, 1979, 63(5):723-760.

[12] 李传亮.压汞过程中的润湿性问题研究[J].天然气工业,1994,(5):84.

[13] 李传亮.用压汞曲线确定油藏原始含油饱和度的方法研究[J].新疆石油地质,2000,21(5):417-418.

[14] 李爱芬,付帅师,张环环,等.实际油藏条件下毛管力曲线测定方法[J].中国石油大学学报(自然科学版),2016,40(3):102-106.

# 致密砂岩气藏井网密度优化与采收率评价新方法

高树生[1,2]　刘华勋[1,2]　叶礼友[1,2]　温志杰[3]　朱文卿[1,2]　张　春[4]

(1. 中国石油集团科学技术研究院有限公司；2. 中国石油勘探开发研究院
渗流流体力学研究所；3. 中国石油吉林油田公司勘探开发研究院；
4. 中国石油西南油气田公司勘探开发研究院)

**摘要**：为了解决目前致密砂岩气藏井网密度优化方法可靠性较低、缺乏井网密度与采收率关系的有效论证等问题，通过建立井间干扰概率的计算方法，绘制出鄂尔多斯盆地苏里格气田目标研究区井间干扰概率曲线，进而建立了1套适用于致密砂岩气藏的井网密度优化与采收率评价新方法，并将该新方法应用于苏里格气田3个加密试验区的井网优化与采收率评价。研究结果表明：(1)致密砂岩气藏井间干扰概率与井网密度密切相关，随着井网密度的增加井间干扰概率呈现逐渐增加的趋势，直至井网密度达到一个相对大的值后，井间干扰概率才达到或接近于1；(2)苏里格气田3个加密试验区的经济最佳井网密度介于2.6~3.1口/km²，对应采收率介于36.6%~39.8%，井间干扰概率介于28%~33%，而经济极限井网密度介于5.2~6.6口/km²，对应采收率介于46.8%~49.8%，井间干扰概率介于83%~89%；(3)苏里格气田致密砂岩气藏经济最佳井网密度对应的井间干扰概率约为30%，经济极限井网密度对应的井间干扰概率约为85%。结论认为，采用新方法可以计算得到经济最佳、经济极限井网密度与对应的采收率，实现致密砂岩气藏的井网优化与采收率评价；该研究成果既可以为苏里格气田的经济高效开发提供理论支撑，也可以为同类型气藏的效益开发提供借鉴。

**关键词**：致密砂岩气藏；干扰概率；采收率；经济最佳井网密度；经济极限井网密度；鄂尔多斯盆地；苏里格气田

中国致密气资源规模巨大，是中国天然气持续增长的重要支柱[1-4]，以鄂尔多斯盆地苏里格气田为典型代表，2017年年产气 $227×10^8 m^3$，探明储量(含基本探明)规模达 $4.77×10^{12} m^3$，是目前中国储量和产量规模最大的天然气气田。但由于致密砂岩气储层渗透率极低，含水饱和度高，非均质性强，致密砂岩气开发普遍存在储量控制程度低、产能低、采收率低的三低问题，储量有效动用面临严峻挑战[5-10]。国内外针对致密砂岩气采收率的研究和开发实践表明：井网密度是影响致密砂岩气田储量动用程度和采收率的关键因素[11-16]，近几年借鉴美国开发致密气密井网开发经验[17]，在苏里格中部开展相应的试验，但由于中国致密砂岩气藏储量丰度低[8]，采用密井网开发时井间干扰严重，单井产量低，经济效益难以保证。因此，合理的井网部署才是开发致密砂岩气藏不断追求的目标。目前，致密砂岩气井网密度优化方法主要有地质分析法、气井泄气半径折算法和经济极限井距法等方法，这些方法从技术、经济两个方面来确定合理的井距及井网密度。其中地质分析法与经济极限井距法为静态分析法，受储层开发地质认识限制，静态分析法可靠性差；气井泄气半径折算法通常只确定合理井网密度，缺乏井网密度与采收率关系有效论证。

为此，笔者通过建立基于致密砂岩气藏井间干扰概率计算方法，绘制出苏里格气田目标研究区井间干扰概率曲线,进而建立了一套适用于致密砂岩气藏的井网密度优化与采收率评价新方法,并在苏里格气田3个加密试验区开展了应用,预测效果与3个加密区的生产动态基本一致。最后论证了致密砂岩气藏经济井网密度与井间干扰概率的关系,认为井间干扰概率在30%左右时,气藏开发最为经济有效,此时对应的井网密度为经济最佳。

# 1 井间干扰概率计算方法

井间干扰现象是指同一油气层中当多井同时生产时,其中任何一口井工作制度的变化所引起的其他井井底压力及产量发生变化的现象[17-19]。对于中、低渗气藏储层中气体流动性强,井间连通性好,井间干扰现象较容易发生,所以开发井网密度比较小,井距一般在1km以上。而对于致密砂岩气藏来说,由于储层渗透率低,非均质性强,井控范围有限且差异大[12,13]。若采用常规中低渗气藏适合井网密度进行开发,将导致气藏的储量有效控制程度低,从而导致气藏采收率低;而若采用密井网开发又会引起井间干扰的发生,使得产气能力低的致密气藏气井产气能力进一步下降,严重影响致密砂岩气藏的效益开发,因此,井间干扰是致密砂岩气藏有效开发必须面对和解决的问题。

李跃刚等[12]提出了井间干扰概率的定义,即干扰井数与总井数比值,并对苏里格气田大量的干扰试井结果进行了统计分析,得到了井间干扰概率与井网密度的关系曲线。但这种方法存在3个问题:(1)干扰试井确定的是试井期间的井间干扰情况,而致密储层渗流能力差,地层压力波传播速度慢,发生井间干扰所需时间长,通常需要几个月甚至几年[19],试井期的井间干扰结果不能代表气井长时间生产情况下的井间干扰情况;(2)干扰试井确定的井间干扰概率受试验井网密度与人为选择因素的影响大,且只能获得试验井网密度下的井间干扰概率数据,同时开展干扰试井时通常选择生产情况相对较好的气井,也缺乏代表性;(3)致密砂岩气藏开展干扰试井需要长时间关井,会对气田生产造成影响,不宜大面积开展。

为此,针对致密砂岩气藏,笔者提出了根据气井动态控制储量与动态控制面积快速确定井间干扰概率的方法。

致密砂岩气藏储层渗透率低,目前常规井网密度下(一般在1口/km²)井间基本无干扰,基本呈现出一井一藏的特征[17]。根据物质平衡方程,单井累计采气量($G_P$)与视地层压力($P/Z$)呈线性关系,表达式如下[17]:

$$G_P = G \frac{P_i/Z_i - P/Z}{P_i/Z_i} \tag{1}$$

其中,$P_i$表示原始地层压力,MPa;$Z_i$表示原始状态下天然气压缩因子,f;$P$表示地层压力,MPa;$Z$表示天然气压缩因子,f;$G_P$表示单井累计采气量,$10^4 m^3$;$G$表示单井动态控制储量,$10^4 m^3$。

将$G_P$与$P/Z$线性段延伸,至地层压力$P=0$时,获取单井动态控制储量$G$。进一步,可根据单井动态控制储量$G$和储层物性参数,求取单井动态控制面积$A$:

$$A = \frac{GB_{gi}}{\phi h S_g} \tag{2}$$

其中,$\phi$ 表示孔隙度,f;$h$ 表示储层厚度,m;$S_g$ 表示含气饱和度,%;$B_{gi}$ 表示原始状态下天然气体积系数,f。

考虑到苏里格气田致密砂岩气藏储层纵向多层叠置,采取多层合采的方式进行开发,可视为纵向连续分布,此时储层厚度、孔隙度和含气饱和度表达式如下:

$$h = \sum_{i=1}^{n} h_i \tag{3}$$

$$\phi = \sum_{i=1}^{n} \phi_i \frac{h_i}{h} \tag{4}$$

$$S_g = \sum_{i=1}^{n} S_{gi} \frac{h_i}{h} \tag{5}$$

其中,$n$ 表示纵向储层数;$i$ 表示各小层层序号。

苏里格气田目标研究区苏 3X 井区一次井网井距 1200m,地质上的单井控制面积 1.44km²,对应井网密度 0.69 口/km²,根据该目标研究区 110 口井的生产动态数据,得到单井的动态控制面积,如图 1 所示,目标区单井动态控制面积 0.1~1.0km²,最大值与最小值相差较大,储层非均质性明显;单井动态控制面积中值 0.31km²,平均值 0.35km²,综合看来,苏里格目标区单井动态控制面积小,井控能力弱,普遍比地质上的单井控制面积(1.44km²)小,因此,可以判断气藏在现有井网密度下井间基本不存在干扰,此时单井动态控制面积可视为单井最大动态控制面积,后期一旦发生井间干扰,单井动态控制面积就会进一步降低。

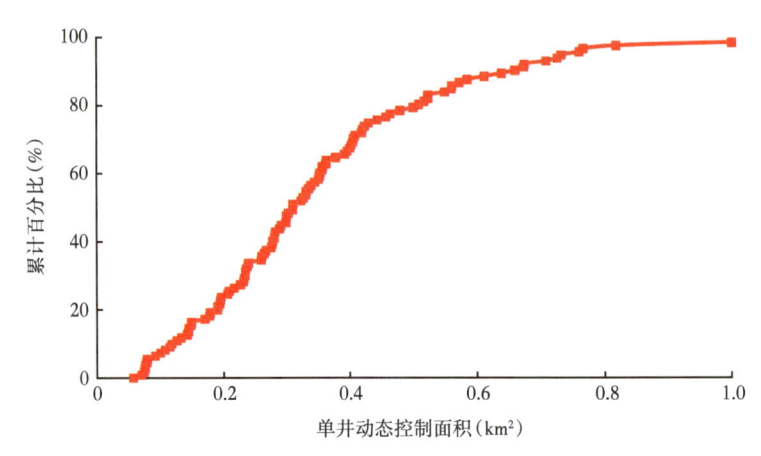

图 1　苏里格气田研究区无井间干扰时单井动态控制面积累计百分比曲线图

井网加密是改善致密砂岩气藏储量动用程度与提高采收率最关键、最有效的手段[12-14],苏里格气田开展了 4 个加密试验区,其中苏 3X 井区加密试验区井网密度达 5.1 口/km²,对应地质上的单井控制面积只有 0.2km²,对比图 1 研究结果,存在部分气井地质上的单井控制面积小于无干扰时的单井动态控制面积,从而发生井间干扰的现象。因此,可以通过对比加密后地质上的单井控制面积与无干扰时的单井动态控制面积对比来判断加密后井间是否发生干扰、发生干扰的井数以及干扰概率,干扰概率表达式如下:

$$F(S) = \frac{n_1}{N} \tag{6}$$

其中，$S$ 表示井网密度，口/km²；$F$ 表示干扰概率，f；$n_1$ 表示无干扰时单井动态控制面积大于 $1/S$ 的井数，口；$N$ 表示总井数，口。

利用目标区 110 口井的生产数据，根据式（6）和图 1 计算，就可以得到的苏里格气田目标研究区的井间干扰概率曲线（图 2）。可以看出，苏里格气田致密砂岩气藏开始发生井间干扰的临界井网密度相对较大，一般大于 1 口/km²，而常规中、低渗气藏由于储层连通性好，井控范围大，在井网密度小于 1 口/km² 时也会发生明显的井间干扰，地层各处压力几乎同步下降[19,20]。致密砂岩气藏开始发生井间干扰后，由于储层非均质性强，井与井之间在生产上存在明显差异，气井之间不是同时发生干扰，而是随井网密度增加井间干扰概率逐渐增加的连续过程，直至井网密度达到一个相对大值后，井间干扰概率才达到或接近 1，基本处于完全干扰状态，发生干扰的概率大小与井网密度密切相关，这一认识与苏里格气田致密砂岩气生产实践认识基本一致[3]，进一步证实本文确定井间干扰概率方法的可行性。

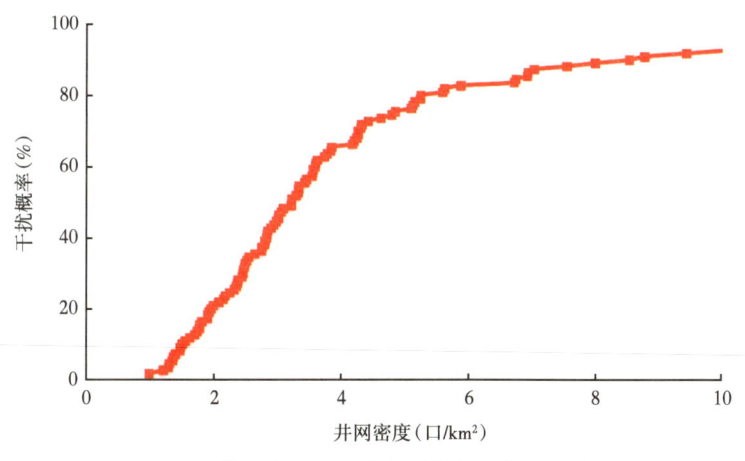

图 2　苏里格目标区井间干扰概率统计曲线

## 2　井网优化与采收率评价模型

根据上述井间干扰理论，由式（6）可推导得到发生干扰的井数 $n_1$ 为：

$$n_1 = NF(S) \tag{7}$$

当单井动态控制面积大于或等于地质上的井控面积，井间才发生干扰，因此 $n_1$ 口干扰井动态控制面积 $A_1$ 大于或等于 $n_1$ 口井地质上的井控面积 $n_1/S$。但受实际井网和地质条件限制，$n_1$ 口井动态控制面积 $A_1$ 只能小于或等于 $n_1$ 口井地质上的井控面积 $n_1/S$，因此 $n_1$ 口干扰井动态控制面积 $A_1$ 只能等于 $n_1$ 口井地质上的井控面积 $n/S$，即：

$$A_1 = \frac{n_1}{S} NF(S)\frac{1}{S} \tag{8}$$

相应的 $n_1$ 口干扰井动态控制储量：

$$G_1 = NF(S)\frac{\phi h S_{gi}}{S B_{gi}} \tag{9}$$

同理,根据井间干扰理论,井网密度 $S$ 时不发生干扰的概率为 $[1-F(S)]$,井数为:

$$n_2 = N[1 - F(S)] \tag{10}$$

根据井间干扰概率的定义,不发生干扰时的气井动态控制面积小于地质上的井控面积,以概率的形式分布在单井最小动态控制面积 $A_{1\min}$ 与地质上的井控面积 $1/S$ 之间,根据概率论,$n_2$ 口不干扰气井单井动态控制面积:

$$A_2 = N[1 - F(S)]\int_S^{1/A_{1\min}} \frac{1}{s} \frac{\partial F}{[1-F(S)]\partial s} ds \tag{11}$$

相应的 $n_2$ 口不干扰气井单井动态控制储量:

$$G_2 = \frac{N\phi h S_{gi}}{B_{gi}}\int_S^{1/A_{1\min}} \frac{1}{s} \frac{\partial F}{\partial s} ds \tag{12}$$

因此,井网密度 $S$ 时,平均单井动态控制储量:

$$\bar{G} = \frac{\phi h S_{gi}}{B_{gi}}\left(\frac{F(S)}{S} + \int_S^{S_{\max}} \frac{1}{s} \frac{\partial F}{\partial s} ds\right) \tag{13}$$

其中,$\bar{G}$ 表示平均单井动态控制储量,$10^4 m^3$。

相应的单井可采气量平均值:

$$\bar{G}_p = \beta \bar{G} \tag{14}$$

其中,$\beta$ 表示单井可采气量与单井动态储量比值,根据气藏工程理论,$\beta$ 主要取决于原始地层压力、废弃地层压力和经济极限产量等参数,以苏里格气田致密砂岩气藏为例,该值一般在 0.6 左右[21]。

根据式(14)计算 $\bar{G}_p$,并与单井极限产气量 $G_{pjx}$ 对比,评价气藏开发是否经济有效,进而确定经济极限井网密度,即:

$$\bar{G}_p|_{S=S_1} = G_{pjx} \tag{15}$$

其中,$S_1$ 表示经济极限井网,口/km²,$G_{pjx}$ 表示单井经济极限产量,$10^4 m^3$,主要受气价、钻井成本和约定等的内部收益率等因素影响[21],以苏里格致密砂岩气藏为例,经济极限产量在 $1300 \times 10^4 m^3$ 左右[12]。

将气井生产成本折合成单井经济极限产量,引入净采气量,定义为区块 $N$ 口井累计采气量减去 $N$ 口气井生产成本折合的气量($NG_{pjx}$),则井网密度 $S$ 时净采气量,表达式如下:

$$G_J = N\bar{G}_p - NG_{pjx} \tag{16}$$

其中,$G_J$ 表示区块净采气量,$10^4 m^3$。

从经济角度来讲,应该追求区块净采气量的最大化,此时的井网密度即为经济最佳井网密

度,根据费马定律[22],式(16)对井数 $N$ 的导数为 0 时,取最大值,因此,经济最佳井网密度满足如下关系:

$$\left.\frac{\partial (N \overline{G}_p)}{\partial N}\right|_{S=S_2} = G_{pjx} \qquad (17)$$

其中,$S_2$ 表示经济最佳井网密度,口/km²。

式(15)、式(17)给出了经济极限和经济最佳井网密度计算公式,因此,矿场可据此计算确定经济极限井网密度 $S_1$ 和经济最佳井网密度 $S_2$ 后,带入式(14)可求取经济最佳和经济极限时单井可采气量平均值,并与区块地质储量对比,求取经济最佳和经济极限采收率,表达式如下:

$$\eta_1 = \left.\frac{N \overline{G}_p}{G}\right|_{S=S_1} \times 100\% \qquad (18)$$

$$\eta_2 = \left.\frac{N \overline{G}_p}{G}\right|_{S=S_2} \times 100\% \qquad (19)$$

其中,$\eta_1$ 表示经济极限井网密度下采收率,%;$\eta_2$ 表示经济最佳井网密度下采收率,%。

需要强调的是,在依据上述模型对致密砂岩气藏进行井网优化与采收率评价时,需要对如图 2 所示的特定目标区的井间干扰概率统计曲线进行拟合,拟合出井间干扰概率 $F(S)$ 表达式,满足数值计算对曲线光滑性的要求。然后给出一系列井网密度 $S$ 值,带入式(14)计算井网密度 $S$ 时的单井可采气量平均值,并根据式(15)和式(16),利用插值法[23]求取经济极限井网密度与经济最佳井网密度,然后再根据式(18)和式(19)计算相应的采收率,完成致密气藏井网优化设计与采收率评价工作。

## 3 试验区井网优化与采收率评价

为了验证前述井网优化与采收率评价模型可靠性,选择苏里格气田 3 个加密试验区开展井网优化与采收率评价。3 个目标区一次性开发井网井距为 1200m 左右,井网密度 0.69 口/km²,选取加密前生产时间较长的气井(生产时间超过 5 年、开发处于中后期的气井)进行生产动态分析,确定试验区单井动态控制面积和井间干扰概率曲线(图 3)。

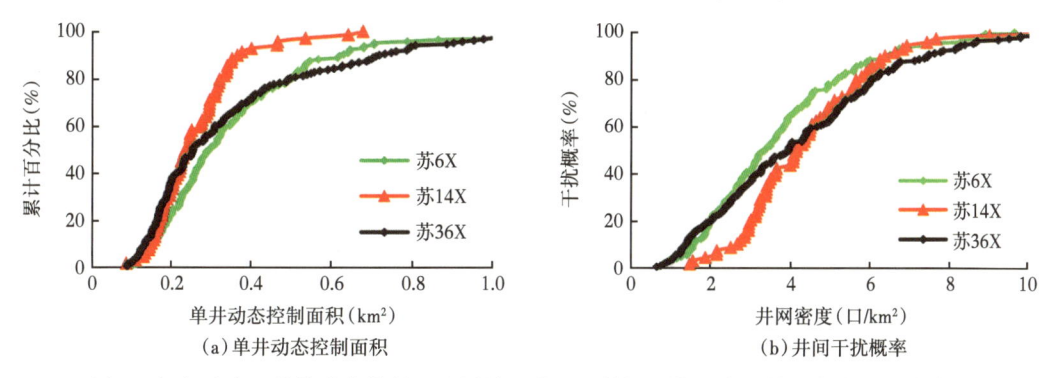

(a) 单井动态控制面积

(b) 井间干扰概率

图 3 加密试验区单井动态控制面积累计百分比及井间干扰概率—井网密度关系曲线图

结果表明,试验区块单井动态控制面积基本小于 1km²,平均只有 0.3km² 左右,加密前井网控制程度低,基本不存在井间干扰,具有加密的潜力。苏 6X、苏 3X 井区单井控制面积相对较大,苏 14X 井区单井控制面积较小,该井区绝大部分气井的动态控制面积小于 0.4km²;如图 3b 所示,3 个试验区的井间干扰概率曲线存在着一定的差异。

以 3 个试验区的储层物性参数为基础(表 1),结合图 3 的统计结果,根据前述方法开展井网优化与采收率评价(表 2),其中单井经济极限产量为 1350×10⁴m³。

表 1 试验区储层物性参数

| 井区 \ 指标 | 含水饱和度 (%) | 厚度 (m) | 孔隙度 (%) | 储量丰度 ($10^8 m^3/km^2$) |
|---|---|---|---|---|
| 苏 6X | 35.2 | 9.3 | 9.4 | 1.65 |
| 苏 4X | 40 | 8.8 | 9.0 | 1.5 |
| 苏 3X | 35 | 9.8 | 10.0 | 1.8 |

表 2 苏里格 3 个目标区块井网优化与采收率评价结果

| 井区 \ 指标 | 井网密度 (口/km²) | 干扰概率 (%) | 单井累计产量 ($10^4 m^3$) | 采收率 (%) |
|---|---|---|---|---|
| 苏 6X | 2.6/6.0 | 33/89 | 2499/1350 | 38.7/49.2 |
| 苏 14X | 3.1/5.2 | 28/83 | 1796/1350 | 36.6/46.8 |
| 苏 36X | 2.8/6.6 | 29/87 | 2658/1350 | 39.8/49.8 |

如图 4 所示,当井网密度达到开始产生井间干扰的井网密度(大约在 1 口/km²)时,随着井网密度的增加,井间干扰概率也会增加,最终导致单井平均产气量逐渐下降,采收率的增加幅度越来越小,最后趋于稳定,存在合理的井网密度。试验区经济最佳井网密度介于 2.6~3.1 口/km²,单井平均产气量(1796~2658)×10⁴m³,采收率 36.6%~39.8%,对应的井间干扰概率都在 30% 左右;极限井网密度 5.2~7.0 口/km²,单井平均产量等于单井极限产量 1350×10⁴m³,采收率 46.8%~49.9%,井间干扰概率 80%~90%,模型计算结果与试验区开发动态基本吻合,其中苏 6X 与苏 14X 试验区加密后的井网密度与本文模型计算的最佳井网密度相当,加密区生产效果良好,加密井平均生产 8.7 年,单井平均产气量 1600×10⁴m³,经济效益好;苏

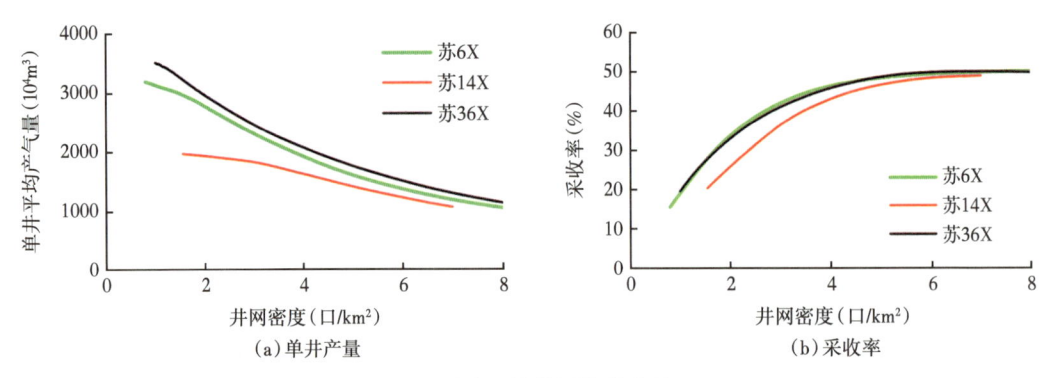

(a)单井产量

(b)采收率

图 4 试验区块井网优化曲线

36X井区加密试验后的井网密度5.1口/km², 与本文模型计算的最佳井网密度与经济极限井网密度的平均值相当, 目前已经生产了3年, 单井累计采气量平均值800×10⁴m³, 根据产气量与油、套压预测, 试验区单井最终可采气量1700×10⁴m³, 与井间干扰概率法预测的单井可采气量相当。研究结果进一步证明致密砂岩气藏井网优化与采收率评价模型的可靠性。

## 4 苏里格气田经济井网密度与干扰概率的关系

苏里格气田3个加密试验区实例计算分析结果表明, 经济井网密度与井间干扰概率存在很好的相关性, 表1说明经济最佳井网密度下井间干扰概率为30%左右, 经济极限井网密度下井间干扰概率为85%左右。因此, 可通过井间干扰概率判断致密砂岩气藏井网密度的合理性, 而且理论上证明, 在一定条件下此种方法在苏里格气田是完全可行的。

以苏里格气田目标区块为例, 储量丰度$1.5×10^8 m^3/km^2$, 从图4b可以看出, 开始发生干扰的井网密度为1口/km²左右, 完全干扰时井网密度为7口/km²左右, 井间干扰概率与井网密度基本呈线性关系, 井间干扰概率可近似用下式表达：

$$F(S) = \frac{S-1}{6} \tag{20}$$

根据式(7)和式(10)可求取不同井网密度的单井平均控制面积：

$$\bar{A} = \frac{1}{S}\frac{S-1}{6} + \int_S^7 \frac{1}{s}\frac{1}{6}ds = \frac{1}{S}\frac{S-1}{6} + \frac{1}{6}\ln\frac{7}{S} \tag{21}$$

式(21)除以地质上的单井控制面积$1/S$, 即可得平面上井控程度, 表达式如下：

$$\alpha = \frac{S-1}{6} + \frac{S}{6}\ln\frac{7}{S} \tag{22}$$

根据式(20)和式(22)可以计算不同井间干扰概率下致密砂岩气藏的平面控制程度(图5)。结果表明: 井网平面控制程度随井间干扰概率增加而增加, 但这种增加为非线性增加, 增加幅度越来越小, 最后趋于稳定; 可以看出, 在干扰概率30%时, 井网平面控制程度即可达到

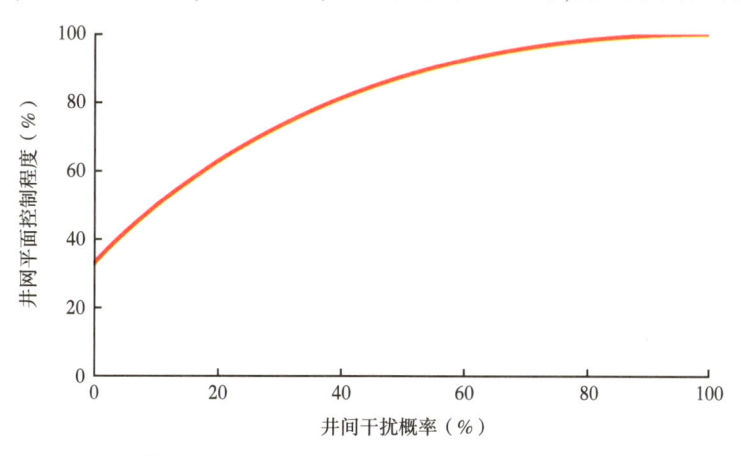

图5 苏里格气田井间干扰概率与井网平面控制程度关系

76%,与致密砂岩气藏开发要求的井网储量控制程度80%的目标相吻合[17];干扰概率85%时,井网平面控制程度达到了90%,控制绝大部分储量,继续增加井网密度,新增控制储量极少,意义不大。

由此可见,苏里格致密砂岩气藏最佳井网密度对应的井间干扰概率约为30%,而经济极限井网密度对应的井间干扰概率约为85%,可通过井间干扰概率来确定苏里格致密砂岩气藏经济合理的井网密度。

## 5 结论

(1)提出了依据致密砂岩气藏气井动态分析统计确定致密砂岩气藏井间干扰概率的新方法,建立了苏里格目标试验区块井间干扰概率曲线,非均质性的存在导致苏里格不同区块井间干扰概率曲线存在一定差异,但曲线形态基本一致。

(2)建立了考虑井间干扰概率的致密砂岩气藏单井可采气量与采收率模型,给出了经济最佳和经济极限井网密度与采收率计算公式,实现了致密砂岩气藏井网优化与采收率评价的目的。

(3)苏里格气田3个加密试验区的经济最佳井网密度介于2.6~3.1口/km$^2$,采收率介于36.6%~39.8%,井间干扰概率30%左右;极限井网密度介于5.2~7.0口/km$^2$,采收率介于46.8%~49.9%,井间干扰概率85%;

(4)苏里格致密砂岩气藏最佳井网密度对应的井间干扰概率约为30%,而经济极限井网密度对应的井间干扰概率约为85%,可通过井间干扰概率快速确定苏里格致密砂岩气藏经济合理的井网密度。

## 参 考 文 献

[1] 邹才能,杨智,何东博,等.常规—非常规天然气理论、技术及前景[J].石油勘探与开发,2018,45(4):575-587.

[2] 贾爱林.中国天然气开发技术进展及展望[J].天然气工业,2018,38(4):77-86.

[3] 胡文瑞.开发非常规天然气是利用低碳资源的现实最佳选择[J].天然气工业,2010,30(9):1-8.

[4] 杨华,刘新社,黄道军,等.长庆油田天然气勘探开发进展与"十三五"发展方向[J].天然气工业,2016,36(5):1-14.

[5] 叶礼友,高树生,杨洪志,等.致密砂岩气藏产水机理与开发对策[J].天然气工业,2015,35(2):41-46.

[6] 马新华,贾爱林,谭健,等.中国致密砂岩气开发工程技术与实践[J].石油勘探与开发,2012,39(5):572-579.

[7] 李奇,高树生,杨朝蓬,等.致密砂岩气藏阈压梯度对采收率的影响[J].天然气地球科学,2014,25(9):1444-1449.

[8] 王峰,田景春,陈蓉,等.鄂尔多斯盆地北部上古生界盒8储层特征及控制因素分析[J].沉积学报,2009,27(2):238-245.

[9] 张吉,史红然,刘艳侠,等.强非均质致密砂岩气藏已动用储量评价新方法[J].特种油气藏,2018,25(3):1-5.

[10] 丁景辰,曹桐生,吴建彪,等.致密砂岩气藏高产液机理研究[J].特种油气藏,2018,25(3):87-91.

[11] 位云生,贾爱林,何东博,等. 中国页岩气与致密气开发特征与开发技术异同[J]. 天然气工业,2017,37(11):43-52.
[12] 李跃刚,徐文,肖峰,等. 基于动态特征的开发井网优化——以苏里格致密强非均质砂岩气田为例[J]. 天然气工业,2014,34(11):56-64.
[13] 何东博,王丽娟,冀光,等. 苏里格致密砂岩气田开发井距优化[J]. 石油勘探与开发,2012,39(4):458-464.
[14] 何东博,贾爱林,冀光,等. 苏里格大型致密砂岩气田开发井型井网技术[J]. 石油勘探与开发,2013,40(1):79-89.
[15] 郭平,景莎莎,彭彩珍. 气藏提高采收率技术及其对策[J]. 天然气工业,2014,34(2):48-55.
[16] 付宁海,唐海发,刘群明. 低渗—致密砂岩气藏开发中后期精细调整技术[J]. 西南石油大学学报(自然科学版),2018,40(3):136-145.
[17] 李熙喆,万玉金,陆家亮,等. 复杂气藏开发技术[M]. 北京:石油工业出版社,2010.
[18] 孔祥言. 高等渗流力学[M]. 合肥:中国科学技术大学出版社,2010.
[19] 孙贺东. 油气井现代产量递减分析方法及应用[M]. 北京:石油工业出版社,2013.
[20] 夏静,谢兴礼,冀光,罗凯. 异常高压有水气藏物质平衡方程推导及应用[J]. 石油学报,2007,28(3):96-99.
[21] 毛美丽,李跃刚,王宏,等. 苏里格气田气井废弃产量预测[J]. 天然气工业,2010,30(4):64-66.
[22] 陈纪修,於崇华,金路. 数学分析(上册)[M]. 北京:高等教育出版社,2003.
[23] 李庆扬,易大义,等. 数值分析[M]. 北京:清华大学出版社,2008.

# 气藏水侵与开发动态的实验综合分析方法

徐 轩[1,2] 梅青燕[3] 陈颖莉[3] 韩永新[1,2] 唐海发[1] 焦春艳[1,2] 郭长敏[1,2]

(1. 中国石油勘探开发研究院；
2. 中国石油天然气集团公司天然气成藏与开发重点实验室；
3. 中国石油西南油气田分公司勘探开发研究院)

**摘要**：传统油气藏开发实验分析多从单因素角度研究气藏水侵及开发动态,不利于发挥物理模拟实验优势。结合水驱气藏开发实验实例,建立了气藏水侵与开发动态实验综合分析方法。从生产动态、水侵程度分析、动态压降剖面、含水饱和度及剩余储量分布5个方面对气藏开发机理和开发动态进行综合描述,论述了单因素分析方法产生偏差的原因,指出实验综合分析方法的优势及必要性。实例研究表明：由于室内模拟气井不受井筒积液影响,裂缝带上的气井长时间带水生产等效实现了排水采气,降低水的影响；存在贯通缝的气藏S3近井地带压力梯度稳定在0.22MPa/cm,不到无贯通缝气藏S1和气藏S2的50%；气藏S3气水同产使水侵后气藏含水饱和度仅增加32.33%,较未带水生产水低10%以上；采收率提高20%以上；气藏不同部位水侵机理不同,水侵气藏由于受水封作用影响,储量动用极不均衡。

**关键词**：水侵程度；开发动态；水驱气藏；压降剖面；剩余储量；分析方法

气藏开发动态监测与分析伴随气藏开发整个过程,水侵识别和水侵量动态分析是开发动态分析的主要内容之一[1-5]。由于油气藏开发的不可逆性,对气藏水侵与开发动态分析不准确、不及时将导致对气藏错误认识,从而制定错误的开发对策,引起难以挽回的损失。因此,在油气藏开发过程中,如何综合利用各种动、静态资料及分析方法,对气藏水侵与开发动态进行准确、及时的分析具有重要意义[6-9]。

气藏开发物理模拟实验是研究气藏开发机理,探索并总结开发规律,创新开发理论的重要手段[10-13]。模拟再现气藏开发过程,开展气藏水侵程度及开发动态的综合分析,既能够最大限度地发挥物理模拟实验的作用,也能为检验气藏开发动态分析技术适用性,创新气藏工程理论方法提供支撑。本文利用自主研发的一套测试气藏内部动态参数的实验装置,开展水驱气藏开发实验,结合实例分析,建立了一套气藏水侵与开发动态实验室综合分析方法。

## 1 建立实验综合分析方法的意义及思路

### 1.1 建立实验综合分析方法的意义

气藏开发实践中,水侵与开发动态分析的及时性和准确性受到多方面影响和制约[14-16]。而且从本质上说,气藏动态分析属于反问题求解,本身具有多解性,这也是开发动态分析需要综合动、静态数据及多种分析方法,以约束多解性的主要原因。以上因素增大了现场气藏开发动态分析难度,为气藏开发动态分析及气藏工程理论发展带来了挑战。

室内物理模拟实验能够模拟不同气藏地质及开发条件,再现气藏开发过程。相对于现场开发动态分析,实验室分析具有如下优势:首先,实验中动态参数测试准确、及时,排除了现场误差影响;其次,实验室模拟地层及开发条件明确,排除了求解反问题多解性的影响;第三,物理模拟具有可重复性,能够检验、对比不同分析方法的准确性和适用性。

然而,传统的开发实验,由于受实验设备限制,通常采用一维常规尺度岩心模拟采气,实验中只能获取气井(夹持器出口)气、水产出数据,无法获得气藏内部压力分布等动态参数,难以应用物质平衡等现场动态分析方法。即使部分实验测试了压力或饱和度等动态参数,也多止步于单因素分析,较少将气藏工程分析方法应用到实验中,忽视了对来之不易的实验数据的进一步挖掘,制约了开发实验作用的发挥[17,18]。

## 1.2 主要研究思路

随着开发实验技术的发展,模型制备及测试手段得到全方位的提升。一方面,物理模型从一维常规尺度向二维、三维大尺度模型跨越发展,模拟的地质及开发条件更为全面;另一方面,压力场动态监测、核磁共振及电阻率测饱和度等技术的应用,使得实验中获取的动态参数更加丰富。实验技术的进步为发展气藏水侵与开发动态实验综合分析方法奠定了基础[19,20]。

主要思路为通过研发物理模拟装置及测试技术,在充分模拟气藏地质及开发条件的基础上,开展气藏开发物理模拟实验。实验过程中,实时监测气井气、水产量的同时,尽可能多地测试气藏动态参数,包括气藏内部压力,气、水饱和度分布等,在此基础上,发挥实验室储层及开发条件明确,动静态参数测试准确、及时等优势,结合多种气藏工程分析方法,对气藏进行多角度水侵与开发动态综合分析。从而为实验数据的充分挖掘与利用,为开发技术政策制定,动态分析技术发展提供实验和理论支撑。

# 2 水驱气藏开发实验实例

为阐述上述综合分析方法,选取一组水驱气藏开发物理模拟实验开展实例研究。结合静态地质模型、气井气水产量、剩余压力及饱和度等动静态参数开展开发动态综合分析。

## 2.1 地质模型与实验设计

裂缝性气藏开发普遍受水侵影响,以川东石炭系龙吊气藏为例,水驱裂缝性气藏地质及生产特征参见文献[21,22]。龙吊气藏具层状裂缝—孔隙性气藏特点,整个构造被多条大断层切割,区域内断层及裂缝极为发育。如图1所示,池39井是位于吊钟坝高点北端的一口高产气井,气井单井控制储量$26.5×10^8m^3$,于1992年3月投产,初期日产气量$35.0×10^4m^3$,1994年3月突然产出地层水,日产水$3m^3$,日产气量迅速降至$7.3×10^4m^3$。研究表明池39井属典型的大裂缝导通型水侵,两条大断裂导通了水体和39井区,断裂带试井渗透率达60mD以上。

实验以龙吊气藏池39井区地质及开发实例为原型,抽提出典型裂缝水驱气藏三种地质模型,设计并开展水侵与治水物理模拟实验。如图1所示,模型左端连接水体,右侧为具有非均质性的,由不同物性裂缝区和基质区组合而成的储层。裂缝区(高渗透区)位于基质区的上下两侧,基质区根据物性不同,分为基质区1和基质区2(图中用不同填充色表示)。三种地质模型主要区别在于右侧基质区 2 的范围(图中所示三种模型边界)和气井位置:模型 1 和模型 2 中气井位于基质区 2,代表了具有低渗透基质区遮挡,裂缝没有完全连通气井和水体的气藏;模型 3 中气井通过高导断裂带直接与裂缝区相连,代表了气井通过裂缝或高渗透区直接连通水体的气藏。

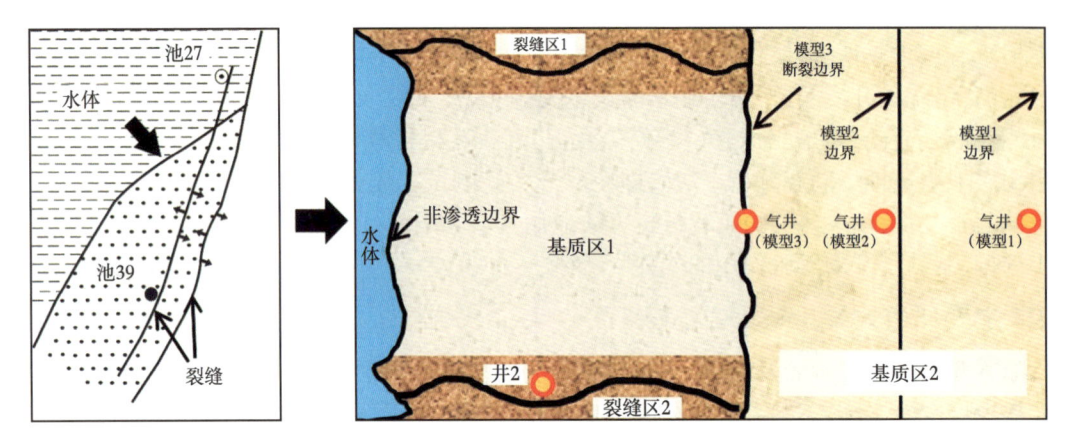

图 1 典型裂缝水驱气藏地质模式图(三种地质模型)

对应地质模型,建立裂缝水驱气藏开发动态物理模拟实验装置及方法。

实验采用单个长 50cm,总长达 200cm 的大尺度岩心模型模拟图中气藏不同区域。模型岩心夹持器采用自主研发的新型多测压点长岩心夹持器[23],可布设多个测压探头实时监测气藏内部压力。

如图 2,表 1 中所示,实验储层主要由四组不同物性的岩心构成。

长岩心 1 和岩心 3 采用基质岩心定向造缝获得沿岩心长轴方向的贯通缝,加围压后裂缝开度减小,裂缝规模主要由渗透率体现。经压裂造缝后岩心 1 覆压 30MPa 下渗透率达到 9.68mD,代表小裂缝带;岩心 3 覆压渗透率 18.08mD,代表大裂缝带。长岩心 2 和岩心 4 模拟基质区。岩心 4 根据所模拟的地质模型选取不同长度:模型 1 取 50cm,模型 2 取 25cm,模型 3

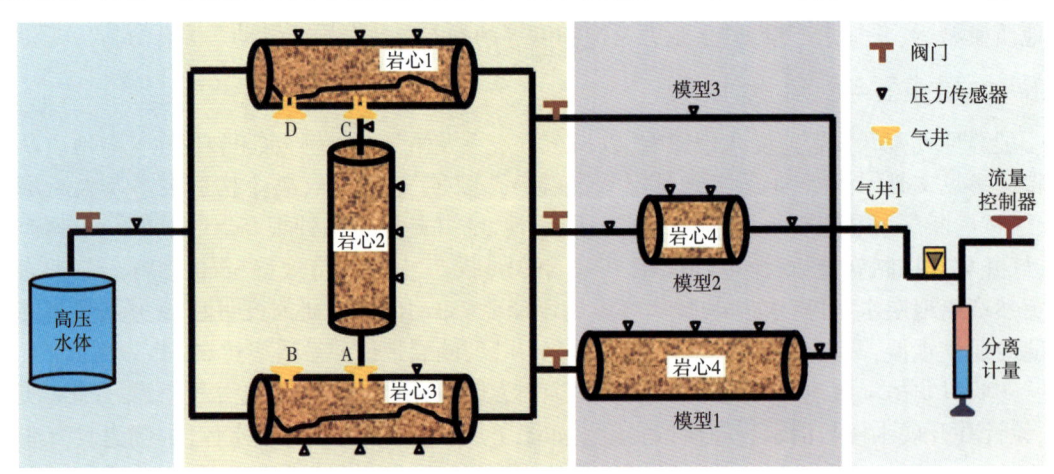

图 2 裂缝水驱气藏实验装置流程图(三种地质模型)

取 0cm,对应于气井模型 1—模型 3 的裂缝穿透率分别为 50%,67% 及 100%(具体参数见表 1)。

表 1 实验中四组岩心基础物性参数

| 岩心编号 | 储层类型 | 长度(cm) | 直径(cm) | 孔隙度(%) | 30MPa 下渗透率(mD) |
|---|---|---|---|---|---|
| 1 | 贯通缝 | 50 | 2.52 | 5.42 | 9.68 |
| 2 | 孔隙型 | 50 | 2.51 | 4.54 | 0.67 |
| 3 | 贯通缝 | 50 | 2.53 | 5.63 | 18.08 |
| 4 | 孔隙型 | 50/25/0 | 2.51 | 5.91 | 2.54 |

针对影响气藏水侵及开发效果的主要地质及生产条件,设计四组实验,实验目的、特点、模型及水体参数见表 2,详述如下:

实验 S1—S3 分别采用模型 1—模型 3 研究气井与裂缝区/水体位置关系对水驱气藏开发的影响,气藏均为单井生产。

实验 S4 为对比实验,采用模型 2 模拟无水体的定容封闭气藏,目的在于与水驱气藏进行对比分析。

表 2 不同组次实验内容及参数

| 编号 | 实验目的 | 实验特点 | 地质模型 | 水体 |
|---|---|---|---|---|
| S1 | | | 模型 1 | |
| S2 | 气井与裂缝区位置关系 | 水驱气藏<br>气井与裂缝区距离不同 | 模型 2 | 有限水体<br>(15 倍) |
| S3 | | | 模型 3 | |
| S4 | 对比实验 | 无水体封闭气藏 | 模型 2 | 无 |

考虑到模拟对象川东石炭系气藏含气储层含气饱和度普遍达 80% 以上,水侵前气井基本不产水,为气相单相流动,实验中采用干岩心模拟气藏。

实验 S1—S3 中气藏连通有限水体,考虑实际气藏岩石及水体均释放弹性能,实验中统一设置等效 15 倍水体。所有实验气井均配产 400mL/min 生产,气井废弃产量设为 10mL/min。

## 2.2 实验流程及步骤

第一步:根据实验方案准备岩心模型,将全部岩心装入岩心夹持器并加围压至 35MPa。

第二步:对岩心模型从两端缓慢饱和气至 30MPa。

第三步:将模拟地层水装入耐高压中间容器,加压至 30MPa,根据需要设置 15 倍水体。

第四步:将中间容器水体与饱和气后的岩心模型连通,根据实验方案,气井按配产 400mL/min 生产,模拟气藏开采。

第五步:开采过程中,通过夹持器上设置的测压探头实时记录岩心沿程压力剖面;通过出口流量计和气水分离器记录瞬时气、水产量、累计气、水产量、见水时间等参数。当气井达到废弃产量或出口端检测不到气流量时,继续观察 30min 以上,看是否复产气,否则停止实验。

第六步:实验结束后,取出岩心称重,获得不同位置处岩心含水饱和度。

## 3 气藏水侵与开发动态综合分析

结合上述水驱气藏开发物理模拟实例,开展气藏水侵与开发动态的实验综合分析,阐述实验综合分析的具体方法及意义。

### 3.1 气藏动态生产参数分析

绘制水驱气藏 S1—S3 及定容封闭气藏 S4 的生产曲线于图 3,主要生产参数见表 3。

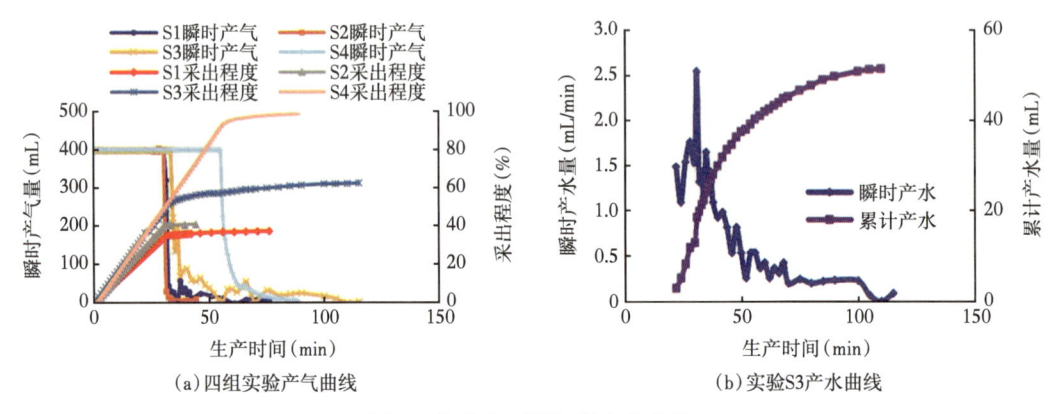

(a) 四组实验产气曲线

(b) 实验S3产水曲线

图 3 实验中不同气藏生产曲线

表 3 实验中不同气藏生产参数统计

| 实验组次 | 4 号岩心长度（cm） | 稳产时间（min） | 稳产期采出程度（%） | 累计生产时间（min） | 采收率（%） | 见水时间（min） | 累计产水量（mL） |
|---|---|---|---|---|---|---|---|
| S1 | 50 | 30 | 34.2 | 76 | 37.9 | 74 | 0.2 |
| S2 | 25 | 30 | 41.4 | 49 | 41.4 | 47 | 0.05 |
| S3 | 0 | 32 | 51.7 | 115 | 62.5 | 22 | 51.7 |
| S4 | 25 | 55 | 91.7 | 88 | 98.6 | 无 | 无 |

实验结果表明气井与裂缝区距离不同,生产动态和采收率差异显著。

气藏 S1 和 S2 中气井均位于低渗透基质区 2,与裂缝区存在遮挡。两气藏气井一旦见水,产量即急速下降至废弃产量:气藏 S1 在 74min 见水,76min 后即水淹至停产;气藏 S2 在 47min 见水,49min 后废弃停产(由于 S1 及 S2 见水即停产,图 3b 中仅绘制 S3 的产水曲线)。而气井直接连通裂缝区的气藏 S3,见水后气井仍能较长时间带水生产。分析原因是由于实验中气井能及时、充分排水,虽然气井开井 22min 后即见水,但见水后仍具有稳产能力,稳产至 32min,此后气水同产至 115min,气水同产期采收率达到 25.6%。

从见水时间上看,气井距连通水体的裂缝区越近,见水越快。气井与裂缝区距离不同,生产受水侵影响程度不同,带水生产能力迥异。三个气藏虽累计产气量接近,但由于气藏储量差异较大,气藏 S1 采收率仅为 37.9%,S2 采收率为 41.4%,而直接连通裂缝区的气藏 S3 采收率则达到了 62.5%。而不受水侵影响的定容封闭气藏 S4,其稳产期采收率就达到了 91.7%,最终采收率达 98.6%,远高于水驱气藏 S1—S3,说明水侵对气藏采收率影响巨大。

实验表明,虽然裂缝区连通水体,但气井距裂缝区越近,采收率仍然越高。分析认为:首先,裂缝作为高速供气通道供气能力强,使得气藏稳产能力强,储量动用快,即使气藏S3见水快,但裂缝依然具备较强渗流能力,见水后依然能向气井供气,使得气井见水后持续稳产,并长期气、水同产。其次,由于实验室条件下,气藏S3裂缝区气井不需要排水采气工艺即能够长时间带水生产,客观实现了排水采气,抑制了水体向基质区侵入;第三,实验模型S1和S2中,气井位于基质区2(岩心4),由于近井带渗透率较低,虽能一定程度延缓水体侵入气井,但也导致远端高渗透区向气井供气速度远低于S3,而一旦水体侵入基质区,形成两相流,近井带基质区渗流阻力将极剧增加,气井迅速达到废弃产量。

需要指出上述水驱气藏S1—S3的生产动态、采收率以及基于此进行的分析均基于文中特定的实验条件。裂缝和基质渗透率比值、水体规模、气井位置和配产等具体配置关系、参数变化均将导致开发效果迥异。有研究即指出虽然适当的裂缝能够提高采收率,但裂缝规模过大反而会导致水沿着裂缝快速流动,导致采收率急剧下降[24,25]。因此,本文实验结论具有前提条件,并不意味着裂缝区部署开发井的开发效果一定优于基质区。

以上分析基于实验室气藏地质模型与水体规模均已知的基础上进行。然而,在实际气藏开发动态分析中,如果不结合地质背景和其他动静态资料综合分析,仅根据气藏S1及S2生产参数进行分析,由于气藏S1、S2生产过程中几乎不产水,很难判断气藏S1和S2为水驱气藏,更难以对其水侵程度和开发动态进行正确的分析和评价。这充分说明了气藏动态分析的复杂性,以及开展综合分析研究的必要性。

## 3.2 气藏水侵程度分析

气藏水侵程度分析是开发动态分析的重要手段,结合气井生产曲线和其他动态分析方法,可在开发早期进行水侵识别,并分析气藏水侵程度。目前主要的水侵识别及水侵程度判别方法均基于气藏物质平衡原理,可细分为压降曲线法、水侵体积系数法和视地质储量法三种主要方法[26,27]。文章以压降曲线法和水侵体积系数法为例,对实验中气藏水侵程度及开发特征进行分析。

### 3.2.1 压降曲线法

压降曲线法又称为压降动态法和视地层压力法。常压水驱气藏的物质平衡方程可表示为:

$$\frac{P}{Z} = \frac{P_i}{Z_i}\left[\frac{G - G_p}{G - (W_e - W_p B_w)/B_{gi}}\right] \quad (1)$$

对于定容封闭气藏,$W_e = 0$、$W_p = 0$,其平衡方程可表示为:

$$\frac{P}{Z} = \frac{P_i}{Z_i}\left(1 - \frac{G_p}{G}\right) \quad (2)$$

实际生产中获取准确平均地层压力较困难,因此基于平均地层压力的气藏工程分析方法具有一定的局限性和误差。实验中,通过20余个测压点,实时精确测量了气藏储层内部压力,为准确获取全气藏平均地层压力提供了基础。

模型模拟裂缝发育的非均质气藏,开发过程中不同区域压力及储层物性参数各异,故需采用与储量有关的物性参数加权平均的方法计算气藏平均地层压力：

$$\overline{P} = \frac{\sum_{k=1}^{n} p_k L_k A_k \phi_k}{\sum_{k=1}^{n} L_k A_k \phi_k} \tag{3}$$

同时，由于气藏开发过程中，气藏各处压力变化范围较大，计算视地层压力（$P/Z$）时需考虑氮气偏差因子在不同压力下的取值。

采用上述方法，绘制实验水驱气藏 S1—S3 及定容气藏 S4 的 $P/Z$—$G_p$ 曲线（图4）。为便于分析将气藏 S3 累计产水量（$W_p$）与累计产气量（$G_p$）关系也绘制于图中。

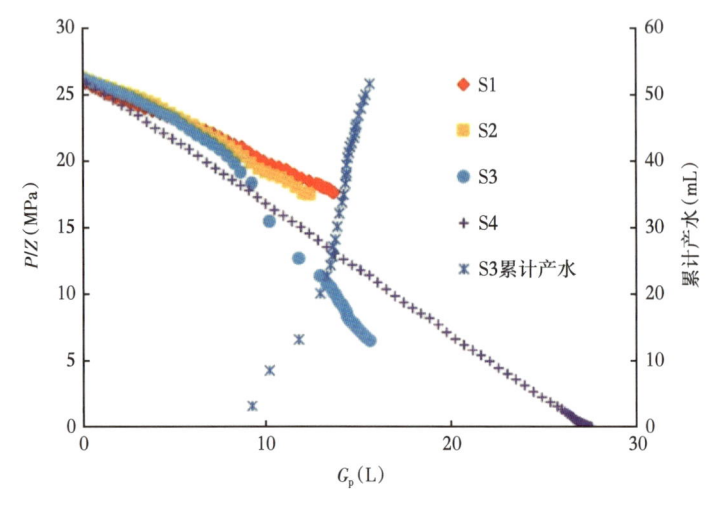

图4 实验中不同气藏视地层压力与累计产气关系图

实验中，由于地质模型和水体条件已知，据图4可进行以下分析判断：(1)气藏 S1—S3 早期 $P/Z$—$G_p$ 关系曲线都上翘，均表现出水驱气藏特征。由于气藏 S1 和 S2 始终未产水，压力曲线较气藏 S3 上倾，表明净水侵量有持续增强趋势；(2)气藏 S3 在气井带水生产后压降曲线开始下倾，显示出强排水特征。图4中累计产水曲线与压降曲线斜率具有较强的对应关系，说明一旦气井开始产水，累计净水侵量即快速变化。(3)视地层压力图4还揭示出水侵对水驱气藏储量动用影响严重，水侵入气藏导致大量剩余气无法采出，气藏 S1 和 S2 剩余视地层压力高达 15MPa 以上，与之对比，定容封闭气藏 S4，采出程度高，剩余视地层压力不到 0.5MPa。

结合生产参数（图3）及水侵程度分析（图4），无疑能更准确地分析气藏水侵程度及动态储量。但实际上由于气藏地质和水侵模型的多解性（同一生产动态可解释为不同地质及水体组合），如果不了解气藏类型及水体背景，仅采用上述方法分析水侵程度、推算气藏储量仍有误判风险。如气藏 S1 及气藏 S2 初期气井均未产水，根据这种生产动态，易将气藏视为定容封闭气藏，在初期（$G_p$<10L 前），如果根据 $P/Z$—$G_p$ 关系外推气藏储量，仍有极大可能高估气藏动储量。实例进一步说明在水侵程度判断和气藏储量分析中，应尽量了解气藏地质背景，包括气藏类型（水驱气藏/定容气藏）、水体大小、生产历史等，并随着开发的深入，紧密结合气藏生产动态，多因素综合分析，滚动评价，逐步落实对气藏储量和水侵的认识。

### 3.2.2 水侵体积系数法

水侵体积系数法是陈元千[26]基于物质平衡方法提出的,相对于压降曲线法,水侵体积系数法采用采出程度与地层相对视压力作为横纵坐标,相当于规整化的压降曲线法,更适合对比分析不同储量和地层压力的气藏水侵。

相对于压降曲线法,水侵体积系数法除需获取准确平均地层压力外,还需有较为准确、可靠的气藏原始地质储量以计算气藏采出程度,这也为气藏开发工作者带了更多挑战。实验中,储层孔隙体积和原始压力已知,故原始储量可准确获取。根据原始地质储量将图 4 中 $P/Z$—$G_p$ 关系曲线转化为 $\theta$—$R$ 关系曲线(图 5)。

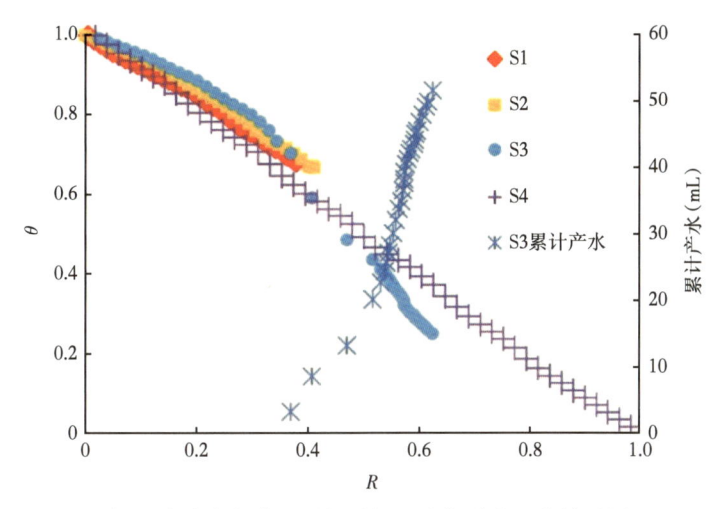

图 5 实验中气藏地层相对视压力与采出程度关系图

由于各项实验地质模型不同,原始储量也不同,转化后的 $\theta$—$R$ 曲线与图 4 中 $P/Z$—$G_p$ 曲线形态具有明显差异。

由图 5 可见,与理论一致,无水体定容气藏 S4,其 $\theta$—$R$ 关系曲线基本符合 45°线。而水驱气藏 S1—S3,生产初期 $\theta$—$R$ 曲线即上倾,均显示出水驱气藏特征。后期三种气藏的 $\theta$—$R$ 关系曲线显示出水侵程度差异:气藏 S1 和 S2,$\theta$—$R$ 曲线斜率变化不大,始终位于 45°线上方,表明地层水持续侵入,未实现有效排水。气井直接连接裂缝区的气藏 S3,在 $R = 0.37$ 左右时相对压降曲线开始向下偏移,对应气井开始产水。随着气藏 S3 累计产水量增加,相对压降曲线在采出程度 $R$ 为 0.4 左右时穿过 45°线,并继续下倾,表明此阶段水体净侵入量持续减小,并由侵入逐渐转化为产出。

## 3.3 气藏动态压降剖面分析

气藏物质平衡法可对气藏整体水侵程度和采收率进行判断,然而要了解气藏内部不同区域储量动用程度和剩余气位置,需进一步了解气藏不同部位剩余压力情况,应用气藏压力等值线图等方法研究气藏水封状况。实验过程中实时监测了气藏内部动态压力,可实时、直观反映气藏内部剩余封存气的位置和储量动用情况。

图 6 为定容封闭气藏 S4 不同时刻压降剖面,图 7—图 9 为水驱气藏 S1、S2 和 S3 不同时刻

压降剖面。为便于分析,根据压力参数,还统计并绘制上述四种气藏不同时刻近井地带(气藏 S1、S2 和 S4 为岩心 4 两端,气藏 S3 为裂缝区/气藏两端)平均压力梯度于图 10。

实验结果表明,定容封闭气藏和水驱气藏压降动态剖面差异显著,反映出丰富的气藏内部信息。

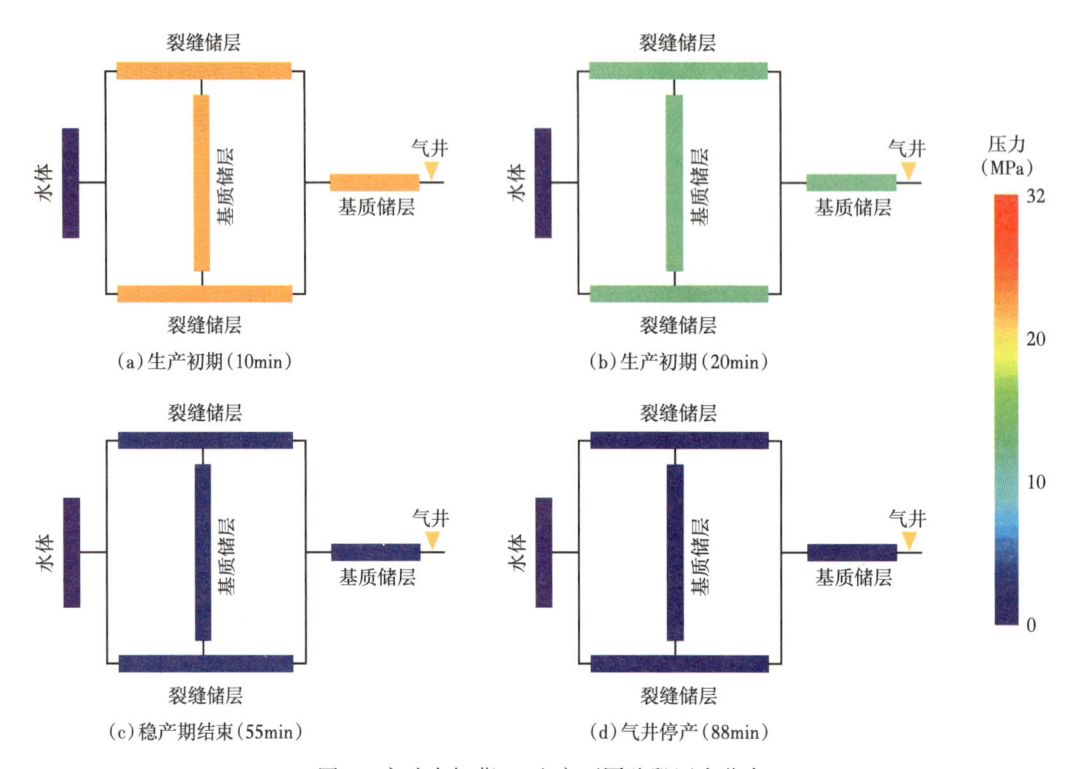

图 6 实验中气藏 S4 生产不同阶段压力分布

压降剖面(图 6)显示定容气藏 S4 生产过程中各部位压力均匀下降,稳产期结束,储量基本均衡动用,至 88min 停产时,各部位剩余压力基本为 0,对应气藏采收率 98%以上。

气藏 S1 基质 2 范围最大,气井距裂缝区最远(50cm)。图 7 中气藏动态压降过程显示,在采气前 20min,整个气藏压力降落相对较同步,表明初期储量动用均衡,储层能向气井持续稳定供气。此阶段气藏供气路径上最大压力梯度区——近井带低渗透率基质区 2(岩心 4)的压力梯度仅为 0.007MPa/cm(图 10)。此后气井虽能维持稳产,但近井地带压力梯度开始迅速增加,在 25min 时达到 0.18MPa/cm,30min 时达到高峰 0.34MPa/cm,稳产随即结束。据此,可以断定,应该是 25min 后地层水开始侵入基质区 2,形成气水两相流,导致流动阻力急剧增加,气井周围形成巨大压降漏斗。大量地层能量损耗在近井带基质区,使得储层中的气体无法流入井底,产量因而急速衰减至配产的 10%左右。至气井废弃停产时,近井带压力梯度最终稳定在 0.38MPa/cm,外围裂缝区和基质区 1(岩心 2)剩余压力高达 19.6~21.6MPa,大量剩余储量被水封无法动用,对应气藏采收率仅 37.9%。

气藏 S2 中气井距裂缝区 25cm,是气藏 S1 距裂缝距离的 50%。如图 8 和图 10 所示,与气藏 S1 类似,初期 S2 压力降落较同步,后期由于地层水侵入近井带基质区 2,使得近井带压力

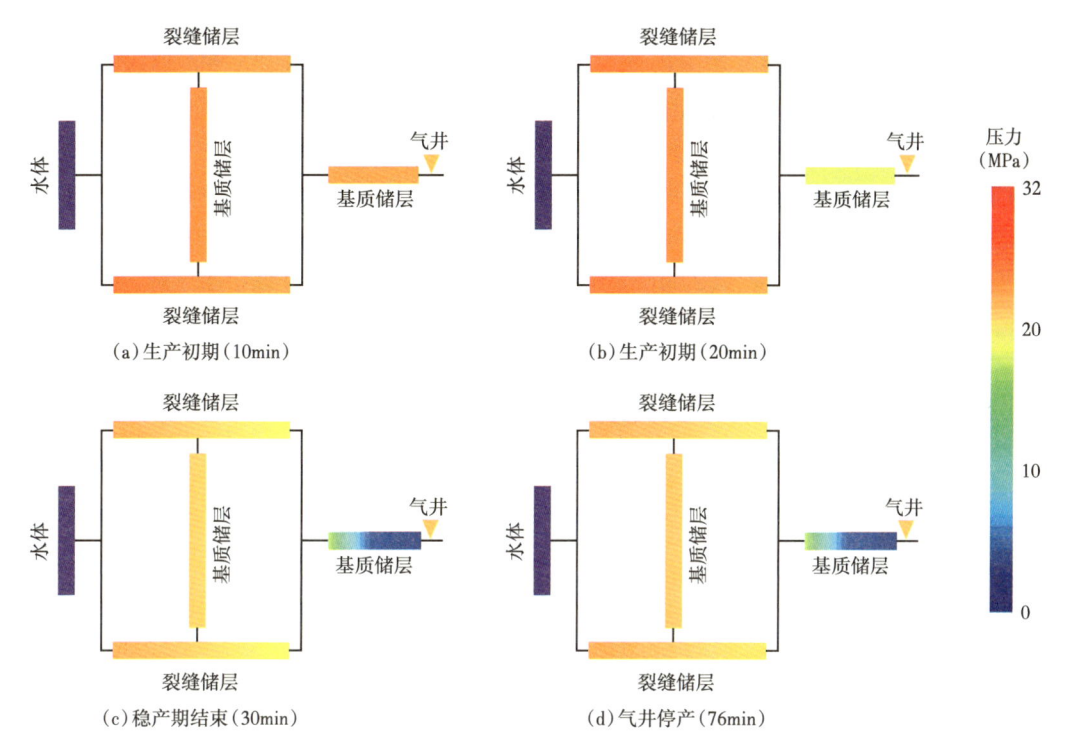

图 7　实验中气藏 S1 生产不同阶段压力分布

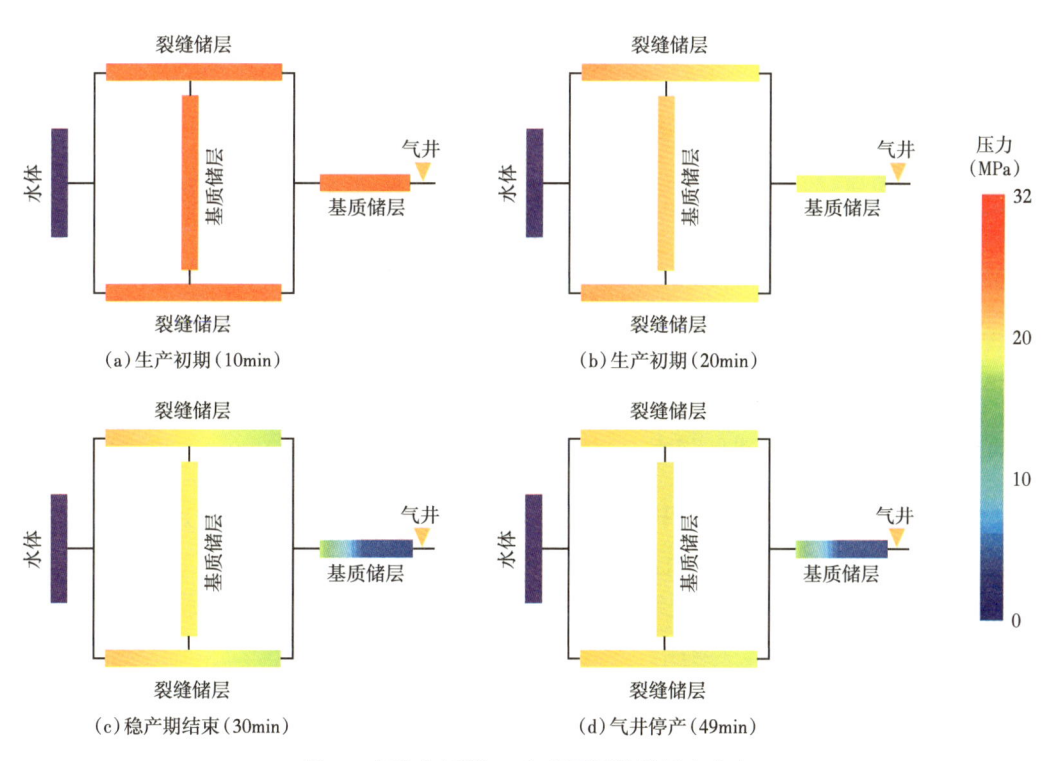

图 8　实验中气藏 S2 生产不同阶段压力分布

梯度迅速增加,在气井产量递减过程中,近井带平均压力梯度稳定在 0.74MPa/cm 左右,约为气藏 S1 的 2 倍,为三种水驱气藏中最高。外围裂缝区和基质区(岩心 2)虽贯通水体,但各区域压力较接近,维持在 17.9~23.2MPa 之间,表明地层能量主要消耗在近井带低渗透基质区,外围大量剩余储量无法动用,对应采收率仅为 41.4%。

气藏 S3 中气井直接连通裂缝区。图 9 显示压降剖面与气藏 S1 和 S2 差异显著,无论是开发初期还是后期,气藏 S3 整体储量动用均较均衡,近井带及外围压力能平缓、同步下降。图 10 也显示,气藏压力梯度仅为 0.22MPa/cm,远小于气藏 S1 和 S2。可见由于气藏 S3 裂缝的高导流作用,气井气、水同产,水体能量消耗的同时,客观降低了水侵危害,气体得以通过裂缝流向气井。最终,气井停产时,水体连同整个气藏压力大幅下降,剩余地层相对视压力仅为 0.25,对应采收率达 62.5%。

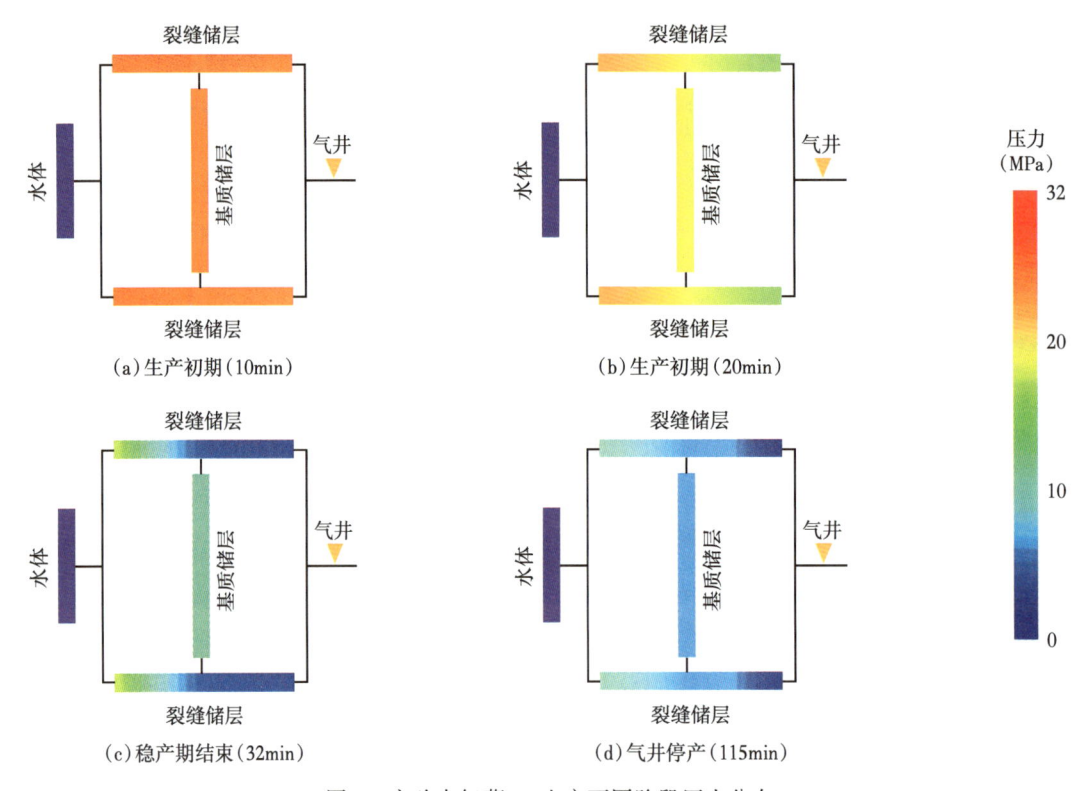

图 9  实验中气藏 S3 生产不同阶段压力分布

气藏压降剖面图还表明,气井生产时,气井远端被裂缝区包围的低渗透基质区 1(岩心 2)储量动用程度取决于外围裂缝区(岩心 1 和岩心 3)。裂缝区动用程度低(气藏 S1 和 S2),则基质区 1 储量被封闭,无法动用;裂缝区动用程度高(气藏 S3),则基质区 1 通过裂缝区向气井供气,其动用程度也较高。

显然,动态压降剖面与上文 3.1 和 3.2 中生产参数、水侵程度分析可互相验证,有效约束气藏工程反问题求解的多解性。

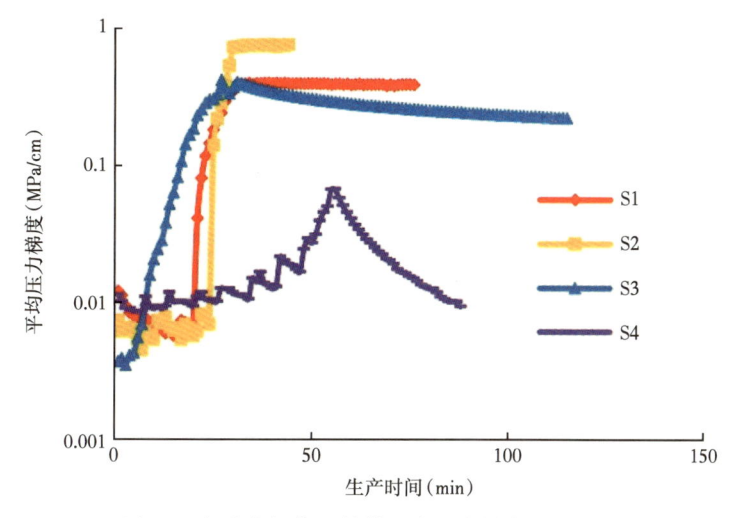

图10 实验中气藏近井带平均压力梯度变化曲线

## 3.4 气藏含水饱和度分析

如前文所述,目前的含水饱和度动态测试手段逐渐趋于丰富,采用 X-射线,电阻率法和核磁共振等方法能够实时检测气藏开发过程中储层内部水饱和度变化,为研究气藏水侵及开发规律,动用剩余储量提供关键依据。由于实验设备限制,本次实验仅在水驱气藏采气实验结束后通过岩心称重方法获得气藏不同部位岩心平均含水饱和度。

表4中气藏不同区域含水饱和度增量统计结果表明气藏 S1 和 S2 生产结束后整体平均含水饱和度增量差异不大,均达到了 40% 以上。气井连通裂缝区的气藏 S3 由于气井见水后能够长期带水生产,排水效果好,含水饱和度仅为 32.33%,较气藏 S1 和 S2 降低约 10%,这与上文3.2节中水侵程度分析一致。

表4 实验结束后水驱气藏不同区域含水饱和度增量　　　　　　单位:%

| 实验组次 | 小缝区1—岩心1 | 基质区1—岩心2 | 大缝区2—岩心3 | 基质区2—岩心4 | 气藏平均 |
| --- | --- | --- | --- | --- | --- |
| S1 | 46.34 | 27.60 | 47.47 | 44.43 | 41.46 |
| S2 | 53.11 | 21.45 | 50.73 | 45.65 | 42.74 |
| S3 | 45.86 | 9.56 | 41.56 | 无 | 32.33 |

分区域比较,三个水驱气藏均表现出直接连通水体的裂缝区水侵最严重,含水饱和度最高,达到 40%~55%;近井带基质区2虽离水体最远,但含水饱和度也达到 45% 左右;而被裂缝区包围的基质区1在各方案中平均含水都最低,气藏 S1、S2 和 S3 分别为 27.60%、21.45% 和 9.56%,远低于其他区域。分析认为,基质区1水侵机理与其他区域不同,主要为渗吸作用,理由在于:首先,基质区1外围具备高导流能力的裂缝区为气、水主要渗流通道,气、水主要流动方向与基质区1垂直;其次,由于基质区1渗透率为 0.67mD,远低于外围裂缝区储层,其供气速度缓慢,储量动用滞后,同期压力始终略高于相连裂缝区(压力剖面可证实)。从渗流力学分析,气、水不可能由低压裂缝区向高压基质区流动。因此,基质区1水侵入的动力只能是毛

细管力引起的渗吸作用。

可见不同地质条件气藏,相同气藏不同区域,气藏水侵机理也可能完全不同,含水饱和度与剩余气分布也差异显著。同时也说明联合气藏压力剖面及气、水饱和度动态分布开展综合分析,对于研究气藏水侵机理,制定控水、治水对策意义重大。

### 3.5 剩余储量分析

剩余储量分析是开发动态分析的关键环节,是气藏后期挖潜,调整开发对策的关键依据。根据含水饱和度换算剩余气饱和度,再考虑气藏各区域剩余压力及孔隙度,可计算不同区域相对剩余储量(或剩余储量占全气藏比例),具体计算方法为:

$$f = \frac{P_{rk} S_{gk} \phi_k}{\sum_{k=1}^{4} P_{rk} S_{gk} \phi_k} \tag{4}$$

采用上式计算三种水驱气藏生产结束后不同区域剩余储量占比见表5。

表5 实验结束后水驱气藏不同区域剩余储量占比　　　　　　　单位:%

| 实验组次 | 小缝区1—岩心1 | 基质区1—岩心2 | 大缝区2—岩心3 | 基质区2—岩心4 |
| --- | --- | --- | --- | --- |
| S1 | 27.93 | 32.08 | 28.51 | 11.48 |
| S2 | 26.93 | 38.05 | 29.56 | 5.46 |
| S3 | 29.48 | 39.69 | 30.83 | 无 |

表5显示,气藏不同区域剩余储量分布并不均匀。三个气藏中基质区1虽孔隙度不高,但由于剩余气饱和度较高,其剩余储量占比均达到30%以上,明显高于其他区域,提示其为气藏后期挖潜的重点区域;气藏S1和S2中近井带基质区2剩余储量最少,储量集中在外围裂缝区和基质区1,气藏储量动用不均衡;气藏S3不同区域剩余储量占比差异不超过10%,气藏储量动用最均衡。

结合上文采收率和剩余储量量化分析可进一步量化得到气藏内部不同区域剩余储量(文中未详细列出),为确定气藏开发挖潜目标,制定开发调整对策提供重要决策依据。

## 4 典型气藏开发动态

上文分析表明水驱气藏开发过程中,对于强非均质性气藏,水易沿裂缝或高渗透带侵入井底,从而形成水封气,使得外围大量剩余储量无法动用(图7—图10)。现实中水侵分割气藏,形成死气区并导致采收率降低的典型实例如威远震旦系灯影组气藏。该气藏1994年由于气藏非均质性及地层水纵窜横侵,被分割成6个互不连通的区块,与实验揭示的机理一致。气井出水后产能下降幅度介于5%~50%,对气藏高效开发带来极大危害[27]。

实验条件下,由于不存在井筒积液,气井位于裂缝带的气藏S3不需要排水工艺即能带水生产,实现了气藏整体排水采气,降低了水侵危害,采收率较未实现排水采气的气藏S1和S2分别提高24.6%、21.1%。以四川盆地石炭系气藏为例,中坝气田须二段裂缝—孔隙型边水气藏是排水采气成功开发的典型,自1978年4月中4井开始产地层水,相继有多口井出水,一些井出水后被水淹不能恢复生产,气井产量、压力快速下降,气藏生产受到严重威胁。为扭转这一被动局面1982年实施整体排水采气,以气举为主的多种排水采气工艺取得了显著的成效[28]。

## 5 结论

气藏物理模拟实验具有可重复性高、数据测量准确、动静态资料齐全、可有效规避多解性等优势。发展气藏水侵与开发动态的实验综合分析方法对于发挥开发实验优势,发展气藏工程分析理论均具有重要意义。

文章结合水驱气藏动态物理模拟实验进行了实例分析,运用动态生产参数、水侵程度分析、动态压降剖面、含水饱和度及剩余储量分布五种不同方法对气藏水侵与开发动态进行了综合分析和对比。研究表明建立的综合分析方法可操作性强,可有效避免单因素分析可能对气藏储量及水侵程度等造成的误判。综合分析方法能够更为准确、全面和精细地分析水侵及开发动态。

文章对开发动态实验综合分析方法进行了有益的探索,结合实例,列举了多种分析方法。在实际操作中,并不局限于文章所述方法。在动、静态参数支持的情况下,可探索将现场众多动态分析及气藏工程方法,如现代产量递减分析、试井分析、物质平衡理论等应用于实验分析中。

### 符号注释

$A_k$——各测点所测岩心横截面积,cm²;$B_w$——地层水体积系数;$B_{gi}$——原始天然气体积系数;$G$——天然气储量,10⁴m³;$G_p$——累计产气量,10⁴m³;$L_k$——各测量点测得的岩心长度,cm;$P$——当前地层压力,MPa;$P_i$——初始地层压力,MPa;$P_{rk}$——各测点测得的岩心剩余压力,MPa;$\bar{P}$——气藏平均地层压力,MPa;$P_k$——各测点测得的岩心平均地层压力,MPa;$R$——地层相对视压力;$S_{gk}$——岩心测量的剩余气饱和度,%;$W_e$——累计水侵量,10⁴m³;$W_p$——累计产水量,10⁴m³;$Z$——压力$P$下的天然气偏差系数;$Z_i$——压力$P_i$下的天然气偏差系数;$\theta$——各测点测得的岩心孔隙度,%;$\omega$——水侵系数;$\phi_k$——各测点测得的岩心孔隙度,%。

### 参考文献

[1] J J Arps. Analysis of Decline Curves[J]. Trans of the AIME,1945,160:228-247.
[2] 陈元千. 气田天然水侵的判断方法[J]. 石油勘探与开发,1978,5(3):51-57.
[3] 李士伦. 天然气工程[M]. 北京:石油工业出版社,2000:99-111.
[4] 陶诗平,冯曦,肖世洪. 应用不稳定试井分析方法识别气藏早期水侵[J]. 天然气工业,2003,2(4):68-70.
[5] 庄惠农. 气藏动态描述和试井[M]. 北京:石油工业出版社,2004:1-20.
[6] 韩永新,庄惠农,孙贺东. 数值试井技术在气藏动态描述中的应用[J]. 油气井测试,2006,(2):9-11,75.
[7] 刘晓华,邹春梅,姜艳东,等. 现代产量递减分析基本原理与应用[J]. 天然气工业,2010,30(5):50-54,139-140.
[8] 贾爱林,闫海军,郭建林,等. 全球不同类型大型气藏的开发特征及经验[J]. 天然气工业,2014,34(10):33-46.

[9] 李熙喆,郭振华,胡勇,等. 中国超深层构造型大气田高效开发策略[J]. 石油勘探与开发,2018,45(1):111-118.

[10] 高树生,熊伟,刘先贵,等. 低渗透砂岩气藏气体渗流机理实验研究现状及新认识[J]. 天然气工业,2010,30(1):52-55.

[11] 刘华勋,任东,高树生,胡志明,等. 边、底水气藏水侵机理与开发对策[J]. 天然气工业,2015,35(2):47-53.

[12] 胡勇,李熙喆,万玉金,等. 裂缝气藏水侵机理及对开发影响实验研究[J]. 天然气地球科学,2016,27(5):910-917.

[13] 徐轩,万玉金,陈颖莉,等. 裂缝性边水气藏水侵机理及治水对策实验[J]. 天然气地球科学,2019,30(10):1508-1518.

[14] 许宁. 储层平面非均质性对气藏开发动态的影响[J]. 天然气工业,2001,21(3):62-65.

[15] 冯曦,贺伟,许清勇. 非均质气藏开发早期动态储量计算问题分析[J]. 天然气工业,2002,22(增刊):87-90.

[16] 高承泰,卢涛,高炜欣,等. 分区物质平衡法在边水气藏动态预测与优化布井中的应用[J]. 石油勘探与开发,2006,33(1):103-106.

[17] 郭平,徐永高,陈召佑,等. 对低渗气藏渗流机理实验研究的新认识[J]. 天然气工业,2007,27(7):86-88,140-141.

[18] 朱华银,朱维耀,罗瑞兰. 低渗透气藏开发机理研究进展[J]. 天然气工业,2010,30(11):44-47,118-119.

[19] 徐轩,杨正明,刘学伟,等. 特低渗透大型露头模型流场测量技术及分布规律研究[J]. 岩土力学,2012,33(11):3331-3337.

[20] 徐轩,刘学伟,杨正明,等. 特低渗透砂岩大型露头模型单相渗流特征实验[J]. 石油学报,2012,33(3):453-458.

[21] 杨江海,邵勇,罗远平,等. 池27井区治水效果分析及川东石炭系气藏治水建议[J]. 钻采工艺,2003(S1):77-82,14.

[22] 苟文安,冉宏,李纯红. 大池干井气田石炭系气藏水侵早期治理现场试验及成效分析[J]. 天然气工业,2002,(4):67-70,4.

[23] 徐轩,胡勇,朱华银,等. 低渗透储层气体渗流启动压力梯度测量装置及测量方法:中国,ZL201410553044.7[P]. 2016-08-13.

[24] Fang F F,Shen W J,Li X Z,et al. Experimental study on water invasion mechanism of fractured carbonate gas reservoirs in LongWangMiao formation,Moxi block,Sichuan Basin[J]. Environmental Earth Sciences,2019,78(316):1-10.

[25] 徐轩,万玉金,陈颖莉,等. 裂缝性边水气藏水侵机理及治水对策实验[J]. 天然气地球科学,2019,30(10):1508-1518.

[26] 陈元千. 油气藏的物质平衡方程式及其应用[M]. 北京:石油工业出版社,1979.

[27] 李士伦,王鸣华,何江川,等. 气田与凝析气田开发[M]. 北京:石油工业出版社,2004.

[28] 刘义成. 中坝气田须二气藏提高采收率研究[J]. 天然气勘探与开发,2000,23(3):12-20.

# 致密砂岩气藏井网加密优化

胡 勇[1]　梅青燕[2]　王继平[3]　陈颖莉[2]　徐 轩[1]　焦春艳[1]　郭长敏[1]

(1. 中国石油勘探开发研究院；2. 中国石油西南油气田分公司；
3. 中国石油长庆油气田分公司)

**摘要**：采用物理模拟实验与数学评价方法相结合，系统研究了井控范围从500m逐步加密至100m（相当于井距从1000m加密至200m）过程中不同渗透率砂岩储层在不同含水饱和度条件时的储量采出程度，揭示了井网加密对提高储量采出程度作用，以采出程度提高5%~10%和大于10%为依据，建立井网加密可行性判识图表，为气藏井网部署和加密方案优化提供了参考依据。实验岩心常规空气渗透率分别为1.63mD、0.58mD、0.175mD、0.063mD，含水饱和度介于30.3%~71.1%之间。研究结果表明：渗透率为1.63mD的储层，采出程度总体均较高，除了在含水饱和度高达69.9%时的采出程度与井控范围有关外，其余含水饱和度条件下，采出程度与井控范围关系不大，可以采用大井距开发；渗透率为0.58mD的储层，采出程度与含水饱和度和井控范围关系密切，随含水饱和度降低、井控范围加密而增加；渗透率为0.175mD的储层，采出程度受含水饱和度的影响十分显著，只有在含水饱和度小于等于52.3%时，井网加密优化可提高储量采出程度，当含水饱和度大于52.3%时，储量采出程度均较低，一般小于等于10%，即使井控范围加密至100m，也难以得到提高；渗透率为0.063mD的储层，总体上采出程度非常低，即使含水饱和度仅有31.6%，井控范围加密至100m，其采出程度最高也只有2.3%，因此，该类储层依靠井网加密难以得到有效动用。

**关键词**：致密砂岩气藏；井网加密；储量动用；物理模拟；开发优化

井网加密是致密砂岩气藏提高储量采出程度的重要技术方法[1-3]，如鄂尔多斯盆地苏里格气田，气藏开发初期的井排距不小于800m，目前部分区块也加密至400m左右，开发井网在不断优化加密调整，国内外学者气藏开发井网优化以及加密调整等开展了大量工作，取得了很多成果认识[4-16]。但是，通过井网加密到底可以提高哪些类型储层的采出程度以及井网加密提高采出程度需要满足的条件，目前尚不清楚，因此，如何在气藏开发早期取得基础资料时认识不同类型储层动用特征，论证合理井网井距，对于气藏科学开发具有重要指导意义。

本文采用物理模拟实验与数学评价方法相结合，系统研究了不同井控范围（从500m逐步加密至100m）下不同渗透率砂岩储层在不同含水饱和度条件时的储量采出程度，揭示了井网加密适应条件，为气藏井网加密方案优化提供了参考依据。

对于实际气井来讲，一般是三维径向流供气，与一维模型供气的最大差异是瞬时产气能力和动用范围波及时间，但对于模拟研究生产过程中压降特征和动用范围大小影响不大。因此，本文采用一维物理模拟实验方法，模拟研究了气井某一供气方向上动用范围和压降特征，对于气井最终动用范围大小评价结果具有借鉴意义。

## 1 实验方法

### 1.1 仪器设备及流程

建立一套长岩心多点测压物理模拟实验方法及流程[1,2](图1),通过在岩心夹持器及胶皮套上布置测压孔,可以在线实时动态检测气衰竭开采过程中岩心内部不同位置孔隙压力变化特征。

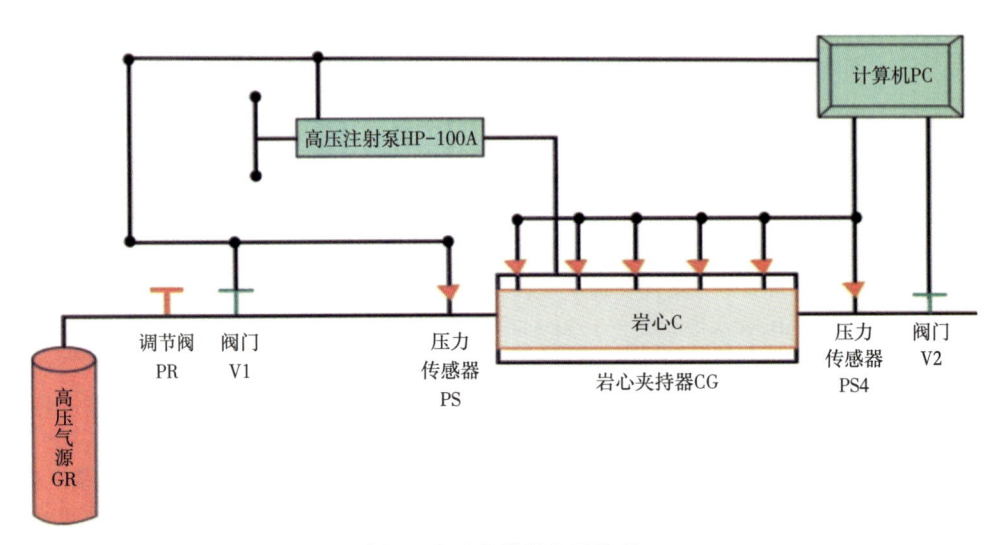

图1 实验仪器设备及流程

### 1.2 实验材料

选用鄂尔多斯盆地苏里格气田和四川盆地须家河组气藏天然岩心(直径×长度 = 2.5cm×50cm)开展实验,常规空气渗透率分别为1.63mD、0.58mD、0.175mD、0.063mD,储层原始含水饱和度介于30%~70%,原始地层压力20MPa。岩心基本参数和实验条件见表1。

表1 岩心基础参数和实验条件

| 孔隙度(%) | 渗透率(mD) | $S_w$(%) | 直径×长度(cm×cm) | 初始孔隙压力(MPa) | 实验配产(L/min) |
|---|---|---|---|---|---|
| 12.7 | 1.63 | 69.9 | 2.5×52.0 | 20 | 0.50 |
|  |  | 40.1 |  | 20 |  |
|  |  | 30.3 |  | 20 |  |
| 10.6 | 0.58 | 71.0 | 2.5×51.5 | 20 | 0.05 |
|  |  | 56.6 |  | 20 |  |
|  |  | 46.3 |  | 20 |  |
|  |  | 32.1 |  | 20 |  |

续表

| 孔隙度（%） | 渗透率（mD） | $S_w$（%） | 直径×长度（cm×cm） | 初始孔隙压力（MPa） | 实验配产（L/min） |
|---|---|---|---|---|---|
| 6.9 | 0.175 | 70.0 | 2.5×24.8 | 20 | 0.05 |
|  |  | 52.3 |  | 20 |  |
|  |  | 41.8 |  | 20 |  |
|  |  | 30.4 |  | 20 |  |
| 5.9 | 0.063 | 71.1 | 2.5×52.2 | 20 | 0.05 |
|  |  | 53.6 |  | 10 |  |
|  |  | 31.6 |  | 10 |  |

## 1.3 实验步骤

(1)选择实验用岩心并建立初始含水饱和度。
(2)将饱和水的岩心装入岩心夹持器并加围压至设定压力值。
(3)通过高压气源对岩心孔隙饱和气至设计地层压力。
(4)饱和气完毕后关闭气源,确保实验岩心处于独立压力系统。
(5)从岩心夹持器一端以一定速度释放孔隙压力(模拟气藏定产量衰竭开采)。
(6)实时记录实验过程中开采时间、各测点压力、瞬时产气量、瞬时产水量、累计产气量和累计产水量等参数。
(7)实验结束条件为检测不到气流量或各测点压力保持基本不变。

# 2 实验结果

通过气藏开采物理模拟实验,测试了产气量与压力变化特征,可以实现全生命周期的评价,为了评价方便,本文仅选取瞬时产气量降为初期配产的10%作为废弃产量条件为例开展评价工作,其他时期的评价方法可以参考。绘制了不同渗透率储层在不同含水饱和度条件下的产气量变化特征曲线(图2)以及废弃产量条件下的压力剖面(图3)。图3中的动用距离是指由近井向远井区压力下降波及区域。

总体上,产气量和压力剖面具有相似特征,从产气量特征曲线上可以看出,各渗透率储层均有以下特征,即相同采气速度下,稳产时间随含水饱和度增加而减少,稳产期结束后产量快速递减,低产周期长。从压力剖面上看,近井区压力下降最快,远井区相对平缓,呈现出"漏斗状"形态。

图 2　产气量变化特征曲线

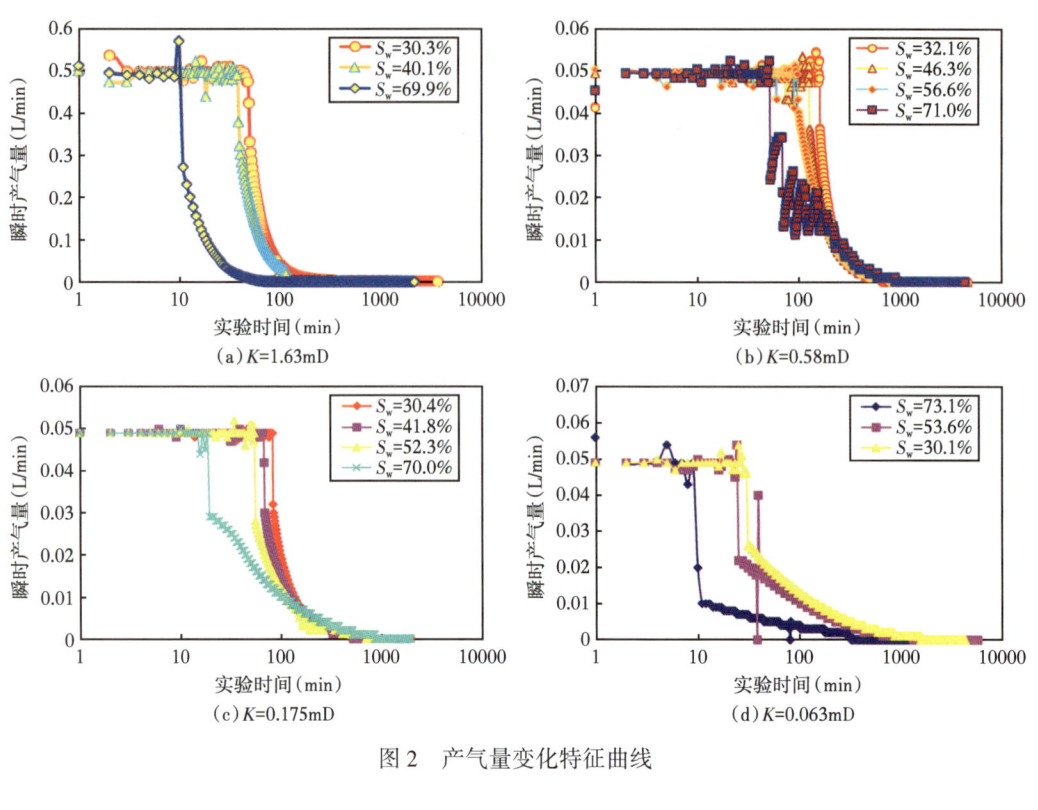

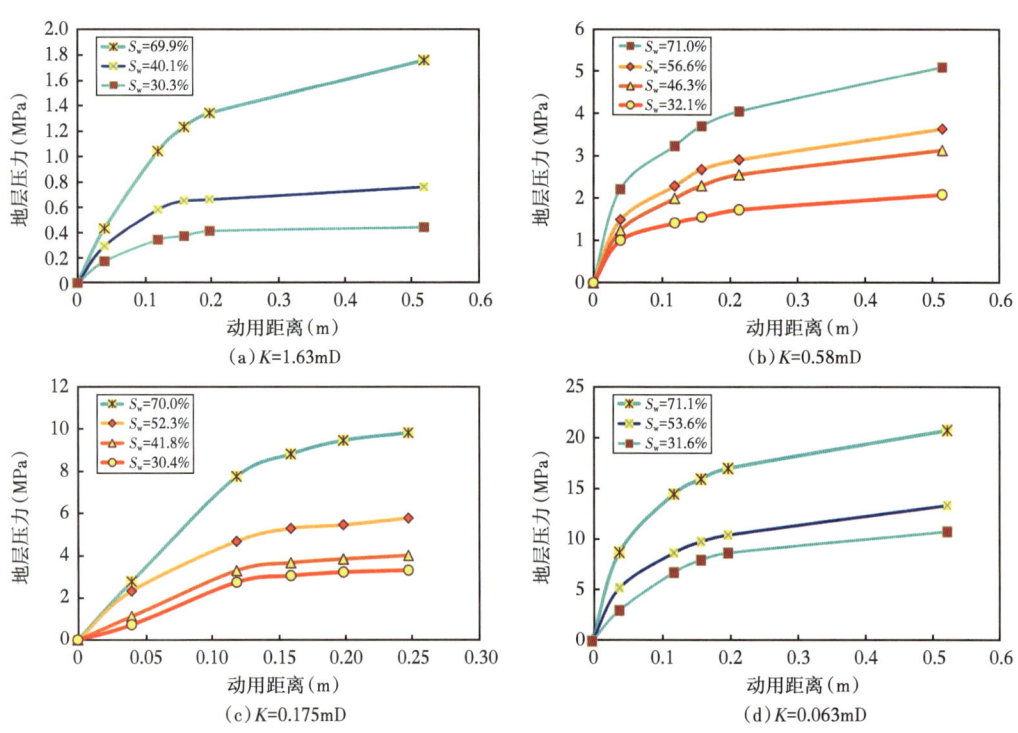

图 3　废弃产量条件下的压力剖面

## 3 动用范围与压力分布预测

### 3.1 幂函数拟合关系

根据图3中废弃产量条件下压力剖面特征,为了排除末端效应对实验的影响,选择离采气端较远的数据点采用幂函数进行拟合(图4),得出地层压力与动用距离数学拟合函数关系。

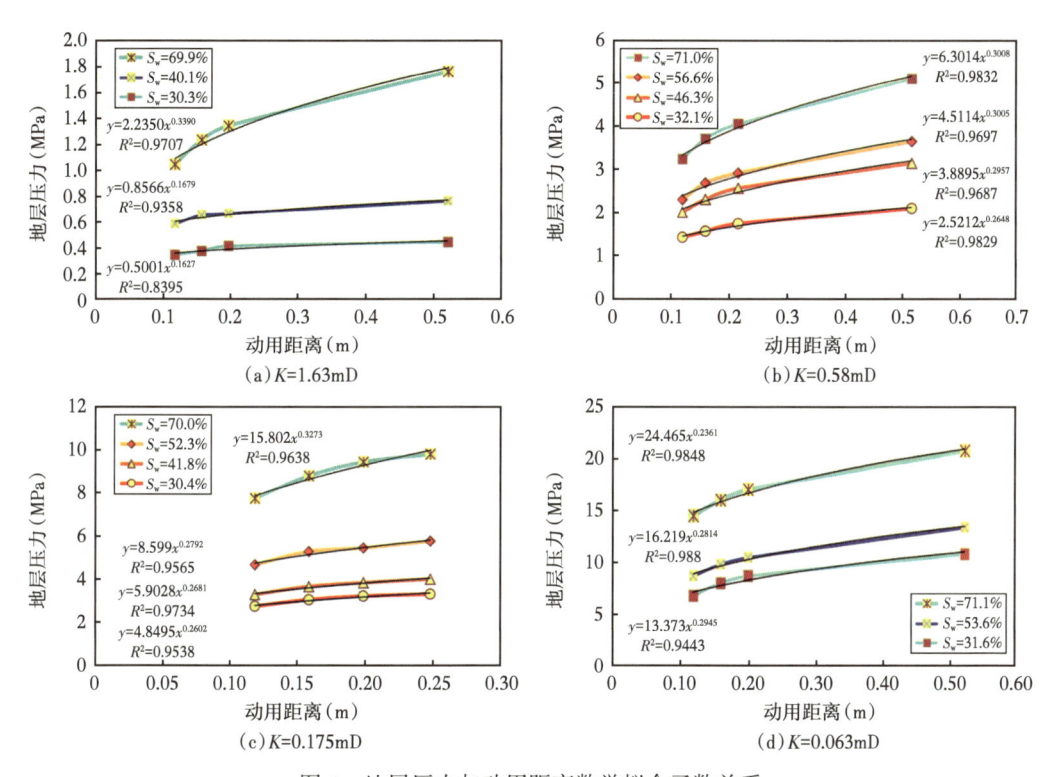

图4 地层压力与动用距离数学拟合函数关系

### 3.2 动用距离与压力分布

根据图4幂函数拟合关系式预测动用距离与压力分布特征(图5)。

## 4 井网加密优化分析

### 4.1 不同井控范围的采出程度评价

采用面积占比法,根据图5中动用距离与压力分布关系曲线,以地层压力30 MPa为例,系统研究了不同井控范围(从500m逐步加密至100m)下不同渗透率砂岩储层在不同含水饱和度条件时的储量采出程度(图6)。

渗透率为1.63mD的储层,采出程度总体均较高,除了在含水饱和度高达69.9%时的采出程度与井控范围有关外,其余含水饱和度条件下,采出程度与井控范围关系不大。当含水饱和度小于等于40.1%时,即使井控范围达500m,其采出程度也大于等于93.0%,井控范围从

图 5　动用距离与压力分布特征

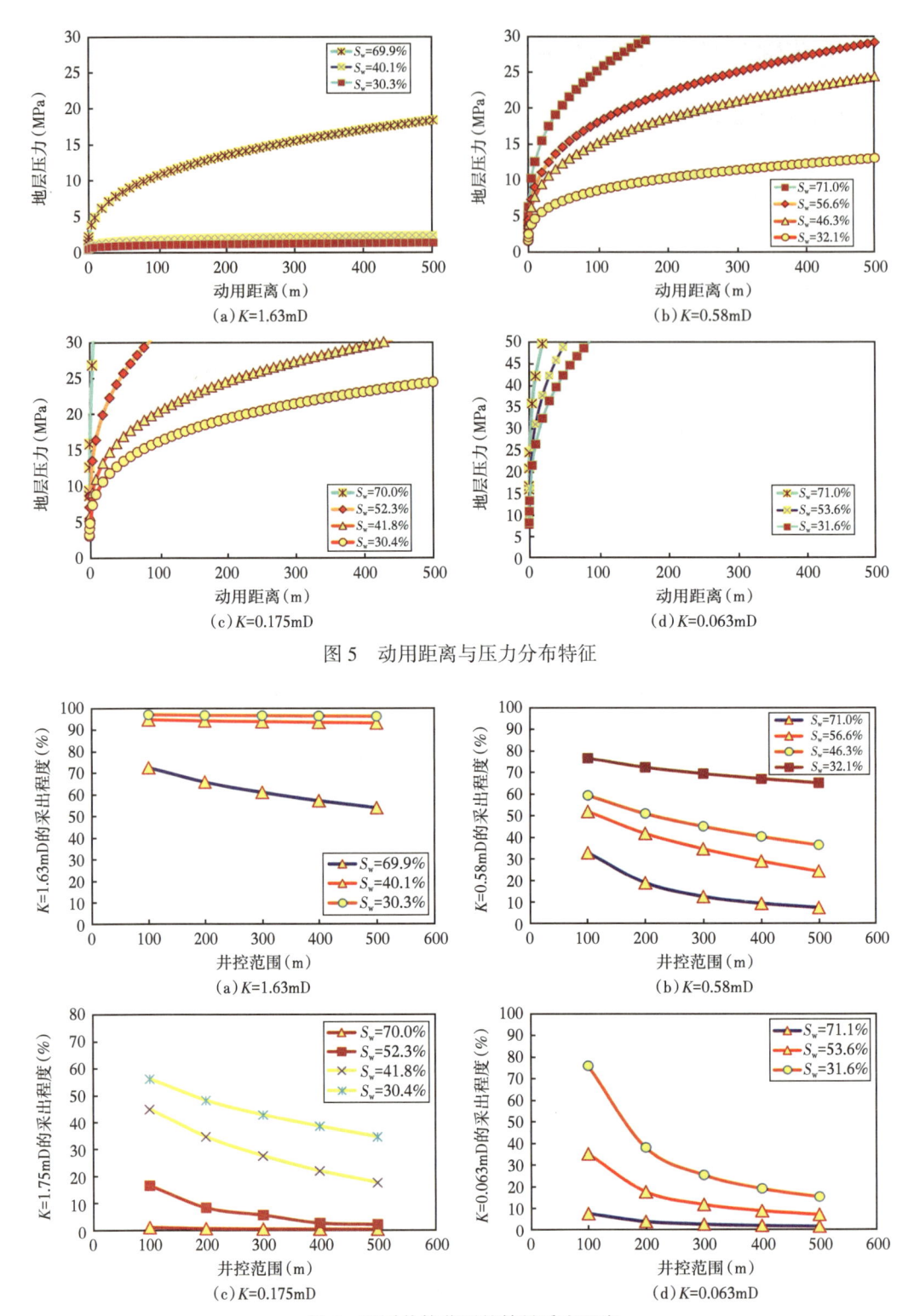

图 6　不同井控范围的储量采出程度

500m 加密至 100m 的采出程度差异不大;当含水饱和度为 69.9% 时,采出程度有所降低,但随井控范围加密而增加,井控范围从 500m 加密至 100m,采出程度可从 53.8% 提高至 72.3%,提高 18.5%。因此,对于这类储层,在含水饱和度不高的条件下,可以采用大井距开发。

渗透率为 0.58mD 的储层,采出程度与含水饱和度和井控范围关系密切,随含水饱和度降低、井控范围加密而增加。井控范围从 500m 加密至 100m,当含水饱和度为 32.1% 时,采出程度可从 65.2% 提高至 76.6%,提高 11.4%;当含水饱和度为 46.3% 时,采出程度可从 36.5% 提高至 59.4%,提高 22.9%;当含水饱和度为 56.6% 时,采出程度可从 24.4% 提高至 52.0%,提高 27.6%;当含水饱和度为 71.0% 时,采出程度可从 7.6% 提高至 32.9%,提高 25.3%。因此,对于这类储层,无论含水饱和度高低,井网加密优化对于提高储量采出程度均会发挥积极作用。

渗透率为 0.175mD 的储层,采出程度受含水饱和度的影响十分显著,只有在含水饱和度小于等于 52.3% 时,井网加密优化可提高储量采出程度,当含水饱和度大于 52.3% 时,储量采出程度均较低,一般小于等于 10%,即使井控范围加密至 100m,也难以得到提高。井控范围从 500m 加密至 100m,当含水饱和度为 30.4% 时,采出程度可从 34.4% 提高至 56.0%,提高 21.6%;当含水饱和度为 41.8% 时,采出程度可从 17.5% 提高至 44.7%,提高 27.3%;当含水饱和度为 52.3% 时,采出程度可从 2.1% 提高至 16.4%,提高 14.3%。因此,对于这类储层,井网加密优化需要清楚认识储层含水饱和度特征,有针对性开展才能发挥效果。

渗透率为 0.063mD 的储层,在该废弃产量条件下的采出程度非常低,即使含水饱和度仅有 31.6%,其采出程度最高也只有 2.3%。因此,该类储层依靠井网加密难以得到有效动用。

## 4.2 井网加密优化适用条件

对比分析井控范围从 400m 加密至 200m 时,不同渗透率储层的采出程度提高幅度(图 7),分别以采出程度提高 5% 和 10% 作为评价依据,给出了各类储层的井网加密适用条件及判识(表 2)。

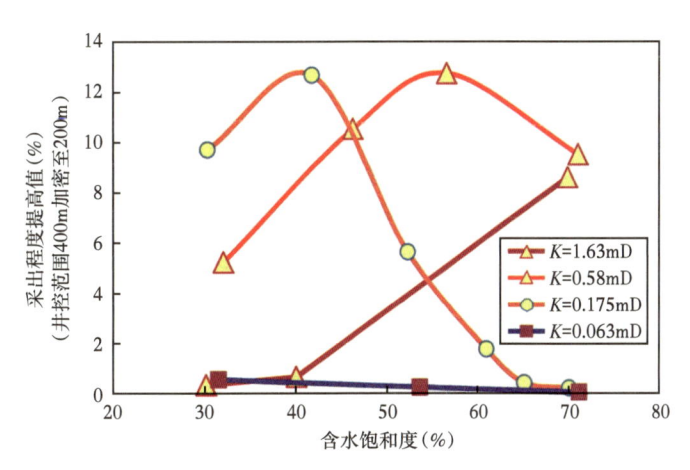

图 7 井控范围 400m 加密至 200m 采出程度提高幅度

表 2  各类储层井网加密优化适用条件及判识

| 序号 | $\phi$（%） | $K$（mD） | $S_w$（%） | 采出程度 200m | 采出程度 400m | 井控范围400m加密至200m提高幅度(%) | 适用条件判识 提高5% | 适用条件判识 提高10% |
|---|---|---|---|---|---|---|---|---|
| 1 | 12.745 | 1.63 | 69.9 | 65.6 | 57.0 | 8.6 | √ | × |
| 2 | 12.745 | 1.63 | 40.1 | 93.9 | 93.2 | 0.7 | × | × |
| 3 | 12.745 | 1.63 | 30.3 | 96.5 | 96.1 | 0.4 | × | × |
| 4 | 10.6 | 0.58 | 71.0 | 19.0 | 9.5 | 9.5 | √ | × |
| 5 | 10.6 | 0.58 | 56.6 | 41.9 | 29.2 | 12.7 | √ | √ |
| 6 | 10.6 | 0.58 | 46.3 | 51.0 | 40.4 | 10.5 | √ | √ |
| 7 | 10.6 | 0.58 | 32.1 | 72.4 | 67.2 | 5.2 | √ | × |
| 8 | 6.9 | 0.175 | 70.0 | 0.4 | 0.2 | 0.2 | × | × |
| 9 | 6.9 | 0.175 | 52.3 | 8.2 | 2.6 | 5.6 | √ | × |
| 10 | 6.9 | 0.175 | 41.8 | 34.5 | 21.8 | 12.7 | √ | √ |
| 11 | 6.9 | 0.175 | 30.4 | 48.1 | 38.3 | 9.7 | √ | × |
| 12 | 5.9 | 0.063 | 71.1 | 0.1 | 0.1 | 0.1 | × | × |
| 13 | 5.9 | 0.063 | 53.6 | 0.5 | 0.3 | 0.3 | × | × |
| 14 | 5.9 | 0.063 | 31.6 | 1.1 | 0.6 | 0.6 | × | × |

## 5 结论与讨论

（1）本文采用物理模拟实验与数学评价方法相结合,系统研究了不同井控范围(从500m逐步加密至100m)下不同渗透率砂岩储层在不同含水饱和度条件时的储量采出程度,揭示了井网加密提高储量采出程度的适用条件和判识图表,为气藏井网部署和加密方案优化提供了参考依据。

（2）明确了井网加密提高储量采出程度作用,对于渗透率为1.63mD高含水饱和度储层、渗透率为0.58mD的储层以及渗透率为0.175mD但含水饱和度小于等于52.3%的储层,通过井网加密可以一定程度提高储量采出程度,其余储层效果不明显。

（3）本文研究了井网加密提高储量采出程度的适应条件,但在生产现场具体实施井网加密方案时,在本文适应条件基础上,还需考虑地质条件和经济效益目标等因素确定最优的加密井距。

## 参 考 文 献

[1] Stephen A Holditch.Tight gas sands[J].SPE J,2006(1):86-93.
[2] 李熙喆,万玉金,陆家亮,等. 复杂气藏开发技术[M]. 石油工业出版社,2010.
[3] 庄惠农. 气藏动态描述和试井[M]. 北京：石油工业出版社,2009.
[4] 马新华,贾爱林,谭健,等.中国致密砂岩气开发工程技术与实践[J].石油勘探与开发,2012,39(5):572-579.

[5] 李熙喆,卢德唐,罗瑞兰,等.复杂多孔介质主流通道定量判识标准[J].石油勘探与开发,2019,46(5):943-949.
[6] 雷群,李熙喆,万玉金,等.中国低渗透砂岩气藏开发现状及发展方向[J].天然气工业,2009,29(6):1-3,133.
[7] 赵文智,卞从胜,徐兆辉.苏里格气田与川中须家河组气田成藏共性与差异[J].石油勘探与开发,2013,40(4):400-408.
[8] 胡勇,李熙喆,万玉金,等.致密砂岩气渗流特征物理模拟[J].石油勘探与开发,2013,40(5):580-584.
[9] 何东博,贾爱林,冀光,等.苏里格大型致密砂岩气田开发井型井网技术[J].石油勘探与开发,2013,40(1):79-89.
[10] 李熙喆,郭振华,胡勇,等.中国超深层构造型大气田高效开发策略[J].石油勘探与开发,2018,45(1):111-118.
[11] 李熙喆,郭振华,胡勇,等.中国超深层大气田高质量开发的挑战、对策与建议[J].天然气工业,2020,40(2):75-82.
[12] 胡勇,李熙喆,李跃刚,等.低渗致密砂岩气藏提高采收率实验研究[J].天然气地球科学,2015,26(11):2142-2148.
[13] 李熙喆,刘晓华,苏云河,等.中国大型气田井均动态储量与初始无阻流量定量关系的建立与应用[J].石油勘探与开发,2018,45(06):1020-1025.
[14] 胡勇.气体渗流启动压力实验测试及应用[J].天然气工业,2010,30(11):48-50,119.
[15] 郭智,贾爱林,冀光,等.致密砂岩气田储量分类及井网加密调整方法——以苏里格气田为例[J].石油学报,2017,38(11):1299-1309.
[16] 徐轩,胡勇,万玉金,等.高含水低渗致密砂岩气藏储量动用动态物理模拟[J].天然气地球科学,2015,26(12):2352-2359.

# 地质应用篇

# 川中合川气田须二段致密砂岩储层"甜点"研究

张满郎[1]　谷江锐[1]　孔凡志[1]　郭振华[1]
钱玮玮[2]　付　晶[1]　郑国强[1]　石　石[1]

(1. 中国石油勘探开发研究院;2. 中国石油华北油田分公司第四采油厂)

**摘要**:以川中合川气田须二段为研究对象,利用大量钻井、测井资料、岩心、铸体薄片、扫描电镜、压汞及物性分析数据,系统研究致密砂岩储层特征,阐明储层形成主控因素并开展储层甜点预测。研究结果表明:须二段被划分为两个半高频旋回、五个砂层组,优质储层主要分布在层序界面之上的基准面上升的早期,须二段下部的X2-1砂层组储层最发育,连续性最好;有利沉积微相为三角洲前缘水下分流河道、河口坝,有效储层主要为中粒、中细粒岩屑长石石英砂岩,发育残余粒间孔、粒间溶孔、粒内溶孔、杂基内微孔和微裂缝;须家河组煤系地层生烃高峰期酸性地层水所引起的长石、岩屑和杂基的溶蚀作用是形成次生溶孔,改善储层物性的主要原因;早期绿泥石衬边胶结提高砂岩的抗压强度并抑制石英加大边的形成,从而使得较多的原生粒间孔隙得以保存;在雷口坡组古残丘顶部的须二段砂岩中,由于差异压实而发育微裂缝,这种微裂缝大幅改善了储层的渗透能力。归纳出七种成岩—储集相组合类型,基于沉积相—成岩相耦合,结合储层物性、有效厚度、天然气测试产量等指标,进行了储层"甜点"定量评价,筛选出三个储层"甜点"发育区。Ⅰ类储层分布于三角洲前缘主河道和汇流区的水下分流河道—河口坝叠置砂体与绿泥石衬边—强溶蚀相或微裂缝—强溶蚀相耦合区域,绿泥石衬边—强溶蚀相呈片状分布,范围较大,而微裂缝—强溶蚀相分布局限,微裂缝发育部位与古残丘有关。

**关键词**:致密砂岩;控制因素;沉积相—成岩相耦合;储层"甜点";须二段;合川气田;川中地区

　　四川盆地上三叠统须家河组是中国致密砂岩气藏的典型代表,其分布面积广、资源量丰富,具有较大的勘探开发潜力。须家河组被划分为六个岩性段,其中,须一、三、五段以泥岩为主夹薄层泥质粉砂岩、煤层或煤线,是须家河组的主要烃源层和盖层,须二、四、六段发育大面积分布的厚层砂岩,是须家河组的主要含气层段。

　　四川盆地在晚三叠世为前陆盆地演化阶段,川西前陆盆地西起龙门山推覆带,北至米仓山—大巴山推覆带,东至达县—重庆—泸州一线,南抵雅安、自贡等地,分布面积约80000km²。须家河组被划分为四个三级层序[1],其中SQ1相当于须一段,SQ2相当于须二段、须三段,SQ3相当于须四段,SQ4相当于须五段、须六段。层序界面为不整合面(须家河组与雷口坡组、自流井组与须家河组分界面均为区域不整合面)、河道冲刷侵蚀面(须二段、须四段底界为冲刷侵蚀面,为SQ2和SQ3的底界面)、进积—退积转换面。最大湖泛面为退积—进积转换面,一般为滨浅湖或前三角洲泥岩沉积。平面上发育具有继承性的盆地边缘西、北、东三个物源方向控制的冲积扇—河流—(扇)三角洲沉积体系[1]。须家河组沉积期古地形具有西、北陡,东、南缓的特点,在盆地西北部的江油—剑阁地区和盆地北部的旺苍地区发育以砂砾岩为主的冲积扇和扇三角洲沉积,在川中地区主要发育中粒、中细粒砂岩组成的三角洲前缘沉积,在盆地中西部发育滨浅湖沉积(图1)。

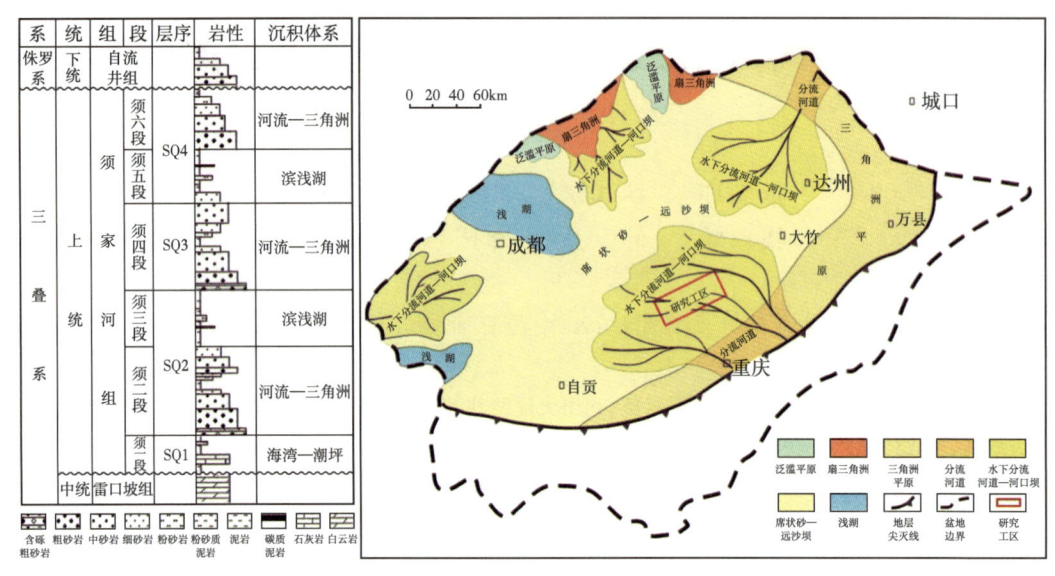

图1 四川盆地须二段沉积体系展布及研究区位置图

川中地区在晚三叠世处于川西前陆盆地的斜坡带和前隆带,沉积了巨厚的、叠置连片的河流、三角洲砂岩,具备大面积岩性气藏形成的有利条件[2,3]。

须二段是川中须家河组的主要含气层段,具有厚砂体、薄气层、纵向多层、低孔致密、裂缝发育、储层非均质性强、气水关系复杂等特点[4]。虽然川中须二段储量规模巨大,但空井、低产井、产水井比例偏大,开发效果欠佳。针对这种大面积、低丰度、气水关系复杂、强非均质性的致密砂岩气藏,优选储层"甜点"和天然气富集区块,提高单井产量是实现气藏高效开发的关键。

本文对川中合川气田须二段开展了储层综合研究。基于16口取心井1000余米岩心精细描述,利用209口井的钻井、测井资料,对须二段进行高分辨率层序地层划分对比,编制12条对比剖面及5个小层的沉积砂体分布图件,明确了层序格架中的有利砂体类型及分布。利用大量铸体薄片、扫描电镜、压汞实验及物性分析数据(共22口井1795样次),研究了合川气田须二段的岩石学特征、孔隙结构、成岩作用及储层物性特征,结合沉积、成岩及裂缝分布综合分析,确定了优质储层形成的主要控制因素,综合考虑沉积相、成岩相、储层物性、含气性、有效储层规模等因素,开展了储层"甜点"评价与预测。

# 1 储层基本地质特征

## 1.1 储层岩石学特征

岩石成分统计结果表明,研究区须二段砂岩中石英含量较高,石英含量大多在49%~74.5%之间,石英/(长石+岩屑)介于1~3之间。须二段砂岩主要岩石类型为岩屑长石石英砂岩,其次为长石岩屑砂岩,而岩屑长石砂岩、岩屑砂岩较少。粒度以中粒为主,次为中—细粒、细粒,分选中等—好,磨圆次圆—次棱,多为点—线接触、颗粒支撑,孔隙式胶结。其总体特征是结构成熟度中等,矿物成分成熟度较低。

## 1.2 储层物性特征

根据岩心分析资料,合川气田须二段砂岩具有低孔、致密特征。孔隙度主要集中分布在 5%~10% 之间,孔隙度 ≥6.0% 的岩样占 86.9%;渗透率主要集中分布在 0.04~0.64mD 之间,占砂岩样品的 77.69%(图2);储层段样品渗透率主要集中在 0.04~2.5mD,占 87.2%,储层段平均渗透率为 0.93mD,产气段渗透率在 0.1~2.5mD 之间。

岩心样品孔隙度与渗透率成正相关,潼南区块样品的孔渗相关性比合川区块好,主要发育孔隙型储层。合川区块孔隙度 4%~8% 的部分样品渗透率偏高,发育裂缝—孔隙型储层。

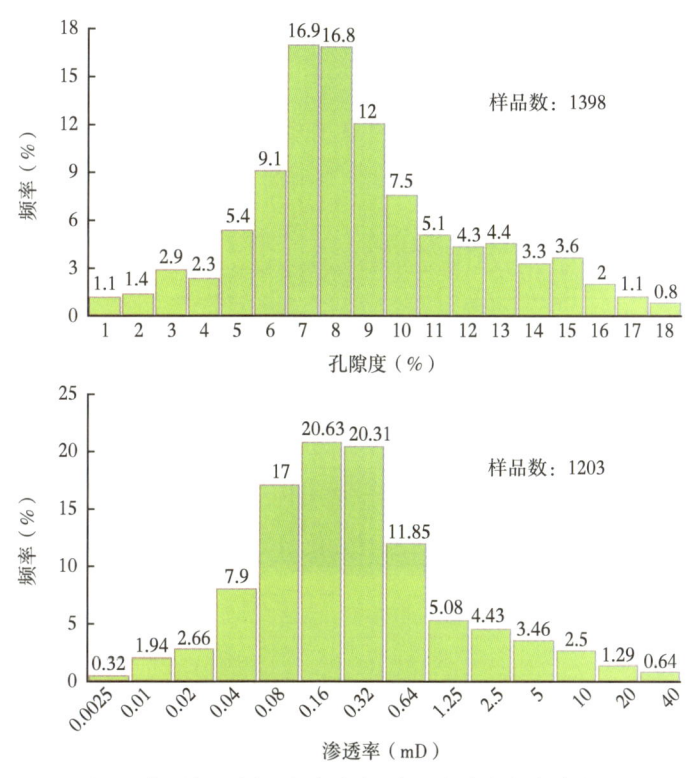

图2 合川气田须二段孔隙度、渗透率分布频率直方图

## 1.3 储层孔隙结构特征

储层孔隙结构特征是指岩石所具有的孔隙和喉道的大小、形状、分布及相互连通关系。研究区须二段储层孔隙类型为残余粒间孔、次生孔隙(粒间溶孔、粒内溶孔、铸模孔、杂基内微孔)及微裂缝,可见到微裂缝溶蚀扩大形成的带状孔隙,发育片状喉道、弯状喉道及管束状喉道(图3、表1、表2)。

— 155 —

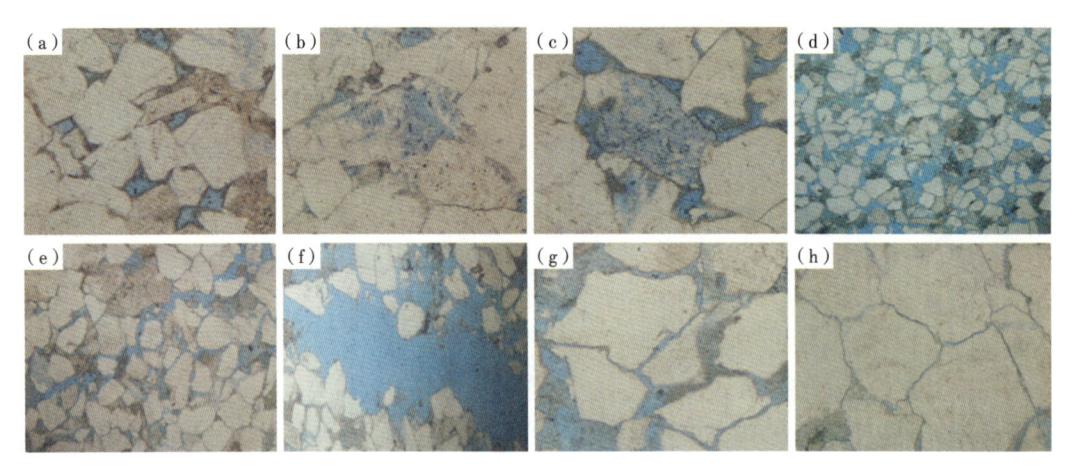

图 3 合川气田须二段砂岩储层显微特征

(a)残余原生粒间孔,三角状、多角状,孔隙连通性差,TN105 井,2227m,单偏光,×50;(b)粒内溶孔,沿长石解理溶蚀,TN107 井,2246m,单偏光,×50;(c)粒内溶孔,粒间杂基溶孔,不规则状、港湾状,颗粒边缘见溶蚀残余,TN105 井,2185m,单偏光,×50;(d)粒间溶孔发育,孔隙连通性好,TN101 井,2228m,单偏光,×50;(e)微裂缝,沟通粒间孔,TN104 井,2215m,单偏光,×50;(f)裂缝溶蚀扩大,形成条带状孔隙,TN101 井,2231m,×50;(g)粒间溶孔、残余原生粒间孔,片状和弯状喉道,TN101 井,2228m,×100;(h)管束状喉道,TN104 井,2193m,×100

表 1 合川气田须二段储层孔隙类型划分表

| 孔隙类型 | 形成机理 | 孔隙基本形状 | 发育程度 |
| --- | --- | --- | --- |
| 残余粒间孔 | 沉积作用 | 三角状、多角状、不规则 | 发育 |
| 微孔隙 | | 受层段控制、形状不规则 | 少 |
| 粒间溶孔 | 溶蚀作用 | 形态多样、多角状、不规则状、港湾状,位于颗粒之间 | 发育 |
| 粒内溶孔 | | 形态多样,多为长石溶孔,少量为岩屑粒内溶孔 | 发育 |
| 铸模孔 | 溶蚀作用 | 形态多样,零星分布 | 少 |
| 杂基内微孔 | 重结晶作用 | 多呈斑状分布,孔隙细小 | 少 |
| 微裂缝 | 差异压实 | 线状,切割颗粒,溶蚀扩大可形成条带状孔隙 | 较发育 |

表 2 合川气田须二段岩心铸体薄片面孔率统计表

| 孔隙度(%) || 样品数(个) | 粒间孔(%) || 粒内溶孔(%) | 铸模孔(%) | 微裂缝 | 平均总面孔率(%) |
| --- | --- | --- | --- | --- | --- | --- | --- | --- |
| 区间 | 平均 | | 原生孔 | 溶蚀孔 | | | | |
| <6 | 5.09 | 59 | — | 3.91 | 1 | 1 | 0.11 | 3.95 |
| 6~9 | 6.78 | 31 | — | 5.41 | 0.88 | 1.25 | 0.15 | 5.68 |
| 9~12 | 10.36 | 18 | 4.60 | 6.69 | 1.50 | — | 0.20 | 8.57 |
| ≥12 | 15.05 | 29 | 3.17 | 13.59 | 1.31 | 1 | 0.25 | 11.80 |

## 2 优质储层发育的主要控制因素

致密砂岩储层的大量研究实例表明,优质储层形成控制因素主要包括层序地层界面、有利沉积相带、次生溶蚀及裂缝发育等[5-8]。沉积作用为储层发育提供物质基础,对储集岩原始孔隙的形成具有控制作用。一般而言,低能薄层砂体,易于遭受强烈压实和碳酸盐胶结,多形成致密砂岩,而三角洲前缘高能厚层河道砂体,抗压实能力强,颗粒粗、分选好,易于保留较多的残余粒间孔隙,且利于孔隙水流动,形成粒间溶孔发育的相对高孔隙度渗透率储层[9]。沉积砂体能否最终成为有效储集体,其在埋藏历史中经历的成岩作用非常关键。成岩作用改造原始孔隙并最终决定现今储层的好坏以及储层的分布状况[10-16]。成岩相是控制储层质量的重要因素之一。基于铸体薄片鉴定、扫描电镜分析等手段,研究储层的成岩特征与孔隙结构,开展单因素成岩相分析和成岩—储集相组合类型分析,对成岩相进行定量表征[17-19]。

## 2.1 层序界面对储层形成的控制作用

川中地区须家河组与下伏的雷口坡组石灰岩呈不整合接触,在雷口坡组侵蚀面上发生填平补齐,沉积了须一段暗色泥岩。在雷口坡组古残丘之上,须一段厚度薄,少数钻井缺失须一段。须二段分布较稳定,厚度在80~140m之间,平面表现为东南薄西北厚的分布格局,其岩性组合为叠置的多套厚层砂岩夹薄层的深灰色泥岩或煤线。

对川中须家河组进行层序地层划分对比,研究层序格架中的砂岩储层分布规律[20,21]。依据高分辨率层序地层学原理,基于测井资料、岩心描述、露头观测研究成果,将须二段划分成两个半旋回、五个砂层组:X2-1砂层组为基准面上升半旋回,水进过程,可容纳空间增加,自下而上岩性由粗变细;X2-2砂层组为基准面下降半旋回,水退过程,可容纳空间减少,自下而上岩性由细变粗。二者构成一个完整的基准面上升—下降旋回。同样,X2-3砂层组和X2-4砂层组构成一个完整的基准面上升—下降旋回。X2-5砂层组为基准面上升半旋回,与须三段湖相泥岩连续沉积。

合川气田须二段纵向多层,一般钻遇气层为4~10层,单层厚度平均3~5m。须二段已获工业气流或低产气流试气井段的统计结果显示,气层主要分布在层序界面之上,在基准面上升的早期可容空间小,厚层的中粒—中粗粒河道砂体分布范围广,侧向连通性好,有利于气藏的形成。最有利储层主要发育于须二段下部的X2-1砂层组,其次为X2-3砂层组,二者均为基准面上升半旋回沉积(图4)。

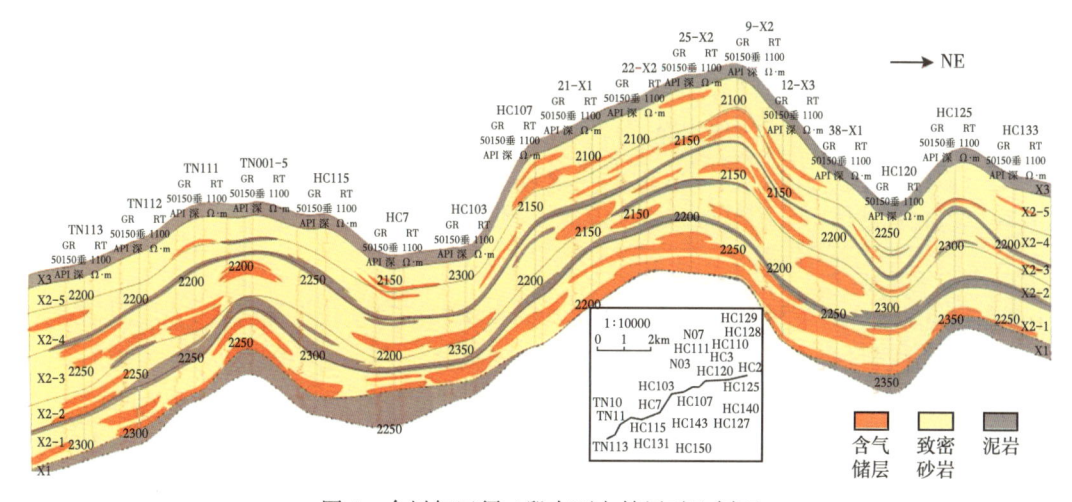

图4 合川气田须二段东西向储层对比剖面

## 2.2 沉积作用对储层形成的控制作用

### 2.2.1 有利沉积微相

研究表明,川中合川气田须二段以三角洲前缘水下分流河道砂体物性最好,河口坝砂体物性较好(表3),而远沙坝、滩坝砂体的物性较差,一般不发育效储层。

表3 合川气田须二段储层沉积微相与物性特征

| 井号 | 井深(m) | 测井曲线形态 | 储层厚度(m) | 孔隙度(%) | 渗透率(mD) | 沉积微相 | 测试成果 |
|---|---|---|---|---|---|---|---|
| HC1 | 2110~2162 | 微齿箱形 | 12.0 | 9.07 | 0.1 | 水下分流河道 | 气:4.44×10$^4$m$^3$/d |
| | 2162~2172 | 齿化漏斗形 | 10.0 | 8.4 | 0.1 | 河口坝 | 未试油 |
| HC3 | 2106~2119 | 微齿箱形 | 4.5 | 8.59 | 0.1 | 水下分流河道 | 气:6.71×10$^4$m$^3$/d |
| | 2134~2141 | 齿化漏斗形 | 6.0 | 10.1 | 0.4 | 河口坝 | 水:3.3m$^3$/d |
| HC106 | 2190~2193.3 | 箱形 | 2.9 | 9.5 | 0.2 | 水下分流河道 | 气:7.1×10$^4$m$^3$/d |
| | 2194.7~2200 | 箱形 | 4.8 | 9.5 | 0.1 | 水下分流河道 | |
| HC109 | 2255~2284 | 齿化箱形 | 14.2 | 9.7 | 0.2 | 水下分流河道 | 气:12.9×10$^4$m$^3$/d |
| | 2180.8~2218.5 | 齿化漏斗形 | 2.7 | 7.9 | 0.1 | 河口坝 | 气:1.3×10$^4$m$^3$/d |
| TN1 | 2190.5~2194 | 箱形 | 3.5 | 6.7 | 0.1 | 水下分流河道 | 气:5.2×10$^4$m$^3$/d |
| | 2247.5~2250.5 | 箱形 | 3.0 | 7.3 | 0.1 | 水下分流河道 | |
| TN101 | 2231.5~2237 | 微齿箱形 | 5.2 | 10.4 | 0.4 | 水下分流河道 | 气:4.1×10$^4$m$^3$/d |
| | 2243.7~2252 | 微齿箱形 | 6.8 | 8.6 | 0.3 | 水下分流河道 | |
| TN111 | 2209.6~2217.1 | 微齿箱形 | 7.5 | 7.2 | 0.1 | 水下分流河道 | 气:10.7×10$^4$m$^3$/d |
| | 2218~2223 | 齿化漏斗形 | 4.2 | 7.8 | 0.2 | 河口坝 | |
| | 2225.6~2233.7 | 箱形 | 8.1 | 13.3 | 1.7 | 水下分流河道 | |

由于沉积相演化序列及砂体成因的差异,可划分出水下分流河道叠置、河口坝叠置以及水下分流河道—河口坝叠置等砂体组合类型。储层物性及含气性均反映出水下分流河道—河口坝叠置砂体组合最优,其复合砂体的厚度大,发育中粒、中粗粒、中细粒长石石英砂岩、岩屑长石石英砂岩,砂岩粒度适中,淘洗干净、分选好,储层物性优良;水下分流河道叠置砂体组合的厚度较大,物性较好,但在该类组合上部的单层砂体厚度变薄、粒度变细、泥质含量增高,物性变差;河口坝叠置砂体组合的单层砂体厚度薄、以细砂岩为主,储层较致密。

### 2.2.2 沉积物组分对储层物性的影响

沉积物组分对储层物性的影响主要表现为碎屑颗粒(石英、长石和岩屑)和填隙物对储层物性的影响。统计结果表明,孔隙度随着石英含量的增加而变大,随着岩屑含量的增加而降低(图5、图6)。在孔隙度不小于6%的有效储层中,孔隙度与石英含量、岩屑含量的相关性显著;但在孔隙度小于6%的致密砂岩中,石英含量、岩屑含量与孔隙度的相关性差,因为除了碎屑颗粒组分外,颗粒粒度及胶结作用影响砂岩物性,即使同样为石英砂岩,晚期的碳酸盐胶结也可使其强烈致密化。石英为刚性颗粒,起骨架支撑和抗压实作用,而软岩屑呈塑性,易被压实,造成孔隙损失。长石含量与孔隙度的相关性不明显,其原因可能是:少量长石可以形成溶蚀孔隙,在酸性地层水条件下,长石易于沿解理发生溶蚀,甚至形成铸模孔;而长石易发生高岭石化,其塑性较强,当长石含量太高时可能堵塞孔隙,而不利于地层水的交换和溶蚀作用的发生。因而区内有利储层为长石石英砂岩和岩屑长石石英砂岩,而岩屑长石砂岩一般物性较差。另外,有效储层中填隙物含量一般为5%~10%,填隙物含量太高将堵塞孔隙,导致孔隙度的下降。

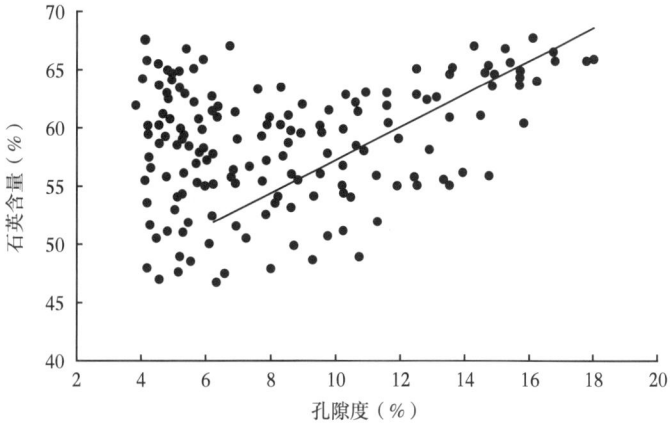

图 5 石英含量与孔隙度交会图

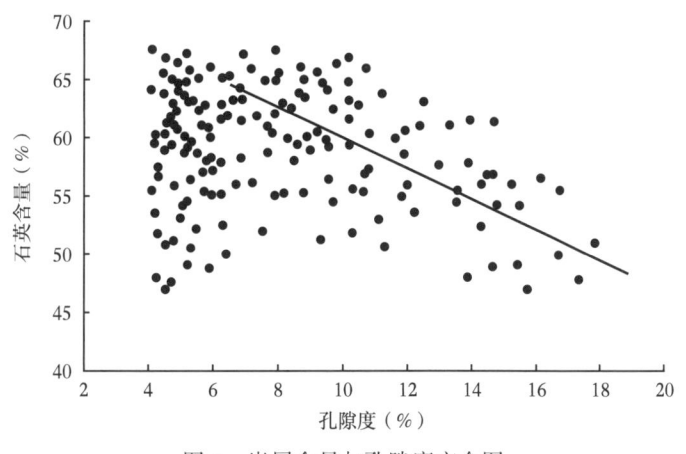

图 6 岩屑含量与孔隙度交会图

## 2.3 成岩作用对储层形成的控制作用

川中上三叠统须家河组具有埋藏深度大(一般2200~2400m)、深埋时间长、成岩演化程度高(干酪根镜质组反射率 $R_o$ 大多在1.3%~2%之间)的特点。通过大量的岩石薄片显微观察,结合前人研究和其他成岩阶段划分标志,发现目前须家河组砂岩储层整体处于晚成岩阶段 B期。在早成岩 A 期,压实作用最为明显,绿泥石和方解石发育,此时孔隙度较大,但随压实而迅速降低,此时孔隙类型为原生孔隙,成岩环境为弱碱性;早成岩 B 期是一个过渡阶段,压实作用继续进行,高岭石、石英次生加大、溶蚀作用开始出现,成岩环境由弱碱性向酸性发展;晚成岩 A1 期，压实作用由强变弱，高岭石、石英次生加大和溶蚀作用大量发育，为次生孔隙的主要形成时期；晚成岩 A2 期，压实作用弱，高岭石、石英次生加大和溶蚀作用逐渐减弱，孔隙度降低快；晚成岩 B 期，成岩环境为碱性环境，压实作用由弱变强，颗粒接触类型为线—凹凸接触，铁方解石和铁白云石大量发育，碳酸盐胶结造成了大量的孔隙损失。

由大量铸体薄片、扫描电镜照片观察可知，合川气田须二段储层经历了压实、压溶、胶结等破坏性成岩作用，也经历了溶蚀、破裂等建设性成岩作用(图7)，最终形成残余粒间孔、粒间溶孔、粒内溶孔、杂基内微孔及微裂缝等储集空间。

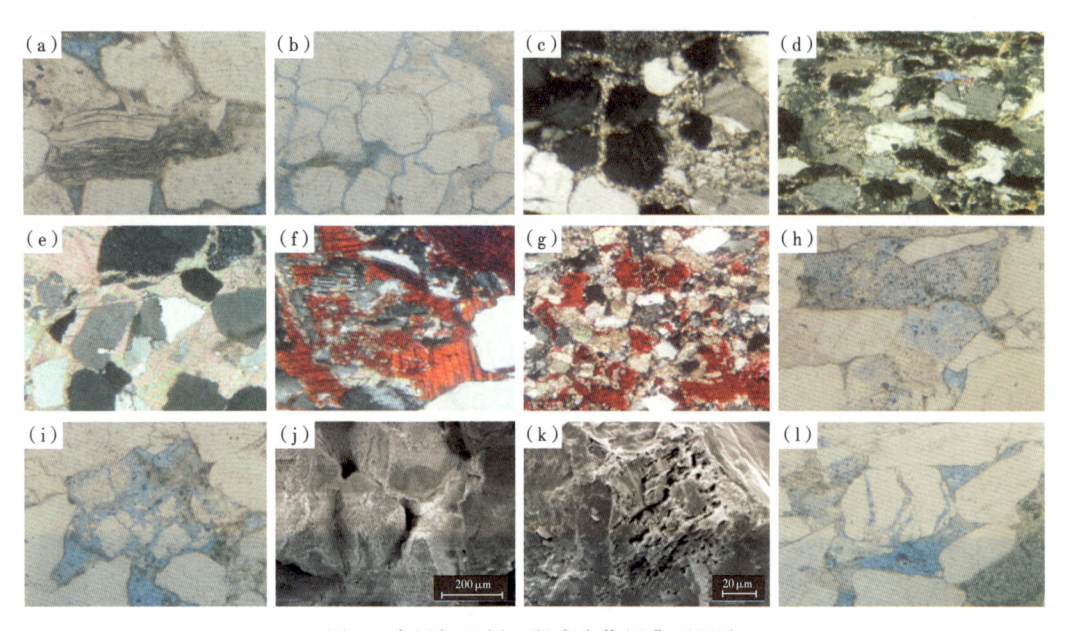

图 7 合川气田须二段成岩作用典型照片

(a)压实作用，塑性岩屑变形，颗粒凹凸接触，TN107 井，2256m，单偏光，×50；(b)压溶作用，石英次生加大，颗粒呈线接触、凹凸接触，TN105 井，2231m，单偏光，×50；(c)黏土矿物充填粒间，HC3 井 2070m，正交偏光，×50；(d)黏土矿物在颗粒边缘胶结，HC5 井 2217.91m，正交偏光，×50；(e)碳酸盐胶结及重结晶，连晶方解石使岩石完全致密化，HC5 井，2212.77m，正交偏光，×50；(f)铁方解石交代长石，HC5 井，2213.73m，正交偏光，×50；(g)方解石和铁方解石交代岩屑，HC3 井，2072m，正交偏光，×50；(h)岩屑溶蚀、杂基溶蚀，形成蜂窝状溶孔，TN105 井，2233m，单偏光，×50；(i)颗粒碎裂，粒内溶孔，杂基溶蚀形成的粒间溶孔，绿泥石衬边发育，TN104 井，2293m，单偏光，×50；(j)绿泥石衬边，残余原生粒间孔，HC106 井，2193.43m，扫描电镜；(k)长石溶蚀孔隙，HC1 井，2158.50m，扫描电镜；(l)粒内裂缝孔，TN104 井，2291m，单偏光，×50

### 2.3.1 压实作用未完全破坏砂岩骨架，保留了部分原生粒间孔

对研究区不同类型孔隙面孔率与孔隙度之间关系分析表明，粒间孔面孔率与孔隙度呈明显正相关关系(图8)，说明砂岩骨架并未因压实作用遭受完全破坏，仍保留了部分原生粒间孔隙，而残余粒间孔是地层水交换和粒间溶孔形成之基础。

### 2.3.2 长石、岩屑和杂基的溶蚀作用是形成次生溶孔，改善储层物性的主要原因

在须家河组砂岩中，长石和岩屑是主要的易溶组分，但并不是二者含量越高孔隙性越好，长石含量介于 15%~25% 的样品孔隙性较好，溶蚀作用使得孔隙性变好时要有一定量的石英

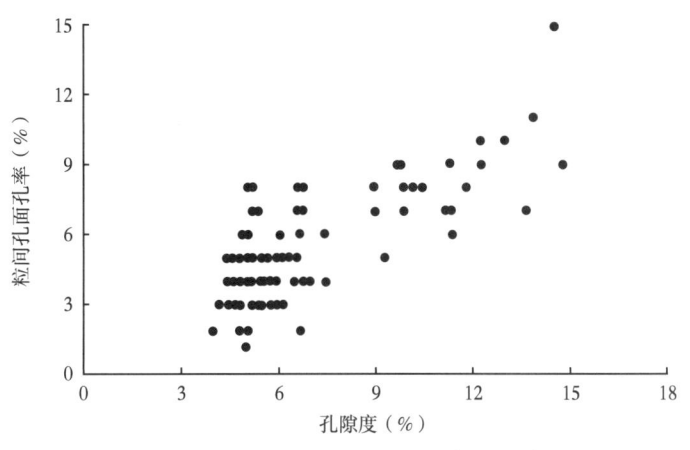

图8 合川气田须二段砂岩粒间孔面孔率与孔隙度关系图

作为岩石的支撑骨架,否则受溶蚀后压实作用影响,岩石孔隙并没有显著增加。

黏土杂基也常被溶蚀,并且在黏土充填物含量较小时容易发生,溶孔呈不规则形状。若黏土杂基过分充填则造成地层水的交换困难,不利于溶蚀孔隙的形成。高岭石交代也可形成一些高岭石晶间孔,但它是一种微孔,在面孔率中所占比例较低。

长石、岩屑和杂基的溶蚀孔隙主要发育在2060~2250m的深度段,溶蚀作用发生在晚成岩A期的酸性成岩环境,与须家河组煤系地层生烃高峰同期或略晚于生烃高峰期,当时的古地温为110~130℃,干酪根镜质组反射率$R_o$在0.7%~1.3%之间。

### 2.3.3 绿泥石含量与孔隙度、渗透率呈正相关,绿泥石衬边的形成是一种建设性成岩作用

前人针对绿泥石胶结物的成因及其与砂岩储层的储集性能的关系进行过大量的研究。鄂尔多斯盆地延长组砂岩储层研究成果认为,早期绿泥石衬边胶结作用有利于粒间孔隙的保存,与储集性能关系密切[9,22,23]。四川盆地须家河组砂岩中也发育绿泥石衬边,其形成对储层原生孔隙的保存起到了积极的作用,它主要通过抑制石英在颗粒表面成核的数量,来抑制硅质胶结物在孔隙内的生长[24-26]。

合川气田须二段储层研究成果表明,绿泥石含量与孔隙度、渗透率呈正相关(图9),早期

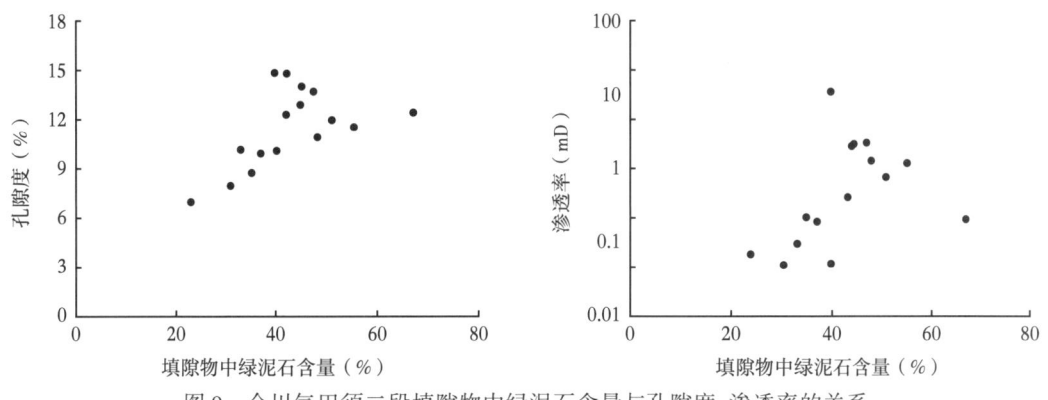

图9 合川气田须二段填隙物中绿泥石含量与孔隙度、渗透率的关系

绿泥石衬边一方面增强了砂岩储层的抗压强度,另一方面抑制了硅质胶结和石英加大边的形成,较多的原生粒间孔隙得以保存。

### 2.3.4 裂缝发育可以大幅度提高储层的渗透性

川中地区雷口坡组侵蚀面之上发育数量较多的古残丘,在沉积作用和差异压实作用下,可在古残丘形成须家河组的原始倾斜和尖灭,并产生同沉积期小断层和裂缝[27]。岩心观察和薄片鉴定表明,研究区须二段储层裂缝为微裂缝,这种微裂缝的形成主要与雷口坡古残丘之上须二段砂岩遭受差异压实作用有关,形成于同沉积期和成岩期,具有成岩缝的特点。该区断裂、褶皱等后期构造作用弱,构造裂缝不太发育。

通过储层横向展布特征研究发现,虽然须二段储层分布在三角洲前缘水下分流河道、河口坝发育区,但在古残丘和沉积微相有利叠合区,即合川构造主体街子坝构造储层最为发育,砂岩储层累计厚度多大于30m,物性较好,Ⅰ+Ⅱ类储层所占比例达50%。须家河组下伏地层为雷口坡组古残丘,尤其是构造主体部位须一段要么缺失要么较薄,最厚不超过5m,古残丘之上须二段砂岩储层发育差异压实形成的微裂缝。统计表明,尽管微裂缝对孔隙度的贡献不大(裂缝面孔率一般为0.1%~0.25%),但裂缝渗透率远大于基质渗透率,对改善储层渗透性具有重要作用。

## 3 沉积—成岩耦合与储层"甜点"预测

### 3.1 成岩—储集相组合类型

成岩—储集相是描述影响储层性质的某种或某几种成岩作用和特有的储集空间的组合。可以选择影响储层孔隙发育的主要成岩作用类型及相应参数,进行成岩—储集相的划分[28]。根据主要成岩作用和孔隙度、渗透率等储层物性参数及孔隙特征,可将须二段储层划分为7类成岩—储集相,即Ⅰ类储层可划分为绿泥石衬边—强溶蚀相、微裂缝—强溶蚀相;Ⅱ类储层可划分为绿泥石衬边—中溶蚀相、弱杂基充填—中溶蚀相和微裂缝—中溶蚀相三种组合类型;Ⅲ类储层可划分为压实—碳酸盐胶结相和压实—高岭石胶结相两种组合类型(表4)。

表4 须二段成岩—储集相组合类型

| 储层分类 | 成岩—储集相组合类型 | 孔隙度(%) | 渗透率(mD) | 孔隙特征 |
| --- | --- | --- | --- | --- |
| Ⅰ | Ⅰa 绿泥石衬边—强溶蚀相 | 12.02~15.25 (13.22) | 0.56~6.13 (2.18) | 粒间溶孔为主,其次为残余粒间孔,绿泥石衬边发育,溶孔面孔率8%~12%,残余粒间孔面孔率2%~4% |
|  | Ⅰb 微裂缝—强溶蚀相 | 11.85~18.15 (14.05) | 1.25~20.28 (4.36) | 粒间溶孔、粒内溶孔,裂缝溶蚀扩大,形成带状孔隙,溶孔面孔率8%~12%,残余粒间孔面孔率2%~4%,裂缝面孔率0.25% |
| Ⅱ | Ⅱa 绿泥石衬边—中溶蚀相 | 9.05~11.78 (10.62) | 0.20~2.13 (0.54) | 粒间溶孔、残余粒间孔,绿泥石衬边发育,溶孔面孔率5%~7%,残余粒间孔面孔率2%~4% |

续表

| 储层分类 | | 成岩—储集相组合类型 | 孔隙度(%) | 渗透率(mD) | 孔隙特征 |
|---|---|---|---|---|---|
| Ⅱ | Ⅱb | 弱杂基充填—中溶蚀相 | 9.02~11.64 (10.16) | 0.21~2.21 (0.52) | 残余粒间孔、粒间溶孔、杂基内微孔、高岭石晶间孔,溶孔面孔率5%~7%,残余粒间孔面孔率2%~4% |
| | Ⅱc | 微裂缝—中溶蚀相 | 8.95~10.68 (9.87) | 0.32~5.21 (0.76) | 粒间、粒内溶孔、残余粒间孔、裂缝发育,溶孔面孔率5%~7%,残余粒间孔面孔率2%~4%,裂缝面孔率0.2% |
| Ⅲ | Ⅲa | 压实—碳酸盐胶结相 | 5.92~8.18 (6.55) | 0.01~0.20 (0.14) | 残余粒间孔,粒内溶孔,溶孔面孔率3%~4%,残余粒间孔面孔率1%~2% |
| | Ⅲb | 压实—高岭石胶结相 | 6.03~8.50 (7.26) | 0.02~0.26 (0.19) | 残余粒间孔,高岭石晶间孔,杂基溶孔,溶孔面孔率3%~4%,残余粒间孔面孔率1%~2% |

## 3.2 沉积相—成岩相耦合

水下分流河道和河口坝处于高能相带,经过强烈的筛选、磨蚀和搬运,形成结构成熟度和成分成熟度均较高的中粒、中细粒岩屑长石石英砂岩。由于其具有较好的颗粒支撑结构,其抗压实能力较强,保存了较多的残余原生粒间孔隙,有利于地层水交换和次生溶蚀孔隙的形成。有利沉积微相和砂岩的初始物性是储层形成的基础,除非遇到特殊的成岩改造作用,如强烈的晚期碳酸盐胶结作用可使原生孔隙丧失殆尽,而裂缝发育可促进地层水交换,沿裂缝发生溶蚀扩大,并沟通孤立孔隙,改善储层的渗透性。将有利沉积相带与有利成岩相带结合,可以更加客观地评价和预测优质储层分布[29,30]。

图10为研究区须二段X2-1小层沉积相—成岩相耦合结果,从中可以看出:(1)合川区块发育由东部、东北部物源控制的河流—三角洲沉积体系,潼南区块发育由南部、西南部物源控制的河流—三角洲沉积体系,储层主要发育于三角洲前缘水下分流河道、河口坝及其叠置砂体中,有利储层受主河道部位(TN116-TN110-TN105、HC129-HC110-HC1-HC106、HC133-HC125-HC120-HC108)及河道汇流区(合川中部的HC1、HC3、HC108井区)控制;(2)Ⅰ类储层分布于水下分流河道—河口坝叠置砂体与绿泥石衬边—强溶蚀相或微裂缝—强溶蚀相耦合区域,绿泥石衬边—强溶蚀相呈片状分布,范围较大,而微裂缝—强溶蚀相分布与古残丘有关;(3)Ⅱ类储层分布于Ⅰ类储层之外围,溶蚀强度中等;(4)Ⅲ类储层分布于河道侧翼、河口坝远端及远沙坝,岩性偏细,压实作用较强,并发育碳酸盐胶结和高岭石胶结。

## 3.3 储层"甜点"预测

沉积相是储层形成和演化的物质基础。成岩相是储层改造和定型的关键因素。除了这两个因素以外,储层"甜点"评价还需考虑储层的物性(孔隙度、渗透率)、含气性(含气饱和度、储量丰度、测试产量等)、有效储层规模(有效厚度、分布范围、储能系数)。综合考虑沉积相、成岩相、孔隙度、有效厚度(有效储层下限参数:孔隙≥6%,动态法确定含气饱和度≥47%)、天然气测试产量等因素,按专家打分法对每个因素进行评价(评分范围2~10分),依据其重要程

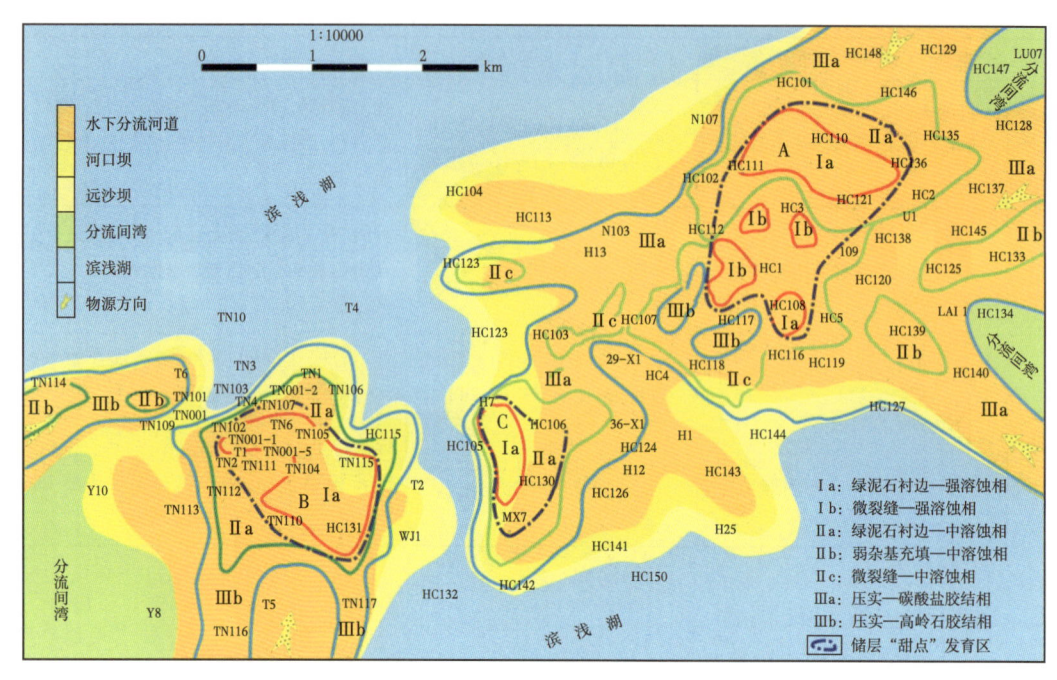

图 10　合川气田须二段 X2-1 小层沉积相—成岩相耦合与储层"甜点"预测

度给他们赋予不同的权重,建立了合川气田须二段致密砂岩储层"甜点"评价标准(表5)。沉积相、成岩相为储层发育的重要控制因素,赋予的权重均为 0.25。测井解释孔隙度作为储层"甜点"评价的重要指标也被赋予 0.25 的权重。勘探开发实践表明,须二段具有厚砂薄储的特点,天然气测试产量的影响因素复杂,与有效储层厚度关系不太密切,分别被赋予 0.15 和 0.1 的权重。储层"甜点"评价得分=沉积相评分×0.25+成岩相评分×0.25+孔隙度评分×0.25+有效厚度评分×0.1+天然气测试产量评分×0.15。

表 5　须二段储层"甜点"评价标准

| 评价指标 | | 评分 | 权重 |
|---|---|---|---|
| 沉积相 | 水下分流河道—河口坝叠置 | 10 | 0.25 |
| | 水下分流河道叠置 | 10 | |
| | 河口坝叠置 | 6 | |
| | 远沙坝或滩坝 | 2 | |
| 成岩相 | 绿泥石衬边—强溶蚀相 | 10 | 0.25 |
| | 裂缝—强溶蚀相 | 10 | |
| | 绿泥石衬边—中溶蚀相 | 6 | |
| | 弱杂基充填—中溶蚀相 | 6 | |
| | 裂缝—中溶蚀相 | 6 | |
| | 压实—碳酸盐胶结相 | 2 | |
| | 压实—高岭石胶结相 | 2 | |

续表

| 评价指标 | | 评分 | 权重 |
|---|---|---|---|
| 孔隙度<br>(%) | ≥12 | 10 | 0.25 |
| | 9~12 | 8 | |
| | 6~9 | 6 | |
| | <6 | 2 | |
| 有效砂岩厚度<br>(m) | ≥20 | 10 | 0.10 |
| | 15~20 | 8 | |
| | 10~15 | 6 | |
| | <10 | 2 | |
| 测试产量<br>($10^4 m^3/d$) | >10 | 10 | 0.15 |
| | 4~10 | 8 | |
| | 1~4 | 6 | |
| | <1 | 2 | |

依据合川气田须二段的沉积相、成岩相分布格局、耦合关系、储层分布及天然气测试产量，初步筛选出三个储层发育有利区(图10)。有利区 A 为 HC110-HC3-HC1 井区，位于合川构造主体的街子坝构造，雷口坡古残丘之上须二段裂缝发育，位于水下分流河道交汇区与微裂缝—强溶蚀相、绿泥石衬边—强溶蚀相叠合区块，孔隙度一般为9%~15%，储层厚度20~25m，天然气测试产量(4.4~14.2)×$10^4 m^3/d$；有利区 B 位于潼南区块中部的 TN105-TN104-TN110 井区，属于水下分流河道—河口坝叠置砂体与绿泥石衬边—强溶蚀相叠合区块，孔隙度一般为9%~13%，储层厚度10~15m，天然气测试产量(3.0~10.7)×$10^4 m^3/d$；有利区 C 为合川区块的 HC7-HC130 井区，位于三角洲前缘河口坝叠置砂体与绿泥石衬边—中、强溶蚀相叠合区块，孔隙度一般为9%~12%，储层厚度10~15m，天然气测试产量(1.2~7.1)×$10^4 m^3/d$。

基于各个井点的测井解释结果(孔隙度、有效储层厚度)、测试产量和沉积相、成岩相划分结果，进行储层"甜点"评价。评价结果表明(表6)，有利区 A 储层"甜点"评价得分9.525，有利区 B 得分8.850，有利区 C 得分7.150。根据储层规模和开发效果，优选有利区 A(HC110-HC3-HC1 井区)和有利区 B(TN105-TN104-TN110 井区)为重点开发区块，有利区 C(HC7-HC130 井区)作为开发接替区块。

表6 合川气田须二段 X2-1 小层储层"甜点"评价结果

| 有利区 | 沉积相 | 成岩相 | 孔隙度(%) | 有效厚度(m) | 测试产量($10^4 m^3/d$) | 总得分 |
|---|---|---|---|---|---|---|
| 有利区 A | 10 | 9 | 10 | 10 | 8.5 | 9.525 |
| 有利区 B | 10 | 9.2 | 9 | 6 | 8 | 8.850 |
| 有利区 C | 6 | 8.6 | 8 | 6 | 6 | 7.150 |

# 4 结论

(1)将须二段划分为两个半旋回、五个砂层组，建立了川中地区须二段的高分辨率层序地层格架。

（2）合川气田须二段储层为中粒、中细粒岩屑长石石英砂岩，发育残余粒间孔、粒间溶孔、粒内溶孔、杂基内微孔和微裂缝，具有低孔隙度、致密特征，局部发育相对高孔隙度渗透率储层。

（3）通过宏观与微观分析相结合，明确了该区砂岩储层形成的主要控制因素：①优质储层主要分布在层序界面之上的基准面上升的早期，须二段下部的 X2-1 砂层组储层最发育，连续性最好；②有利沉积微相为水下分流河道及河口坝，主要储集体为水下分流河道叠置砂体和水下分流河道—河口坝叠置砂体；③长石、岩屑和杂基的溶蚀作用是形成次生溶孔，改善储层物性的主要原因；④早期绿泥石衬边胶结可以提高砂岩的抗压强度并抑制石英加大边的形成，使得较多的原生粒间孔隙得以保存；⑤雷口坡古残丘顶部须二段储层裂缝发育，大幅改善储层的渗透能力。

（4）归纳出 7 种成岩—储集相组合类型，基于沉积相—成岩相耦合，结合储层物性、有效厚度、测试产量等指标，进行储层"甜点"定量评价，筛选出 3 个储层"甜点"发育区。

## 参 考 文 献

[1] 田继军,姜在兴,李熙喆,等. 川西前陆盆地上三叠统层序地层学研究[J]. 天然气工业,2008,28(2):30-33.

[2] 施振生,杨威. 四川盆地上三叠统砂体大面积分布的成因[J]. 沉积学报,2011,29(6):1058-1068.

[3] 田继军,姜在兴,李熙喆,等. 川西前陆盆地上三叠统岩性地层圈闭勘探前景分析[J]. 油气地质与采收率,2009,16(1):22-25.

[4] 谢武仁,杨威,李熙喆,等. 四川盆地上三叠统砂岩储层特征研究[J]. 天然气地球科学,2008,29(5):623-629.

[5] 杨晓萍,赵文智,邹才能,等. 川中气田与苏里格气田甜点储层对比研究[J]. 天然气工业,2007,27(1):4-7.

[6] 杨晓萍,赵文智,邹才能,等. 低渗透储层成因机理及优质储层形成与分布[J]. 石油学报,2007,28(4):57-61.

[7] 李熙喆,张满郎,谢武仁,等. 鄂尔多斯盆地上古生界层序格架内的成岩作用[J]. 沉积学报,2007,25(6):923-933.

[8] 孙雨,于海涛,马世忠,等. 致密砂岩储层物性特征及其控制因素——以松辽盆地大安地区白垩系泉头组四段为例[J]. 中国矿业大学学报,2017,46(4):809-819.

[9] 周勇,徐黎明,纪友亮,等. 致密砂岩相对高渗储层特征及分布控制因素研究——以鄂尔多斯盆地陇东地区延长组长 82 为例[J]. 中国矿业大学学报,2017,46(1):106-120.

[10] 朱如凯,邹才能,张鼐,等. 致密砂岩气藏储层成岩流体演化与致密成因机理——以四川盆地上三叠统须家河组为例[J]. 中国科学(D 辑:地球科学),2009,39(3):327-339.

[11] 林承焰,王文广,董春梅,等. 储层成岩数值模拟研究现状及进展[J]. 中国矿业大学学报,2017,46(5):1084-1101.

[12] Abercrobie H J, Hutcheon I E, Blochu J D, et al. Silica activity and the smectite-illite reaction [J]. Geology, 1994, 22(6): 539-542.

[13] Karen E H, Horst Z, Agnes G R, et al. Diagenesis, porosity evolution, and petroleum emplacement in tight gas reservoirs, Taranaki basin, New Zealand[J]. Journal of Sedimentary Research, 2007, 77(12):1003-1025.

[14] Geoffrey T, Bernard P B, Mogens R, et al. Simulation of potassium feldspar dissolution and illitization in the Statfjord formation, north Sea[J]. AAPG Bulletin, 2001, 85(4) : 621-635.

[15] Lander R H, Walderhang O. Predicting porosity through simulating sandstone compaction and quartz cementation[J]. AAPG Bulletin, 1999, 83(3) : 433-449.

[16] Tobin R C, Mcclain T, Lieber R B, et al. Reservoir quality modeling of tight-gas sands in Wamsutter field: Integration of diagenesis, petroleum systems, and generation data [J]. AAPG, 2010, 94(8) : 1229-1266.

[17] 杜业波,季汉成,吴因业,等. 前陆层序致密储层的单因素成岩相分析[J]. 石油学报,2006,27(2):48-52.

[18] 陈桂菊,姜在兴,田继军,等. 成岩相对磨溪气田上三叠统致密储层的控制作用[J]. 大庆石油地质与开发,2007,26(2):41-45.

[19] 马宝全,杨少春,张鸿,等. 基于DEA定量表征低渗透砂岩储层成岩相——以鄂尔多斯盆地演武地区延长组81段为例[J]. 中国矿业大学学报,2018,47(2):357-366.

[20] 田继军,姜在兴,李熙喆,等. 川中上三叠统须二段厚层砂岩的成因及其对储层、气藏的控制[J]. 大庆石油地质与开发,2008,27(2):34-37.

[21] 谢武仁,李熙喆,张满郎,等. 川中地区上三叠统须四段厚层砂体成因及油气运移通道分析[J]. 石油学报,2008,29(4):504-508.

[22] 姚泾利,王琪,张瑞,等. 鄂尔多斯盆地华庆地区延长组长6砂岩绿泥石膜的形成机理及其环境指示意义[J]. 沉积学报,2011,29(1):72-79.

[23] 丁晓琪,张哨楠,葛鹏莉,等. 鄂南延长组绿泥石环边与储集性能关系研究[J]. 高校地质学报,2010,16(2):247-254.

[24] 孙治雷,黄思静,张玉修,等. 四川盆地须家河组砂岩储层中自生绿泥石的来源与成岩演化[J]. 沉积学报,2008,26(3):459-468.

[25] 刘金库,彭军,刘建军,等. 绿泥石环边胶结物对致密砂岩孔隙的保存机制——以川中—川南过渡带包界地区须家河组储层为例[J]. 石油与天然气地质,2009,30(1):53-58.

[26] 孙全力,孙晗森,贾趵,等. 川西须家河组致密砂岩储层绿泥石成因及其与优质储层关系[J]. 石油与天然气地质,2012,33(5):751-757.

[27] 朱仕军,黄继祥,李先荣,等. 川中—川南过渡带古残丘的地震解释及其对香溪群油气藏的控制作用[J]. 西南石油学院学报,1994,16(4):7-10.

[28] 邹才能,陶士振,周慧,等. 成岩相的形成、分类与定量评价方法[J]. 石油勘探与开发,2008,35(5):526-540.

[29] 李熙喆,张满郎,周兆华,等. 河包场地区须家河组储层有利区带预测[J]. 天然气工业,2009,29(9):24-27.

[30] 季汉成,翁庆萍,杨潇. 鄂尔多斯盆地安塞——神木地区山西组成岩与沉积相耦合关系[J]. 石油勘探与开发,2009,36(6):709-717.

# 致密砂岩气藏储渗单元研究方法与应用
## ——以鄂尔多斯盆地二叠系下石盒子组为例

郭建林　贾成业　闫海军　季丽丹　李易隆　袁　贺

(中国石油勘探开发研究院)

**摘要**：精细表征储层特征和储层结构是致密砂岩气藏开发中后期的主要技术需求。基于不同类型砂体的相似孔隙度、渗透率特征将辫状河沉积体系中河道充填和心滩砂体聚类为储渗单元，提出了储渗单元研究概念和研究思路，开展了辫状河沉积体系储渗单元发育模式研究。通过露头观测和测井相标志，识别出辫状河沉积体系中储渗单元发育心滩叠置型、河道充填叠置型、心滩和河道充填叠置型、心滩或河道充填孤立型四种储渗单元叠置模式；基于储渗单元发育模式，提出了河流相致密砂岩气藏开发井型的适应性，指出辫状河沉积体系中河道叠置带是叠置型储渗单元发育的有利部位，是水平井开发的有利目标，辫状河沉积体系中的过渡带和洼地主要发育孤立型储渗单元，适合直井或丛式井组开发。鄂尔多斯盆地二叠系下石盒子组野外露头研究和苏里格气田加密井区井间干扰试验表明，辫状河沉积体系中储渗单元发育规模为顺古水流方向长600m和垂直水流方向宽400m左右；表明证实苏里格气田具备进一步加密到400m×600m的条件,预计可提高采收率15%~20%。

**关键词**：致密砂岩气藏；辫状河沉积体系；储渗单元；叠置模式；开发井型；采收率

　　致密砂岩气藏是指地层条件下覆压渗透率小于0.1mD(不包含裂缝渗透率)的砂岩储层,一般情况下没有自然产能或自然产能低于工业标准,需要采用增产措施或特殊工艺井才能获得商业气流[1]。致密砂岩气藏是中国主要的天然气藏类型之一,在天然气产业地质储量和年产量中占相当大的比重。截至2016年12月,致密砂岩气藏探明储量占全国天然气总探明储量的35%,致密砂岩气藏年产量占国内天然气总年产量的比例为22%。近十年来,随着苏里格、大牛地、榆林、子洲、神木、米脂等一批致密砂岩大气田投入开发,致密砂岩气探明储量和年产量实现了快速双增长[2,3]。但上产生产高峰期之后,主力致密砂岩气田将相继进入开发中后期,开发调整、稳产挖潜和提高采收率是该期致密砂岩气藏开发中面临的主要技术问题。与国外海相—海陆过渡相致密砂岩储层不同,河流相砂体是中国致密砂岩气藏的主要储层类型,进一步精细表征河流相沉积体系特征,开展河流相沉积体系中沉积微相和微相组合精细描述是致密砂岩气藏开发中后期的主要研究方向。

　　Martin(1993)、Collinson(1996)和Miall(1996)对河流相沉积体系的研究表明[4-6],辫状河砂体具有较好的渗透率、孔隙度和较高的净毛比,是品质较好的油气藏储集体。同时,由于辫状河体系内部渗透性差异形成的低渗单元是阻碍流体流动、制约波及系数的主要因素,是储层表征和油气藏开发中面临的主要技术挑战[7-12]。Hearn(1984)提出了流动单元的概念[13],流动单元研究以一致的岩石学和水动力学特征为基础,将具有不同特征的沉积微相划分为不同级别的流动单元,预测剩余油分布规律。Miall(1996)依据河流沉积层序中的不同级次界面和结构单元建立了储层内部建筑结构[6],将不同岩性界面划分为五级界面,为表征河流相储层非均质性和河流沉积学研究提供了有益的研究思路。在前人对河流相沉积体系储层构型和流动单元划分的基础上,结合多年的研究与实践,笔者提出了储渗单元研究思路,以河流相沉积边界和储层非均质性差异为标志,针对河流相沉积体系中高渗透率、低渗透率储层单元开展识别和分析,建立不同渗透性特征的储集体空间分布模式,指导河流相致密砂岩气藏开发实践。储渗单元是指受岩性或物性边界约束的、具有相似储集性能和渗流特征的沉积亚(微)相或亚(微)相组合。由于天然气的流动性远高于原油,通常气藏开发中压降波及范围内的天然气可采储量均可实现商业开发,因此储集体内部储集和渗流特征评价是气藏开发评价的研究重点。与流动单元不同,储渗单元研究以阻流边界(通常为岩性或物性边界)识别为基础,将阻流边界控制范围以内,分布连续、具有相似物性特征的沉积微相和微相组合划分为不同品质的储渗单元。从本质上,流动单元研究是对沉积微相按流动特征的分级分类,而储渗单元研究是将不同类型的沉积微相按渗透性聚类,通过不同沉积微相的叠置关系建立储渗单元内部结构模式。本文通过鄂尔多斯盆地二叠系下石盒子组辫状河沉积体系露头和实钻储层特征分析,识别储渗单元和建立发育模式,以期为致密砂岩气藏开发中后期加密部署和水平井开发提供技术思路。

# 1 储渗单元特征与发育模式

## 1.1 储渗单元边界类型

Lynds和Hajek(2006)对美国内布拉斯加州奈厄布拉勒河和北卢普河现代河流沉积的研究表明[10],受水深的控制,同期河道沉积的顶界为泛滥平原或溢岸沉积;随着河道改道作用的影响,河道沿侧向移动,河道对泛滥平原或溢岸沉积的泥岩产生切割,在下一期河道底部形成于泥岩衬里。由于泛滥平原或溢岸沉积是河道滞留沉积的横向而不是纵向伴生亚相,所以其与河道沉积属同一期河流沉积的侧向沉积物。因此,不同期次河道沉积界限为泛滥平原或溢岸泥岩相,即辫状河储层建筑结构中的四级构型界面[6],多期河道侧向往复改道,形成纵向和横向上多期河道砂体、泛滥平原或溢岸泥岩相互叠置形成复合河道带,鄂尔多斯盆地二叠系下石盒子组气藏即为多期辫状河河道叠置而成的大型复合河道带。

储渗单元研究的首要任务是识别储渗单元内部和外部边界。储渗单元识别的基础是不同储渗单元与内、外部边界的岩性和物性差异。河流相致密砂岩气藏有效储层成因与岩石组构、成岩作用密切相关。以鄂尔多斯盆地二叠系下石盒子组气藏为例,辫状河沉积体系中心滩和河道充填底部粗砂岩分选差、大粒径矿物颗粒形成岩石骨架结构,石英类刚性矿物含量高、抗压实能力强,有利于原生孔隙的保存和孔隙流体的流动,溶蚀作用相对发育,粒径大于0.5mm的粗砂岩孔隙度普遍大于5%,渗透率与孔隙度呈正相关关系,整体上粗砂岩平均孔隙度为9.1%、平均覆压渗透率为0.098mD[14-17],渗流条件好(图1),是有效储层发育的有利岩相;河道充填中、上部的中细砂岩中火山岩屑等塑性颗粒含量高、分选好,呈致密压实相,不利于孔隙流体的流动和溶蚀作用的发生,孔隙度和渗透率均较低,有效储层不发育。同时,废弃河道、泛滥平原、心滩内部不同期次的单元坝间泥岩夹层和落淤层等泥岩相渗透率极低。储渗单元研究中将心滩、河道底部充填等渗透率高、物性条件好的沉积微相或微相组合归为储渗单元,而心滩和河道底部充填受岩性和物性边界的隔档,形成阻流边界。溢岸、心滩侧向加积的坝内粉砂质泥岩夹层和落淤层是储渗单元研究中的内部边界,废弃河道、泛滥平原是储渗单元研究中的外部边界。

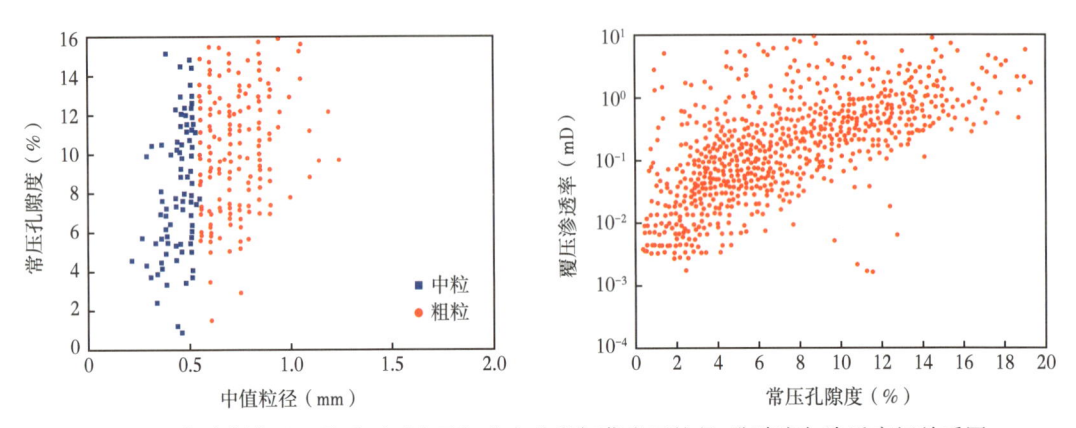

图1 鄂尔多斯盆地二叠系下石盒子组致密砂岩气藏岩石粒径、孔隙度与渗透率间关系图

泛滥平原和溢岸泥岩(图2),即同期河道复合砂体的顶界面,是储渗单元的标志性顶底界面,也是储渗单元在纵向上的主要识别标志。泛滥平原和溢岸泥岩厚度一般为数十厘米到数米不等,延伸范围较广。废弃河道泥岩位于储渗单元顶面附近,同属于储渗单元的外部边界,是河道水流侧向运动或河流水动力减小形成,一般河流上游部位废弃河道泥岩规模发育较小,河流下游由于水动力减弱废弃河道泥岩发育规模增大,但整体上废弃河道泥岩位于同期河道充填沉积范围以内。河道充填沉积顶部由于水流的沉积分异作用,沉积物分选好,粒径相对较小,通常在细粉砂级别,受压实作用影响大渗透性较差。由于河道顶部充填物性较差,形成储渗单元的外部物性边界。河道充填沉积顶部与底部间的岩性界面是河流相储层建筑结构的三级界面[6,7,18]。

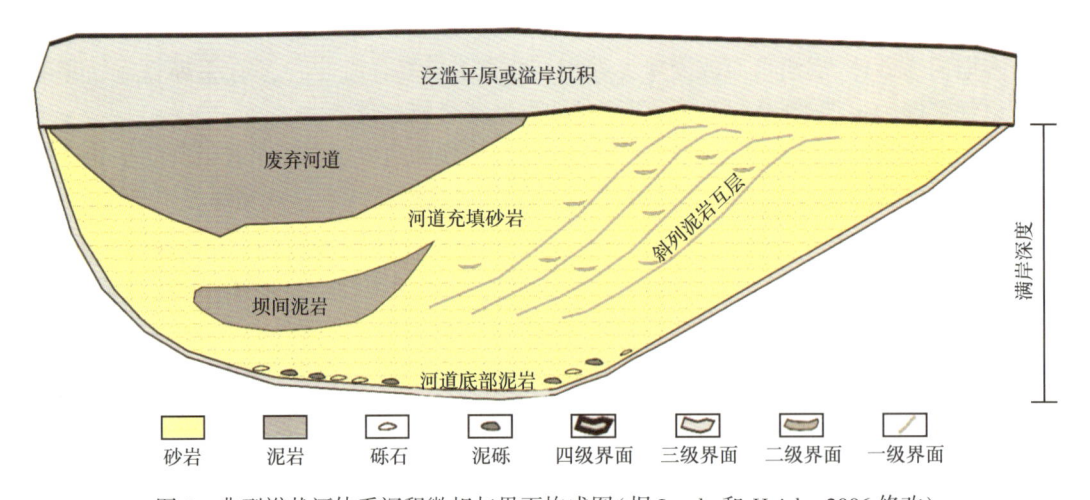

图2 典型辫状河体系沉积微相与界面构成图(据 Lynds 和 Hajek,2006 修改)

单元坝是构成心滩的基本单元,河道水流受多个心滩单元坝阻隔发育坝间次级水道,或称串沟,次级水道水动力相对较弱,粉砂质泥岩、泥岩在该区域易沉降,形成坝间泥岩。坝间泥岩为储渗单元内部边界,通常规模相对较小、物性差,一般厚度小于0.5m,是河流相储层建筑结构的二级界面。

坝内泥岩是心滩侧向加积作用产生的粉砂质、泥质夹层,通常厚度较薄、纹层或夹层级,分选好;落淤层泥岩则形成于季节性洪水期的泥质沉积物,一般侧向延伸宽度有限,发育规模较小。坝内泥岩、落淤层或统称为斜列泥岩互层(图2),同属储渗单元的内部边界,是层系组或单个层系界面,属建筑结构中的一级界面。

## 1.2 储渗单元发育模式

野外露头观测和致密砂岩气藏开发中密井网区精细解剖可识别不同类型的储渗单元,从而建立储渗单元叠置模式。笔者采用储渗单元研究思路,对鄂尔多斯盆地南缘柳林地区二叠系下石盒子组露头剖面观测,结合气井钻遇砂体分析,将储渗单元划分为:心滩与河道底部充填叠置型、河道充填叠置型、心滩叠置型和心滩或河道底部充填孤立型四种发育模式(图3)。

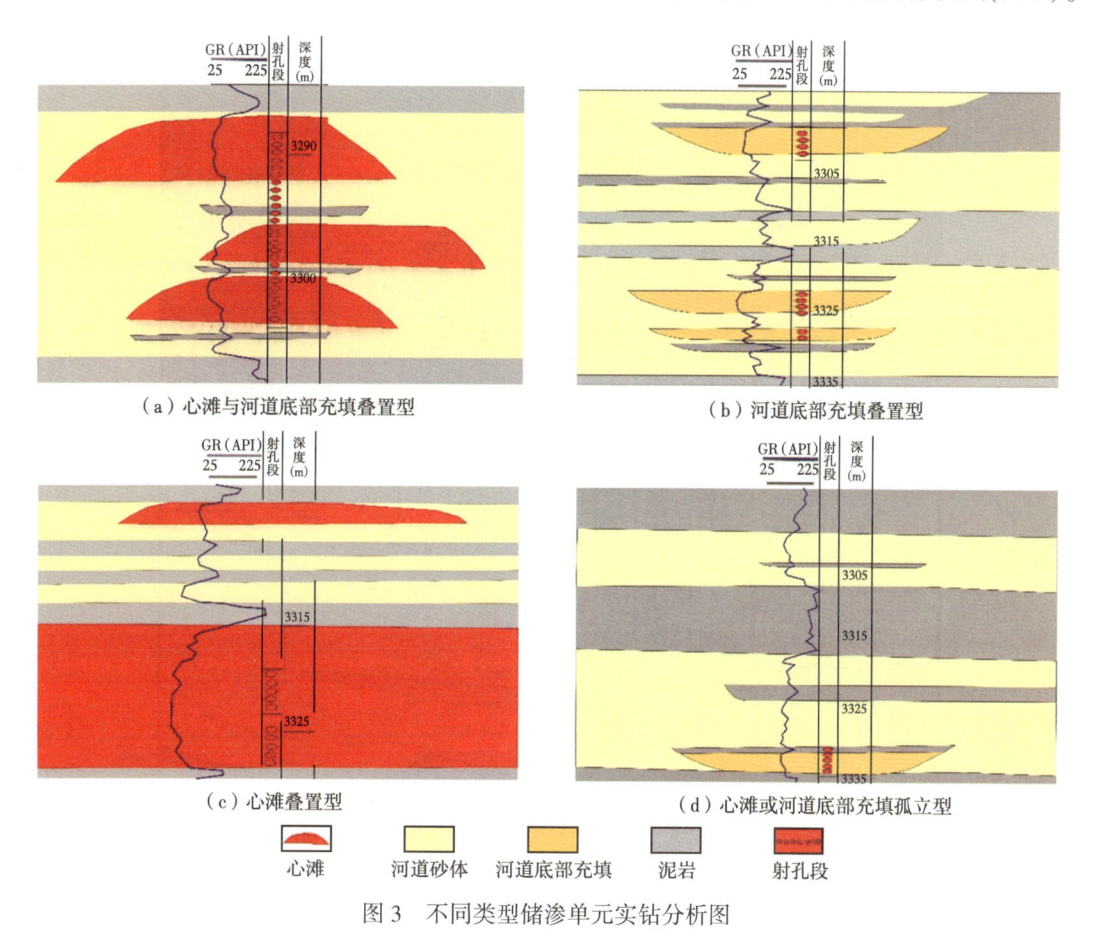

(a)心滩与河道底部充填叠置型

(b)河道底部充填叠置型

(c)心滩叠置型

(d)心滩或河道底部充填孤立型

心滩　河道砂体　河道底部充填　泥岩　射孔段

图3　不同类型储渗单元实钻分析图

### 1.2.1 心滩与河道底部充填叠置型

心滩整体上一般为正粒序,底部为砾岩或粗砂岩相,向上过渡为粗、中砂岩相。由于心滩边部水动力作用较弱,通常为斜层状的泥岩或粉砂质泥岩(落淤层),同时侧向加积作用剧烈,形成心滩复合砂体内部夹杂泥岩或粉砂质泥岩夹层,即储渗单元的内部阻流边界。该阻流边界通常规模较小,呈纹层状,从露头剖面和水平井实钻轨迹中可识别出该类型储渗单元内部阻流边界。心滩与河道底部充填型储渗单元内部落淤层阻流边界纵向发育规模较小,一般小于0.5m(图4),水平井钻井过程中通常沿心滩侧向钻进过程中钻遇薄层状泥质夹层即为落淤层边界。心滩与河道底部充填叠置型储渗单元底部边界为上期河道消亡时沉积的泛滥平原或废弃河道泥岩、粉砂质泥岩,属岩性边界;其上部边界为河道顶部充填沉积形成的中、细砂岩相,分选较好,物性较差,形成物性边界。该类型储渗单元通常为厚层状,纵向上厚度较大,一般在10~15m。

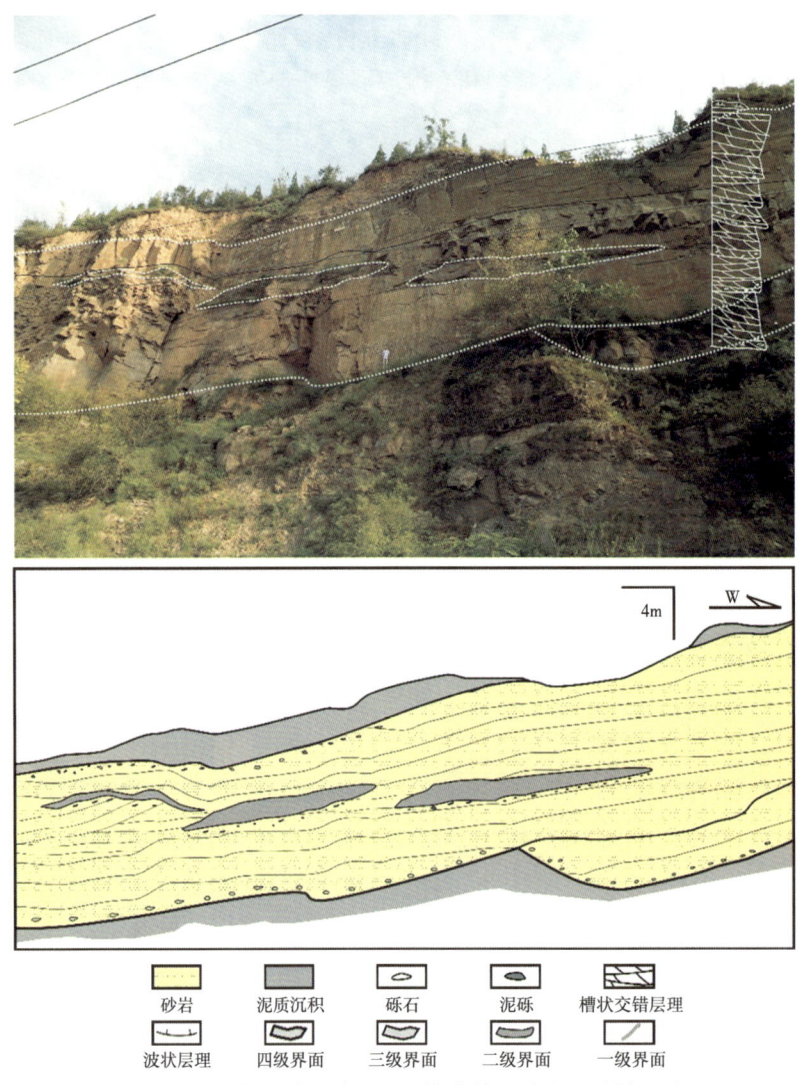

图4 心滩与河道充填叠置型储渗单元露头观测剖面图

## 1.2.2 河道底部充填叠置型

两期或多期河道底部充填呈垂向叠置,河道带砂体底部发育明显的冲刷界面,一般呈不规则下凹状,底部以含砾粗砂岩、粗砂岩为主,单期河道砂体内部呈正粒序旋回。受河道迁移、改道作用的影响,在不同期次河道充填的顶部形成废弃河道或泛滥平原泥岩、粉砂质泥岩的互层,是该类型储渗单元的内部岩性边界(图5)。由于河道底部充填的冲刷作用,泥岩或粉砂质泥岩互层较薄,通常在0.5~1m。单个河道带砂体厚度4~7m,河道底部充填叠置型储渗单元通常由3~5个河道砂体垂向叠置而成。因此,该类型储渗单元厚度较大(6~10m)。

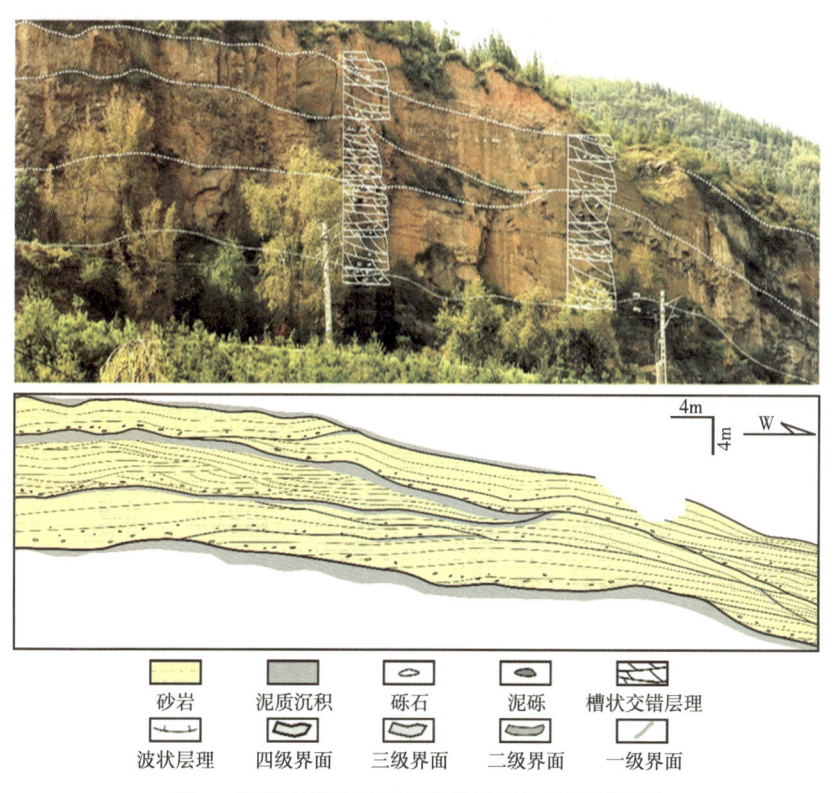

图5 河道充填叠置型储渗单元露头观测剖面图

## 1.2.3 心滩叠置型

辫状河发育带中两期或多期心滩垂向叠置形成规模较大的储渗单元,该类型储渗单元厚10~20m(图6)。不同期次心滩砂体间由于辫状河道的改道作用频繁,通常夹薄层状泛滥平原或废弃河道泥岩,一般厚度在3~5m,形成储渗单元内部岩性边界,整体上该类型储渗单元发育频率较低。心滩砂体一般呈块状,具有纵向上粒度逐渐变细的特征,但心滩砂体内部由于侧向加积作用,内部常见倾斜状泥岩夹层(落淤层)。

## 1.2.4 心滩或河道充填孤立型

与上述叠置型储渗单元不同,该类型储渗单元由于改道作用的影响,不同期次辫状河道砂体侧向变化距离较大,在辫状河体系过渡带或体系间洼地,心滩或河道充填砂体沉积频率较低,从而纵向上心滩或河道底部充填砂体呈孤立状(图7、图8)。心滩砂体顶部偶见侧向加积

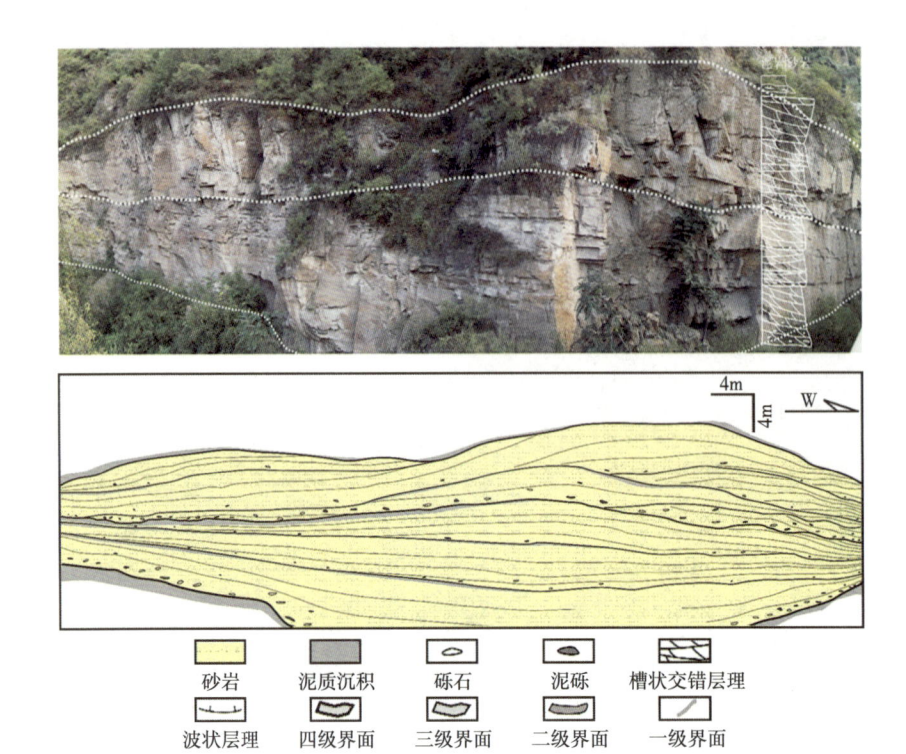

图 6　心滩叠置型储渗单元露头观测剖面图

图 7　心滩孤立型储渗单元露头观测剖面图

形成的倾斜状泥岩,呈薄层状,通常在0~0.5m。河道充填孤立型储渗单元由于河道体系过渡带或河道体系间水动力较弱,储渗单元顶部通常为河道顶部充填砂体,粒度较细,呈中、细砂岩,渗透性较差;同时,随着该期河道的改道或消亡,河道充填顶部向上为废弃河道或泛滥平原沉积泥岩或粉砂质泥岩。单个心滩厚度一般为5~8m,河道底部充填砂体厚度一般为3~5m,因此该类型储渗单元与叠置型相比,具有发育规模较小、侧向上连续性和连通性差的特点。

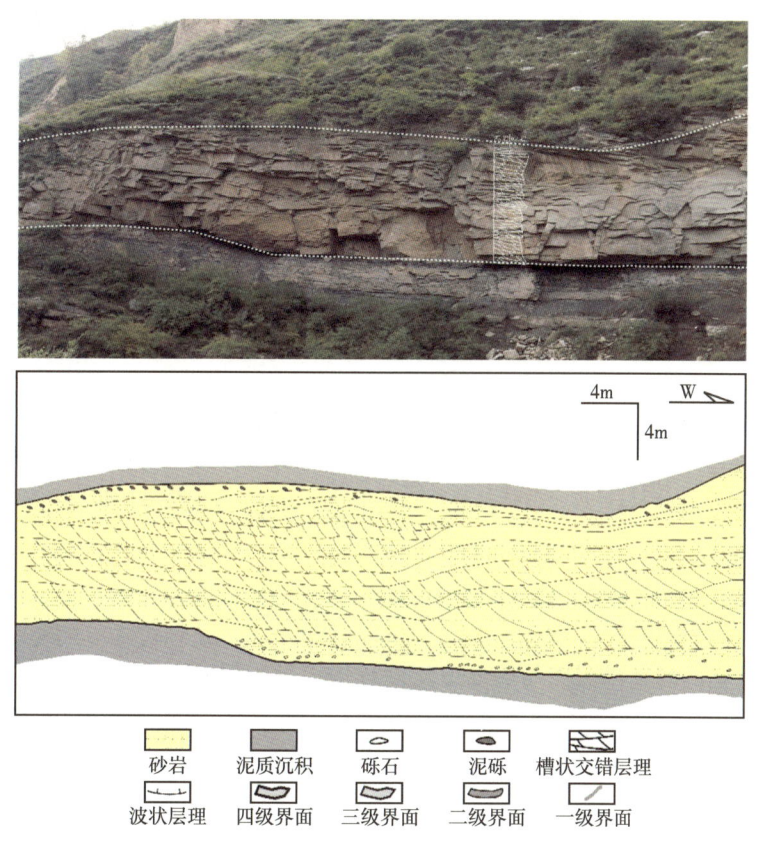

图8 河道充填孤立型储渗单元露头观测剖面图

## 1.3 储渗单元发育规模

### 1.3.1 储渗单元井间识别

通过密井网井间干扰试验可进一步精细识别和表征井间储渗单元形态和边界类型。以密井网开发试验区早期投产井苏38-16-5和加密井苏6-j21为例,两口井均射开二叠系下石盒子组盒8段砂体,纵向上射孔层段对应一致,因此相应层位连续性和连通性可对比性强。苏38-16-5井射孔层段测井曲线形态底部呈齿化箱形、上部呈平滑箱形,储渗单元类型为河道与心滩叠置型;苏6-j21射孔层段测井曲线形态底部为平滑箱形与齿化箱形垂向分布,储渗单元类型为心滩与河道充填叠置型,上部为平滑箱形零星分布,储渗单元类型为心滩孤立型。苏38-16-5井投产于2003年10月,经过6年的生产,到2009年9月地层压力已降至6MPa,而苏6-j21加密井在2009年9月投产时,地层压力仍高达30.45MPa,仍维持原始地层压力。结合两口井射孔层段的一致性,可判断出苏38-16-5井泄流范围小于两口井间距即对应层段储渗单元间存在外部边界,根据对应的部位分别为心滩和河道充填砂体,两口井储渗单元边界类型为坝间泥岩相形成的岩性边界(图9)。

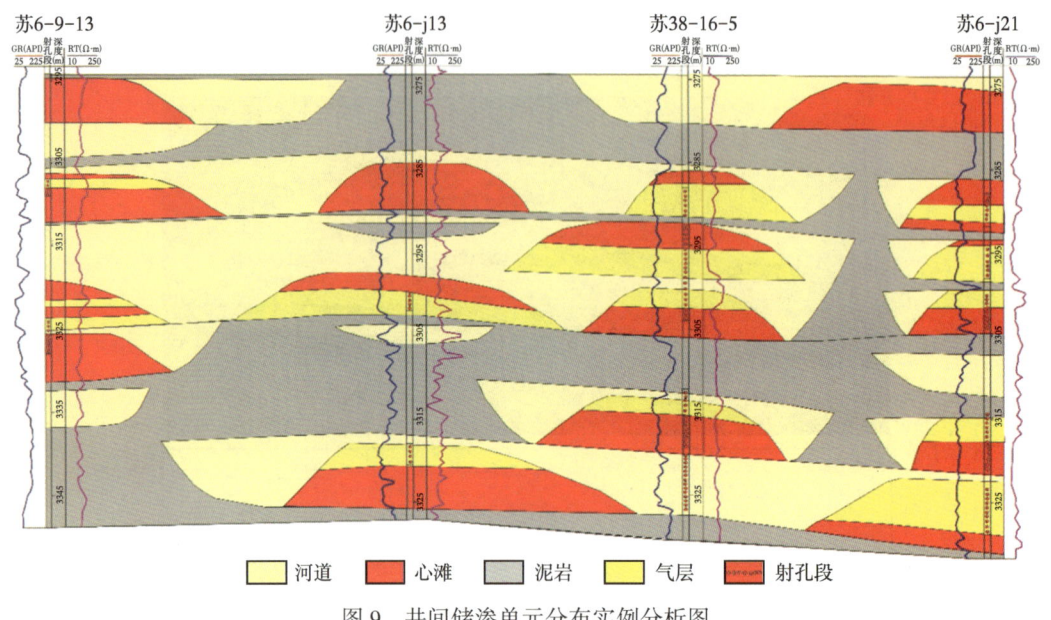

图9 井间储渗单元分布实例分析图

### 1.3.2 储渗单元发育规模

同一期河道充填的满岸深度与储渗单元发育规模呈正相关。沉积序列完整的心滩微相代表了河流的满岸深度,山西柳林地区二叠系下石盒子组露头剖面观测表明,单个心滩厚度为3~6.5m。据此,苏里格地区辫状河体系的河流满岸深度为3~6.5m。而河流相沉积的宽厚比通常在40~70[19,20],因此河流相沉积体系中储渗单元宽度范围在200~400m。不同学者对现代河流沉积砂体的研究表明:高坡降辫状河的心滩坝微相不发育,主要微相单元是河道亚相;低坡降辫状河的心滩微相发育明显,顺物源方向心滩砂体长宽比一般为2~5,长宽比最大可达到10∶1[21-25]。据此,苏里格气田下石盒子组心滩砂体长度分布范围为400~1500m,由于顺河道水流方向心滩砂体形态通常呈弯曲状,沿南北向心滩砂体长度为400~900m[26-28]。

密井网开发试验是井间砂体解剖的最直接方式,同时配合干扰试井分析是判识井间渗透性和砂体连通性的最有效途径。通过大量加密井井间地层压力测试,对比不同批次加密井原始地层压力,若投产前地层压力接近原始地层压力,则表明储渗单元不连通或连通性较差,同时,通过储渗单元测井相分析识别出储渗单元类型,从而建立储渗单元发育规模知识库。

通过对投产时间长的气井井底压力和新投产井原始地层压力的监测,可判断出井间储渗单元发育规模。以苏里格气田苏6加密井区为例,该井区开发层位均为二叠系下石盒子组致密砂岩储层,如图10所示,作为苏里格气田最早的开发试验区该井区在初期一次骨架井网1200m×800m的基础上,通过不断加密部署,逐渐形成井距366~800m、排距约600m的加密井网。骨架井网原始地层压力约30MPa,随着投产时间的不断增加,投产井井底压力不断降低,最早投产的苏6井、苏38-16、苏38-16-2、苏38-16-3、苏38-16-4和苏38-16-5井井底压力已分别下降至4.99MPa、6.02MPa、9.35MPa、13.78MPa、6.03MPa及9.67MPa。同一井排上加密井井距分布366~466m不等，投产前原始地层压力除苏6-j5井外，均呈现较大幅度的压降；而另外两排加密井井距分布范围为417~820m，新投产井原始地层压力23.60~30.72MPa。表明当井距低于400m时，井间已出现明显压力干扰；当井距超过400m，井间压力干扰的概率大幅度降低，储渗单元井间东西向规模小于400m。同时，对相邻两排井投产前原始地层压力监测表明，上下两排气井与首批投产井排间排距约600m，投产前地层压力均未发现压力降低，未受到早期投产气井生产的影响。因此，储渗单元南北向发育规模范围小于600m。

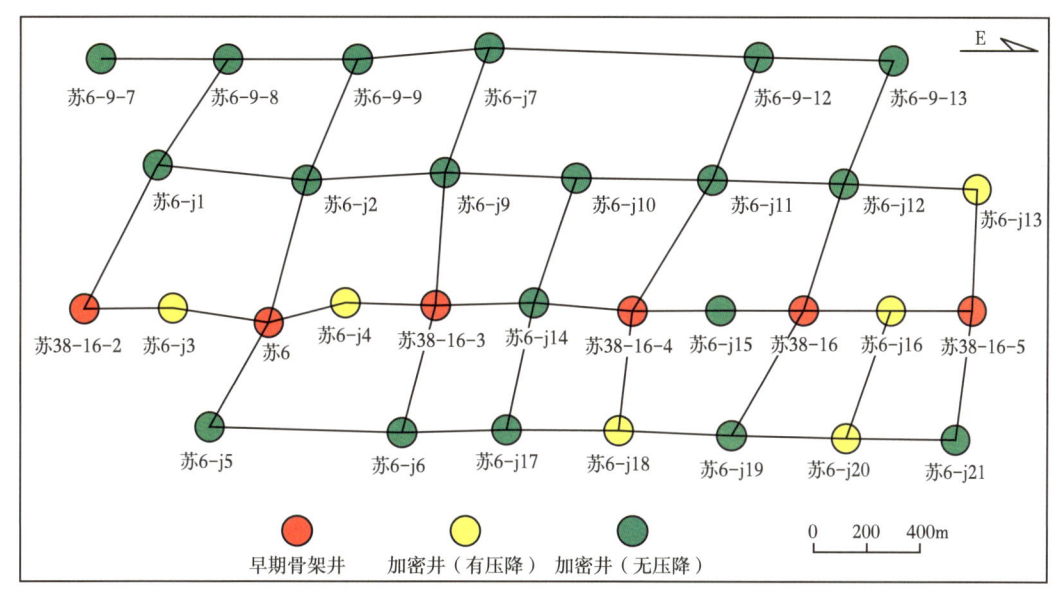

图10 苏里格气田苏6加密井区井位图

## 2 储渗单元研究与开发部署

### 2.1 储渗单元与气田加密部署

井网加密是大规模致密砂岩气藏开发中后期提高采收率的主要技术手段之一。位于美国得克萨斯州西南部的奥卓拉气田(Ozona)是致密砂岩气藏开发的典型气田。该气田于1960年代投入开发，开发初期一次部署井网面积为1.29km²；1995年以后，通过加密部署和井间干扰试验，采用地质统计分析，开展气井泄流面积和井间剩余储量评价，以井间压力干扰为加密部署极限的判识标准。随着地质认识逐渐加深，砂体规模小于井网密度、井间砂体连续性有限，进一步对储渗单元泄流能力评价，气井泄流面积0.16~0.32km²，平均0.24km²；将单井控制面积逐步加密到0.65km²和0.32km²，主力开发区单井控制面积0.16km²(占总井数52%)、井网密度平均6口/km²，最大井网密度可达12口/km²[29]。

苏里格气田自2003年投入开发，规模开发初期开展"甜点"区预测、富集区优选和地质建模研究，基于辫状河体系主河道带复合砂体长1000~1800m、中值1200m，宽600~1000m，在气田中部富集区采用800m×1200m骨架井网,气井平均可采储量2000×10⁴m³,实现了气田有效开发;到2009年,通过井位优化和加密部署在原有一次井网基础上,通过主河道带内单砂体精细刻画和一次井网下剩余储量分布研究,局部加密到600m×800m,在保障气井可采储量不下降的前提下,进一步扩大气田产能;2010年以后,气田实现整体开发,通过滚动评价、扩边生产,开发区域逐步扩展到东区、中区和西区,直井开发区块内全面实现600m×800m井网,气田开发规模进一步上升,且较一次井网下采收率提高幅度为10%,成为中国储量和产能规模最大的天然气田[2,3,26,27]。通过储渗单元研究,鄂尔多斯盆地辫状河沉积体系中心滩和河道充填砂体形成的储渗单元呈现不同叠置模式,不同类型储渗单元南北向(顺古水流方向)和东西向(垂直古水流方向)发育规模分别为600m和400m以内,表明苏里格气田具备进一步加密的条件,达到400m×600m井网,预计提高气田采出程度15%~20%。

## 2.2 储渗单元与水平井部署

致密砂岩气藏储渗单元研究以不同类型砂体间叠置关系研究为主体,将具有相似渗流特征的砂体聚类为储渗单元,结合复合砂体的形态和辫状河体系发育特征,辫状河体系主河道叠置带内心滩和河道充填砂体富集,心滩叠置型,心滩与河道叠置型,以及河道充填叠置型储渗单元发育。整体上辫状河体系砂体发育规模较小,不适合大面积采用水平井开发;但辫状河主河道叠置带河道继承性发育,较强的水动力条件有利于孔渗条件好的粗砂岩形成和富集,砂地比普遍大于70%,纵向上多期叠置型储渗单元富集,叠置型储渗单元总厚度通常超过10~15m,空间分布稳定,侧向上储渗单元连续性好(图11)。因此,叠置型储渗单元发育带即辫状河体系的主河道叠置带是水平井部署的有利目标区,可优选井位、部署水平井。

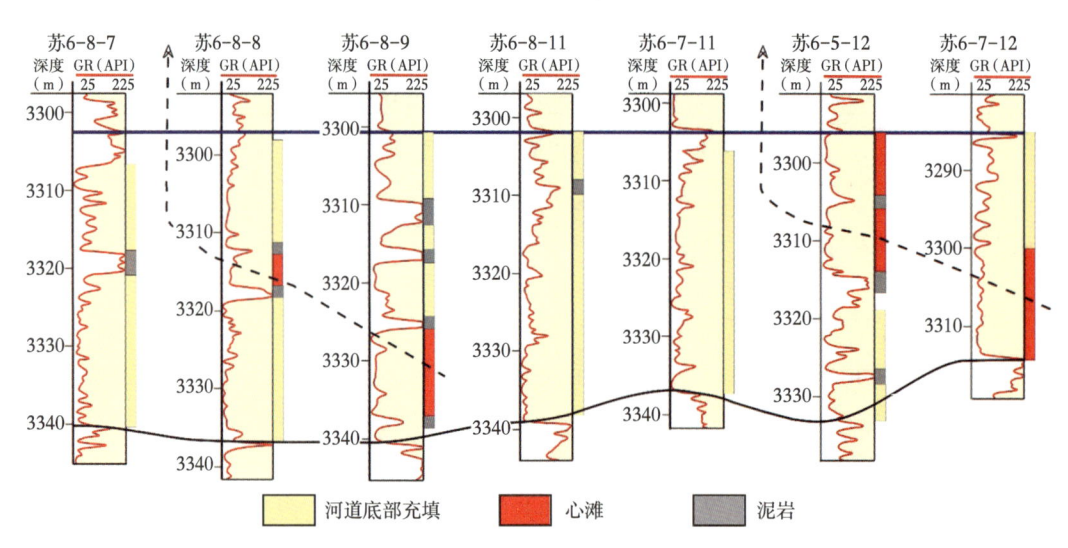

图11 辫状河沉积体系河道叠置带储渗单元发育特征

对于辫状河体系过渡带和体系间洼地,辫状河体系过渡带虽然发育因洪水期改道短期内形成的河道充填,沉积速率快、粒度粗,形成单层厚度较大的粗砂岩,但整体上砂地比较低,一般在30%~70%(图12);辫状河体系间洼地,砂地比更低一般小于30%,以泥岩、粉砂质泥岩为主,夹薄层状中细砂岩(图13)。辫状河体系过渡带和体系间洼地储渗单元类型主要为零星分布的心滩或河道充填孤立型，叠置型储渗单元零星分布，不利于水平井整体部署，宜采用直井或直井丛式井组开发。

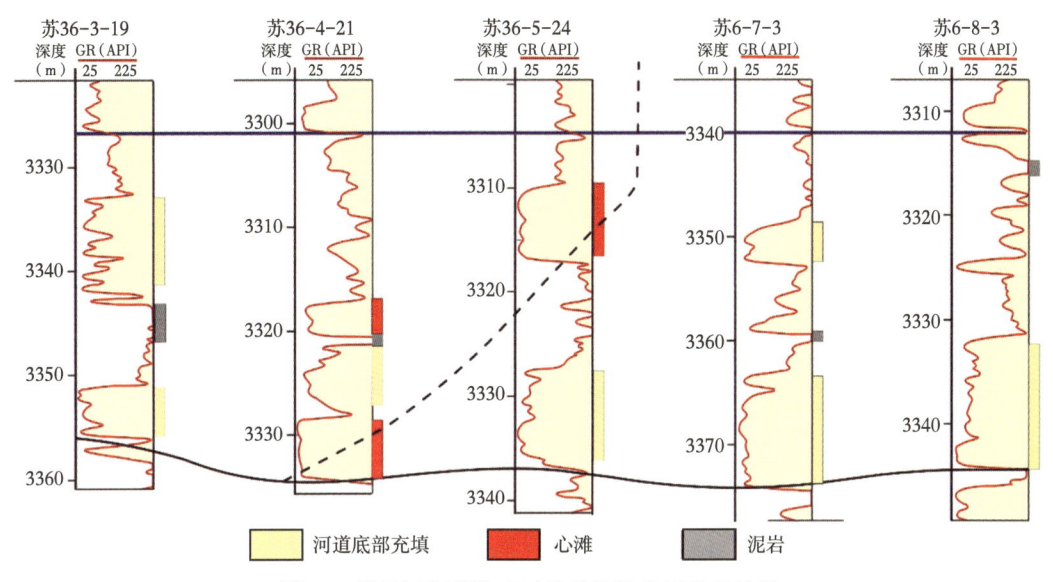

图12 辫状河沉积体系过渡带储渗单元发育特征

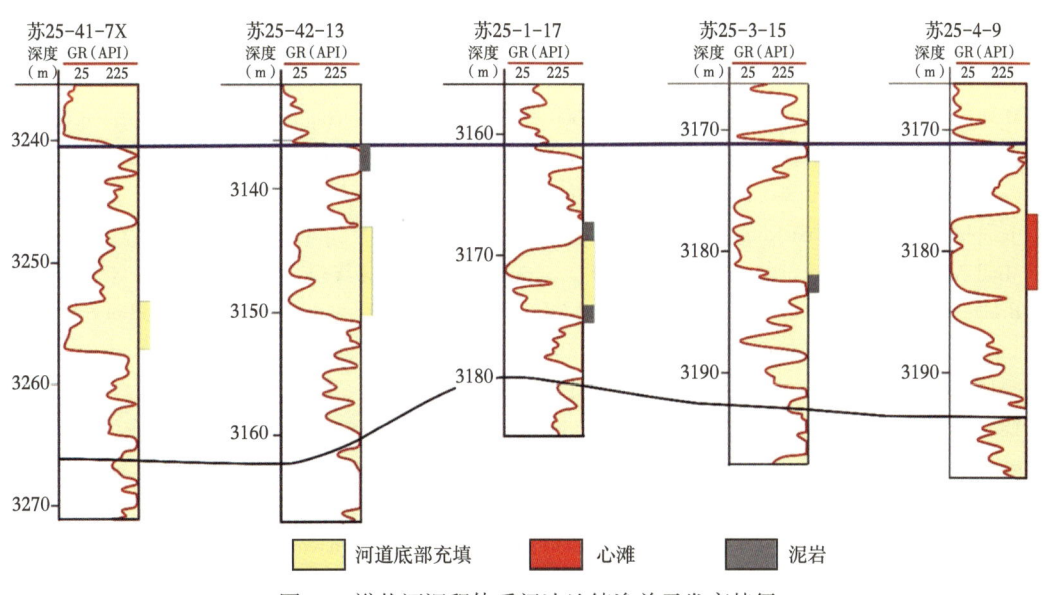

图13 辫状河沉积体系洼地储渗单元发育特征

## 3 结论

（1）将具有相似储渗特征的地质体或沉积微相聚类为储渗单元，通过储渗单元发育特征、叠置模式、露头识别研究，建立储渗单元空间分布和发育规模等地质认识，明确不同类型储渗单元分布规律，可有效指导天然气藏开发中井网和井型、井距优选、开发中后期加密部署等关键开发技术政策的制定,为天然气藏高效开发和提高采收率提供技术支撑。

（2）采用储渗单元研究技术思路和方法,根据不同类型砂体的储渗特征,将孔隙度和渗透率高的粗砂岩相砂体聚类为储渗单元,心滩和河道充填砂体是主要的储渗单元类型；溢岸、心滩侧向加积的坝内粉砂质泥岩夹层和落淤层是储渗单元研究中的内部边界,废弃河道、泛滥平原形成储渗单元研究中的外部边界。

（3）河流相致密砂岩气藏储渗单元发育四种叠置模式:心滩叠置型、河道充填叠置型、心滩或河道充填叠置型,以及心滩或河道充填孤立型。叠置型储渗单元主要发育在辫状河沉积体系的河道叠置带内,是致密砂岩气藏水平井开发的有利目标;孤立型储渗单元主要发育在辫状河沉积体系过渡带和体系间洼地,适宜采用直井或直井丛式井组开发。通过露头观测和加密井区井间干扰试验,对储渗单元发育规模定量开展表征,储渗单元空间发育规模为顺古水流方向600m和垂直古水流方向400m以内;苏里格气田在当前600m×800m井网的基础上,仍具备井网加密的条件,预计进一步加密到400m×600m井网可提高采收率15%~20%。

## 参 考 文 献

[1] 国家能源局. 中华人民共和国石油和天然气行业标准（SY/T 6832—2011）：致密砂岩气地质评价方法. 北京：石油工业出版社, 2011.

[2] 戴金星,倪云燕,吴小奇. 中国致密砂岩气及在勘探开发上的重要意义[J]. 石油勘探与开发, 2012. 39 (3)：257-264.

[3] 马新华,贾爱林,谭健,等. 中国致密砂岩气开发工程技术与实践[J]. 石油勘探与开发,2012,39(5)：572-579.

[4] Martin J H. Areview of braided fluvial hydrocarbon reservoirs：The petroleum engineer's perspective[G]//Best J L and Bristow, C,S, Braided rivers, Geological Society (London) Special Publication,1993,75, 333-368.

[5] Collinson, J D. Alluvial Sediments[G]// Reading H G. Sedimentary Environments：Processes, facies and stratigraphy, Oxford, United Kingdom, Blackwell Science, 1996:37-82.

[6] Miall A D. The Geology of Fluvial Deposits[M]. Springer Verlag, New York, 1996:75-178.

[7] Bridge, J. S., Mackey S. D. A theoretical study of fluvial sandstone body dimensions[G]// Flint S S and bryant I D et al. Geological modeling of hydrocarbon reservoirs. International Association of Sedimentologists, Special Publication 15, Utrecht, Netherlands,1993:213-236.

[8] Leclair S F, Bridge J S, Wang F,et al. Preservation of crossstrata due to migration of subaqueous dunes over aggrading and non-aggrading beds：comparison of experimental data with theory[J]. Geoscience Canada, 1997,24 (1)：55-66.

[9] Lane S N. Approaching the system-scale understanding of braided river behavior[G]// Smith G S, Best J, Bristow C et al. Braided Rivers：Process, deposits, ecology and management, blackwell publishing, London, 2006：107-135.

[10] Lynds R., Hajek E. Conceptual model for predicting mudstone,dimensions in sandy braided-river reservoirs [J]. AAPG Bulletin, 2006,90(8)：1273-1288.

[11] Labourdette R. Stratigraphy and static connectivity of braided fluvial deposits of the Lower Escanilla Formation, South Central Pyrenees, Spain[J]. AAPG Bulletin, 2011,95(4)：585-617.

[12] Lunt I A, Smith G H, Best J L,et al. Deposits of the sandy braided South Saskatchewan River Implications for the use of modern analogs in reconstructing channel dimensions in reservoir characterization[J]. AAPG Bulle-

tin,2013,97(4):553-576.
- [13] Hearn C L,Ebanks W J,Tye R S,et al. Geological factors influencing reservoir performance of the Hartzog Draw Field, Wyoming[J]. Journal of Petroleum Technology,1984,36(8):1335-1344.
- [14] 何东博,贾爱林,田昌炳,等. 苏里格气田储集层成岩作用及有效储集层成因[J]. 石油勘探与开发,2004,31(3):69-71.
- [15] 李易隆,贾爱林,何东博,等. 致密砂岩有效储层形成的控制因素[J]. 石油学报,2013,34(1):71-82.
- [16] 王国亭,何东博,王少飞,等. 苏里格致密砂岩气田储层岩石孔隙结构及储集性能特征[J]. 石油学报,2013,34(4):660-666.
- [17] Jia C Y,Jia A L,Zhao X,et al. Architecture and quantitative assessment of channeled clastic deposits, Shihezi sandstone (Low Permian), Ordos Basin, China[J]. Journal of Natural Gas Geoscience,2017,2(1):11-20.
- [18] Kelly S. Scaling and hierarchy in braided rivers and their deposits: examples and implications for reservoir modeling[G]// Smith G S. Best J, Bristow C et al. Braided rivers: Process, deposits, ecology and management, blackwell Publishing, London:2006:75-106.
- [19] 廖保方,张为民,李列,等. 辫状河现代沉积研究与相模式——中国永定河剖析[J]. 沉积学报,1998,16(1):34-39.
- [20] 吴胜和. 储层表征与建模[M]. 北京:石油工业出版社,2010.
- [21] 钟建华,马在平. 黄河三角洲胜利Ⅰ号心滩的研究[J]. 沉积学报,1998,16(2):38-42.
- [22] 于兴河,马兴祥,穆龙新,等. 辫状河储层地质模式及层次界面分析[M]. 北京:石油工业出版社,2004.
- [23] 何顺利,兰朝利,门成全,等. 苏里格气田储层的新型辫状河沉积模式[J]. 石油学报,2005,26(06):25-29.
- [24] 金振奎,杨有星,尚建林,等. 辫状河砂体构型及定量参数研究——以阜康、柳林和延安地区辫状河露头为例[J]. 天然气地球科学,2014,25(3):311-317.
- [25] 雷卞军,李跃刚,李浮萍,等. 鄂尔多斯盆地苏里格中部水平井开发区盒8段沉积微相及砂体展布[J]. 古地理学报,2015,17(1):91-105.
- [26] 何东博,王丽娟,冀光,等. 苏里格致密砂岩气田开发井距优化[J]. 石油勘探与开发,2012,39(4):458-464.
- [27] 何东博,贾爱林,冀光,等. 苏里格大型致密砂岩气田开发井型井网技术[J]. 石油勘探与开发,2013,40(1):79-89.
- [28] 卢涛,刘艳侠,武力超,等. 鄂尔多斯盆地苏里格气田致密砂岩气藏稳产难点与对策[J]. 天然气工业,2015,35(6):43-52.
- [29] Cipolla C. L,Mayerhofer M. Infill drilling and reserve growth determination in lenticular tight gas sands[C]// SPE Annual Technical Conference & Exhibition,Denver, Colorado: Society of Petroleum Engineers,1996:533-554.

# 高磨地区灯四段岩溶古地貌分布特征及其对气藏开发的指导意义

闫海军[1]　彭　先[2]　夏钦禹[1]　徐　伟[2]　罗文军[2]
李新豫[1]　张　林[1]　朱秋影[1]　朱　迅[2]　刘曦翔[2]

(1. 中国石油勘探开发研究院；2. 中国石油西南油气田分公司勘探开发研究院)

**摘要**：基于四川盆地高磨地区三维地震和完钻井资料，结合高分辨率法、层拉平方法、残厚法和印模法的优点，采用双界面法对高磨地区震旦系灯四段岩溶古地貌进行恢复。灯四段岩溶古地貌包括岩溶低地、岩溶斜坡和岩溶台地三种一级古地貌单元以及陡坡、缓坡、台坡、台面、洼地、沟谷和残丘七种二级古地貌单元。灯四段岩溶古地貌表现为"两沟三区、北缓南陡"的特征，同时高石梯发育台坡、台面和残丘三种古地貌单元，磨溪地区主体发育缓坡和台面，台地内部发育台面、洼地和残丘。台坡、台面和斜坡微地貌单元岩溶发育程度好，优质储层发育，完钻井效果较好，高石梯和磨溪地区差异性明显。结果表明，岩溶古地貌对高产井控制作用明显，下一步需要精细刻画沟谷分布，论证断层及古沟槽对古地貌分布的控制作用，分区建立古地貌划分标准，评价台地内部优势微地貌单元分布，支撑建产有利区筛选和目标开发井位优化，从而为高磨地区震旦系气藏快速建产和长期稳产提供保障。

**关键词**：古地貌；双界面；岩溶；灯四段；高磨；四川盆地

　　四川盆地高磨地区震旦系气藏三级储量规模已达万亿立方米，是四川盆地天然气上产的重要领域[1-3]。震旦系灯影组气藏为岩溶风化型碳酸盐岩气藏，岩溶古地貌单元的不同导致后期成岩环境中形成差异性储层单元，从而影响开发有利区筛选和井位部署。虽然在地质历史时期中四川盆地经历了频繁构造运动，但是在川中稳定基底的控制作用下，自震旦纪以来四川盆地总体仍是以下沉为主。自基底开始四川盆地先后经历了六期主要的构造旋回，扬子旋回、加里东旋回、海西旋回、印支旋回、燕山旋回和喜马拉雅旋回，震旦系顶部风化壳由加里东旋回第二幕桐湾运动造成，导致上扬子地区整体抬升，灯四段广泛遭受剥蚀，灯四段丘—滩相白云岩遭受风化剥蚀及大气淡水溶蚀淋滤，形成大量的溶蚀孔洞，成为四川盆地灯影组天然气藏的有效储层。对于四川盆地灯影组古地貌分布特征，李启桂等[4]利用全盆地地震格架大剖面解释资料和钻井资料，结合厚度印模法，恢复了桐湾不整合面古地貌，古地貌特征呈现西北高、中部斜坡、东南洼的格局；汤济广等[5]采用地震资料分析不整合类型，井震结合对乐山—龙女寺古隆起进行古地貌恢复，恢复结果显示灯影组沉积末期古岩溶地貌类型可划分为2个岩溶斜坡、3个岩溶高地和2个岩溶盆地；汪泽成等[6]通过对桐湾运动性质、期次的分析，利用地震、钻井、露头等资料，采用残余厚度法和印模法刻画盆地震旦系顶部岩溶古地貌形态；罗思聪[7]应用钻井资料以及四川盆地周缘150余条野外露头剖面资料，结合区域地震资料，采用印模法恢复震旦系灯影组古岩溶地貌，恢复结果显示整个盆地南北向呈现"三隆"（镇巴、川中、黔江—正安)"两坳"(阆中—通江、重庆—开县)特征,而东西向被分割为相对独立的两个古隆起体系。金民东[8]等以高磨地区三维地震和钻井资料,采用印模法进行灯影组四段岩溶古地貌的恢复,将灯四段岩溶古地貌进一步划分为岩溶台面、斜坡和叠合斜坡3种地貌单元,在地貌单元的划分中引入意大利马尔凯地区岩溶台面水系模式图的概念。这些古地貌恢复通常以整个盆地为恢复工区[4,6,7],或者以构造单元为研究工区[5],整体上全盆地都表现为西北高、东南低的古地貌分布特征,但是盆地内部地貌单元分布存在差异或者完全相悖的结论,同时在古地貌恢复方法上多采用印模法或者印模法、残厚法两者结合,由于印模法和残厚法在古地貌恢复过程中有自己的不足[9],制约了古地貌恢复的精度。

笔者针对四川盆地震旦系气藏高石梯—磨溪建产区块,利用近60口钻井资料和三维地震资料采用"双界面法"开展古地貌恢复,恢复结果可有效指导开发有利区筛选和井位部署,对于震旦系气藏快速上产具有实际意义。

## 1 气藏概况

研究区位于四川省中部资阳市、重庆市潼南县境内,区域构造上位于盆地中部川中古隆起平缓构造区威远至龙女寺构造群(图1)。乐山—龙女寺古隆起是四川盆地形成最早、规模最大、延续时间最长的巨型隆起,轴线西起乐山,东至龙女寺,其形成演化对震旦系灯影组油气藏具有重要影响和明显控制作用[10,11]。震旦系灯影组沉积期到下寒武统沉积期,上扬子地区长

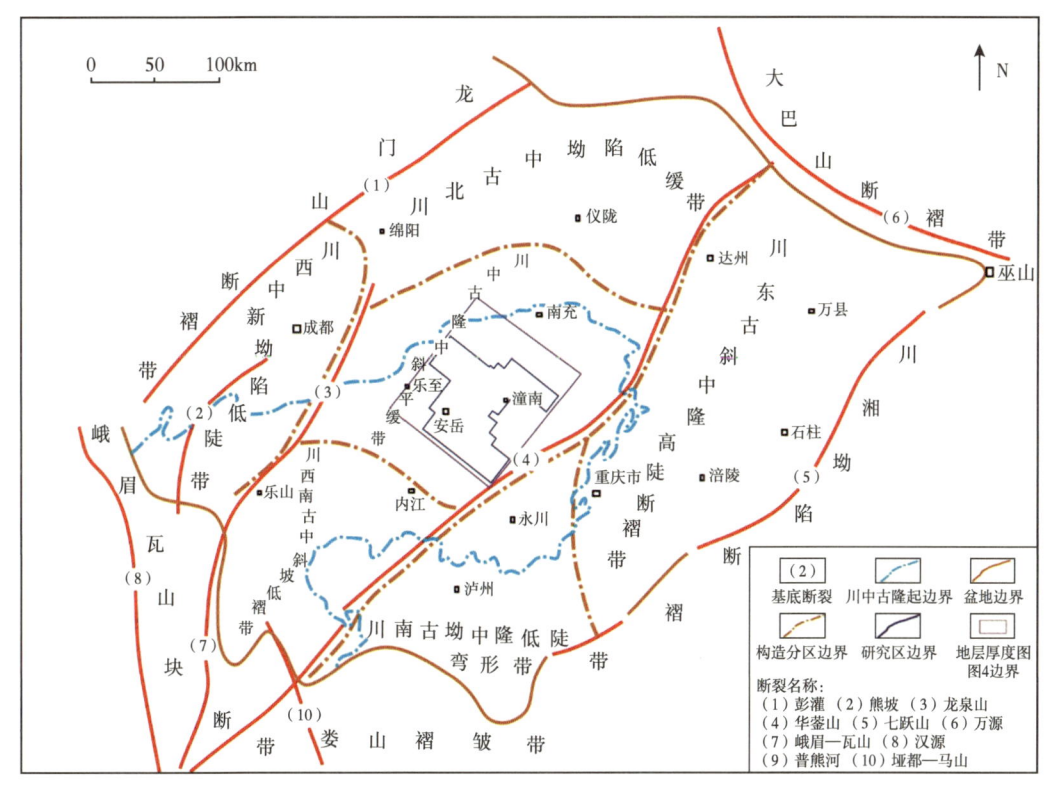

图1 高磨地区构造位置

期处于拉张环境,四川盆地表现为克拉通内裂陷盆地特征。震旦纪伸展作用在四川盆地的直接响应是形成德阳—安岳台内裂陷。受该裂陷形成与演化的影响,四川盆地震旦系并非铁板一块,而是具有隆凹相间的构造格局,灯一—灯二段沉积期表现为"一隆四凹"的古地理背景;在灯三—灯四段沉积期,由于德阳—安岳台内裂陷持续张裂并与长宁裂陷贯通,将四川盆地分割成东西两个部分,从"一隆四凹"演化为"两隆四凹"的古地理格局[12]。灯影组分为4个层段,其中灯一段和灯三段为海侵域,灯二段和灯四段为高位域,有效储层主要发育在灯二段和灯四段。受安岳—德阳裂陷槽控制,高磨地区发育开阔台地相沉积,台地边缘发育高能丘滩复合体,台地内部发育低能丘滩复合体。受古环境、古构造、古水深、古气候等特征影响,丘、滩体发育在纵向上表现为自上而下由孤立状向侧向叠置再向垂直叠置型发育,台地边缘向台地内部丘、滩体发育程度降低,丘、滩体连续性连通性变差。构造演化研究表明[13],乐山—龙女寺古隆起是张应力背景下形成的受基地和断裂共同控制的继承性隆起。在震旦系灯影组沉积期,形成同沉积隆起兼剥蚀隆起雏形,为低隆起时期;灯影组沉积末期,桐湾Ⅱ幕差异抬升作用导致古隆起发生较大幅度的相对隆升,灯四段遭受不同程度的淋滤和剥蚀,形成优质储层。

高磨地区灯四段与下伏混积潮坪相灯三段整合接触,其上与下寒武统筇竹寺组呈平行不整合接触。研究区内岩性较为复杂,藻凝块云岩、藻叠层云岩、藻砂屑云岩为主,灯四段厚度介于158~380m,总体呈西厚东薄、南北厚度分布相对稳定的特征。灯四段沉积时期经历两期快速海侵缓慢海退的旋回,在中部发育一套具有高GR低能相碳酸盐岩,依据该旋回可将灯四段分为灯四$_1$(第一旋回)和灯四$_2$(第二旋回),有效储层主要发育在灯四$_2$中上部的藻云岩和砂屑云岩中。储层以次生粒间溶孔、晶间溶孔、中小溶洞为主。数字岩心重构分析表明,溶洞形状多为扁圆形、条带状顺层分布。储层柱塞样孔隙度平均值为3.87%,渗透率平均值为0.51mD,全直径孔隙度平均值为3.97%,水平渗透率平均值为2.89mD,表现为低孔隙度低渗透率特征。

高磨地区震旦系气藏储量规模大,目前气藏开发正处于快速上产阶段,完钻井试气特征和开发井试采特征研究表明有效储层发育受沉积+岩溶控制,特别是岩溶古地貌对于缝洞型储层的发育程度及发育规模控制作用明显,导致优势古地貌单元高产井比例大,因此岩溶古地貌特征在一定程度上制约气藏快速建产和高效开发。目前对于四川盆地和高磨地区震旦系顶部岩溶风化壳古地貌的恢复结果存在极大的差异[4-7]。因此,基于前人研究成果结合高磨地区特征,论证高磨震旦系岩溶古地貌恢复方法,开展古地貌恢复对于震旦系气藏产能建设发挥重要支撑作用。

## 2 古地貌恢复方法筛选

### 2.1 古地貌恢复方法综述

目前常用的古地貌恢复方法[14]主要包括盆地分析回剥法[15]、层拉平方法、地震古地貌学方法[16]、层序地层恢复方法[17]、沉积学分析法[18]、残厚法[19,20]、印模法[21]等。但这些方法均需要借助钻井、录井、岩心、地震资料等,实现在某一构造级别、某一精度范围内的古地貌恢复。每一种方法在进行古地貌恢复的过程中有自己的优势,但也存在一些不足(表1)。

## 表 1 不同古地貌恢复方法对比[14-20]

| 方法 | 技术要点 | 应用资料 | 优缺点 |
| --- | --- | --- | --- |
| 沉积学方法 | 关键是古地形、古环境、古构造三者的有机统一,其基本内容是:<br>①利用沉积前古地质图、地层等厚图、砂岩等厚图、岩相古地理图分析古地形;<br>②结合岩相、成因相、古流向分析古环境;<br>③依据构造演化史分析古构造 | 钻井、岩心、薄片、录井资料 | 优点:综合性强,定性—半定量化<br>缺点:影响因素复杂,基础图件多,工作量大 |
| 印模法 | 关键是选取上覆标志层,其基本要求是:<br>①是全区范围内分布的等时界面,能够代表当时的海平面;<br>②该沉积界面离风化壳越近越好;<br>③地层厚度要有地震解释成果图作为依据 | 钻井、录井资料 | 优点:易操作,半定量化,地层厚薄的变化能够迅速反映出古地势背景信息<br>缺点:上覆标志层不易确定,地层去压实矫正难度较大,没有地震资料约束,误差较大 |
| 残厚法 | 关键是选取下伏地层基准面,其基本要求是:<br>①所选基准面必须为等时界面,不能发生穿层现象;<br>②基准面距离风化壳越近越好;<br>③地层厚度要有地震解释成果图作为依据 | 钻井、录井资料 | 优点:直观真实,易操作,半定量化<br>缺点:未考虑沉积前地形及剥蚀差异的影响,误差大 |
| 层拉平方法 | 关键是盆地大量研究基础之上的界面选取,其基本流程是:<br>①对盆地的古地质背景和古构造特点进行分析;<br>②选定对比层序的参照顶底面,利用多井合成记录对参照面标准层进行精细解释;<br>③利用相关的物探软件进行顶面层序拉平操作,此时得到的底面形态就是该层序的沉积前的相对古地貌 | 地震、钻井资料 | 优点:可以对较大工区范围内(盆地级或者是断陷级)进行古地貌恢复<br>缺点:恢复精度不够,误差较大 |
| 高分辨率学方法 | 关键是高分辨率层序等时地层格架的建立以及上覆对比参照面的选取,其基本流程是:<br>①建立高分辨率层序地层格架,正确划分基准面旋回级次;<br>②进行井间对比,同时为了提高恢复精度,应用压实系数进行厚度矫正;<br>③选取上覆等时界面(最大洪泛面或者是层序边界),求取参照面到不整合面的厚度,进而分析古地貌分布特征 | 测井、录井、岩心、薄片、钻井、地震资料 | 优点:选择的上覆基准面相对等时性强,理论上更接近原始古地貌,精度更高<br>缺点:工区范围广,恢复误差较大;没有地震资料约束,平面上恢复精度较低 |
| 双界面法 | 综合高分辨率方法、层拉平方法和印模法筛选上覆标志面和下伏基准面开展古地貌恢复:<br>①建立高分辨率层序地层格架,多因素筛选上覆标志面;<br>②上覆标志面拉平,得到震旦系顶部岩溶古地貌;<br>③依据工区范围筛选下伏标志面,将高度值转变为厚度值,为定量刻画奠定基础 | 测井、录井、岩性、薄片、地震、钻井资料 | 优点:选择的上覆基准面相对等时性强,理论上更接近原始古地貌,精度更高;以下伏基准面为基准,实现古地貌值定量化,为定量刻画微地貌奠定基础<br>缺点:恢复结果受地震资料精度限制 |

## 2.2 "双界面"岩溶古地貌恢复方法及其步骤

经典的古地貌恢复方法包括残厚法和印模法,印模法地层厚度的变化趋势能够迅速反馈出研究区古地势背景信息,但上覆标志层的选取容易存在穿时性,造成恢复精度不够;残厚法中残留厚度的厚薄反应古地貌相对位置的高低,但是不能反映古构造背景。因此高磨地区震旦系岩溶古地貌恢复采用"双界面"法[14],该方法恢复岩溶古地貌原理本质上仍旧等同于印模法,依靠上覆地层对风化壳界面的填平补齐实现对古地貌相对位置高低的刻画和表征。"双界面"法综合高分辨层序地层学方法、印模法、残厚法以及层拉平方法各自的优点,采用上覆标志面来表征风化壳古地貌起伏,采用下伏基准面作为表征风化壳古地貌的基准,实现恢复范围古地貌恢复结果的横向可对比性,为微地貌单元的定量刻画奠定了基础。

"双界面"岩溶古地貌恢复法操作步骤包括以下3个方面:(1)该方法吸收高分辨率层序地层恢复方法优点,划分上覆地层高分辨率层序地层,依据工区构造级别及工区规模、范围大小,筛选可对比追踪上覆等时地层界面,多因素分析优选上覆标志面,理论上认为在上覆标志面沉积期全区范围内实现对风化壳界面的填平补齐;(2)钻井和地震资料相互结合,采用层拉平手段,风化壳上覆地层界面高度大小即代表风化壳古地貌高低,该过程同印模法原理相同,但在这一过程中采用地震和钻井资料相互结合,采用上覆标志面拉平的手段,更加准确刻画古地貌差异;(3)在工区范围内选择下伏基准面,将古地貌高度值转换为厚度(相对)值,以厚度大小来描述古地貌差异,为微地貌单元的定量刻画奠定了基础。需要说明的是,下伏基准面不是一个几何面,具有其物理意义,该界面平行于本工区范围内矫正原始构造相对位置高低幅度、灯三段差异沉积厚度和灯二段顶部原始古地貌高低之后上一期构造填平补齐界面。由此可以看出,"双界面"法综合现有古地貌恢复方法的优点,规避了现有方法的各种不足,能够更加精确实现对原始古地貌的恢复(表1)。

# 3 上覆标志面选取

上覆标志面的选取是"双界面"法岩溶古地貌恢复的核心,其选取结果关系到古地貌恢复的精度。

## 3.1 上覆地层标志面选取原则

上覆标志面的选取主要有以下原则:(1)实现填平补齐:必须在恢复区范围内实现对下伏灯影组顶面风化壳的填平补齐;(2)具备等时属性:该界面必须是等时界面,大致平行于当时的海平面;(3)拥有最近距离:该标志面距离风化壳面越近越好,越接近风化壳受后期构造活动影响越小,该标志面与风化壳面间的地层厚度越能反映古地貌形态;(4)容易识别追踪:该界面地震反射波特征明显,在地震剖面上容易识别和对比。

## 3.2 上覆地层层序特征

震旦系上覆地层为寒武系,四川盆地寒武系以台地相沉积为特点[21],地层层序完整。高磨地区下寒武统自下而上发育麦地坪组、筇竹寺组、沧浪铺组和龙王庙组。高分辨率层序地层分析表明,下寒武统划分为一个长期基准面旋回,4个中期基准面旋回,4个中期基准面旋回分别对应麦地坪组、筇竹寺组、沧浪铺组和龙王庙组(图2)。高磨地区大部分完钻井不发育麦地坪组,仅发育3个中期基准面旋回。4个中期基准面旋回转换面和海泛面具有明显的岩性、电性识别标志,完钻井之间可以对比追踪。

### 3.3 上覆地层发育特征

通过对下寒武统(麦地坪组至龙王庙组)层序特征进行分析,以中期基准面旋回转换面、海泛面为界,灯影组上覆地层存在5个等时界面(图3):筇竹寺组内部最大海泛面、筇竹寺组与沧浪铺组岩性转换面、沧浪铺组内部次级海泛面、沧浪铺组与龙王庙组岩性突变面及龙王庙组顶界面,该5个界面在全区基本可以对比追踪,为上覆标志面的选取奠定了基础。

#### 3.3.1 界面发育特征

(1)筇竹寺组内部最大海泛面。筇竹寺组内部最大海泛面在测井曲线上表现为异常高GR特征,特征明显,井间易于对比。但由于磨溪区块较高石梯区块和台内区块整体地势低,磨溪区块在筇竹寺组沉积早期接受沉积,发育上升半旋回,而高石梯—磨溪区块和台内区域未能接受沉积,不发育基准面上升半旋回。由于距灯影组顶部风化壳较近,未能实现填平补齐,因此不能作为标志层。

(2)筇竹寺组与沧浪铺组岩性转换面、沧浪铺组内部次级海泛面。筇竹寺组与沧浪铺组为一砂泥岩转换面,筇竹寺组以泥岩为主,沧浪铺组以三角洲和碎屑滨岸相砂岩为主,因此在GR曲线上有较好的响应特征。沧浪铺组内部次级海泛面表现为GR值由低到高再变低,GR高值对应次级海泛细粒沉积,这两个界面完钻井特征明显,发育较为稳定,可以作为标志面选取。

(3)沧浪铺组与龙王庙组岩性突变面、龙王庙组顶界面。沧浪铺组与龙王庙组为碎屑岩与碳酸盐岩岩性突变界面,GR曲线表现为龙王庙组低GR和沧浪铺组高GR呈突变式接

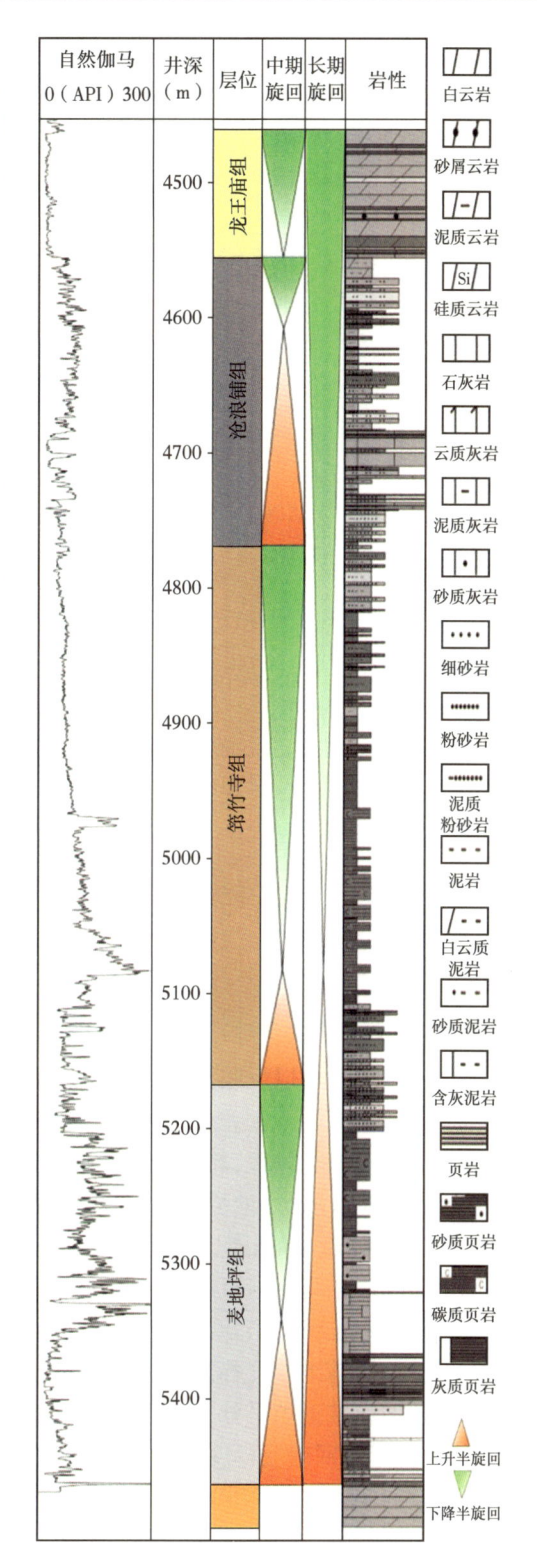

图2 高磨地区下寒武统层序特征

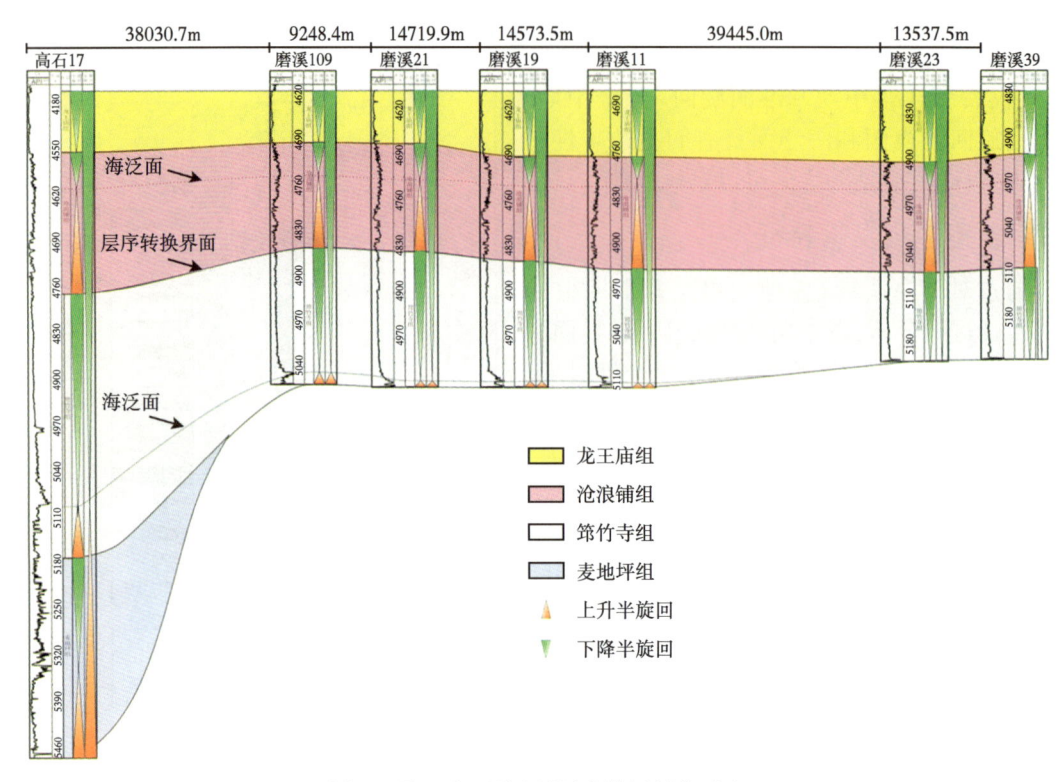

图 3　震旦系上覆地层层序旋回划分对比

触,特征十分明显。龙王庙组与上覆高台组为平行不整合接触,岩性、岩相及测井曲线突变接触,容易识别划分与对比。这两个界面在工区内发育稳定,特征明显易于识别,可作为标志面选取。

### 3.3.2　上覆地层厚度分析

地层厚度差异可以反映出在该地层沉积前原始古地貌的高低差异,如果厚度差异越大表明沉积前古地貌高低幅度差异越大,如果厚度差异越小代表沉积前古地貌高低幅度差异越小。麦地坪组仅在海槽区发育,筇竹寺组沉积时期厚度介于 60~400m,由海槽向台地地层厚度差异较大,地层西厚东薄,地层正处于填平补齐阶段;沧浪铺组沉积时期地层厚度介于 120~230m,龙王庙组沉积时期地层厚度介于 52~129m,平面上龙王庙组厚度分析差异较小,表明该沧浪铺组时期海槽与台地沉积基本实现了填平补齐(图4)。

龙王庙组和沧浪铺组地层厚度统计结果表明,相比沧浪铺组,龙王庙组厚度分布更为集中。完钻井沧浪铺组厚度标准差为 20.7,龙王庙组厚度标准差为 14.1,因此沧浪铺组顶面比筇竹寺组顶面更接近填平补齐(图5)。沧浪铺组内部上升与下降旋回地层厚度统计结果表明(图6),上升半旋回地层厚度标准差为 19,厚度分布分散,下降半旋回地层厚度标准差为 8.8,厚度分布较为集中。下降半旋回厚度分布的均一化表明,与其他界面相比,沧浪铺组内部海泛面最接近填平补齐。

地质应用篇

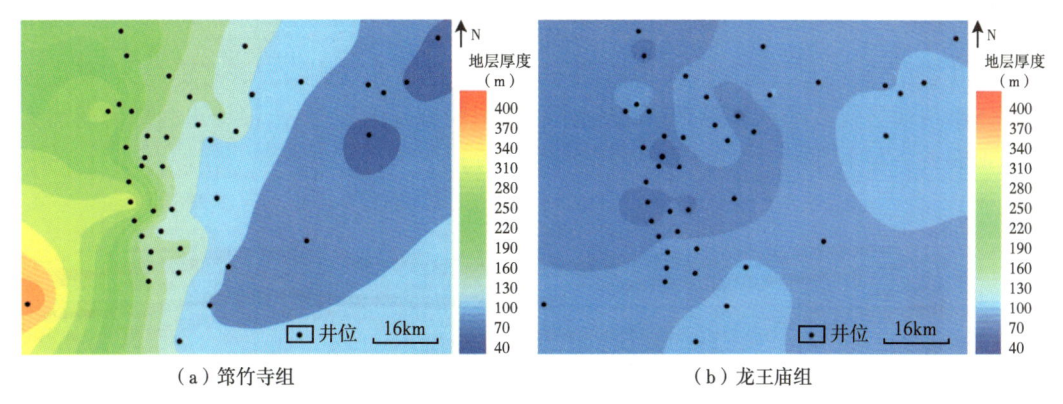

图 4　灯影组上覆地层厚度平面图
（a）筇竹寺组
（b）龙王庙组

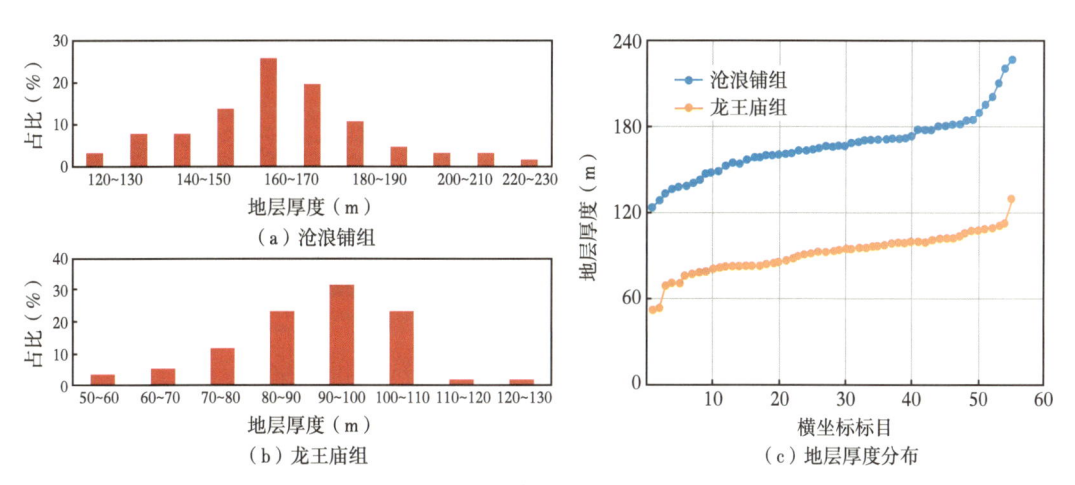

图 5　沧浪铺组和龙王庙组沉积地层厚度分布特征
（a）沧浪铺组
（b）龙王庙组
（c）地层厚度分布

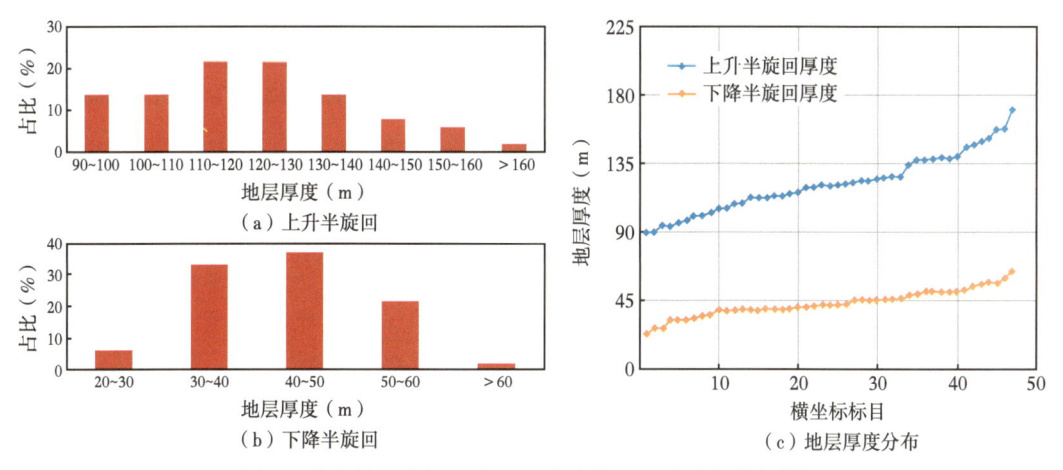

图 6　沧浪铺组内部上升和下降半旋回地层厚度分布特征
（a）上升半旋回
（b）下降半旋回
（c）地层厚度分布

— 189 —

## 3.4 地震特征分析

通过井震结合,标定层面对应同相轴,在侧向对同相轴进行追踪。研究发现,灯影组顶面和龙王庙组顶面呈强波峰反射特征;沧浪铺组顶面呈中强反射波谷;沧浪铺组内部海泛面为沧浪铺组顶面波谷与下部强反射波峰之间转换面(图7)。

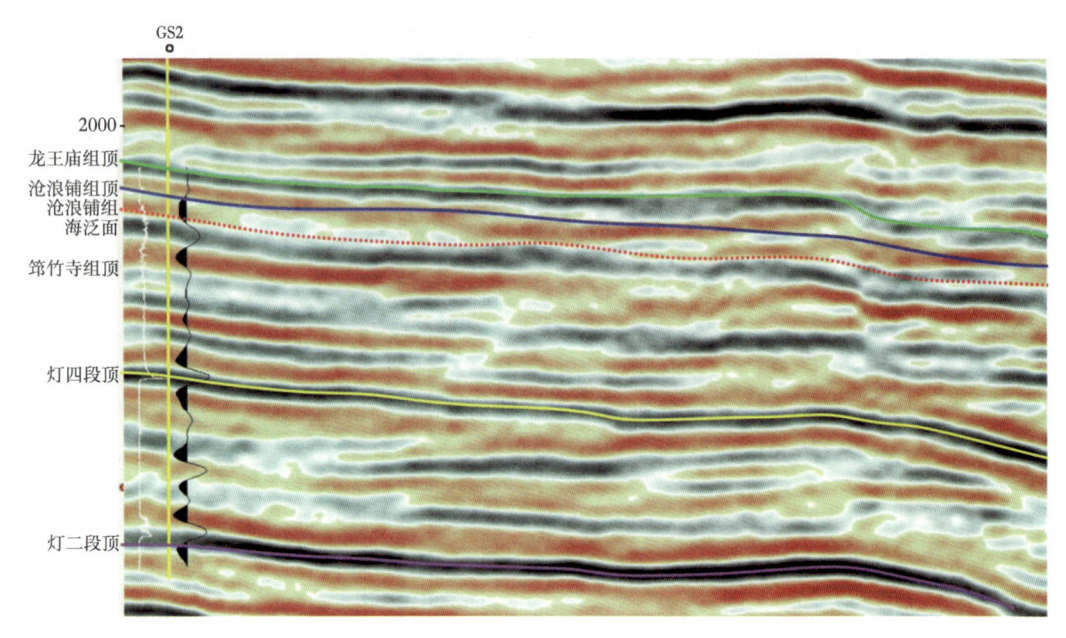

图7 高石2井合成地震记录层位标定及各界面侧向追踪结果

筇竹寺组顶面呈弱波峰反射,部分地区特征较为明显,但侧向连续性较差,难以大范围对比追踪。沧浪铺组内部海泛面位于沧浪铺组顶面红轴之下,介于红轴与下部强反射黑轴之间,区域可追踪性较好。沧浪铺组顶面及龙王庙组顶面能在研究区内稳定追踪,可靠程度较高。

综合考虑上覆标志面选取的各项指标,结合上覆标志面选取的原则选取沧浪铺组内部海泛面为上覆标志面(表2)。沧浪铺内部海泛面在高磨地区尚未对其进行追踪解释,在古地貌恢复过程中前人也没有对其进行有效利用。一方面选取沧浪铺内海泛面作为上覆标志面进行

表2 上覆标志面综合选取评价

| 备选界面 | 区域稳定性 | 单井特征 | 填平补齐 | 距风化壳距离 | 地震可追踪性 | 地震追踪工作现状 | 标志面选取 |
|---|---|---|---|---|---|---|---|
| 龙王庙组顶面 | 好 | 明显 | 是 | 远 | 好 | 部分追踪 | 否 |
| 沧浪铺组顶面 | 好 | 明显 | 是 | 较远 | 好 | 部分追踪 | 否 |
| 沧浪铺组内部海泛面 | 好 | 明显 | 是 | 较近 | 好 | 没有 | 是 |
| 沧浪铺组泥岩顶面 | 较差 | 不明显 | 基本 | 近 | 较好 | 部分追踪 | 否 |
| 筇竹寺组顶面 | 好 | 明显 | 基本 | 近 | 差 | 部分追踪 | 否 |
| 筇竹寺组内部海泛面 | 差 | 明显 | 否 | 近 | 差 | 没有 | 否 |

古地貌恢复更加科学合理,对解释结果更加精细,另一方面也是对古地貌恢复的另一种尝试,本次古地貌恢复结果与其他标志面所做结果的对比能够验证"双界面"岩溶古地貌恢复方法的适用性,也可以通过分析差异,增加对该地区岩溶古地貌恢复方法的深层次理解及对高磨岩溶古地貌特征的认识。在具体的实施过程中,沧浪铺内部海泛面为三级海泛面,其地震、测井识别标志比较明显,测井上为沧浪铺内部高 GR 值处,地震上位于沧浪铺组顶面红轴之下,介于红轴与下部强反射黑轴之间,区域可追踪性较好(图3、图7),能够进行全区域的有效对比追踪和精细解释。

## 4 古地貌恢复及分布特征

### 4.1 古地貌恢复

#### 4.1.1 上覆标志层对比追踪

确定上覆地层标志面之后,完成工区范围内 58 口完钻井的合成记录,建立骨架井剖面,完成骨架剖面层位追踪。由于工区范围内完钻井较少,井覆盖区域以井点约束为主,无完钻井区域以沉积模式为指导,结合同相轴反射特征开展层位解释。通过多地震剖面对比和地质分析,沧浪铺组内部海泛面之上发育三角洲和滨岸相砂体前积,通过该特征可准确定海泛面位置,从而完成对上覆标志面的全区层位解释。

#### 4.1.2 下伏基准面选取刻画岩溶古地貌

上覆标志面追踪之后,通过时深转换,将上覆标志面时间域转化成深度域,转换后深度与井点深度误差相比,平均误差小于 0.1m,97% 的井误差小于 1m,准确度较高。根据镜像原理,对沧浪铺组内部海泛面进行层拉平,灯影组顶面即为寒武系沉积前古地貌,此时灯影组顶面高低起伏是通过以上覆标志面为 0 值的深度表征,对于在具体的微地貌表征过程中采用深度值不方便,有必要将深度值转化成厚度值,通过厚度值来反映岩溶古地貌的高低起伏。通过工区范围大小,优选灯影组顶面下部通过最低古地貌位置且平行于拉平后的上覆标志面的界面为下伏基准面(图8),实现了古地貌值(厚度值)来刻画古地貌的高低,实现古地貌定量刻画。

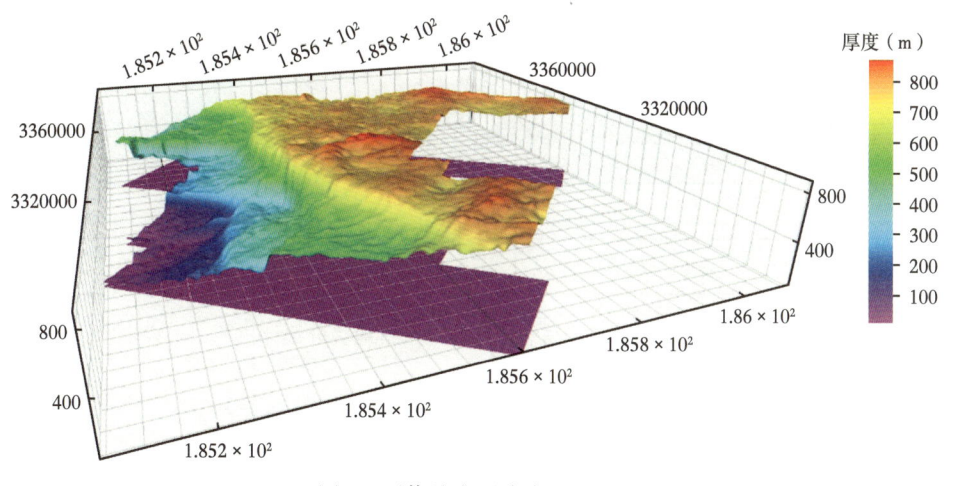

图 8 下伏基准面选取

## 4.2 古地貌分布

### 4.2.1 古地貌分级评价

碳酸盐岩岩溶古地貌平面上分区特征明显,一般来说,岩溶古地貌在平面上存在岩溶高地、岩溶斜坡和岩溶盆地3个单元。古地貌单元的分级涉及古地貌的范围,如果以盆地级别为古地貌恢复范围,综合前人研究成果[22],结合古地貌学词典对于各单元的定义[23],笔者认为与盆地级别对应的古地貌单元应该划分为岩溶高地、岩溶斜坡、岩溶低地、岩溶洼地和岩溶盆地5个一级古地貌单元。岩溶高地整体处于岩溶地貌的高部位,长期处于风化剥蚀及大气淡水淋滤状态,以地表岩溶作用为主,伴生风化残积物。地表水以径流为主,岩溶形态以漏斗、溶沟等为主,多被后期沉积的泥砾和角砾等混合充填,储层质量差,不利于油气的聚集。岩溶斜坡是岩溶高地与岩溶洼地之间的过渡带,呈环状发育于岩溶高地周缘,是大气淡水垂直渗流和水平潜流溶蚀最强烈的部位,发育厚层的垂直渗流带和水平潜流带,岩溶斜坡一般会依据坡度大小分为缓坡和陡坡2个二级古地貌单元。岩溶台地由坡度较陡的台坡和坡度较缓的台面组成,台面水平投影面积大于台坡投影面积,是由构造运动上升而明显比周围地区地面高并主要遭受剥蚀的"正地貌"单元,一般包括残丘、台坡、台面和洼地4个二级古地貌单元。岩溶低地往往与地质构造有关,是构造运动下降而明显比周围地区地面低的"负地貌"单元,岩溶低地一般被岩溶斜坡包围。岩溶盆地是地下水的汇聚泄流区,水流以地表径流和停滞水为主,并与广海相连。

高磨地区古地貌单元仅仅存在岩溶台地、岩溶斜坡和岩溶低地3个一级地貌单元,同时依据一级地貌单元内地貌高低、坡度等差异将岩溶台地和岩溶斜坡划分为台坡、台面、洼地、残丘、沟谷、陡坡和缓坡等二级古地貌单元(图9)。

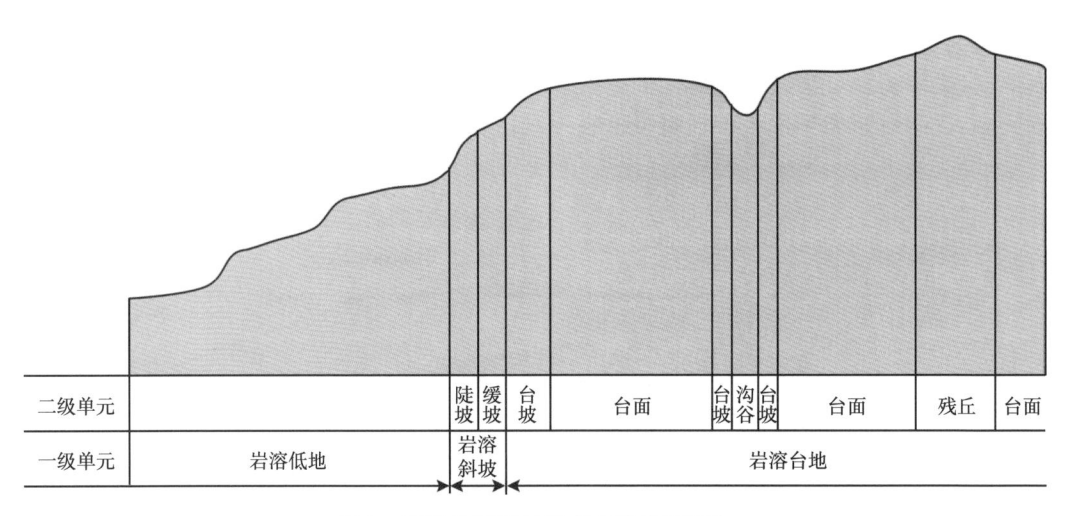

图9 高磨地区震旦系古地貌分级评价

### 4.2.2 古地貌分布特征

对于整个四川盆地震旦系顶部岩溶古地貌恢复结果,岩溶高地主要分布在四川盆地西侧,岩溶盆地主要分布于四川盆地东部及东北部,四川盆地大部分处于岩溶斜坡位置,局部发育南北向岩溶低地和岩溶台地。针对于高磨地区,受裂陷槽影响,自东向西发育岩溶低地、岩溶斜坡和岩溶台地,这与本次高磨地区恢复古地貌认识相符(图10)。高石梯磨溪古地貌分布存在较大差异,高石梯地区以陡坡为主,磨溪地区以缓坡为主。同时高石梯古地貌差异较大,残丘、台面普遍发育,而磨溪古地貌差异较小,以台面和缓斜坡为主。本次恢复结果表明,高磨基本格局为西低东高,与前人的研究成果基本一致。但本次古地貌的恢复结果更加精细:(1)经过综合地质分析表明,高石梯和磨溪之间的沟谷受断层发育影响,该断层较为古老,同时控制高磨震旦系原始沉积和震旦系顶部岩溶风化壳格局;(2)高石梯和磨溪古地貌格局存在差异,高石梯整体地貌更高,磨溪整体地貌更低,究其原因受原始沉积格局及后期差异升降作用影响,高石梯构造幅度差异大,斜坡以陡坡为主,磨溪地区构造幅度差异较小,斜坡以缓坡为主;(3)磨溪台地内部和台缘带古地貌差异明显,台缘带古地貌更多受控于差异沉降作用,台内古地貌更多受控于沉积特别是岩性的差异性;(4)古地貌恢复结果更加精细,体现不同区域、不同井区微地貌分布特征的差异,下一步可尝试定量刻画岩溶微地貌,探索微地貌单元同高产井的定量、半定量关系。

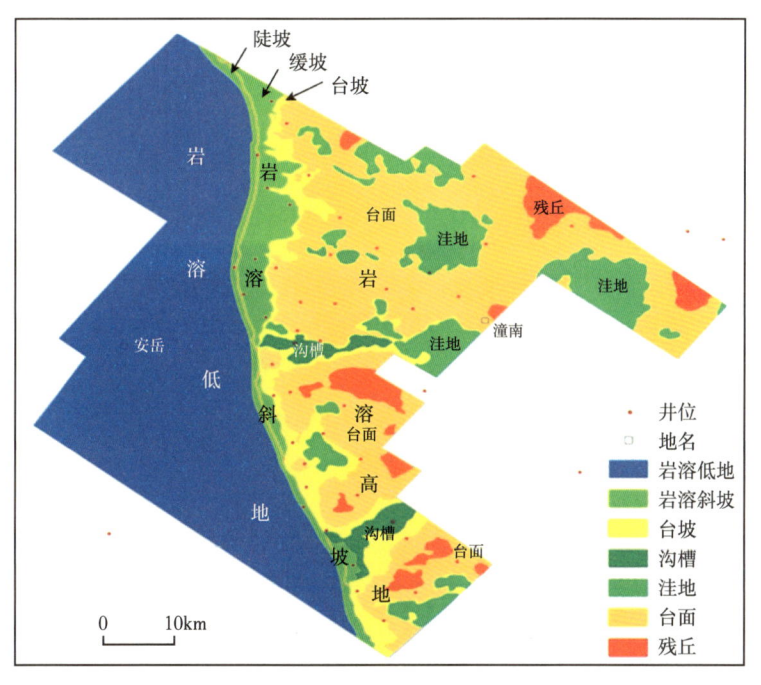

图10 高磨地区古地貌分布特征

## 4.3 古地貌分布模式

受多期构造运动影响,四川盆地震旦系顶部风化壳古地貌同鄂尔多斯下古风化壳模式不同,鄂尔多斯下古生界风化壳为典型的岩溶高地、岩溶斜坡和岩溶坡地模式[24-28],优质储层主要发育在岩溶斜坡上,可以说是教科书式的古地貌模式。但是四川盆地震旦系基地发育受多期次拉张运动影响,发育多期次裂陷槽,后期受桐湾运动影响,虽然整体上仍然表现为西部岩溶高地,中部岩溶斜坡和东部岩溶盆地的特征,但是局部地区发育岩溶低地和岩溶台地,这就

— 193 —

造成四川盆地整个盆地的岩溶模式同局部地区的岩溶模式不同或者相反的现象,以高磨地区为例,古地貌表现为东高西低的特征。

对于高磨地区,自西向东为岩溶低地、岩溶斜坡和岩溶高地的分布特征,同时整体格局表现为"两沟三区、北缓南陡"的特征。南北两个沟谷将高磨地区古地貌划分为高石梯南区、高石梯区和磨溪区块3个单元(图11)。在分区上,高石梯南和高石梯分布特征相似,均表现为较陡的斜坡,同时古地貌幅度差异较大,发育残丘和洼地微地貌。而磨溪台缘表现为较大部分的缓坡和台面,内部发育台面、残丘和洼地微地貌单元。

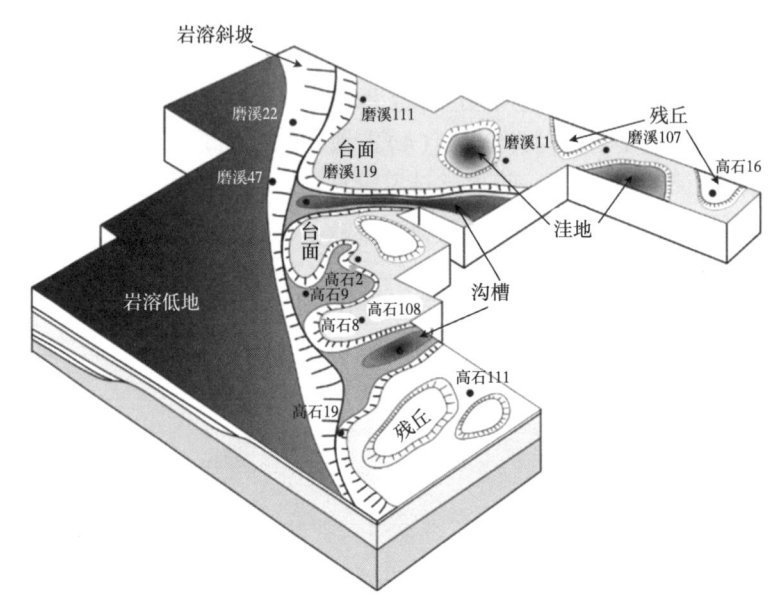

图11 高磨地区震旦系岩溶古地貌模式

## 5 古地貌对高产井的控制作用

按照无阻流量大于$100×10^4m^3$定为高产井的标准。平面上,高产井主要分布在台地边缘。整体上高磨地区除陡坡、残丘、洼地和沟谷之外,缓坡、台坡和台面3个微地貌单元溶蚀作用强烈,有效储层发育,测试井无阻流量高。对于各微地貌单元,岩溶台坡溶蚀作用最强,形成储层质量最好,古地貌单元平均无阻流量为$98×10^4m^3$,高产井占比为27%,受原始沉积体规模、物性和早期成岩作用影响,台坡内部完钻井测试产量差异较大(图12)[29-35]。岩溶缓坡溶蚀作用较强,形成的储层质量相对较好,完钻井平均无阻流量达$72×10^4m^3$,高产井占比44%。岩溶台面溶蚀作用也较强,但高石梯和磨溪差异较大,该单元平均无阻流量为$79×10^4m^3$,高产井占比为43%。在分区块上,高石梯好于磨溪,高石梯完钻井集中分布在缓坡、台坡和台面,磨溪地区完钻井仅仅分布在缓坡,台坡完钻井数少不具有代表性,台面完钻井数多,效果较差。以台面微地貌单元为例,高石梯台面古地貌位置相对较高,淋滤风化条件优越,岩溶储层发育程度和发育规模较大,完钻井测试平均无阻流量达$128×10^4m^3$,而磨溪台面古地貌位置相对较低,风化条件较差,岩溶储层发育不充分,完钻井测试平均无阻流量仅仅$10×10^4m^3$。

依据古地貌划分结果,结合优势沉积发育相带、高产井地震影响模式对高磨地区部署试采井及开发建产井位 9 口,4 口井无阻流量大于 $120\times10^4 \text{m}^3/\text{d}$,只有 2 口井无阻流量低于 $60\times10^4 \text{m}^3/\text{d}$,9 口完钻井平均无阻流量达到 $105\times10^4 \text{m}^3/\text{d}$,比前期提高 78%,开发效果较好。

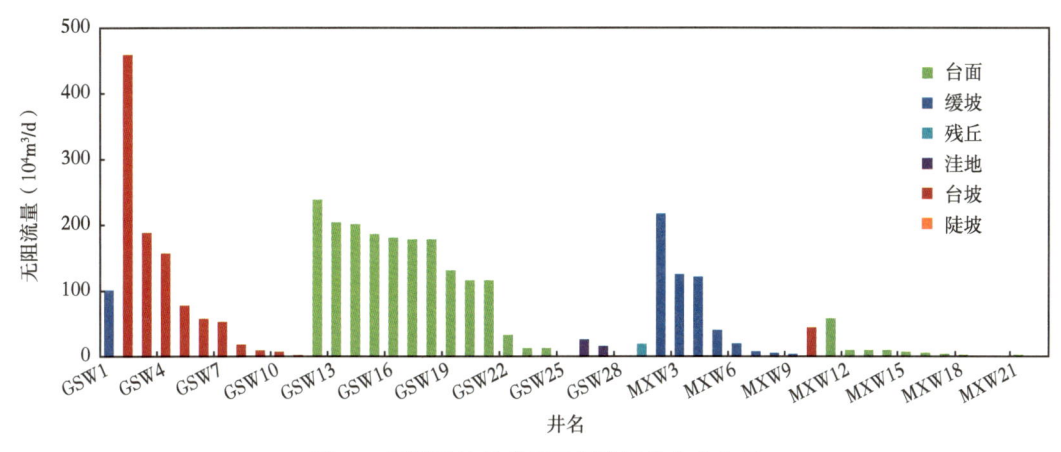

图 12 不同微地貌单元无阻流量分布直方图

## 6 对气藏开发的指导意义

(1)古地貌对高产井控制作用研究表明,除去沟谷、洼地之外,高磨地区台地边缘的缓坡、台坡和台面古地貌单元有效储层较发育,完钻井无阻流量较高,是高磨地区建产主体,也是高磨外围滚动增储的潜力区。但是需要精细刻画沉积期的水道和剥蚀期的沟谷分布,避免开发井落空。

(2)高石梯和磨溪区块古地貌控制作用差异明显,高石梯地区高产井主要分布在台面、台坡和缓坡,其中台面差异较小,台坡无阻流量较高,但差异较大,缓坡仅仅完钻 1 口井,不具有代表性。磨溪地区高产井主要分布在缓坡,占更大面积的台面有效储层发育较差,完钻井效果较差。高石梯的古地貌划分标准是否适合磨溪地区值得探讨。构造、沉积、岩溶等的差异性特征,要求分区建立相对的古地貌划分标准,从而指导不同区块开发有利区筛选和井位部署。

(3)高磨地区除去台地边缘之外,大面积的台地内部勘探程度低,整体完钻井效果较差。综合沉积、古地貌和微裂缝研究,在广大的台内地区寻找有利建产区块对于高磨地区震旦系的规模建产和长期稳产具有重要的意义。

## 7 结论

(1)综合残厚法、印模法、高分辨率法和层拉平法,采用"双界面"法对高磨地区开展古地貌恢复,综合筛选沧浪铺组内部海泛面作为上覆标志面对高磨地区进行古地貌恢复,同时确定下伏基准面,将古地貌高度值转化为厚度值,为定量刻画奠定基础。

(2)高磨地区古地貌自西向东分为岩溶低地、岩溶斜坡和岩溶台地 3 个一级地貌单元,同时将岩溶斜坡划分为缓坡和陡坡 2 个二级古地貌单元,将岩溶台地划分为台坡、台面、洼地和沟谷 4 个二级古地貌单元。高磨地区古地貌分布呈现"两沟三区、北缓南陡"的特征,高石梯地区大面积分布台坡、台面和残丘,磨溪地区仅发育斜坡和台面,台内发育台面、残丘和洼地。高产井主要分布在台坡、台面和缓坡上,高石梯和磨溪之间存在较大差异。

(3)古地貌分布特征及对高产井控制作用分析表明,高磨地区台地边缘完钻井数多、勘探开发程度深,是目前建产的最可靠地区。高石梯和磨溪地区的差异性特征,要求分区建立古地貌划分标准,同时随着动静态资料的逐渐丰富,综合沉积、裂缝发育特征可以优选台地内部建产有利区,从而支撑四川盆地高磨地区震旦系气藏快速建产和长期稳产。

## 参 考 文 献

[1] 邹才能,杜金虎,徐春春,等.四川盆地震旦系——寒武系特大型气田形成分布、资源潜力及勘探发现[J].石油勘探与开发,2014,41(3):278-293.

[2] 魏国齐,王志宏,李剑,等.四川盆地震旦系、寒武系烃源岩特征、资源潜力与勘探方向[J].天然气地球科学,2017,28(1):1-13.

[3] 魏国齐,杨威,杜金虎,等.四川盆地高石梯—磨溪古隆起构造特征及对特大型气田形成的控制作用[J].石油勘探与开发,2015,42(3):257-265.

[4] 李启桂,李克胜,周卓铸,等.四川盆地桐湾不整合面古地貌特征与岩溶分布预测[J].石油与天然气地质,2013,34(4):516-521.

[5] 汤济广,胡望水,李伟,等.古地貌与不整合动态结合预测风化壳岩溶储集层分布——以四川盆地乐山—龙女寺古隆起灯影组为例[J].石油勘探与开发,2013,40(6):674-681.

[6] 汪泽成,姜华,王铜山,等.四川盆地桐湾期古地貌特征及成藏意义[J].石油勘探与开发,2014,41(3):305-312.

[7] 罗思聪.四川盆地灯影组岩溶古地貌恢复及意义[D].成都:西南石油大学硕士论文,2015.

[8] 金民东,谭秀成,童明胜,等.四川盆地高石梯—磨溪地区灯四段岩溶古地貌恢复及地质意义[J].石油勘探与开发,2017,44(1):58-68.

[9] 闫海军,何东博,许文壮,等.古地貌恢复及对流体分布的控制作用——以鄂尔多斯盆地高桥区气藏评价阶段为例[J].石油学报,2016,37(12):1483-1494.

[10] 杨跃明,文龙,罗冰,等.四川盆地乐山—龙女寺古隆起震旦系天然气成藏特征[J].石油勘探与开发,2016,43(2):179-188.

[11] 杜金虎,邹才能,徐春春,等.川中古隆起龙王庙组特大型气田战略发现与理论技术创新[J].石油勘探与开发,2014,41(3):268-277.

[12] 金民东.高磨地区震旦系灯四段岩溶型储层发育规律及预测[D].成都:西南石油大学博士论文,2017.

[13] 许海龙,魏国齐,贾承造,等.乐山—龙女寺古隆起构造演化及对震旦系成藏的控制[J].石油勘探与开发,2012,39(4):406-416.

[14] 黄捍东,罗群,王春英,等.柴北缘西部中生界剥蚀厚度恢复及其地质意义[J].石油勘探与开发,2006,33(1):44-48.

[15] 加东辉,徐长贵,杨波,等.辽东湾辽东带中南部古近纪古地貌恢复和演化及其对沉积体系的控制[J].古地理学报,2007,9(2):155-166.

[16] 赵俊兴,陈洪德,向芳.高分辨率层序地层学方法在沉积前古地貌恢复中的应用[J].成都理工大学学报:自然科学版,2003,30(1):76-81.

[17] 赵俊兴,陈洪德,时志强.古地貌恢复技术方法及其研究意义——以鄂尔多斯盆地侏罗纪沉积前古地貌研究为例[J].成都理工学院学报,2001,28(3):260-266.

[18] 王敏芳,焦养泉,任建业,等.沉积盆地中古地貌恢复的方法与思路——以准噶尔盆地西山窑组沉积期为例[J].新疆地质,2006,24(3):326-330.

[19] 何自新,郑聪斌,陈安宁,等.长庆气田奥陶系古沟槽展布及其对气藏的控制[J].石油学报,2001,22(4):35-38.

[20] 庞艳君,代宗仰,刘善华,等.川中乐山—龙女寺古隆起奥陶系风化壳古地貌恢复方法及其特征[J].石油地质与工程,2007,21(5):8-10.

[21] 杜金虎.古老碳酸盐岩大气田地质理论与勘探实践[M].北京:石油工业出版社,2015:18.

[22] 刘宏,罗思聪,谭秀成,等.四川盆地震旦系灯影组古岩溶地貌恢复及意义[J].石油勘探与开发,2015,42(3):283-293.

[23] 周成虎.地貌学词典[M].北京:中国水利水电出版社,2006.

[24] 马振芳,付锁堂,陈安宁.鄂尔多斯盆地奥陶系古风化壳气藏分布规律[J].海相油气地质,2000,5(1/2):98-102.

[25] 王雪莲,王长陆,陈振林,等.鄂尔多斯盆地奥陶系风化壳岩溶储层研究[J].特种油气藏,2005,12(3):32-35.

[26] 兰才俊,徐哲航,马肖琳,等.四川盆地震旦系灯影组丘滩体发育分布及对储层的控制[J].石油学报,2019,40(9):1069-1084.

[27] 杨跃明,杨雨,杨光,等.安岳气田震旦系、寒武系气藏成藏条件及勘探开发关键技术[J].石油学报,2019,40(4):493-508.

[28] 魏国齐,杨威,谢武仁,等.四川盆地震旦系—寒武系天然气成藏模式与勘探领域[J].石油学报,2018,39(12):1317-1327.

[29] 李熙喆,郭振华,胡勇,等.中国超深层大气田高质量开发的挑战、对策与建议[J].天然气工业,2020,40(2):75-82.

[30] 李熙喆,郭振华,胡勇,等.中国超深层构造型大气田高效开发策略[J].石油勘探与开发,2018,45(1):111-118.

[31] 李熙喆,郭振华,万玉金,等.安岳气田龙王庙组气藏地质特征与开发技术政策[J].石油勘探与开发,2017,44(3):398-406.

[32] 闫海军,贾爱林,冀光,等.岩溶风化壳型含水气藏气水分布特征及开发技术对策——以鄂尔多斯盆地高桥区下古气藏为例[J].天然气地球科学,2017,28(5):801-811.

[33] 贾爱林,闫海军.不同类型典型碳酸盐岩气藏开发面临问题与对策[J].石油学报,2014,35(3):519-527.

[34] 贾爱林,闫海军,郭建林,等.不同类型碳酸盐岩气藏开发特征[J].石油学报,2013,34(5):914-923.

[35] 贾爱林,闫海军,郭建林,等.全球不同类型大型气藏的开发特征及经验[J].天然气工业,2014,34(10):33-46.

# 四川盆地九龙山气田珍珠冲组砂砾岩储层评价及有利区优选

张满郎　孔凡志　谷江锐　郭振华　付　晶　郑国强　钱品淑

(中国石油勘探开发研究院)

**摘要**：九龙山气田珍珠冲组气藏为非均质性很强的裂缝性致密砂砾岩储层。为明确开发层系,优选有利开发区块,利用岩心描述、铸体薄片观察、成像测井解释、全直径岩心分析等手段,对珍珠冲组开展了高分辨率层序地层划分,系统分析了储层沉积微相、岩石学特征、储集空间类型及裂缝分布特征,实现了对珍珠冲组裂缝性砂砾岩储层的精细描述。结果表明:(1)珍珠冲组可划分为3个中期旋回和6个短期旋回,有利储层为Ⅱ旋回扇三角洲前缘辫状水下分流河道多层叠置形成的巨厚砂砾岩体;(2)珍珠冲组为裂缝—孔隙型储层,发育砾间溶孔、砾内溶孔及裂缝扩溶孔,储层孔隙度主峰为2.0%~4.5%,水平渗透率为1~100mD,垂直渗透率为0.1~10.0mD,低角度缝和网状缝显著改善了储层的渗流能力;(3)Ⅱ3和Ⅱ1小层裂缝发育,渗透率较高,为2个有利开发层系;(4)通过单层评价,多层叠合,优选出2个一类有利区和3个二类有利区。该研究成果对九龙山气田开发部署具有指导意义。

**关键词**:致密气;砂砾岩储层;储层评价;珍珠冲组;九龙山气田;四川盆地

九龙山气田为极具特色的裂缝性致密砂砾岩储层。储层主体为厚层块状砾岩,砾径大,基质物性极差,裂缝发育且复杂多样。实践证明,裂缝可以有效改善储层物性,增加孔喉连通性,提高渗流能力,是砂砾岩油气藏的重要储集空间和渗流通道,因此开展储层特征研究对油气高效开发具有十分重要的指导作用[1-3]。前人对致密砂岩储层裂缝的发育特征、控制因素及其渗流规律作了大量研究;刘树根等[4]和卢虎胜等[5]关于厚层砂岩和砂泥岩互层岩石力学性质和裂缝形成机理研究;张惠良等[6]、王俊鹏等[7]、张博等[8]对储层裂缝定量评价和裂缝描述方法研究;张博等[9]、李世川等[10]、王珂等[11,12]对库车前陆盆地典型气田超深层致密砂岩储层裂缝发育特征及其形成主控因素研究;何鹏等[13]、白斌等[14]对九龙山气田须二段致密砂岩裂缝评价及其主控因素也进行了细致研究。在砾岩储层裂缝发育控制因素及分布预测方面也曾进行了不同程度的研究。砾岩、砂砾岩储层在准噶尔盆地西北缘、四川盆地西部、渤海湾断陷盆地等地的冲积扇相带发育[15-17],但裂缝性砾岩、裂缝性砂砾岩储层比较少见,以四川九龙山珍珠冲组、剑阁区块须三段砾岩储层为典型代表,前人在珍珠冲组砾岩裂缝地震预测[18]、岩心、露头裂缝描述、裂缝对储层渗透性的贡献[19-21]、储层敏感性[22]等方面开展了研究。但是对于九龙山气田珍珠冲组储层,前人笼统以一个大层进行研究,对该气藏的主力含气层系认识不清楚,对裂缝的研究手段相对单一[18-21]。

为了提高开发效果,对珍珠冲组进行了高分辨率层序地层划分对比,分析了储层的沉积微相、岩石学特征及储集空间类型,结合岩心描述、铸体薄片观察、全直径岩心分析和成像测井解释等手段,对珍珠冲组裂缝性砂砾岩储层开展精细评价,以期确定天然气开发层系和有利开发区块。

# 1 层序格架中的砂砾岩体展布

## 1.1 建立高分辨率层序地层格架

九龙山气田位于四川盆地北部米仓山前缘隆起带与川西坳陷北部梓潼凹陷的交会部位,其主要含气层位于下侏罗统珍珠冲组,钻厚130~206m,其中下部为石英砂岩砾石、石英岩质砾石和少量燧石砾石、石灰岩砾石组成的砾岩、砂砾岩地层,局部夹中细砂岩及薄层泥岩;其上部由数个浅灰或黄灰色细粒石英砂岩与黄绿色砂质泥岩、页岩韵律互层组成,该岩性段普遍夹薄煤层和煤线,富含菱铁矿结核,产有较丰富的陆相双壳类化石。

对于巨厚的冲积扇、扇三角洲砂砾岩地层,首先必需开展高分辨率层序地层划分,确定短期旋回与沉积微相及储层物性的对应关系,明确有利开发层系[16]。项目组依据钻测井资料,结合岩心描述成果,参考地震资料,完成了81口井的单井层序地层划分,编制9条层序地层对比剖面,建立了珍珠冲组的高分辨率层序地层格架。将珍珠冲组划分为3个中期旋回(Ⅰ、Ⅱ、Ⅲ)、6个短期旋回(Ⅰ1、Ⅱ1、Ⅱ2、Ⅱ3、Ⅲ1、Ⅲ2),每个短期旋回又细分为基准面上升或下降2个时期(图1)。

层序Ⅰ的底界为珍珠冲组与上三叠统须家河组之间的低角度不整合,地震剖面上具有上超充填或削截特征。九龙山地区须四段、须五段全遭剥蚀,须三段亦遭受不同程度的剥蚀。由于古隆起的影响,层序Ⅰ仅发育在九龙山东南部的龙17—龙112—龙111—龙108井区,地层厚度为10~20m,自然伽马曲线为箱形和钟形,底部发育一套白色砾岩和灰白色石英砂岩,中上部为褐灰色中—细砂岩夹薄层泥岩。层序界面之下为须三段暗色泥岩,具有高伽马、低电阻率特征。

层序Ⅱ为珍珠冲组砾岩、砂砾岩的主要发育层位,可细分出3个短期旋回(Ⅱ1、Ⅱ2、Ⅱ3),每个短期旋回的底部均发育砂砾岩底部冲刷面。元素俘获测井解释及岩心观察表明,层序Ⅱ2普遍发育钙质胶结物和石灰岩砾石,与层序Ⅱ1、Ⅱ3的硅质胶结、石英质砾石形成明显区别。Ⅱ1主要为厚层砾岩、砂砾岩夹薄层砂岩,Ⅱ2、Ⅱ3的下部为砾岩、砂砾岩,上部为含砾砂岩、细砂岩夹薄层泥岩。巨厚砾岩发育在Ⅱ1、Ⅱ3及Ⅱ2的中下部。

层序Ⅲ对应于珍珠冲组上部的大套泥岩、砂质泥岩夹细砂岩,可将其细分为Ⅲ1、Ⅲ2两个短期旋回,短期旋回的层序界面为加积(退积)或进积转换面,短期旋回内部发育湖泛面(进积或退积转换面),将其分割为基准面上升半旋回和基准面下降半旋回。

层序Ⅲ的顶界,珍珠冲组与上覆的下侏罗统东岳庙组为整合接触关系。东岳庙组底部发育一套厚度为几米至数十米的细—中粒岩屑砂岩,底部发育冲刷面,其底界比较稳定和容易识别。层序Ⅲ的底界为沉积体系转换面,界面之上为层序Ⅲ的大套滨浅湖相泥岩、砂质泥岩夹细粒石英砂岩,界面之下为层序Ⅱ的扇三角洲相巨厚砾岩、砂砾岩。

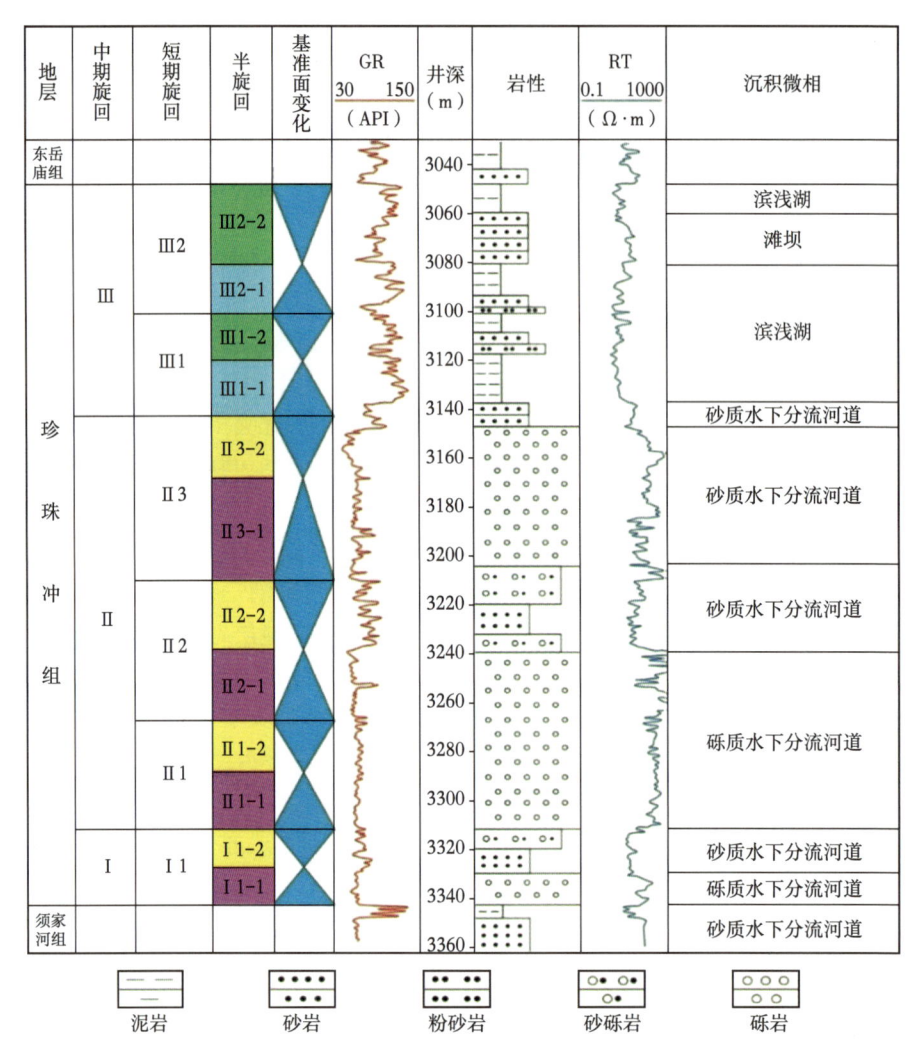

图 1　九龙山气田珍珠冲组层序地层划分柱状图

## 1.2　沉积相类型及分布

川西北地区珍珠冲组发育冲积扇—扇三角洲—滨浅湖沉积,在九龙山和金子山地区发育2个大型的扇三角洲,存在明显的地形坡折带,沉积物越过坡折带后快速卸载,形成巨厚块状砾岩。与准噶尔盆地西北缘相似,研究区发育同沉积逆断裂坡折[17,18],其折控制了扇三角洲辫状水下分流河道砂砾岩体的分布。

在盆地边缘珍珠冲组露头发育湿地冲积扇沉积,形成巨厚砾岩层,显示泥石流沉积特征。以广元樊家岩白田坝剖面为例,冲积扇辫状河道之间主要为深灰色泥岩、黑色碳质泥岩、薄煤层、煤线组成的河间沼泽沉积,砂砾岩粒间主要为硅质胶结及少量黏土杂基充填,黏土杂基主要矿物成分为高岭石,反映冲积扇形成于潮湿—半潮湿气候。巨厚砾岩向上演变为砂砾岩、含砾砂岩、中细砂岩、泥质砂岩夹薄层泥岩,发育冲刷面、粒序层理、槽状、板状交错层理及平行层理,常见黄铁矿结核、菱铁矿结核及水下变形构造,显示水下还原环境扇三角洲前缘沉积特征(图2)。

**图 2 珍珠冲组露头及岩心地质特征**

(a)白田坝剖面,珍珠冲组底部砾岩,冲刷充填;(b)白田坝剖面,珍珠冲组,砂砾岩,槽状交错层理;(c)龙 113 井,3342.23~3342.53m,砾质水下分流河道,冲刷面;(d)龙 119 井,3369.09~3369.31m,珍珠冲组砾岩特征;(e)龙 113 井,3323.7~3323.9m,砾质水下分流河道,砾石杂乱排列;(f)龙 118 井,4040.3~4041.2m,砂质水下分流河道之底部砾岩;(g)龙 113 井,3355.84~3355.99m,砂质水下分流河道,小型板状交错层理;(h)龙 118 井,4036.07~4036.12m,分流间湾,泥质砂岩;(i)龙 113 井,3318.84~3319.04m,分流间湾,黑色碳质泥岩

珍珠冲组Ⅰ、Ⅱ旋回为扇三角洲前缘沉积,发育厚层砾岩、砂砾岩夹中细砂岩。Ⅲ旋回为滨浅湖泥岩、粉砂质泥岩夹薄层中细砂岩,发育少量滩坝砂体。主要产气层为Ⅱ旋回砂砾岩储层,发育扇三角洲沉积,包含砾质及砂质水下分流河道、河口坝、分流间湾等沉积微相类型,辫状水下分流河道多层叠置形成巨厚的砾岩、砂砾岩体。

短期旋回Ⅱ1~Ⅱ3发育受北部物源控制的继承性较强的扇三角洲沉积体系,砾岩厚度受地形坡折带和主河道控制,Ⅱ2旋回泥岩夹层较厚,其砾岩厚度比Ⅱ1、Ⅱ3旋回薄,Ⅱ1、Ⅱ3旋回为砾岩储层主要发育层段(图3)。

基于高分辨率层序地层划分对比、单井沉积相分析、岩心描述,以及砾岩厚度、砂岩厚度、砾地比、砂砾地比等单因素做图,并参考地震预测成果,编制了九龙山地区珍珠冲组各短期旋回的沉积相分布图(图4)。

研究成果表明,砾岩厚度受地形坡折带和主河道控制,坡折带为北东—南西向,位于龙105—龙104—龙002-4—龙102—龙115—龙17井一带,坡折带下部及主河道带砾岩厚度相对较大,Ⅱ1~Ⅱ3发育2个受北部物源控制的继承性较强的扇三角洲沉积体系。

主河道带的砂砾地比>0.8,砾地比>0.6,至下游可降至0.5~0.6,发育厚层砾岩30~50m,属于砾质辫状河水下分流河道叠置,偶夹薄层砂质水下分流河道砂体。分流间湾的砂砾地比<0.4,砾地比<0.2,发育厚层泥岩,夹薄层粉细砂岩,砂岩属于薄层砂质水下分流河道及砂质河口坝。河道侧翼介于二者之间,砂砾地比为0.4~0.7,砾地比为0.3~0.5,发育砾质及砂质水下分流河道砂体。

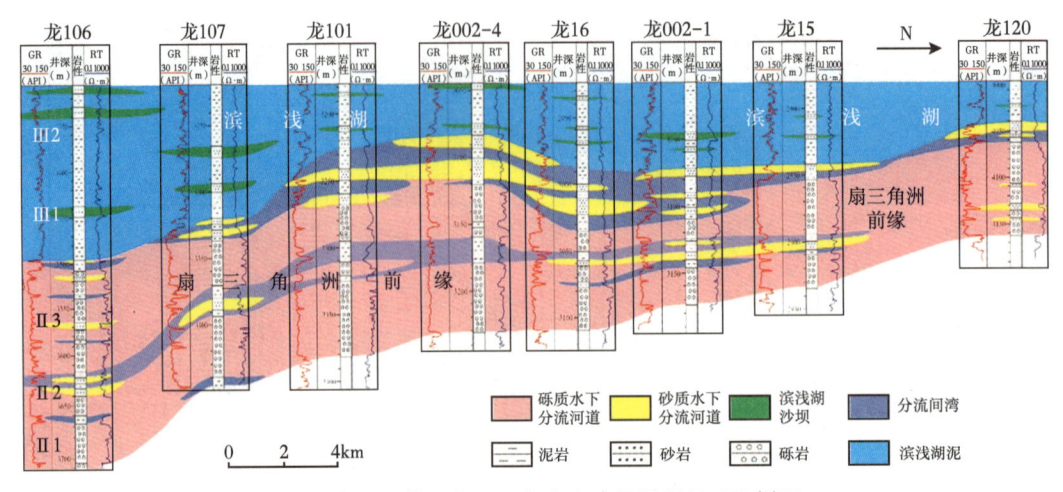

图 3　龙 106 井—龙 120 井珍珠冲组沉积相对比剖面

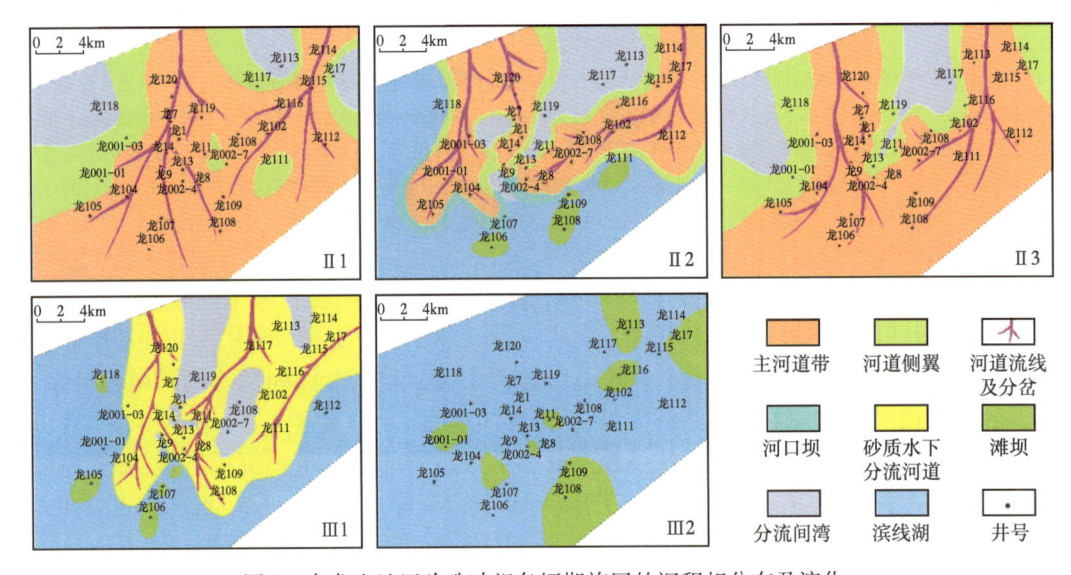

图 4　九龙山地区珍珠冲组各短期旋回的沉积相分布及演化

Ⅲ1 旋回三角洲急剧萎缩，局部发育砂质水下分流河道及滨浅湖沙坝。Ⅲ2 旋回为湖相沉积，滨浅湖沙坝零星分布。

## 2　储层特征及物性分布

### 2.1　储层岩石学特征与储集空间类型

九龙山气田珍珠冲组主要产气层为Ⅱ旋回的砾岩、含砾砂岩及砂岩储层。砾岩储层砾石体积分数约为 75%，砾石成分单一，多为石英砂岩砾石，少量燧石砾、石英岩砾及泥砾，局部见碳酸盐砾石；砾石最大砾径为 85mm，平均为 35～50mm，呈次棱角—次圆状，砾间为不等粒碎屑颗粒和黏土杂基充填，颗粒充填物成分主要为石英、燧石及岩屑。含砾砂岩储层以含砾岩屑砂岩为主，砾石体积分数为5%~20%，成分以石英砂岩砾石为主，平均砾径5~8mm；除砾石外的碎屑颗粒成分以石英为主，约占65%，岩屑体积分数为20%~30%，长石体积分数为1%~2%；颗粒粒间胶结致密，其成分以硅质为主，还有少量黏土杂基。砂岩储层主要为细到中粒岩屑石英砂岩、岩屑砂岩，碎屑颗粒以石英、岩屑为主，石英体积分数为55%~80%，含有少量长石；岩屑成分主要为变质砂岩、变质石英岩、千枚岩、白云岩及燧石等，颗粒多成线接触、压溶缝合接触等，颗粒胶结紧密，以硅质胶结和碳酸盐胶结为主。

九龙山气田珍珠冲组砂岩致密，有效储层主要为裂缝性砾岩、砂砾岩储层，储集空间类型为砾间溶孔（洞）、砾内溶孔、裂缝及裂缝扩容孔、晶间微孔、砾内原生孔等，可见裂缝连通砾内孔、砾间孔和裂缝相互连通现象（图5）。

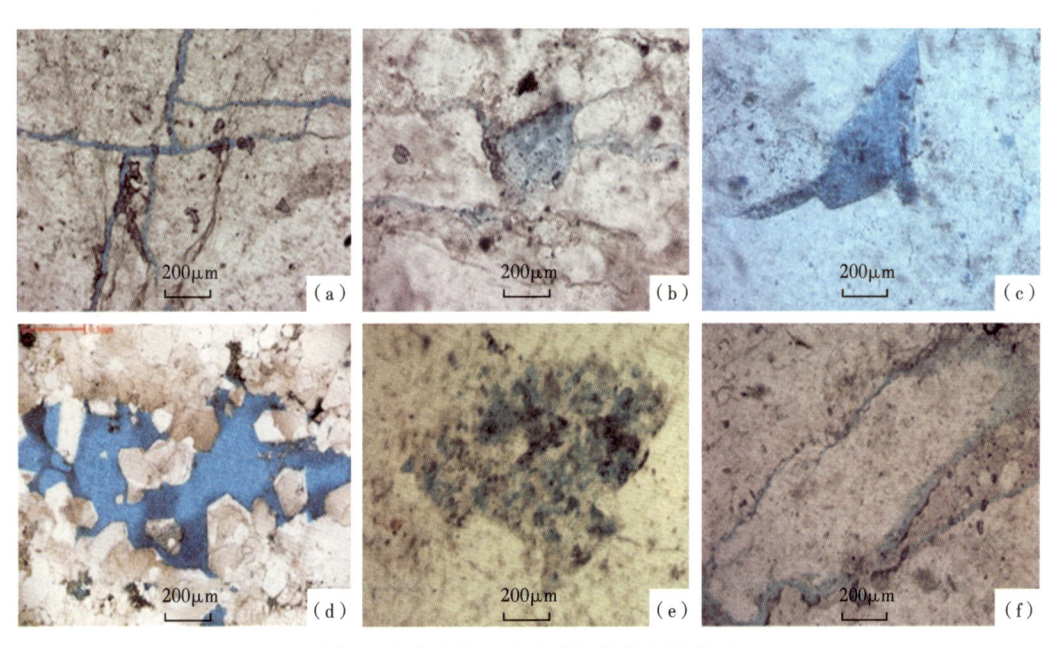

图5 九龙山气田珍珠冲组储集空间类型
（a）裂缝相互连通，网状缝，龙105井，3279.3m，铸体薄片，单偏光；（b）溶蚀小孔洞，与裂缝连通，龙107井，3356.9m，铸体薄片，单偏光；（c）砾内残余原生孔隙，发育微裂缝，龙118井，4033.87m，铸体薄片，单偏光；（d）粒间溶蚀孔洞，龙14井，3043.5m，铸体薄片，单偏光；（e）砾内岩屑溶孔，龙118井，4039.34m，铸体薄片，单偏光；（f）砾缘缝遭受溶蚀，龙14井，3018.2m，铸体薄片，单偏光

压汞分析资料表明，珍珠冲组储层排驱压力为0.0275~10.318MPa，排驱压力大于5MPa的占比为64.54%，反映珍珠冲组储层孔隙度、渗透率整体较差，孔隙喉道偏细。最大孔喉半径为0.071~26.76μm，最大孔喉半径小于0.1μm的样品占比为50%，最大孔喉半径小于1μm的样品占比为91.67%，样品最大孔喉半径均较小，渗流能力较差。最大进汞饱和度分布区间分散，进汞饱和度小于60%的样品占比为53.85%，反映有效孔隙体积相对较低，储集性能较差。分选系数为0.95~4.86，峰值为1~2，分选系数小于2的样品占比为84.62%，反映珍珠冲组砂砾岩储层颗粒分选程度较差。

## 2.2 储层物性特征

珍珠冲组储层具有低孔隙度、低渗透率特征,裂缝发育有效改善了储层物性,主要为裂缝—孔隙型储层,局部发育孔隙—裂缝型储层。

### 2.2.1 孔隙度分布

通过对 148 块全直径岩心样品的实测物性分析可以看出,岩心分析孔隙度相对较低(图 6),为 1.5%~6.5%,主峰为 2%~4.5%,孔隙度大于 2.5% 的样品占 56.03%。Ⅱ1 旋回孔隙度较低,孔隙度大于 3% 的样品仅占样品总数的 22.53%,Ⅱ2、Ⅱ3 旋回孔隙度大于 3% 的样品分别占 88.1% 和 82.86%(图 6)。

(a)珍珠冲组全直径岩心样品孔隙度分布

(b)Ⅱ1、Ⅱ2、Ⅱ3 小层孔隙度特征

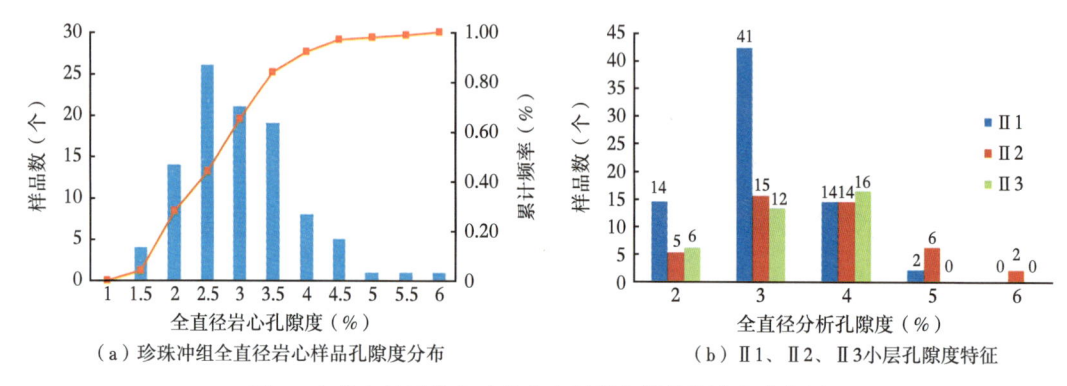

图 6 九龙山气田珍珠冲组全直径岩心样品孔隙度分布图

### 2.2.2 渗透率分布

通过对 138 个全直径岩心样品水平渗透率分析和 142 个全直径岩心样品垂直渗透率分析可知(图 7a),水平渗透率为 0.0006~517.98mD,主峰值为 1~100mD,水平渗透率大于 1mD 的样品占比为 84.67%;垂直渗透率为 0.000029~81.33mD,主峰值为 0.01~1mD,垂直渗透率大于 1mD 的样品占比仅为 29.2%,储层的水平渗透率明显高于垂直渗透率,分析认为,水平渗透率较高与砂砾岩低角度层理及低角度裂缝的发育有关。

Ⅱ1 旋回储层全直径岩心样品的水平渗透率为 0.1~50mD,水平渗透率大于 1mD 的样品占样品总数的 69.57%;垂直渗透率为 0.01~5mD,垂直渗透率大于 1mD 的样品占比为 28.57%。Ⅱ2 旋回全直径岩心样品水平渗透率为 1~500mD,水平渗透率大于 1mD 的样品占 90.24%;垂直渗透率为 0.01~5.00mD,垂直渗透率大于 1mD 的样品占比为 35.71%。Ⅱ3 旋回全直径岩心样品水平渗透率为 1~500mD,水平渗透率大于 1mD 的样品占比为 81.48%;垂直渗透率为 0.1~10.0mD,垂直渗透率大于 1mD 的样品占比为 33.33%(图 7b、c)。

总体而言,珍珠冲组储层孔隙度主峰为 2%~4.5%,水平渗透率多为 1~100mD,远大于垂直渗透率。

### 2.2.3 孔隙度—渗透率关系

珍珠冲组全直径岩心样品孔隙度—渗透率关系图(图 7d)反映出水平渗透率、垂直渗透率与孔隙度略呈正相关,但相关性较差,表明储层物性不完全受孔隙度控制,裂缝发育显著改善储层物性,使其在较低的孔隙条件下仍然具有较好的渗透性。

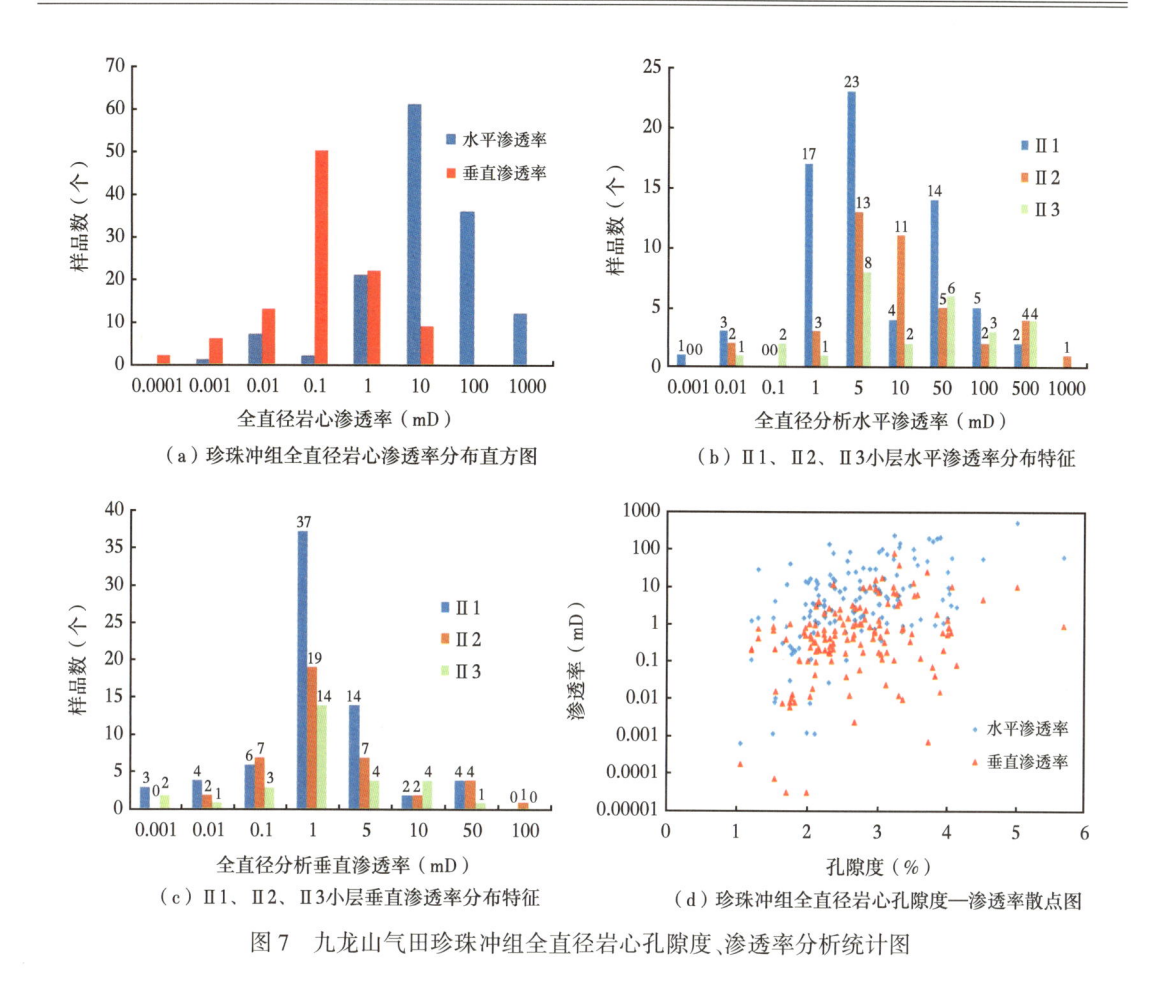

(a) 珍珠冲组全直径岩心渗透率分布直方图

(b) Ⅱ1、Ⅱ2、Ⅱ3小层水平渗透率分布特征

(c) Ⅱ1、Ⅱ2、Ⅱ3小层垂直渗透率分布特征

(d) 珍珠冲组全直径岩心孔隙度—渗透率散点图

图7 九龙山气田珍珠冲组全直径岩心孔隙度、渗透率分析统计图

## 3 裂缝特征及分布

### 3.1 岩心裂缝特征

根据岩心以及FMI成像测井资料综合分析,珍珠冲组裂缝成因类型包括构造裂缝、成岩裂缝及原岩裂缝3种,主要类型为构造剪切裂缝[11]。裂缝按产状可划分为低角度缝(包括水平缝)、斜交缝、高角度缝、网状缝、穿砾缝、砾内缝、砾缘缝7种类型,主要发育低角度缝、网状缝和穿砾缝(图8)。此3种类型裂缝均为构造成因的剪切裂缝,其规模相对较大,延伸较长,不受砾石限制。多种类型裂缝共同组成的网状系统极大提高了砂砾岩储层的渗流能力。

对11口井珍珠冲组第Ⅱ旋回岩心裂缝的观察和统计结果表明(图9),裂缝主要发育在石英砾石中,燧石砾石中裂缝不发育。岩心裂缝以低角度缝为主,裂缝密度为4.75~281.25条/m,主要集中在10~50条/m,密度小于40条/m的占低角度缝总数的81.25%。高角度缝和斜交缝发育程度相对较差,高角度缝密度为0.93~67.86条/m,密度小于20条/m的占82.35%。斜交缝密度为3.75~71.79条/m,密度小于20条/m的占79.27%。分别对Ⅱ旋回3个小层进

图 8　九龙山气田珍珠冲组岩心裂缝特征

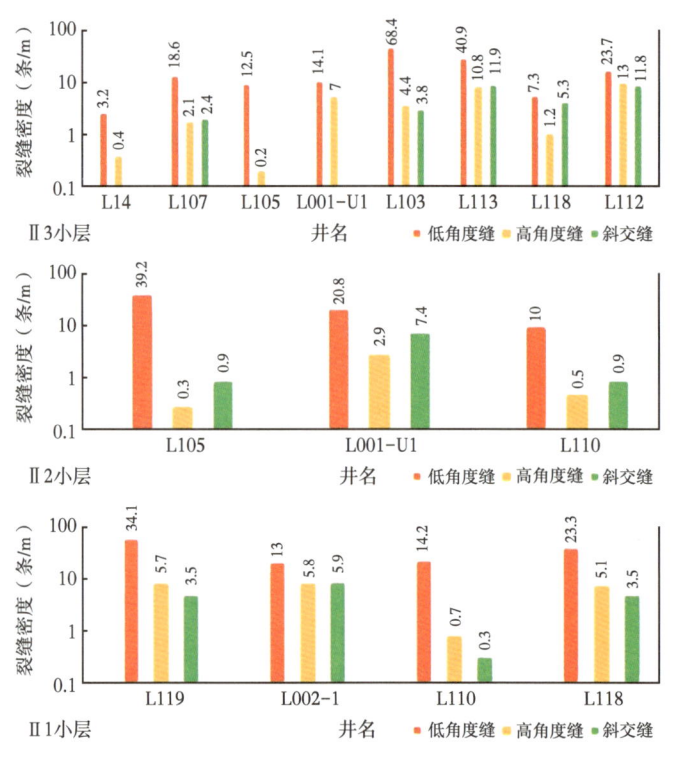

(a) 龙 119 井, 3368.27~3368.36m, 斜交裂缝; (b) 龙 001-U2, 3113.72~3113.85m, 水平缝, 网状缝;
(c) 龙 105 井, 3359.15~3359.26m, 网状缝, 砾缘缝, 穿砾缝, 砾内缝; (d) 龙 105 井, 3274.67~3274.74m,
水平缝, 网状缝, 穿砾缝, 砾缘缝; (e) 龙 110 井, 3301.78~3302.03m, 低角度缝, 穿砾, 砾缘缝, 砾内缝;
(f) 龙 110 井, 3280.6~3280.8m, 低角度缝, 穿砾缝, 高角度缝

图 9　九龙山气田珍珠冲组Ⅱ旋回岩心裂缝密度直方图

— 206 —

行裂缝密度分析,发现Ⅱ3旋回的岩心裂缝密度高于Ⅱ1、Ⅱ2旋回,低角度缝密度大于高角度缝与斜交缝密度。

## 3.2 成像测井裂缝渗透率解释

岩心观察、钻井显示(气侵、井漏)、成像测井及试气成果等方面均反映九龙山气田珍珠冲组储层裂缝发育,但分布不均,裂缝对气井产量具有重要控制作用。

利用FMI成像测井数据解释计算珍珠冲组Ⅱ旋回15口井的裂缝渗透率,计算得到裂缝渗透率为0.13~8.10mD,渗透率大于1mD的样品占43.75%。基于曲率和相干属性,对九龙山气田珍珠冲组Ⅱ旋回开展裂缝分布预测(图10),并将单井成像测井裂缝渗透率解释成果与裂缝分布预测图叠合,其中的暗色区域为裂缝发育区,红色柱子显示裂缝渗透率高低。从中可以看出,裂缝发育主要受北北东向断裂控制,局部发育与之共轭的北北西向剪切断裂,九龙山构造主体部位比其东部的倾伏端裂缝更为发育。

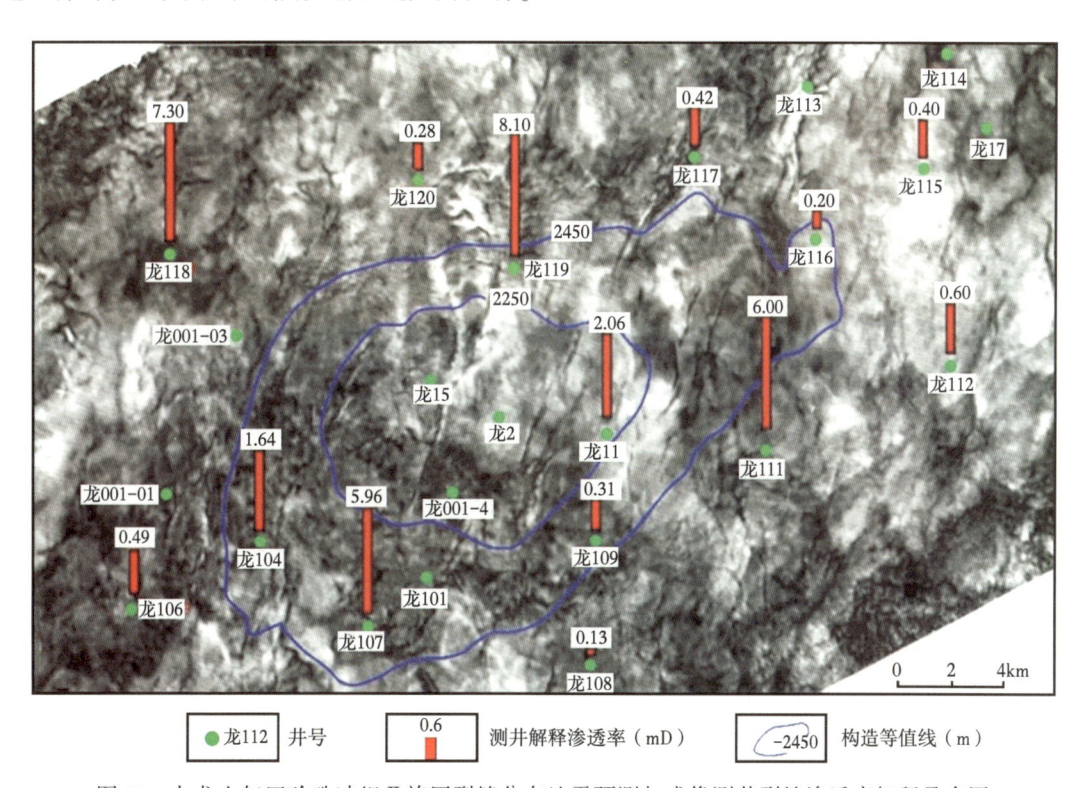

图10 九龙山气田珍珠冲组Ⅱ旋回裂缝分布地震预测与成像测井裂缝渗透率解释叠合图

Ⅱ旋回3个小层的成像测井裂缝渗透率计算结果表明(图11):Ⅱ1小层裂缝渗透率小于0.6mD,龙107井、龙119井、龙104井及龙111井裂缝渗透率达3~8mD;Ⅱ2小层的计算渗透率小于0.4mD,仅龙110井达1.24mD;Ⅱ3小层裂缝发育程度较高,大部分井计算裂缝渗透率大于1mD,最高达7.3mD;Ⅱ3小层的裂缝渗透率高于Ⅱ1小层,Ⅱ2小层裂缝渗透率较低。

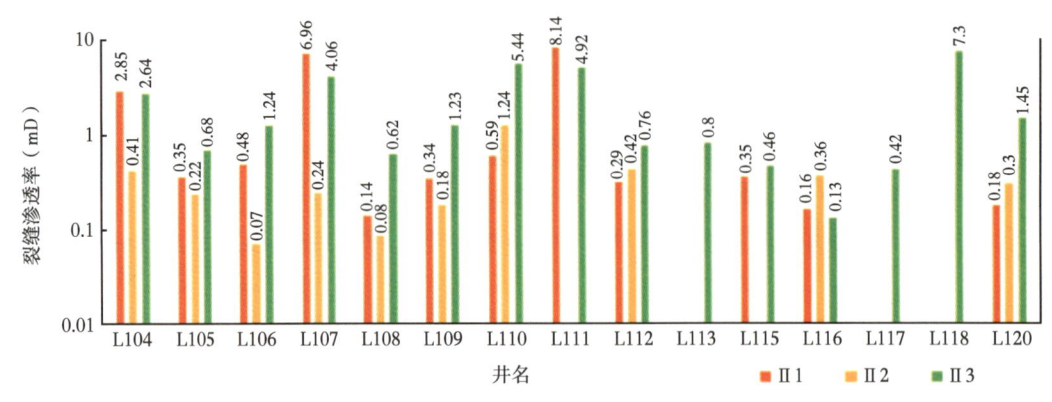

图 11 珍珠冲组Ⅱ1、Ⅱ2、Ⅱ3 小层成像测井计算裂缝渗透率统计直方图

## 4 有利区块评价

从前面的沉积旋回划分、储层特征及物性分布可以看出：九龙山气田珍珠冲组储层类型主要为裂缝性砂砾岩储层。中期旋回Ⅰ的地层厚度 10~20m，仅发育在九龙山构造东南部的局部井区，砂砾岩厚度薄，分布局限。中期旋回Ⅲ为滨浅湖及三角洲前缘沉积，发育薄层的砂质水下分流河道、河口坝及滩坝砂体，岩性为中细粒岩屑砂岩、岩屑石英砂岩，遭受了强烈的压实和碳酸盐胶结作用，岩性致密，物性较差，仅局部发育较薄的有效储层。中期旋回Ⅱ发育巨厚的扇三角洲前缘砂砾岩体，也是珍珠冲组的主要储层发育层段。由于短期旋回Ⅱ2 的泥岩夹层较厚，其砾岩厚度比Ⅱ1、Ⅱ3 旋回薄，且短期旋回Ⅱ2 普遍发育钙质胶结物和石灰岩砾石，其储层物性比Ⅱ1、Ⅱ3 旋回差。Ⅱ1、Ⅱ3 旋回的砾石成分主要为石英砂岩砾石、石英质砾石，粒间发育硅质胶结和少量黏土杂基充填，Ⅱ1、Ⅱ3 旋回的石英质砾石和硅质胶结砾岩比Ⅱ2 旋回的石灰岩砾石和碳酸盐胶结砾岩脆性更大，裂缝更发育，渗透率更高，为珍珠冲组气藏最有利的开发层系。

根据储层渗透率、孔隙度、裂缝密度、有效厚度、构造背景、试气产量等因素，建立了珍珠冲组各短期旋回的有利区块评价标准(表1)，对九龙山气田珍珠冲组进行短期旋回开发有利区块评价与优选。Ⅱ1、Ⅱ2、Ⅱ3 及Ⅲ1 短期旋回共评价出 4 个一类有利区块，计算储量约为 $65.98×10^8m^3$；8 个二类有利区，计算储量约 $149.52×10^8m^3$，一类、二类有利区计算总储量为 $215.50×10^8m^3$。从表2可以看出，有效储层和天然气储量主要分布在Ⅱ3 小层和Ⅱ1 小层，且Ⅱ3 小层优于Ⅱ1 小层。Ⅱ2 小层储量较小，仅有二类储量 $6.26×10^8m^3$，Ⅲ1 小层仅有二类储量 $0.45×10^8m^3$。

表 1 九龙山气田珍珠冲组短期旋回有利区评价标准

| 类型 | 渗透率（mD） | 孔隙度（%） | 裂缝密度（条/m） | 试气产量（$10^4m^3/d$） | 有效厚度（m） | 构造背景 | 评价 |
| --- | --- | --- | --- | --- | --- | --- | --- |
| 一类 | ≥5 | 2.5~6.5 | 20~90，一般40左右 | >5 | >15 | 构造主体 | 最有利 |
| 二类 | 0.5~5 | 2.5~3.5 | 20~40，一般30左右 | 2~5 | 10~15 | 构造主体及外围斜坡 | 有利 |
| 三类 | 0.02~0.5 | 2~2.5 | 10~30 | 1~2 | 5~10 | 构造主体及外围斜坡 | 有潜力 |

## 地质应用篇

**表 2　九龙山气田珍珠冲组短期旋回有利区评价结果**

| 层位 | 有利区 | 井区 | 面积(km²) | 有效厚度(m) | 孔隙度(%) | $S_g$(%) | 体积系数 | 储量($10^8$m³) |
|---|---|---|---|---|---|---|---|---|
| Ⅲ1 | 二类有利区 | 龙002-4井区 | 0.8 | 7.9 | 3.6 | 0.58 | 0.00294 | 0.45 |
| Ⅱ3 | 一类有利区 | 龙002-4井区 | 22.1 | 17.7 | 4.4 | 0.65 | 0.00294 | 38.05 |
| | | 龙105井区 | 4.9 | 16.4 | 2.7 | 0.62 | 0.00294 | 4.58 |
| | 二类有利区 | 龙001-U2井区 | 21.3 | 22.9 | 3.2 | 0.59 | 0.00294 | 31.32 |
| | | 龙112井区 | 20.5 | 28.9 | 2.7 | 0.64 | 0.00294 | 34.82 |
| | | 龙113井区 | 14.6 | 12.2 | 2.9 | 0.65 | 0.00294 | 11.42 |
| Ⅱ2 | 二类有利区 | 龙105井区 | 10.3 | 10.1 | 2.6 | 0.68 | 0.00294 | 6.26 |
| Ⅱ1 | 一类有利区 | 龙002-4井区 | 15.4 | 18.7 | 3.3 | 0.63 | 0.00294 | 20.36 |
| | | 龙105井区 | 4.4 | 12.3 | 2.5 | 0.65 | 0.00294 | 2.99 |
| | 二类有利区 | 龙002-10-1井区 | 26.3 | 18.9 | 3.1 | 0.58 | 0.00294 | 30.40 |
| | | 龙102井区 | 13.6 | 21.7 | 2.8 | 0.72 | 0.00294 | 20.24 |
| | | 龙113井区 | 14.6 | 15.8 | 2.7 | 0.69 | 0.00294 | 14.62 |

通过单层评价,多层叠合,综合优选九龙山气田珍珠冲组的有利开发区块,共优选出 2 个一类有利区和 3 个二类有利区,计算储量约为 244.69×$10^8$m³(表3,图12)。

一类有利区的分布受九龙山构造主体和珍珠冲组砾质水下分流河道双重控制,发育于构造高点或裂缝发育部位,而二类有利区分布于外围斜坡,受岩性及裂缝控制。

**表 3　九龙山气田珍珠冲组开发有利区评价结果**

| 层位 | 有利区 | 井区 | 面积(km²) | $h$(m) | $\phi$(%) | $S_g$(%) | 体积系数 | 储量($10^8$m³) |
|---|---|---|---|---|---|---|---|---|
| 珍珠冲组 | 一类有利区 | A. 龙002-4井区 | 19.1 | 38.9 | 3.5 | 0.64 | 0.00294 | 56.61 |
| | | B. 龙105井区 | 3.7 | 35.9 | 2.9 | 0.67 | 0.00294 | 8.78 |
| | 二类有利区 | C. 龙002-10-1井区 | 34.6 | 39.3 | 3.3 | 0.61 | 0.00294 | 93.10 |
| | | D. 龙102井区 | 23.8 | 39.3 | 2.8 | 0.69 | 0.00294 | 61.47 |
| | | E. 龙113井区 | 13.9 | 28.5 | 2.7 | 0.68 | 0.00294 | 24.74 |

一类有利区 A:分布在构造主体部位的龙14—龙15—龙10—龙002-4井区,其中龙10井日产气 112.8×$10^4$m³、龙002-4井日产气 61.6×$10^4$m³、龙002-9-1井日产气 60×$10^4$m³,有利区块面积约为 19.1km²,计算储量为 56.61×$10^4$m³。

一类有利区 B:构造主体外围的龙105井区,龙105井日产气 13×$10^4$m³,面积约为 3.7km²,计算储量为 8.78×$10^4$m³。

二类有利区 C:分布在构造主体部位的龙002-10-1—龙001-U2—龙1—龙10—龙101—龙110井区,面积约为 34.6km²,计算储量为 93.10×$10^4$m³。

二类有利区 D:主体构造外围的龙112—龙102—龙116井区,其中龙112井日产气 3.4×$10^4$m³、龙116井日产气 3.3×$10^4$m³,有利面积 23.8km²,计算储量为 61.47×$10^4$m³。

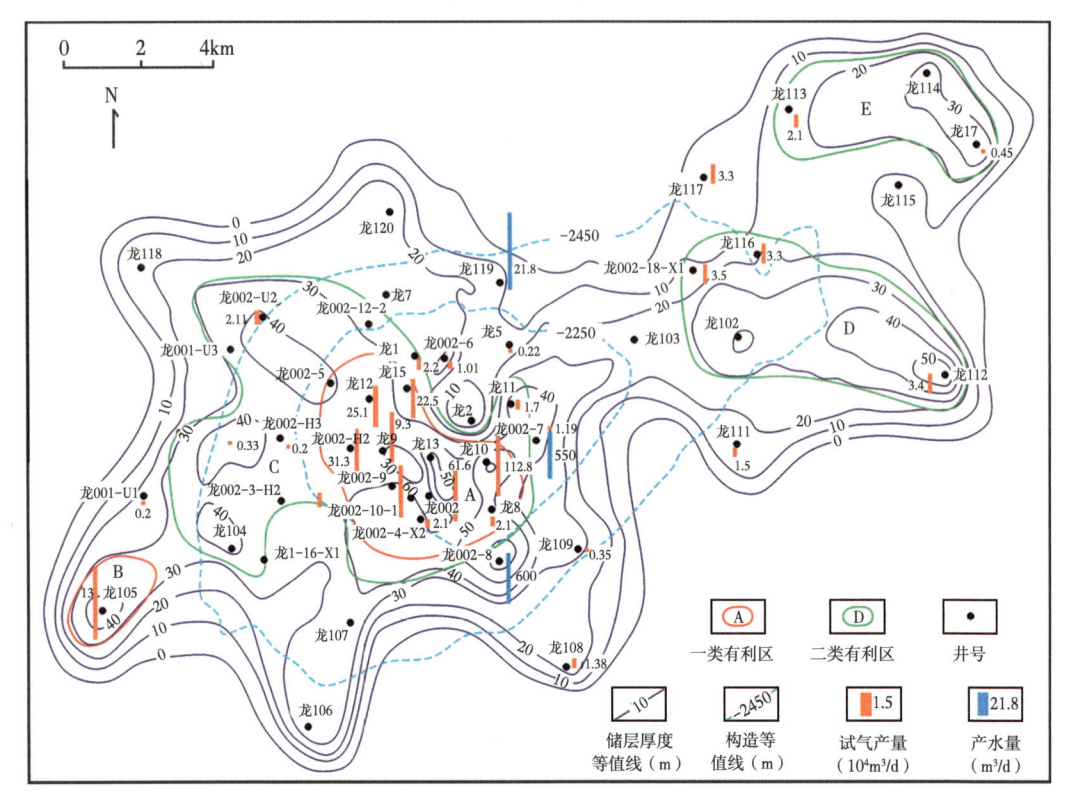

图 12　九龙山气田珍珠冲组有效储层厚度分布及有利区块评价图

二类有利区 E：龙 17—龙 114—龙 113 井区，龙 113 井日产气 $2.1×10^4 m^3$，有利区面积约为 $13.9 km^2$，计算储量为 $24.74×10^4 m^3$。

基于井间干扰、压力分布、构造背景、储层及裂缝发育情况，初步将九龙山气田划分为 4 个压力系统：(1)构造主体区(包括有利区 A 和 C)，原始地层压力为 50.22~53.93MPa，压力系数为 1.74~1.75，位于主河道带，以 Ⅰ、Ⅱ 类储层为主，有效储层厚度 30~50m；(2)龙 105 井区(有利区 B)，原始地层压力为 64.94MPa，压力系数为 2.03，明显高于其他井区，发育 Ⅰ 类储层，位于水下分支河道，储层厚度为 20~35m，分布范围相对局限；(3)龙 102—龙 112 井区(有利区 D)，龙 112 井压力为 58.99MPa，压力系数为 1.71，位于主河道带、发育 Ⅱ、Ⅲ 类储层，储层厚度为 20~30m；(4)龙 113—龙 17 井区(有利区 D)，位于主河道带及其侧翼，主要为 Ⅱ、Ⅲ 类储层，储层厚度为 10~20m。

在九龙山气田珍珠冲组气藏开发部署方面，建议立足构造主体，逐步向外围扩展：优先开发构造主体(龙 002-4 井区)，向构造外围(龙 001-U2—龙 104 井区)拓展；工区东部龙 102—龙 112 井区储层及含气性较好、储层厚度较大，属于东部主河道带，为现实的接替区块；西南部的龙 105 井区储层好、单井产量和气藏压力高、值得重视。

## 5　结论

(1)珍珠冲组被划分为 3 个中期旋回、6 个短期旋回，发育受北部物源控制的继承性较强的扇三角洲沉积体系,有利储层为Ⅱ旋回扇三角洲前缘辫状水下分流河道多层叠置形成的巨厚砂砾岩体,砂砾岩厚度受地形坡折带和主河道带控制。

(2)珍珠冲组砂岩致密,有效储层主要为裂缝性砾岩、砂砾岩储层,砾石成分以石英砂岩砾石为主,含少量燧石砾、石英岩砾及泥砾;为裂缝—孔隙型储层,储集空间为砾间溶孔(洞)、砾内溶孔、裂缝及裂缝扩容孔、晶间微孔、砾内原生孔等。

(3)发育低角度缝、斜交缝、网状缝、穿砾缝、砾缘缝等多种裂缝,显著改善了储层的渗流能力。全直径岩心分析表明,珍珠冲组储层孔隙度主峰为2.0%~4.5%,水平渗透率主峰为1~100mD,垂直渗透率主峰值为0.01~1.00mD,水平渗透率高于垂直渗透率。

(4)对Ⅱ旋回3个小层进行了岩心裂缝密度分析和成像测井裂缝解释,裂缝发育程度、储层物性及含气性均反映出Ⅱ3、Ⅱ1小层较好,为2个有利开发层系,且Ⅱ3小层优于Ⅱ1小层。

(5)根据储层渗透率、孔隙度、裂缝密度、有效厚度、构造背景、试气产量等因素,建立了有利区块评价标准,通过单层评价,多层叠合,优选出2个一类有利区和3个二类有利区,计算储量约为244.69×10$^8$m$^3$。有利区分布受九龙山构造主体、水下分流河道砂砾岩体及裂缝发育程度三重因素控制。

## 参 考 文 献

[1] 曾联波,漆家福,王永秀.低渗透储层构造裂缝的成因类型及其形成地质条件[J].石油学报,2007,28(4):52-56.

[2] 丁文龙,樊太亮,黄晓波,等.塔中地区中—下奥陶统古构造应力场模拟与裂缝储层有利区预测[J].中国石油大学学报:自然科学版,2010,34(5):1-6.

[3] 申本科,胡永乐,田昌炳,等.陆相砂砾岩油藏裂缝发育特征分析——以克拉玛依油田八区乌尔禾组油藏为例[J].石油勘探与开发,2005,32(3):41-44.

[4] 刘树根,何鹏,邓荣贵,等.川西九龙山构造须二段岩石力学性质与裂缝发育规律研究[J].天然气工业,1996,16(2):1-4.

[5] 卢虎胜,张俊峰,李世银,等.砂泥岩间互地层破裂准则选取及在裂缝穿透性评价中的应用[J].中国石油大学学报:自然科学版,2012,36(3):14-19.

[6] 张惠良,张荣虎,杨海军,等.超深层裂缝—孔隙型致密砂岩储集层表征与评价:以库车前陆盆地克拉苏构造带白垩系巴什基奇克组为例[J].石油勘探与开发,2014,41(2):158-167.

[7] 王俊鹏,张荣虎,赵继龙,等.超深层致密砂岩储层裂缝定量评价及预测研究:以塔里木盆地克深气田为例[J].天然气地球科学,2014,25(11):1735-1744.

[8] 张博,李江海,吴世萍,等,大北气田储层裂缝定量描述[J].天然气地球科学,2010,21(1):42-46.

[9] 张博,袁文芳,曹少芳,等.库车坳陷大北地区砂岩储层裂缝主控因素的模糊评判[J].天然气地球科学,2011,22(2):250-253.

[10] 李世川,成荣红,王勇,等.库车坳陷大北1气藏白垩系储层裂缝发育规律[J].天然气工业,2012,32(10):24-27.

[11] 王珂,张荣虎,戴俊生,等.库车坳陷克深2气田低渗透砂岩储层裂缝发育特征[J].油气地质与采收率,2016,23(1):1-6.

[12] 王珂,张惠良,张荣虎,等.塔里木盆地克深2气田储层构造裂缝多方法综合评价[J].石油学报,2015,36(6):673-685.

[13] 何鹏,王允诚,刘树根,等.九龙山气田须二下亚段气藏裂缝的多信息叠合评价[J].成都理工学院学

报,1999,26(3):225-227.
[14] 白斌,邹才能,朱如凯,等. 四川盆地九龙山构造须二段致密砂岩储层裂缝特征、形成时期与主控因素[J]. 石油与天然气地质,2012,33(4):526-535.
[15] 李啸,刘海磊,王学勇,等. 坡折带砂体成因及分布规律:以准噶尔盆地车排子地区下白垩统清水河组为例[J]. 岩性油气藏,2017,29(1):35-42.
[16] 冯有良,胡素云,李建忠,等. 准噶尔盆地西北缘同沉积构造坡折对层序建造和岩性油气藏富集带的控制[J]. 岩性油气藏,2018,30(4):14-25.
[17] 李晨,樊太亮,高志前,等. 冲积扇高分辨率层序地层分析:以辽河坳陷曙一区杜84块SAGD开发区馆陶组为例[J]. 岩性油气藏,2017,29(3):66-75.
[18] 郭鸿喜,梁大庆,易斌,等. 九龙山构造珍珠冲砾岩裂缝地震预测研究[J]. 四川地质学报,2015,35(2):280-284.
[19] 李跃纲,巩磊,曾联波,等. 四川盆地九龙山构造致密砾岩储层裂缝特征及其贡献[J]. 天然气工业,2012,32(1):1-5.
[20] 巩磊,曾联波,张本健,等. 九龙山构造致密砾岩储层裂缝发育的控制因素[J]. 中国石油大学学报(自然科学版),2012,36(6):6-11.
[21] 王婷,侯明才,王文楷,等. 川西北九龙山—剑阁地区珍珠冲段储层特征与主控因素探讨[J]. 山东科技大学学报,2014,33(1):20-26.
[22] 朱华银,蒋德生,安来志,等. 川西地区九龙山构造砾岩储层敏感性实验分析[J]. 天然气工业,2012,32(9):40-43.

# Study on the Effects of Fracture on Permeability with Pore-Fracture Network Model

Chunyan Jiao[1]　Yong Hu[1]　Xuan Xu[1]　Xiaobing Lu[2]
Weijun Shen[2]　Xinhai Hu[1]

(1. PetroChina Research Institute of Petroleum Exploration & Development, Langfang;
2. Institute of Mechanics, Chinese Academy of Sciences)

**Abstract**: Reservoir quality and productivity of fractured gas reservoirs depend heavily on the degree of fracture development. The fracture evaluation of such reservoir media is the key to quantify reservoir characterization for the purposes such as well drilling and completion as well as development and simulation of fractured gas reservoirs. In this study, a pore-fracture network model (PFNM) was constructed to understand the effects of fracture on permeability in the reservoir media. The microstructure parameters of fractures including fracture length, fracture density, fracture number and fracture radius, were analyzed. Then two modes and effects of matrix and fracture network control were discussed. The results indicate that the network permeability in the fractured reservoir media will increase linearly with fracture length, fracture density, fracture number and fracture radius. When the fracture radius exceeds 80 μm, the fracture radius has a little effect on network permeability. Within the fracture density less than 0.55, it belongs to the matrix control mode while the fracture network control mode is dominant in the fracture density exceeding 0.55. The network permeability in the matrix and fracture network control modes is affected by fracture density and the ratio of fracture radius to pore radius. There is a great change in the critical density for the matrix network control compared with the fracture network control. This work can provide a better understanding of the relationship between matrix and fractures, and the effects of fracture on permeability so as to evaluate the fluid flow in the fractured reservoir media.

**Keywords**: fractured gas reservoirs; matrix; fracture; network permeability; fracture parameter; pore-fracture network

## 1 Introduction

As a special complex gas reservoir, fractured gas reservoir takes a large proportion of gas reservoirs found at home and abroad, which is playing an increasingly important role among energy sources and has attracted wide attention[1-3]. With the further exploration and development of natural gas in China, the proven reserve and production of fractured gas reservoirs have also increased year by year[4]. Unlike other gas reservoirs, fractured gas reservoir is characterized by extremely low permeability in matrix, well developed fractures, strong heterogeneity and active edge and bottom water[3,5]. During the development of gas reservoirs, water will flow to gas well along the fracture channel owing to the complex fractures existing in reservoir media, and gas production will greatly reduce and even stop with the part gas separated by water, which will give rise to great difficulties in the development of gas reservoirs[6-8].

Fractured gas reservoir is the heterogeneous and anisotropic media, which is a mere mixture of two distinct populations of fracture and pore voids, and they consist of matrix blocks and fractures[9-11]. The ultra-low permeability matrix system acts as a source of the fluids while the fractures serve as the main pathway of the fluids towards production wells. Generally, the fractures are embedded in reservoir matrix with micro pores, which plays an important effect on the flow channel, and randomly distributed fractures dominate the flow channel in the reservoir media[11-13]. In the low-permeability reservoir rocks, the fluid flow will occur in highly preferred flow pathways within a limited quantity of fractures[14,15]. These randomly distributed fractures are often connected to form irregular networks, and the permeability evolution in the fractured reservoir media has the significant effect on the fluid flow[11]. Therefore, understanding the questions such as the relationship between matrix and fractures, and the effects of fracture on permeability in the reservoir media is crucial for the fluid flow evaluation and for forecasting gas production in fractured gas reservoirs.

Fractures in reservoir rocks are usually random and disordered, which is always a difficult problem to quantify them in the reservoir development. Over the past few decades, many investigators studied the flow characteristics of fracture networks and proposed several mathematical models. Fatt[16] proposed the pore network model (PNM) to idealize the pore structure of porous media as pore bodies and pore throats, but the model is not able to understand the micro-mechanism of fluid flow in reservoirs with great heterogeneity, fractures, pores, and throats. Barenblatt et al[17] and Warren[18] proposed the dual pore-fracture network model (PFNM) to describe the fluid flow and the exchange between the matrix and the fracture. According to parallel plane model, Snow[19] proposed an analytical method for permeability of fracture networks. Dreuzy et al[20] used a numerical and theoretical methods to study the permeability of randomly fractured networks, and verified the validity of the model with naturally fracture networks. Zheng[21] established a fractal permeability model to describe the relationship between gas permeability and pore structure parameters. Noetinger et al[22] studied the fluid flow and the exchange between the matrix and the fracture using the dual pore-fracture model. Based on truncated octahedral support, Jivkov et al[23] proposed a novel bi-regular network model to calculate the evolution of permeability in porous media. Gong[24] studied the effects of fracture aperture distribution in naturally fractured reservoirs with the numerical method. However, these above studies did not provide a quantitative relationship among the permeability of fracture networks, fracture density and microstructure parameters of fractures, such as fracture length, fracture number and fracture radius etc. The pore-fracture network model is an effective method to understand the micro-mechanism of fluid flow in the fractured reservoir media with great heterogeneity and a multi-scale pore structure consisting of fractures, pores, and throats. Therefore, it is extremely necessary to understand the relationship between matrix and fractures, and the effects of fracture on permeability so as to understand the fluid flow in the fractured reservoir media.

In this study, a pore-fracture network model was constructed to understand the effects of fracture on permeability in the fractured reservoir media. The microstructure parameters of fractures including fracture density, fracture length, fracture number and fracture radius, were analyzed and discussed. Then the critical density of the matrix and fracture network control mode was determined. Moreover, the effects of matrix and fracture network control modes were performed to understand the evolution of permeability in the fractured reservoir media. The results of these work can provide a better understanding of the relationship between matrix and fractures, and the effects of fracture on fluid flow in the fractured reservoir media.

## 2 Pore-fracture network model

### 2.1 Structure of pore-fracture network model

In the study, a 2-D quasistatic pore-fracture network model (PFNM) was constructed to understand the effects of fracture parameters in fractured reservoir media, which consists of the pore, throat and fracture units. The basic unit (a) and structure of pore-fracture network model (b) were shown in Fig. 1. The pore-throat unit is composed of four small spheres and four cylindrical pipes while the fracture unit is made up of a cylindrical pipe. The pore is a larger pore space and the throat is a relatively longer and narrower connective space, while the length of the fracture is much larger than the throat. The main parameters characterizing the pore-fracture network model are dimensions of the network, coordination number, shape factor, pore throat radius, and pore throat distribution etc[25]. The coordination number is used to represent the connectivity between pores and throats, which is a microscopic parameter to describe the connectivity degree of the reservoir media. The larger the coordination number is, the greater the connectivity is. The shape factor is used to describe the irregular pore shape. The smaller the shape factor is, the more irregular the shape is. The ratio between pore and throat is the ratio of pore radius to throat radius with connection, which demonstrates the pore throat alternating variation.

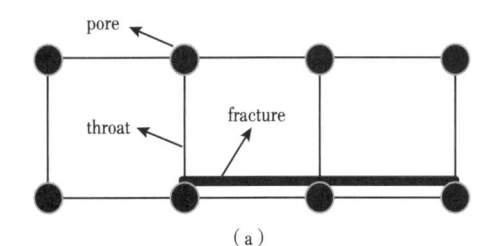

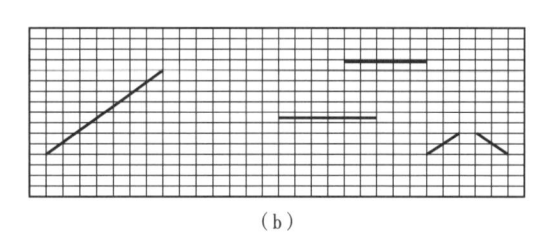

Fig. 1 A 2-D quasistatic pore-fracture network model
(a) Basic unit of PFNM; (b) Structure of PFNM

In the study, the size of the pore-fracture network model is 90 nodes ×30 nodes, which corresponds to the horizontal and vertical direction, respectively, as illustrated in Fig. 2. The parameters of the basic model are listed as follows. The pore radius obey the average distribution between 0.1 $\mu$m and 20 $\mu$m while the throat pore is the normal distribution with the mean value of 2.3 $\mu$m. The fracture length is 5 network step sizes and the corresponding radius is 20 $\mu$m. The network porosity and permeability under the initial condition are 10 % and 3.1 mD, respectively. The simulation is run on the basic model, and on models in which parameters are individually varied.

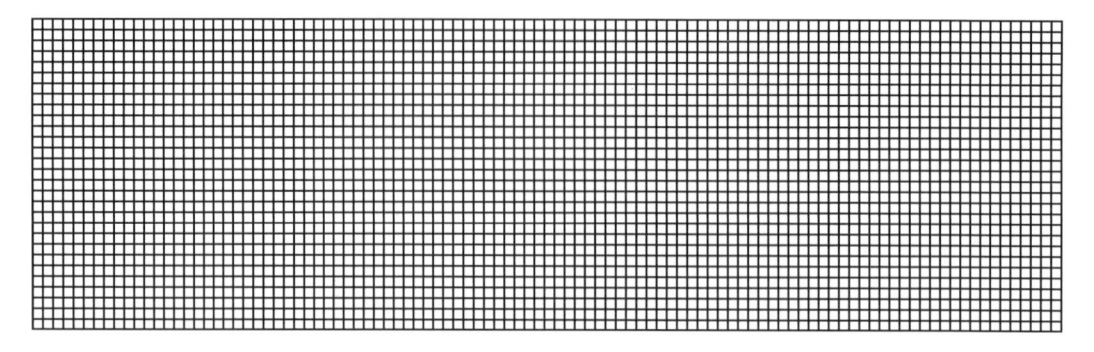

Fig. 2  A pore-fracture network model with 90 nodes ×30 nodes

## 2.2  Flow governing equation

In the pore-fracture network model, the pores acts as a source of storage fluid, and the throats are the relatively narrower and longer connective flow paths. While the fractures provide the longest and widest flow paths, which serves as the main pathway of the fluids. For the single phase flow, the flow in the throats and fractures is assumed to satisfy the Poiseuille equation[26]. The flow governing equation in the throats and fractures can be written as

$$q_{ij} = -\frac{r_{ij}^2}{8\mu_{ij}l_{ij}}(p_j - p_i) \tag{1}$$

where $q_{ij}$ is the flow rate; $r_{ij}$ is the radius of the throat (fracture); $\mu_{ij}$ is the dynamic viscosity of the fluid; $l_{ij}$ is the length of the throat (fracture); $p_j$ and $p_i$ are the pressures at the node $i$ and node $j$, respectively.

The pressure at each node can be obtained by flow governing equation, and then the permeability can be calculated by the total flow $Q$ under the pressure difference $\Delta P$. Based on the pressure of arbitrary cross section in the pore network model, we can obtain the total flow $Q$. Thus the absolute permeability can be calculated by Darcy's law as follows

$$K = \frac{\mu Q L}{A \Delta P} \tag{2}$$

where $K$ is the absolute permeability; $\mu$ is the fluid viscosity; $Q$ is the total flow; $L$ is the length of the network; $A$ is the area of the cross section; $\Delta P$ is the pressure difference.

## 3  Results and discussion

### 3.1  Effect of fracture length and density

Fracture length and fracture density are the important parameters in the characterization of the permeability evolution in the fractured reservoir media, which plays a key role in estimating elastic rock properties, fracture porosity, path length and connectivity for fluid flow of fractured rock[27]. In the study, the effects of fracture length from 5 μm to 40 μm and fracture density from 0 to 0.55 on network permeability have been chosen to investigate the network permeability in the fractured reservoir media, respectively. Fig. 3 (a) shows the variation of network permeability versus fracture length in the model. From the result of Fig. 3 (a), it can be seen that the fracture length affects network permeability in the fractured reservoir media. With the fracture length increasing, the network permeability increases linearly. The variation of network permeability versus fracture density is shown in Fig. 3 (b). From the Fig. 3(b), we can see that the network permeability increases slowly at the low fracture density, and then increases dramatically with the increasing of fracture density. The reason is that the increasing fracture density in such reservoir media is favor of the connection between the pores and the fractures. Thus, the increasing fracturing length and density is beneficial to the fluid flow in the fractured reservoir media.

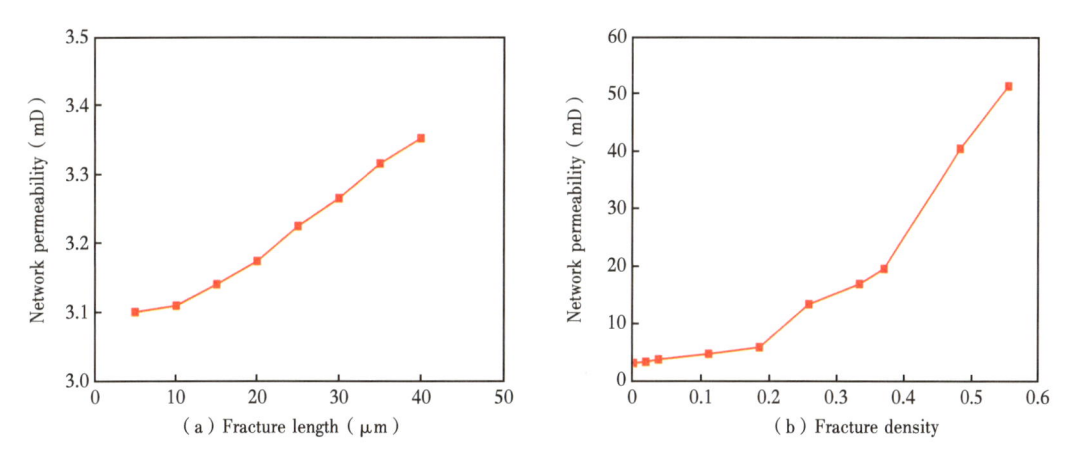

Fig. 3 Variation of network permeability versus fracture length (a) and fracture density (b)

## 3.2 Effect of fracture number and radius

The fractures are embedded in the porous matrix with micro pores, and fracture number and fracture radius are the crucial features controlling the fluid flow behavior in the fractured reservoir media[11,24]. In this study the values of the fracture number from 3 to 30 and fracture radius from 30 μm to 300 μm are conducted to investigate the network permeability in the reservoir media, respectively. The variation of network permeability versus fracture number (a) and fracture radius (b) is illustrated in Fig. 4. From the result of the Fig. 4(a), we can see that the network permeability increases with the increase of the fracture number in the fractured reservoir media. As shown in Fig. 4 (b), it is see that the network permeability increases linearly at the fracture radius less than 80 μm. When the fracture radius exceeds 80 μm, there is very little change in the network permeability. It implies that the fracture radius has a little effect on the increasing of network permeability when the fracture radius is more than 80 μm.

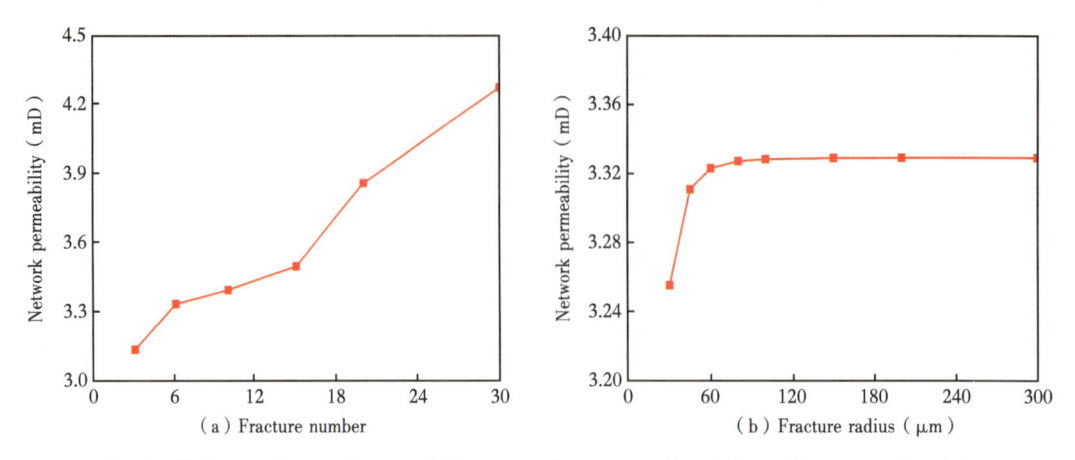

Fig. 4 Variation of network permeability versus fracture number (a) and fracture radius (b)

## 3.3 Matrix and fracture network control modes

In the fractured reservoir media, the matrix is the main storage space while the fracture provides the flow path[2,28]. Thus there is of great significance to understand the relationship between matrix and fractures in such reservoir media. In the study, the modes of matrix and fracture network control are conducted to understand network permeability in the fractured reservoir media. Fig. 5 (a) presents the variation of permeability sensitivity versus fracture density in the pore-fracture network model. It is noted that the critical density of the matrix and fracture network control mode is about 0.55. From the result of Fig. 5 (a), it can be observed that the permeability sensitivity decreases with the increase of the fracture density at the matrix mode. While the permeability sensitivity at the fracture mode increases as the fracture density increases. And the critical density versus the ratio of fracture radius to pore radius is illustrated in Fig. 5 (b). From the Fig. 5 (b), we can see that the

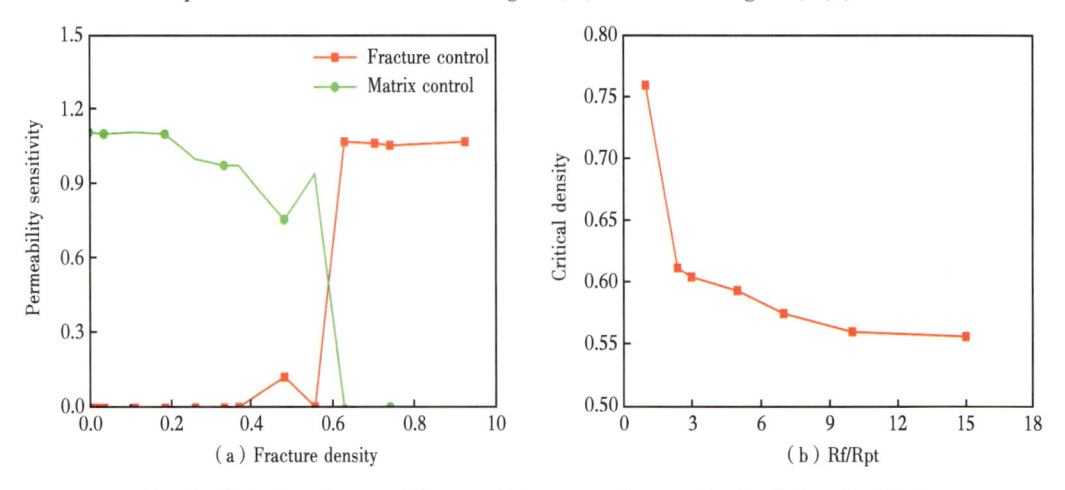

Fig. 5 Variation of permeability sensitivity versus fracture density (a) and critical density versus the ratio of fracture mean radius to pore mean radius (b)

critical density decreases with the increasing of the Rf/Rpt value (Rf and Rpt are fracture mean radius and pore mean radius, respectively). As the Rf/Rpt value increases, the decreasing degree decreases slowly.

## 3.4 Effect of pore and fracure radius

Permeability is a property of the reservoir rock that measures the capacity of the formation to transmit fluid[26]. The matrix pore size, fracture length and radius have the significant influences on permeability in the reservoir media[29]. In order to analyze the effect of pore and fracture radius, the values of the different ratios of fracture radius to pore radius are considered in the pore-fracture network model. Fig. 6 shows the variation of permeability evaluation coefficient versus Rf/Rpt at matrix control (a) and fracture control (b), respectively. From the result of Fig. 6 (a), with the fracture density increasing at the matrix control mode, the permeability evaluation coefficient increases linearly, and then tend to stabilize with the Rf/Rpt value. It indicates that there is a limit that the permeability increases with the Rf/Rpt value at the matrix control. As illustrated in Fig. 6 (b), there is little change in the permeability evaluation coefficient when the Rf/Rpt value is less than 4 at the fracture control mode. If the Rf/Rpt value exceeds 4, the permeability evaluation coefficient increases fast. The reason is that the bigger fracture radius can favor fluid flow in such reservoir media.

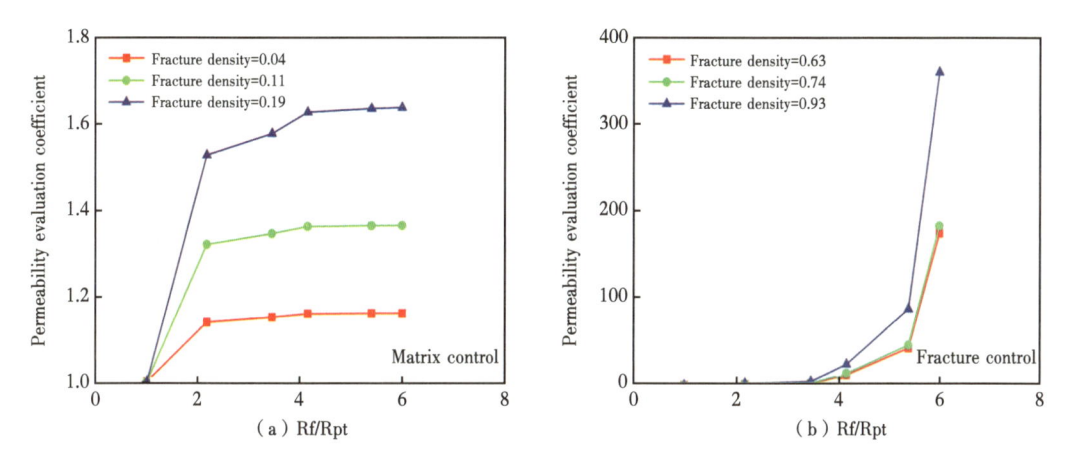

Fig. 6 Variation of permeability evaluation coefficient versus Rf/Rpt at matrix control(a) and fracture control(b)

## 3.5 Effect of fracture density on control mode

Fracture density is one of the main fracture parameters to evaluate the reservoir quality in such reservoir media, which has a great effect on the fluid flow in the fractured reservoir media[29]. In the pore-fracture network model, the fracture density is defined as the fracture number divided by the node number. In order to understand the effect of fracture density on different control modes, the different values of fracture density are considered in the pore-fracture network model. Fig. 7 shows the variation of permeability sensitivity versus fracture density at matrix control (a) and fracture control (b), respectively. From the result of Fig. 7 (a), it can be seen that the permeability sensitivity in the pore-fracture media decreases with the fracture density increasing at the matrix control mode. And the permeability sensitivity increases with the increasing Rf/Rpt value. As illustrated in Fig. 7 (b), there is a contrary trend at the fracture control mode. When it is the fracture control mode, the fracture density has little impact on the permeability sensitivity among the matrix and fractures. Thus the fracture density at the matrix control mode can greatly affect the permeability sensitivity in such media.

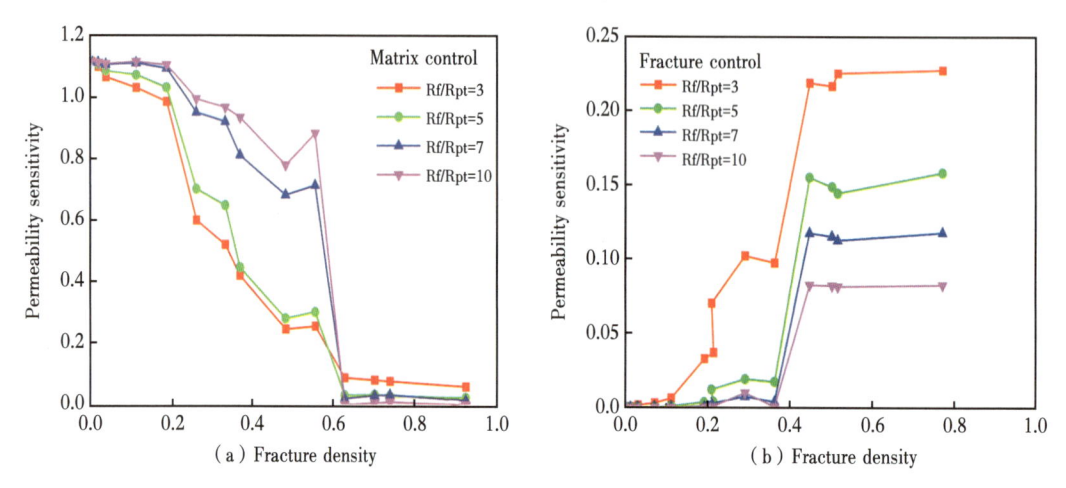

Fig. 7  Variation of permeability sensitivity versus fracture density at matrix control mode (a) and fracture control mode (b)

## 4  Conclusions

In this study, a pore-fracture network model was presented to understand the microscopic flow mechanism in fractured porous media. The microstructure parameters of fractures including fracture density, fracture length, fracture number and fracture radius were analyzed. Then two modes and effects of matrix and fracture network control were discussed. According to the above results, the following conclusions can be drawn: (1) With fracture length, fracture density, fracture number and fracture radius increasing, the network permeability in the pore-fracture media will increase linearly, and favor fluid flow in the fractured reservoir media. While the fracture radius has a little effect on network permeability when the fracture radius exceeds 80 μm. (2) The control modes of matrix and fracture network are affected by the fracture density. Within the fracture density less than 0.55, it belongs to the matrix control mode while the fracture network control mode is dominant in the fracture density exceeding 0.55. (3) Fracture density and the ratio of fracture radius to pore radius have a great effect on the network permeability in the matrix and fracture network control modes. There is a great change in the critical density for the matrix control mode compared with the fracture network control modes. These results can provide a better understanding of the relationship between matrix and fractures, and the effects of fracture on fluid flow in the fractured reservoir media, which is significant for optimizing the extraction condition in fractured gas reservoirs.

## References

[1] Shen WJ, Li XZ, Li XZ, et al. Physical simulation of water influx mechanism in fractured gas reservoirs. Journal of Central South University (Science and Technology), 2014,45(9): 3283-3287.

[2] Shen WJ, Liu XH, Li XZ, et al. Investigation of water coning mechanism in Tarim fractured sandstone gas reservoirs. Journal of Central South University (English Edition), 2015,22(1): 344-349.

[3] Li CH, Li XZ, Gao SS, et al. Experimental on gas-water two-phase seepage and inflow performance curves of gas wells in carbonate reservoirs: A case study of Longwangmiao Formation and Dengying Formation in Gaoshiti-Moxi Block, Sichuan Basin, SW China. Petroleum Exploration and Development, 2017,44(6): 983-992.

[4] Li XZ, Guo ZH, Hu Y, et al. Efficient development strategies for ultra-deep giant structural gas fields in China. Petroleum Exploration and Development, 2018,45(1): 1-8.

[5] Jiang TW, Teng XQ, Yang XT. Integrated techniques for rapid and highly-efficient development and production of ultra-deep tight sand gas reservoirs of Keshen 8 Block in the Tarim Basin. Natural Gas Industry, 2017,4(1): 30-38.

[6] Ould-amer Y, Chikh S. Transient behavior of water-oil interface in an upward flow in porous media. Journal of Porous Media, 2003,6(2): 1-12.

[7] Ould-amer Y, Chikh S, Naji H. Attenuation of water coning using dual completion technology. Journal of Petroleum Science and Engineering, 2004,45: 109-122.

[8] Hu Y, Li XZ, Lu XG. Varying law of water saturation in the depletion-drive development of sandstone gas reservoirs. Petroleum Exploration and Development, 2014,41(6): 723-726.

[9] Hao Y, Yeh TCJ, Xiang J, et al. Hydraulic tomography for detecting fracture zone connectivity. Ground Water, 2008,46(2): 183-192.

[10] Lei G, Dong PC, Mo SY, et al. Calculation of full permeability tensor for fractured anisotropic media. Journal of Petroleum Exploration & Production Technology, 2015,5(2): 167-176.

[11] Miao T, Yu B, Duan Y, et al. A fractal analysis of permeability for fractured rocks. International Journal of Heat & Mass Transfer, 2015,81(81): 75-80.

[12] Cacas, Ledoux, Marsily, et al. Modeling fracture flow with a stochastic discrete fracture network: Calibration and validation: 1. The flow model. Water Resources Research, 1990,26(3): 479-489.

[13] Tsang CF, Neretnieks I. Flow channeling in heterogeneous fractured rocks. Reviews of Geophysics, 1998,36(2): 275-298.

[14] Tsang YW, Tsang CF. Flow channeling in a single fracture as a two-dimensional strongly heterogeneous permeable medium. Water Resources Research, 1989, 25(9):2076-2080.

[15] Goc RL, Dreuzy JRD, Davy P. An inverse problem methodology to identify flow channels in fractured media using synthetic steady-state head and geometrical data. Advances in Water Resources, 2010,33(7): 782-800.

[16] Fatt I. The network model of porous media. Petroleum Trans. AIME, 1956,207, 144-181.

[17] Barenblatt G, Zheltov IP, Kochina I. Basic concepts in the theory of seepage of homogeneous liquids in fissured rocks. Journal of Applied Mathematics and Mechanics, 1960,24: 1286-1303.

[18] Warren J, Root PJ. The behavior of naturally fractured reservoirs. Journal of Petroleum Science and Engineering, 1963,3: 245-255.

[19] Snow DT. A parallel plate model of fractured permeable media. University of California, Berkeley,1965.

[20] De Dreuzy J, Davy P, Bour O. Hydraulic properties of two-dimensional random fracture networks following a power law length distribution: 1. Effective connectivity. Water Resources Research, 2001, 37(37): 2079-

2096.

[21] Zheng Q, Yu B. A fractal permeability model for gas flow through dual-porosity media. Journal of Applied Physics, 2012,111: 024316.

[22] Noetinger B, Jarrige N. A quasi steady state method for solving transient Darcy flow in complex 3D fractured networks. Journal of Computational Physics, 2012,231: 23-38.

[23] Jivkov AP, Hollis C, Etiese F, et al. A novel architecture for pore network modelling with applications to permeability of porous media. Journal of Hydrology, 2013,486(4): 246-258.

[24] Gong J, Rossen WR. Modeling flow in naturally fractured reservoirs: effect of fracture aperture distribution on critical sub-network for flow. Proceedings of the 1st International Conference on Discrete Fracture Network Engineering, Vancouver, Canada, 2014,19-22 October.

[25] Zhang JC, Bian XB, Zhang SC, et al. Research of seepage in artificial fracture using pore network model. Science China (Technological Sciences), 2013,56(3): 756-761.

[26] Shen WJ, Li XZ, Xu YM, et al. Gas Flow Behavior of Nanoscale Pores in Shale Gas Reservoirs. Energies, 2017, 10(6): 1-12.

[27] Mauldon M, Rohrbaugh MB, Dunne WM, et al. Mean fracture trace length and density estimators using circular windows. The 37th U.S. Symposium on Rock Mechanics, Vail, Colorado, 1999,7-9 June.

[28] Shen WJ, Xu YM, Li XZ, et al. Numerical Simulation of Gas and Water Flow Mechanism in Hydraulically Fractured Shale Gas Reservoirs. Journal of Natural Gas Science and Engineering, 2016,35, 726-735.

[29] Saboorian-Jooybari H, Dejam M, Chen Z, et al. Comprehensive evaluation of fracture parameters by dual laterolog data. Journal of Applied Geophysics, 2016,131: 214-221.

# 塔里木盆地克拉2气田储层综合定量评价

徐艳梅[1]　刘兆龙[1]　张永忠[1]　郭振华[1]　肖　鑫[2]　孙　迪[3]

(1. 中国石油勘探开发研究院；2. 北京阿什卡技术开发有限公司；
3. 中国石油塔里木油田勘探开发研究院)

**摘要**：为解决塔里木盆地克拉2气田储层定性评价过程中评价指标互相矛盾的问题，应用灰色关联度分析法确定了储层评价指标的权系数，计算出储层综合评价指标，重新对气藏圈闭内巴什基奇克组储层进行综合定量评价，并给出了储层分类结果，应用产能数据对储层评价结果进行了验证。结果表明，巴什基奇克组一段储层定量分类符合率为87%，巴什基奇克组二段储层定量分类符合率为85%，分类结果具有一定的可靠性，可为气田开发决策提供依据。

**关键词**：克拉2气田；巴什基奇克组；灰色关联度分析；储层定量评价；储层分类

塔里木盆地克拉2气田天然气探明地质储量超过$2000×10^8 m^3$，目前已进入开发中期。通过开展储层评价研究，对储层进行合理分类，有助于科学有效地开发气藏，提高勘探开发效益[1-4]。对储层评价可分为定性评价和定量评价两类[5-7]，克拉2气田储层评价以定性评价为主。朱如凯等人通过研究储层的岩石学、成岩演化、储集空间等特征，制定了不同的分类评价标准，将塔里木盆地北部白垩系—古近系储层定性地分为四类[8]。影响储层评价的因素较多，且各评价参数间的关系极其复杂，在储层评价过程中，常常会出现评价结果互相矛盾的情况。为了解决储层定性评价过程中遇到的问题，本文采用灰色关联度分析法，重新对克拉2气田中气藏圈闭内巴什基奇克组储层进行定量分类评价。

## 1 储层特征及评价存在的问题

### 1.1 地质概况

克拉2气田位于塔里木盆地库车坳陷克拉苏构造带东段，构造轴向呈东西向分布。库车坳陷的构造变化主要经历了三个阶段，分别是前碰撞造山、碰撞造山和陆内造山。库车地区烃源岩主要形成于中生代，克拉2构造圈闭形成和油气成藏阶段主要发生于新生代[9-11]。

克拉2气田主要储层为下白垩统巴西改组、巴什基奇克组和古近系库姆格列木群，岩性为白云岩和砂砾岩[12-13]。其中巴什基奇克组一段、二段为辫状河三角洲前缘亚相沉积，主要发育水下分流河道、河口坝、水下分流间湾微相。岩石类型主要为岩屑砂岩和长石岩屑砂岩，其中巴什基奇克组一段以岩屑砂岩为主，二段以长石岩屑砂岩为主。

## 1.2 储层特征

克拉 2 气田碎屑岩储层物性差异较大,常规物性分析表明,孔隙度最大为 22.39%,最小值 3.06%,主值区间为 8%~20%,峰值为 15%;渗透率主要为 1~1000mD,最大为 1770.15mD,平均为 49.42mD。孔隙度和渗透率均呈正态分布,表现为单峰状,孔渗关系为线性分布(图 1),表明储层孔喉分布较均匀,物性较好。其中,巴什基奇克组一段和二段孔渗相关性好。

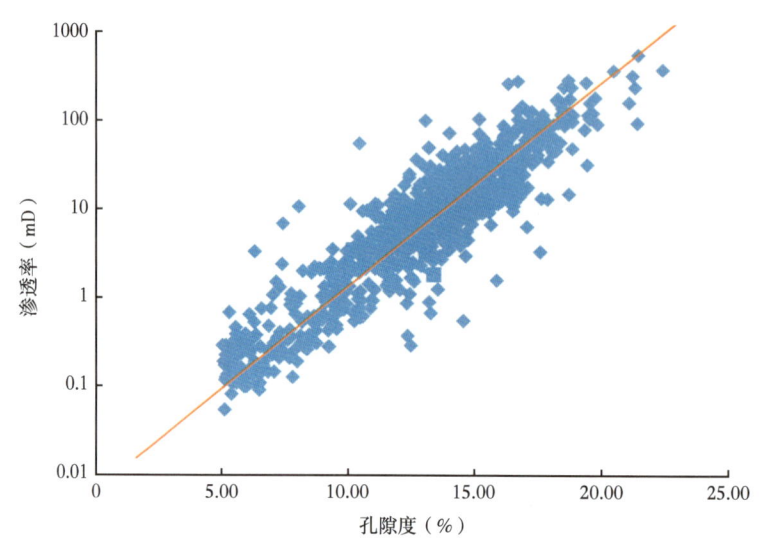

图 1 克拉 2 气田碎屑岩储层物性特征

对克拉 2 气田 528 块砂岩铸体薄片样品进行分析,结果表明孔隙类型以粒间溶孔为主,占总面孔率 74.7%,其次为原生粒间孔,占总面孔率 17.2%。巴什基奇克组一段面孔率为 3.0%~8.0%;巴什基奇克组二段面孔率较好,多数大于 5.0%。砂岩喉道类型主要为缩颈型,片状孔喉较少。巴什基奇克组一段连通性较差,孔喉配位数为 0~2,巴什基奇克组二段连通性好,孔渗性好,孔喉配位数主要为 1~3。

克拉 2 气田储层排驱压力较低,孔喉分选较好,总体属中粗孔中喉—中细孔中小喉的孔隙结构特征。巴什基奇克组一段、二段的毛细管压力曲线中值压力和排驱压力相对较低,平均孔喉半径大,对油气的渗流能力强。

## 1.3 存在问题

前人对克拉 2 气田的储层评价基本上都是通过定性评价方法来分类的,评价参数通常包含孔隙度、渗透率、有效厚度等,通过划分不同级别,制定该区的储层分类评价标准[14]。根据该评价标准,对克拉 2 气田巴什基奇克组储层进行了定性评价分类(表 1)。从表中可以看出:

(1)同时存在多个评价参数时,无法确定哪些参数作为主要评价指标;
(2)确定了多个评价参数时,储层分类的界限相对模糊;
(3)确定分类评价标准非常依赖于专家经验,主观性较强;
(4)评价结果互相矛盾,无法确定单个储集单元的最终评价结果。

表1 克拉2气田巴什基奇克组储层定性评价分类

| 井号 | 砂层组 | 渗透率 | 有效厚度 | 含气饱和度 | 孔隙度 | 泥质含量 |
|---|---|---|---|---|---|---|
| KL203 | $K_1bs_1$ | Ⅲ | Ⅱ | Ⅲ | Ⅲ | Ⅱ |
|  | $K_1bs_2$ | Ⅲ | Ⅰ | Ⅲ | Ⅲ | Ⅲ |
| KL204 | $K_1bs_1$ | Ⅲ | Ⅱ | Ⅲ | Ⅲ | Ⅲ |
|  | $K_1bs_2$ | Ⅰ | Ⅲ | Ⅲ | Ⅱ | Ⅲ |
| KL205 | $K_1bs_1$ | Ⅱ | Ⅲ | Ⅲ | Ⅰ | Ⅲ |
|  | $K_1bs_2$ | Ⅲ | Ⅰ | Ⅱ | Ⅲ | Ⅲ |
| KL2-1 | $K_1bs_1$ | Ⅲ | Ⅱ | Ⅱ | Ⅲ | Ⅲ |
|  | $K_1bs_2$ | Ⅰ | Ⅱ | Ⅰ | Ⅰ | Ⅲ |
| KL2-2 | $K_1bs_1$ | Ⅲ | Ⅲ | Ⅱ | Ⅲ | Ⅲ |
|  | $K_1bs_2$ | Ⅱ | Ⅰ | Ⅰ | Ⅱ | Ⅲ |
| KL2-3 | $K_1bs_1$ | Ⅱ | Ⅰ | Ⅰ | Ⅰ | Ⅲ |
|  | $K_1bs_2$ | Ⅱ | Ⅰ | Ⅰ | Ⅰ | Ⅲ |

## 2 综合定量评价

为了解决单因素定性评价结果相互矛盾的问题,本次研究采用储层综合定量评价方法,提高储层评价的客观性和准确性。

### 2.1 基本原理与计算步骤

储层综合定量评价方法公式为

$$R = \sum_{i=1}^{n} a_i x_i \tag{1}$$

式中 $R$——储层综合评价指标;
$a_i$——储层评价参数的权系数;
$X_i$——储层评价参数;
$n$——储层评价参数的数量。

由式中可以看出,储层评价参数是已知的,只要确定了储层评价参数的权系数,就可以计算出储层综合评价指标。本次研究中使用灰色关联度分析法来确定权系数[15,16]。

灰色关联度分析法是衡量因素间关联程度的一种方法,根据因素之间发展趋势的相似或相异程度,达到分清因素的主次及其影响大小的目的[17]。该方法的优势有:对数据的要求不高,既不需要有大量的数据支持也不需要数据具有典型的正态分布规律,在一定程度上可以忽略数据不对称带来的误差;同时该计算方法较为简便,计算工作量不大。将灰色关联度分析法应用于储层综合评价中,通过分析各个地质评价参数的主要关系,找出影响各个评价参数的重要因素,进而能够快速确定储层评价参数的权系数。

灰色关联度分析法的具体步骤如下:
(1)确定母序列和子序列。

母序列为

$$\{X_t^{(0)}(0)\} \quad t = 1,2,3,\cdots,n \tag{2}$$

子序列为

$$\{X_t^{(0)}(i)\} \quad i = 1,2,3,\cdots,n; i = 1,2,3,\cdots,m \tag{3}$$

式中　$n$——评价样品个数；

　　　$m$——评价参数个数。

（2）构建原始数据矩阵，其矩阵为

$$\boldsymbol{X}^{(0)} = \begin{Bmatrix} X_1^{(0)}(0) & X_1^{(0)}(1) & X_1^{(0)}(2) & \cdots & X_1^{(0)}(m) \\ X_2^{(0)}(0) & X_2^{(0)}(1) & X_2^{(0)}(2) & \cdots & X_2^{(0)}(m) \\ X_3^{(0)}(0) & X_3^{(0)}(1) & X_3^{(0)}(2) & \cdots & X_3^{(0)}(m) \\ \cdots & \cdots & \cdots & & \cdots \\ X_n^{(0)}(0) & X_n^{(0)}(1) & X_n^{(0)}(2) & \cdots & X_n^{(0)}(m) \end{Bmatrix} \tag{4}$$

（3）将各序列进行无量纲化处理，确定标准化后的数据。

（4）计算关联系数。将变换后的母序列和子序列分别记为 $X_t^{(1)}(0)$ 和 $X_t^{(1)}(i)$，则同一观测时刻各子因素与母因素之间的绝对差值为

$$\Delta_t(i,0) = |X_t^{(1)}(i) - X_t^{(1)}(0)| \tag{5}$$

同一观测时刻(观测点)各子因素与母因素之间的绝对差值的最大值为

$$\Delta_{\max} = \max_i \max_t |X_t^{(1)}(i) - X_t^{(1)}(0)| \tag{6}$$

同一观测时刻(观测点)各子因素与母因素之间的绝对差值的最小值为

$$\Delta_{\min} = \min_i \min_t |X_t^{(1)}(i) - X_t^{(1)}(0)| \tag{7}$$

母序列与子序列的关联系数 $L_t(i,0)$ 为

$$L_t(i,0) = \frac{\Delta_{\min} + \xi \Delta_{\max}}{\Delta_t(i,0) + \xi \Delta_{\max}} \tag{8}$$

式中　$\xi$——分辨系数，其作用是为了平衡由于最大绝对差数值较大而出现失真，改善关联系数之间的差异显著性。

这样，我们计算确定各子因素对母因素之间的关联度为

$$r_{i,0} = \frac{1}{n} \sum_{t=1}^{n} L_t(i,0) \tag{9}$$

（5）计算权系数。在计算确定关联系数后，通过归一化处理求得各影响因子的权系数：

$$a_i = \frac{r_i}{\sum_{t=1}^{n} r_i} \tag{10}$$

## 2.2 评价过程

在储层定量评价过程当中,首先筛选出主要评价参数,然后利用灰色关联分析法确定权系数,计算评价指标。

评价参数的选取一般应满足以下要求:(1)代表性:所选评价参数能反映储层的特点且具有代表性;(2)综合性:储层评价参数影响因素复杂,各参数相互影响,因此在选取有利参数的同时,也应选取对储层物性不利的参数;(3)目标性:评价参数与评价目标要一致,防止相互冲突。本次克拉2气田储层定量评价中选取的参数包含孔隙度、渗透率、有效厚度、含气饱和度和泥质含量。

在油田开发阶段,相对于其他评价参数,渗透率是主要影响因素[18,19],因此在本次研究中将渗透率作为主因素,其余4个参数作为子因素。将计算的关联度结果进行归一化处理,得到各指标的权重系数:$a$ =(0.233,0.193,0.192,0.192,0.190)。依照权重系数及关联度大小对全部指标进行排序,即得到关联程度顺序:渗透率>有效厚度>含气饱和度=孔隙度>泥质含量,关联结果与实际情况吻合,能够反映各评价参数对储层的影响程度。最后对每个评价参数进行最大值标准化处理,即可确定综合定量评价指标。

地质特征越相近,评价指标值越相近,不同分布特征的参数会形成不同斜率的直线段。具有相近特征的储层位于同一斜率的直线段上,因此可以利用累积概率曲线来确定分类区间(分类标准),不同直线段的交点即为分类的界限值。这样最终确定了克拉2气田储层综合定量分类的标准(表2)。

表2 储层分类标准

| | Ⅰ类储层 | Ⅱ类储层 | Ⅲ类储层 |
| --- | --- | --- | --- |
| 综合定量评价指标 | ≥0.55 | 0.45~0.55 | ≤0.45 |

## 2.3 评价结果

根据表2的储层综合定量分类标准,对本次研究的储层单元进行了分类(表3)。Ⅰ类储层为有利储层,为克拉2气田圈闭内较好的储层,是天然气富集和勘探开发的有利储层,占储集层单元总数的30.0%,平均孔隙度为15.41%,平均渗透率为94.52mD,有效厚度为129m;Ⅱ类储层为中等储层,占储层单元总数的52.5%,平均孔隙度为14.09%,平均渗透率为42.37mD,有效厚度为102m;Ⅲ类储层为不利储层,占储层单元总数的17.5%,平均孔隙度为12.40%,平均渗透率为8.21mD,有效厚度为59m。

表3 克拉2气田巴什基奇克组储层综合定量评价分类(部分)

| 井号 | 砂层组 | 渗透率(mD) | 有效厚度(m) | 含气饱和度(%) | 孔隙度(%) | 泥质含量(%) | 综合定量评价指标 | 储层分类 |
| --- | --- | --- | --- | --- | --- | --- | --- | --- |
| KL203 | $K_1bs_1$ | 7.20 | 51.50 | 50.12 | 10.54 | 10.57 | 0.360 | Ⅲ |
| | $K_1bs_2$ | 5.99 | 152.50 | 49.58 | 11.87 | 9.41 | 0.494 | Ⅱ |
| KL204 | $K_1bs_1$ | 31.25 | 119.30 | 60.36 | 13.56 | 9.99 | 0.520 | Ⅱ |
| | $K_1bs_2$ | 257.73 | 0.00 | 14.99 | 14.41 | 9.44 | 0.502 | Ⅱ |

续表

| 井号 | 砂层组 | 渗透率（mD） | 有效厚度（m） | 含气饱和度（%） | 孔隙度（%） | 泥质含量（%） | 综合定量评价指标 | 储层分类 |
|---|---|---|---|---|---|---|---|---|
| KL205 | $K_1bs_1$ | 4.49 | 67.80 | 62.85 | 11.95 | 9.33 | 0.439 | Ⅲ |
|  | $K_1bs_2$ | 21.30 | 137.10 | 70.52 | 12.94 | 10.24 | 0.546 | Ⅱ |
| KL2-1 | $K_1bs_1$ | 33.98 | 121.00 | 70.60 | 13.39 | 10.33 | 0.545 | Ⅱ |
|  | $K_1bs_2$ | 137.13 | 143.30 | 76.06 | 15.11 | 10.16 | 0.696 | Ⅰ |
| KL2-2 | $K_1bs_1$ | 29.12 | 88.80 | 68.13 | 13.03 | 10.27 | 0.497 | Ⅱ |
|  | $K_1bs_2$ | 99.15 | 130.40 | 73.88 | 14.55 | 10.81 | 0.629 | Ⅰ |
| KL2-3 | $K_1bs_1$ | 83.62 | 135.30 | 70.11 | 15.82 | 9.56 | 0.642 | Ⅰ |
|  | $K_1bs_2$ | 92.13 | 148.90 | 74.55 | 15.68 | 11.43 | 0.648 | Ⅰ |

## 2.4 结果验证

分类结果直接反映在产能等动态指标上,本文选取产能数据对储层分类结果进行验证。根据不同射孔层段的地层系数(有效厚度×渗透率)权重大小,将单井产能劈分到每个储层单元上,结合每个储层单元标定的产能和储层综合定量评价分类结果,确定了克拉2气田产能标定储层分类标准(表4)。通过对比分析产能标定储集分类和综合定量评价储层分类结果(表5),巴什基奇克组一段储层定量分类与产能标定分类符合率为87%,巴什基奇克组二段储层分类符合率为85%,证明本次储层综合定量评价具有一定的可靠性。分类结果可以在生产中实际应用。

**表4　产能标定储层分类标准**

|  | Ⅰ类储层 | Ⅱ类储层 | Ⅲ类储层 |
|---|---|---|---|
| 产能($10^4m^3/d$) | ≥400 | 150~400 | ≤150 |

**表5　巴什基奇克组储层产能和综合定量评价对比表(部分)**

| 井号 | 砂层组 | 产能($10^4m^3/d$) | 产能分类 | 综合定量评价指标 | 储层分类 |
|---|---|---|---|---|---|
| KL203 | $K_1bs_1$ | 73.918 | Ⅲ类 | 0.360 | Ⅲ类 |
|  | $K_1bs_2$ | 182.082 | Ⅱ类 | 0.494 | Ⅱ类 |
| KL204 | $K_1bs_1$ | 138.063 | Ⅱ类 | 0.520 | Ⅱ类 |
|  | $K_1bs_2$ | 135.240 | Ⅱ类 | 0.502 | Ⅱ类 |
| KL205 | $K_1bs_1$ | 44.241 | Ⅲ类 | 0.439 | Ⅲ类 |
|  | $K_1bs_2$ | 424.759 | Ⅱ类 | 0.546 | Ⅱ类 |
| KL2-1 | $K_1bs_1$ | 171.314 | Ⅱ类 | 0.545 | Ⅱ类 |
|  | $K_1bs_2$ | 818.686 | Ⅰ类 | 0.696 | Ⅰ类 |
| KL2-2 | $K_1bs_1$ | 270.567 | Ⅱ类 | 0.497 | Ⅱ类 |
|  | $K_1bs_2$ | 1352.800 | Ⅰ类 | 0.629 | Ⅰ类 |
| KL2-3 | $K_1bs_1$ | 665.321 | Ⅰ类 | 0.642 | Ⅰ类 |
|  | $K_1bs_2$ | 806.679 | Ⅱ类 | 0.648 | Ⅰ类 |

## 2.5 储层评价分区

将分类结果标定到巴什基奇克组一段和二段上,在平面上生成储层分类区。从图 2 可以看出,Ⅰ类和Ⅱ类储层主要分布在气藏圈闭的中部和东部,Ⅲ类储层主要分布在气藏圈闭的西部。从图 3 可以看出,Ⅰ类储层主要分布在气藏圈闭的中部,Ⅱ类储层主要分布在气藏圈闭的东部,气藏圈闭内没有Ⅲ类储层,说明纵向上巴什基奇克组二段储层物性整体要好于一段。受井控的限制,将气藏圈闭内Ⅰ类、Ⅱ类、Ⅲ类储层以外区域的储层预测为Ⅳ类储层,可能为差气层或干层,是今后需要进一步评价的区域。通过划分克拉 2 气藏的储层分类平面图,确定储层的有利区域,对以后的开发井位部署具有一定的指导作用。

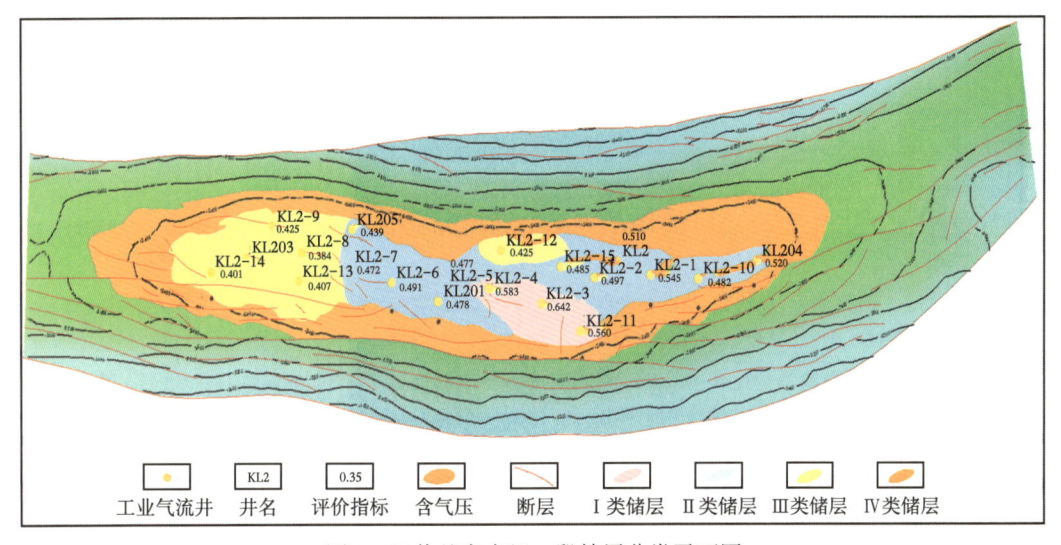

图 2 巴什基奇克组一段储层分类平面图

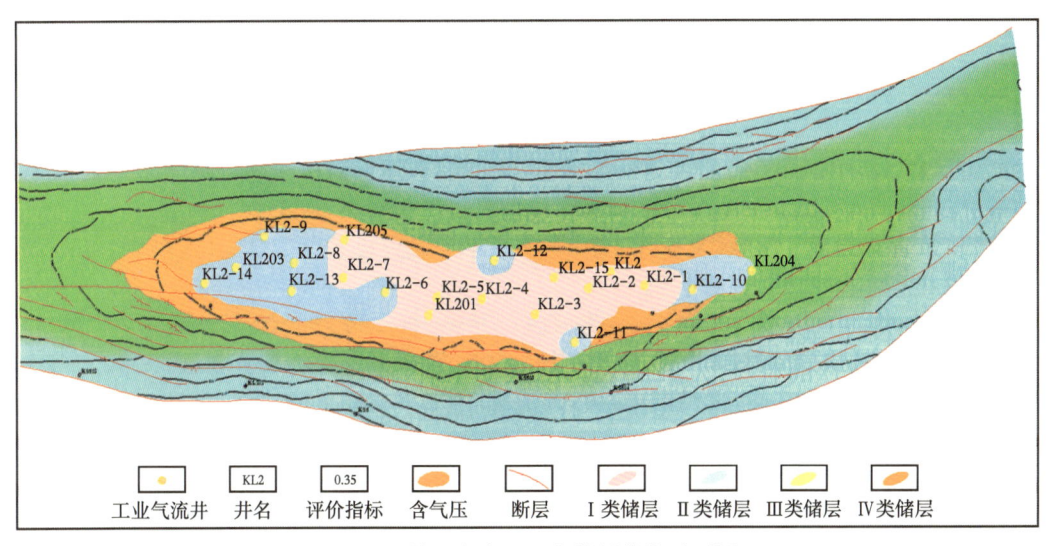

图 3 巴什基奇克组二段储层分类平面图

## 3 结论

(1)本文的储层评价方法解决了储层定性评价中遇到的单一参数评价结果互相矛盾的问题。

(2)对比产能和储层定量评价结果,验证了本文储层综合定量评价分类结果有一定的可靠性。

(3)平面上,巴什基奇克组一段的Ⅰ类和Ⅱ类储层主要分布在中部和东部,Ⅲ类储层主要分布在西部;巴什基奇克组二段的Ⅰ类储层主要分布在中部,Ⅱ类储层主要分布在东部。纵向上,巴什基奇克组二段储层质量整体好于巴什基奇克组一段。

### 参 考 文 献

[1] 宋子齐,谭成仟,曲政.利用灰色理论精细评价油气储层的方法[J].石油学报,1996,17(1):25-31.

[2] 闫明,张云峰,李易霖.基于K-均值聚类与贝叶斯判别的储层定量评价——以大安油田泉四段储层为例[J].深圳大学学报理工版,2016,33(2):211-220.

[3] 涂乙,谢传礼,刘超,等.灰色关联分析法在青东凹陷储集层评价中的应用[J].天然气地球科学,2012,23(2):381-386.

[4] 张琴,朱筱敏.山东省东营凹陷古近系沙河街组碎屑岩储层定量评价及油气意义[J].古地理学报,2008,10(5):465-472.

[5] 何琰.基于模糊综合评判与层次分析的储集层定量评价——以包界地区须家河组为例[J].油气地质与采收率,2011,18(1):23-29.

[6] 赖锦,王贵文,郑新华,等.大北地区巴什基奇克组致密砂岩气储层定量评价[J].中南大学学报(自然科学版),2015,46(6):2285-2298.

[7] 谭秀成,丁熊,陈景山,等.层次分析法在碳酸盐岩储层定量评价中的应用[J].西南石油大学学报(自然科学版),2008,30(2):38-40.

[8] 朱如凯,郭宏莉,高志勇,等.塔里木盆地北部地区白垩系——古近系储集性与储层评价[J].中国地质,2007,34(5):837-842.

[9] Groves, J R, P L Brenckle. Graphic correlation in frontier petroleum provinces: Application to Upper Palaeozoic sections in the Tarim basin, western China: AAPG Bulletin, 1997,81(8):1259-1266.

[10] Wilson, J E, J S Chester, et al. Microfracture analysis of fault growth and wear processes, Punchbowl Fault, San Andreas system, California: Journal of Structural Geology, 2003,25:1855-1873.

[11] 顾家裕,贾进华,方辉.塔里木盆地储层特征与高孔隙度、高渗透率储层成因[J].科学通报,2002,47(增刊):9-15.

[12] 贾承造,王招明,等.克拉2气田石油地质特征[J].科学通报,2002,47(增刊):91-96.

[13] 孙洪志,刘吉余.储层综合定量评价方法研究[J].大庆石油地质与开发,2004,23(6):8-10.

[14] 徐忠波,杨辉廷,何光芒,等.塔里木盆地柯克亚凝析气田西五二段储层定量评价[J].2012,26(2):68-70.

[15] 邓聚龙.灰色系统理论教程[M].武汉:华中理工大学出版社,1990:33-34.

[16] 杨德永,李超.储层评价中的大样本聚类方法[J].石油实验地质,1991,13(3):281-286.

[17] 赵加凡,陈小宏,张勤.灰色关联分析在储层评价中的应用[J].勘探地球物理进展,2003,26(4):282-286.

# 鄂尔多斯盆地东部奥陶系古岩溶型碳酸盐岩致密储层特征、形成机理与天然气富集

王国亭[1] 程立华[1] 孟德伟[1] 朱玉杰[2] 孙建伟[2] 黄锦袖[2] 彭艳霞[3]

[1. 中国石油勘探开发研究院；2. 中国石油长庆油田分公司第二采气厂；
3. 中国地质大学(北京)能源学院]

**摘要**：目前鄂尔多斯盆地东部地区古岩溶型碳酸盐岩储层的研究相对薄弱，因此开展奥陶系古岩溶型储层形成机理与天然气富集潜力研究可为天然气储量规模增加和开发前景评价奠定基础。通过对鄂尔多斯盆地东部奥陶系岩溶储层特征、有效储层形成控制因素、天然气富集主控因素与富集潜力的综合分析，认为盆地东部碳酸盐岩风化壳储层较为发育，气源供给充足，源—运—储配置关系良好，具备天然气大规模富集的条件。鄂尔多斯盆地东部岩溶储层总体表现为低孔、致密的特征，将孔隙度为3%、渗透率为0.05mD确定为有效储层物性下限标准，将孔隙直径为30μm、喉道直径为5μm界定为储层孔、喉尺度下限标准。有利沉积微相或岩相组合、高效岩溶作用、综合成岩作用等共同影响有效储层形成，半充填型硬石膏结核溶孔云岩为最重要的储层类型。总结了天然气富集主控因素：(1)有效储层发育是天然气富集的基础物质条件；(2)良好的源—运—储配置关系是天然气富集的关键；(3)岩溶储层的强非均质性影响着气、水的分布格局。总体而言，盆地东部下古生界碳酸盐岩岩溶储层具备较大的勘探开发潜力。

**关键词**：古岩溶储层；奥陶系；物性下限；天然气富集；鄂尔多斯盆地东部

中国古岩溶型碳酸盐岩储层广泛发育并取得了一系列重大勘探开发突破，重点以塔里木盆地塔河油田及塔里木油田奥陶系石灰岩油气藏、鄂尔多斯盆地靖边下古生界奥陶系气藏、渤海湾盆地任丘油田奥陶系古岩溶油气藏等为代表[1-3]。鄂尔多斯盆地是中国重要的含油气盆地，蕴含丰富的天然气资源，目前天然气探明(含基本探明)储量已达$5.7\times10^{12}m^3$，发现了苏里格、靖边、榆林、大牛地、神木、子洲等多个探明储量超千亿立方米的气田。目前盆地绝大部分探明储量都集中于上古生界山西组至下石盒子组碎屑岩储层中，下古生界探明储量主要集中于古岩溶型碳酸盐岩储层发育的靖边气田，其储量规模仅占目前盆地总探明储量规模的10%。目前针对盆地中部靖边地区的岩溶型碳酸盐岩储层研究较为系统深入[4-9]，而盆地东部地区相关研究比较薄弱且没有储量发现。本文以盆地东部神木地区为依托，深入开展盆地东部碳酸盐岩岩溶储层特征分析、有效储层形成机理及天然气富集潜力评价，以明确盆地东部下古生界的勘探开发前景，从而为盆地东部下古生界古岩溶储层天然气储量规模增加和开发潜力评价奠定基础。

## 1 奥陶系构造沉积与岩溶古地貌格局

古生代时期，鄂尔多斯盆地属于华北克拉通盆地的一部分，南北两侧分别为古秦岭洋和古兴蒙洋。早古生代盆地演化主要受控于南侧古秦岭洋的演化，伴随古洋盆的形成、扩张、俯冲消减及最终闭合消亡，盆地内部经历早期陆表海盆地、后期陆缘海盆地、洋盆闭合并整体抬升

遭受剥蚀的演化过程[4-8]。鄂尔多斯盆地奥陶系沉积期表现为"两隆两鞍两坳陷"的古地貌特征,两隆指北部伊盟隆起和西南部中央古隆起,两鞍指两个隆起间的衔接部位,两坳陷指盆地东部米脂坳陷及盆地西、南部的秦祁海槽[9],上述古地貌控制着奥陶系沉积厚度的变化和相带的展布,决定了鄂尔多斯盆地奥陶纪岩相古地理格局。

盆地中东部奥陶系马家沟组沉积期经历 3 次海进、海退旋回,沉积了一套以碳酸盐岩为主夹蒸发岩的地层,自下而上可换分为马一至马六段等 6 个岩性段。目的层马五段自上而下细分为马五$_1$ 至马五$_{10}$,形成于海退期,主要为以膏岩、盐岩、白云岩为特色的沉积组合[10]。中奥陶世末,华北地块因晚加里东运动整体抬升,经历了约 130—150Ma 的沉积间断,盆地主体缺失晚奥陶—早石炭世沉积,中奥陶统马家沟组经历了长期岩溶作用。该期盆地总体表现为西高东低的岩溶古地貌格局,表现为岩溶高地、岩溶斜坡、岩溶盆地的岩溶古地貌格局(图 1、图 2)。岩溶储层主要发育于马五$_1$ 至马五$_4$ 亚段,是盆地中部靖边地区下古生界的天然气主要的储层与产层。

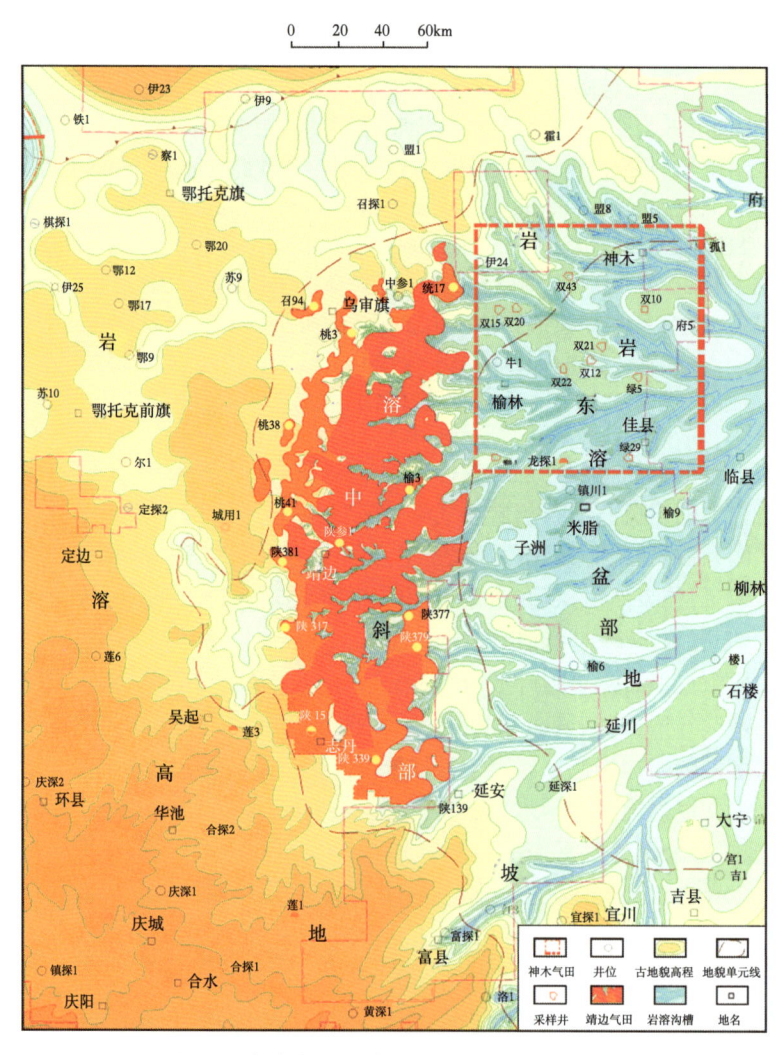

图 1　鄂尔多斯盆地前石炭系岩溶古地貌格局

图 2 鄂尔多斯盆地前石炭系岩溶古地貌剖面

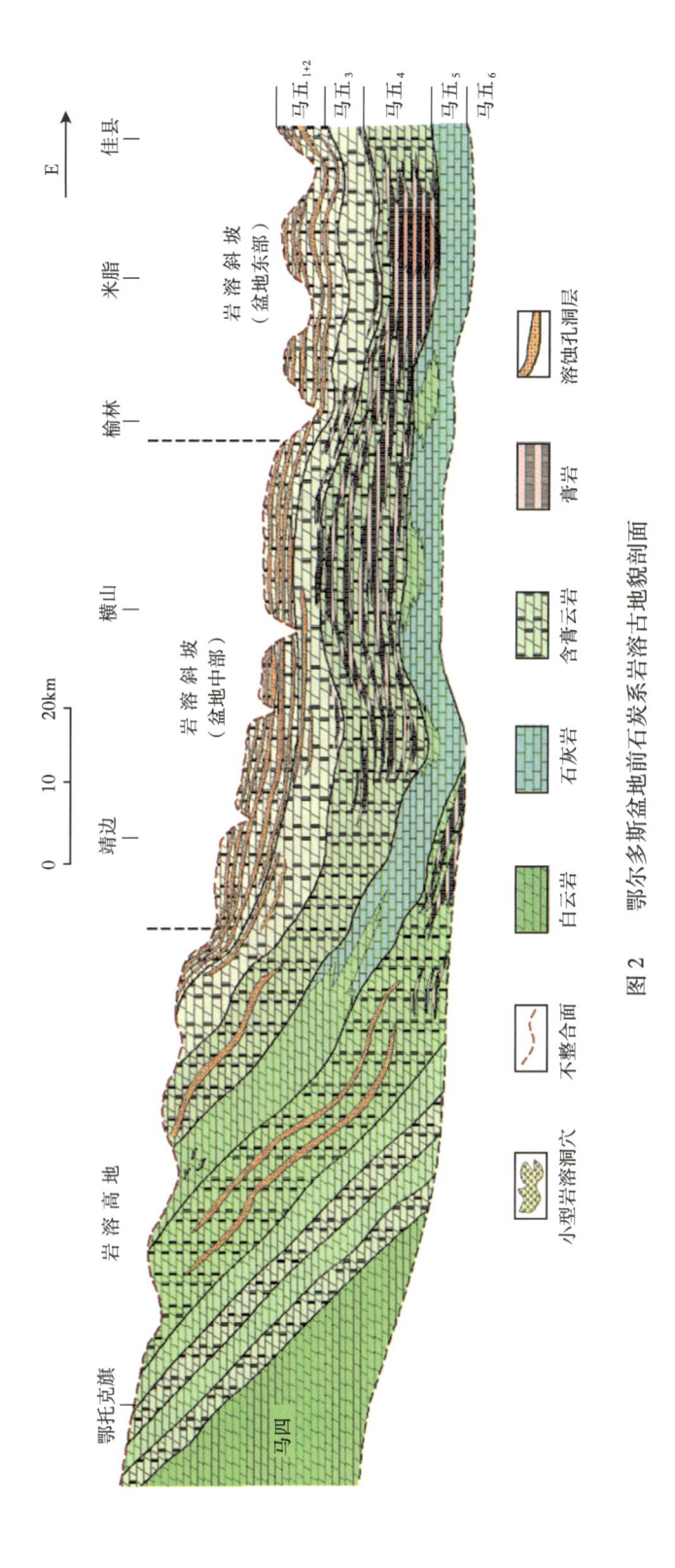

## 2 盆地东部古岩溶储层特征

### 2.1 古岩溶储层岩石学及物性特征

碳酸盐岩抬升裸露地表或近地表后受到各种复杂物理化学风化营力作用影响，并伴随各种机械、重力、化学等沉积作用，沉积期形成的原始地层结构发生改变，最终形成多种复杂类型岩溶岩，可细分为岩溶建造岩和岩溶改造岩两大类[2]。岩溶建造岩为岩溶溶洞中沉积并固化的机械沉积物、化学沉淀物及其他搬运至溶洞再堆积的物质，原始地层结构被彻底改变，可细分为残积岩、塌积岩、冲积岩、填积岩和淀积岩，此类岩石物性普遍较差，难以形成储层（图3a、b、c）。经历过岩溶作用而仍保持一定原始沉积结构的岩石称为岩溶改造岩，根据岩溶作用方式及引起的物理化学变化，可划分为岩溶溶蚀岩、岩溶变形岩及岩溶交代岩，前两种利于储层形成。盆地东部奥陶系顶部碳酸盐岩地层发育的岩石类型包括硬石膏结核云岩、白云岩、泥云

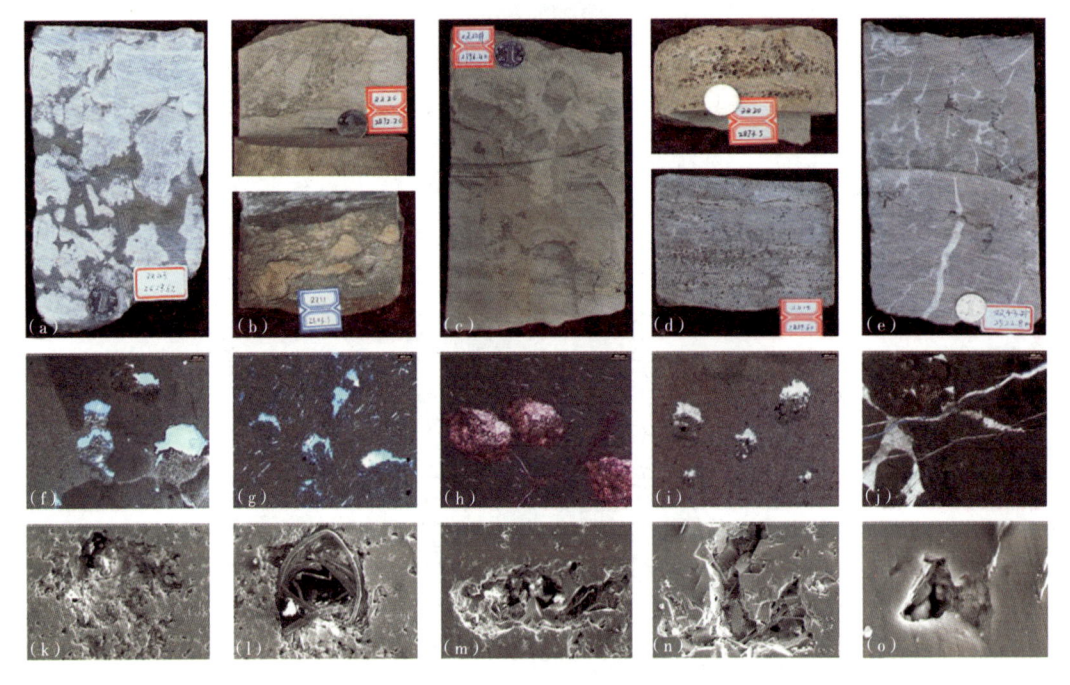

图3 鄂尔多斯盆地东部奥陶系岩溶储层马五$_1$至马五$_4$亚段储层特征

(a)双43井，马五$_2^1$，2329.6m，岩溶建造残积岩，可见溶解残余砾石组分，内部被细粒碎屑充填；(b)双20井，马五$_1^4$，2872.2m，双21井，马五$_1^4$，2623.1m，岩溶建造塌积岩，可见塌积搬运过程中形成的磨蚀、圆化边界；(c)双22井，马五$_1^4$，2794.4m岩溶建造填积岩，可见近垂向充填的棱角状砾石组构；(d)双20井，马五$_1^3$，2873.5m，双15井，马五$_1^4$，2859.6m，岩溶改造溶蚀岩，可见硬石膏结核溶蚀孔；(e)双43井，马五$_1^4$，2522.8m，岩溶改造变形岩，可见大量岩溶裂缝；(f)双20井，马五$_1^3$，硬石膏结核云岩，球状溶发育，白云石粉砂半充填；(g)双43井，马五$_1^3$，硬石膏结核白云岩，球状溶孔、柱状溶孔发育，白云石粉砂半充填；(h)陕267井，马五$_1^2$，硬石膏结核云岩，球状溶孔，方解石+白云石粉砂全充填；(i)双12井，马五$_1^2$，硬石膏结核云岩，球状溶孔发育，白云石粉砂+高岭石全充填；(j)榆82井，马五$_1^3$，白云岩，岩溶裂缝，方解石全充填；(k—o)双15井，马五$_1^2$，2860.5m，泥云岩，石膏高结核溶蚀孔、白云石晶间孔发育，孔隙直径分布于几微米至几百微米

岩、云膏岩、膏云岩、膏岩、云灰岩、灰云岩及石灰岩等多种类型，总体为潮坪沉积环境的产物。发育硬石膏结核及柱状晶体的白云岩在裸露风化期因大气淡水淋滤而形成溶模孔的岩溶溶蚀岩是研究区最重要的储集岩(图3d、f、i)。发生岩体张裂或假角砾化而形成的岩溶变形岩因发育溶滤缝和卸载缝也具有一定的储集性能，取决于裂缝系统的后期充填程度(图3e、m)。

盆地东部储集岩主要为硬石膏结核云岩和白云岩，并以前者为主，与盆地中部靖边气田储集岩性基本类似。盆地东部地区50余口探井密集取样的物性分析表明，下古生界岩溶储层总体表现出低孔、致密的特征。孔隙度主要分布于1%～7%，在此范围的样品比例为76.9%，大于7%的比例为11.7%，平均为3.8%。渗透率主要分布于0.005～1mD，在该区间的样品比例为74.56%，大于1mD的比例为12.57%，平均为0.82mD(图4)。盆地中部靖边地区储层平均孔隙度为6.70%，渗透率为3.80mD，与其相比，东部地区储层品质有所降低。

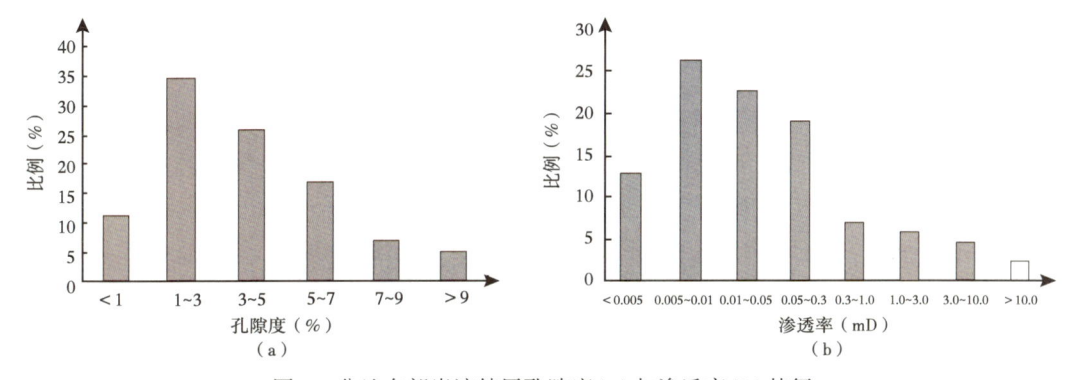

图4 盆地东部岩溶储层孔隙度(a)与渗透率(b)特征

鄂尔多斯盆地上古生界碎屑岩储层孔隙度、渗透率表现出明显线性相关的特征[11]，受研究区碳酸岩储层孔喉结构、残余裂缝及孔洞的影响，下古生界储层物性线性相关性总体偏差，数据分布较为分散(图5a)。不发育裂缝、孔洞的基质储层孔隙度、渗透率正相关性明显，但线性相关性一般，随着孔隙度增加渗透率逐渐增加，但数据点呈现分散、不集中的分布特征。部分样品表现出高孔隙度低渗透率特征，主要原因为结核溶蚀孔虽然发育，但孔隙呈孤立状分布，连通性差；部分样品表现出低孔隙度高渗透率特征，主要受裂缝影响，渗透性虽好，但储集性差。

(a) 孔隙度与渗透率关系

(b) 有效储层物性下限确定

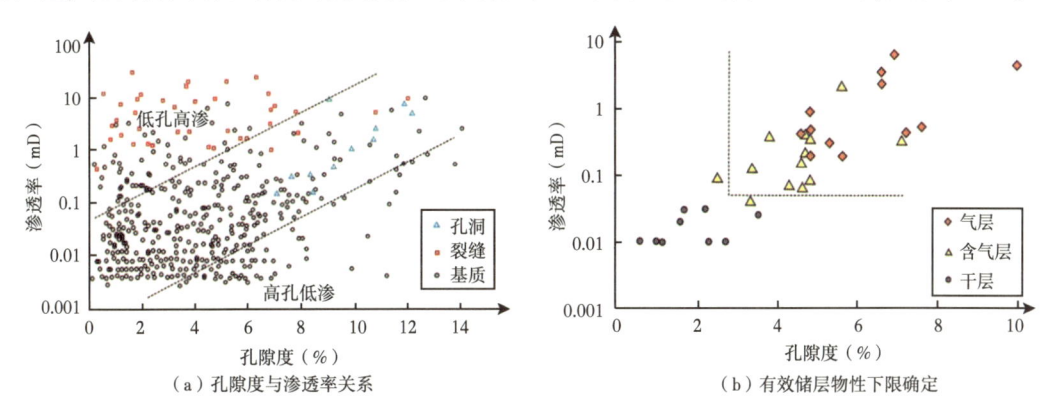

图5 鄂尔多斯盆地东部岩溶储层孔渗关系及物性下限确定

## 2.2 储集空间特征

孔、洞、缝等三大类型储集空间在盆地东部碳酸盐岩风化壳岩溶储层中均有发育,以孔隙为主(图 3d,f—i),偶可见溶洞,且多被后期填充,受岩溶与后期构造作用影响,裂缝体系早期较发育,后期多被填充(图 3e、m)。孔隙类型主要为硬石膏结核球状溶孔、白云石晶间微孔、岩溶角砾间孔等类型,并以前两种为主。硬石膏结核球状溶孔呈圆形、椭圆形,直径范围 0.1~5mm 不等,并以 1.5~3.0mm 为主,多被部分充填或全充填。白云石晶间微孔发育于细粉晶、泥晶白云岩中,也发育于硬石膏结核云岩基质中,直径范围几微米至近百微米(图 3q—r)。总体而言,盆地东部风化壳岩溶储层孔隙直径分布于微米级至毫米级范围,呈连续状分布特征。

铸体薄片孔喉图像分析表明,随着硬石膏结核球状溶孔填充程度的增加或发育程度的降低,对储集性能贡献作用较大的大孔隙发育比例逐渐降低,而微小孔隙发育比例逐渐升高。球状溶孔充填程度低的白云岩储层孔隙总体小于 5mm,占总孔隙体积 50%($P_{50}$)以上的较大孔隙对应的孔隙直径下限为 2mm,即直径为 2~5mm 的大孔隙占总孔隙的 50%,而球状溶孔充填程度高的白云岩、纯云岩、泥云岩等 $P_{50}$ 对应的孔隙直径下限分别为 20μm、15μm、7μm。喉道也表现出类似特征,低球状溶孔充填程度的云岩储层喉道总体小于 50μm,占总喉道 50%($P_{50}$)以上的较大喉道对应的喉道下限直径为 10μm,即直径 10~50μm 的大喉道占总喉道的 50%,而球状溶孔充填程度高的白云岩、纯云岩、泥云岩等 $P_{50}$ 对应的喉道直径下限分别为 4μm、3μm、2μm(图 6)。泥云岩为非储层类,分析表明其孔隙直径总体小于 30μm、喉道直径总体小于 5μm,可以此作为储层与非储层的孔—喉直径分界。球状溶孔充填程度低的白云

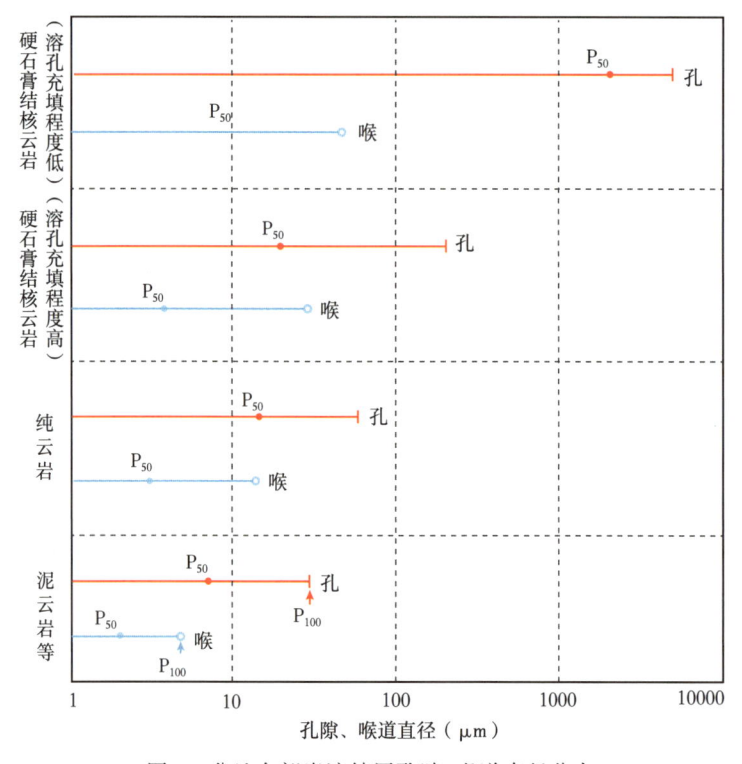

图 6 盆地东部岩溶储层孔隙、喉道直径分布

岩、球状溶孔充填程度高的白云岩、纯云岩等储层无效孔隙占的比例分别为10%、65%、80%，无效喉道占的比例分别为10%、55%、60%。总体而言，随着球状溶孔填充程度的增加或发育程度的降低，无效孔喉占的比例逐渐增加。

## 3 古岩溶储层形成机理

结合大量试气、生产动态数据进行了储层物性下限分析，结果表明，孔隙度低于3%、渗透率低于0.05mD的储层难以形成有效产层。孔隙度为3%、渗透率为0.05mD可界定为有效储层的物性下限标准，高于物性下限标准的储层产气能力较强，为有效储层类型(图5b)。有效储层形成过程极为复杂，综合受到有利沉积微相或岩相组合、高效岩溶作用、建设性及破坏性成岩作用共同影响。

### 3.1 有利沉积岩相

盆地中东部地区马家沟组为海平面周期性升降交替形成的碳酸盐岩与膏岩交互沉积。盆地东部马五$_{1-4}$亚段主体属于潮坪沉积环境的产物，主要亚相类型为潮上带、潮间带，岩相类型主要为含硬石膏结核云岩、白云岩、泥云岩、云灰岩、灰云岩、膏云岩及云膏岩等。盆地东部马五$_{1-4}$亚段在加里东期遭受较强烈的岩溶淋滤作用[12-15]，原始沉积产物受到岩溶作用改造。有利沉积相为储层形成的提供了物质基础，有利岩相的发育是有效储层形成的关键。

潮间带及潮上带发育的含硬石膏结核云岩是盆地东部最有利的沉积相或岩相组合。硬石膏结核云岩由于含有大小适宜且易溶的硬石膏结核组分，便于在古岩溶作用中接受改造而形成百微米至毫米级溶孔，且此类溶孔又相对易于保存，因此硬石膏结核云岩为盆地东部地区最有利的沉积岩相(图3d)。白云岩、泥云岩、云灰岩、灰云岩、石灰岩等不含易溶组分，难以受到岩溶改造作用的影响，不利于有效储层的形成。膏云岩、云膏岩、膏岩等岩相易溶组分含量较高，在强岩溶淋滤作用下可能会因缺乏有效支撑而发生垮塌、填充而最终不利于储集空间的保存。

### 3.2 高效岩溶改造

有利沉积岩相的存在为有效储层的形成创造了基础条件，但如果缺乏高效岩溶改造作用，难以转变为有效储层，因此高效岩溶改造是有效储层形成的关键。加里东末期，盆地整体抬升，马家沟组遭受长达130—150Ma的风化淋滤剥蚀。盆地前石炭纪古地貌分布以近南北向的中央古隆起为中心，古地势逐渐向东西两侧降低，经历剥蚀改造的奥陶系顶部界面呈现为以鄂托克旗—定边—庆阳一线为中心向东西倾伏的特点，并直接影响了盆地不同地区岩溶作用的发育强度[13-17]。

根据盆地岩溶古地貌宏观格局，中东部岩溶古地貌可划分为岩溶高地、岩溶斜坡、岩溶盆地等三种地貌单元(图2)。岩溶高地古地势较高，侵蚀强度大，地层缺失严重，岩溶作用以垂向渗滤为主，形成垂向溶蚀带、落水洞等岩溶形态，非均质性较强。岩溶斜坡地带岩溶作用方式以水平状慢速扩散流溶蚀为主，有利于良好溶蚀性储层的形成，储层均质性较好，靖边气田即位于岩溶斜坡部位[16-19]。岩溶盆地为岩溶斜坡以东大片地区古地势平坦开阔区，岩溶作用以沿地表侵蚀带的溶蚀及浅层地下径流带的岩石溶解为主，层状溶蚀作用偏弱，非均质性更强，此外，该区处于水流的汇水排泄区，充填、淀积作用强，岩溶空间充填作用高[13]。整体而言，盆地东部大部分地区都处于岩溶盆地范围,岩溶作用强度不如靖边气田所处的岩溶斜坡部位。

虽然盆地东部岩溶盆地区岩溶作用强度整体不如盆地中部区,但仍存在有效储层发育的条件。依据神木地区奥陶系风化壳上覆地层石炭系厚度及分布趋势、风化壳残余厚度及残余边界分布将岩溶盆地内古地貌刻画为丘台、坡地、沟槽—洼地等次级微地貌单元。丘台、坡地是古地形相对较高、马五段顶部(马五$_1^4$以浅)地层保留相对完整的地带,其上发育有利沉积岩相的部位可受到较强岩溶作用的改造,是盆地东部地区有利于有效储层发育的沉积岩相—微地貌组合单元。有限的试气资料表明,产气井主要分布于丘台、斜坡部位,沟槽—洼地部位因岩溶作用过于强烈,导致地层缺失严重,为汇水排泄区,充填作用较强,不利于储层的保存发育(图7),孔隙度平面等值线图也证明了这一点(图8)。同盆地中部靖边地区相比,神木地区岩溶作用总体偏弱且非均质性变强,有效储层连通性较差,西部、南部地区比东部地区有效储层发育。

### 3.3 成岩作用

鄂尔多斯盆地中东部下古生界碳酸盐岩地层经历了极其复杂的成岩作用过程,主要成岩作用类型包括溶蚀作用、白云石沉淀、干化脱水、机械压实、压溶、去白云石化、岩溶化和角砾化、胶结等众多类型[18]。

沉积作用之后淋滤剥蚀期之前的浅埋藏期主要发生大气淡水溶蚀、白云石沉淀、干化脱水、机械压实、胶结等成岩作用,淋滤剥蚀期之后的埋藏期主要发生埋藏溶蚀、压溶、胶结、白云石化等成岩作用。

影响最大的成岩作用发生于中石炭世之后的后期埋藏阶段,在淋滤剥蚀期岩溶作用改造了有利岩相储集物性的基础上,后期建设性与破坏性成岩作用的综合叠加决定有效储层的最终形成。埋藏溶蚀和白云石化对储层的改造是建设性的,在一定程度上提高了储层储集性能,而胶结充填作用是后期最主要的破坏性成岩作用,也是影响作用最大的成岩作用,此作用过程堵塞了岩溶改造阶段形成的溶蚀孔隙,使储层品质大幅降低。

盆地东部岩溶储层发育多种胶结充填类型,包括方解石(含铁方解石)、白云石(含铁白云石)、石英、高岭石、黄铁矿、萤石等。这些胶结充填物以单种、两种或多种组合的方式充填于岩溶孔隙空间中,其中白云石与其他充填物相组合的胶结充填方式最为普遍(图3f—i)。根据充填程度的强弱可分为半充填型和全充填型。全充填型的硬石膏结核溶孔虽然具备有利沉积岩相基础,也曾受到高效岩溶作用的改造,但因后期溶孔被完全充填堵死,储层品质变差,最终难以成为有效储层,而半充填型硬石膏结核溶孔云岩的溶蚀孔隙得以部分保存,储层品质较好,最终成为盆地东部地区最重要的有效储层类型(图3f,g)。

### 3.4 裂缝发育影响

盆地东部下古生界顶部风化壳地层受岩溶改造、构造运动、成岩收缩等多种因素影响,裂缝系统较为发育,主要为溶蚀缝、构造缝和成岩缝等三种类型。裂缝镜下宽0.01~1mm,岩心观察宽度一般在0.5~3mm,最宽可达1cm(图3m)。裂缝系统能够有效改善储层储集性能,尤其是渗流性能,物性分析也表明,残存裂缝系统的发育使部分储层具有低孔隙度、高渗透率的特征(图5a)。在后期成岩作用阶段裂缝网络大都被胶结物充填,因此对储层储集与渗流性能的最终改善作用有限,仅少量残存裂缝系统可局部改善储层物性。总体而言,裂缝系统对盆地

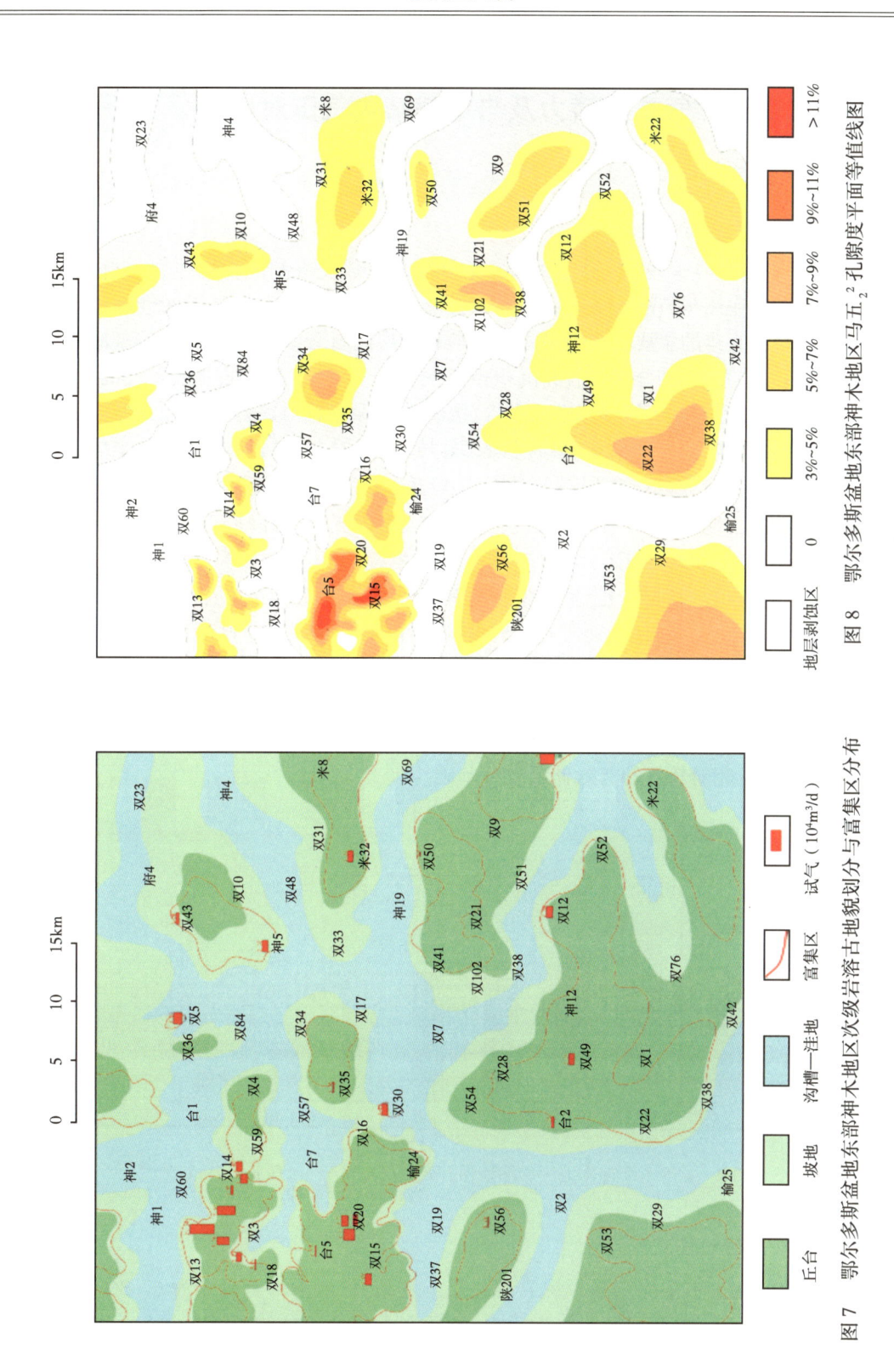

图 8 鄂尔多斯盆地东部神木地区马五$_2^2$孔隙度平面等值线图

图 7 鄂尔多斯盆地东部神木地区次级岩溶古地貌划分与富集区分布

东部地区下古生界碳酸盐岩风化壳型有效储层的形成起较为有限的影响作用。

## 4 盆地东部天然气富集潜力及与中部靖边气田对比

### 4.1 天然气富集潜力

#### 4.1.1 储层发育与分布

盆地东部下古生界碳酸盐岩岩溶地层有效储层的形成受有利沉积相或岩相组合、高效岩溶改造、后期成岩作用与裂缝发育等因素的综合影响。有效储层发育是天然气能够富集的基本条件,是天然气富集的主要控制因素之一。半充填型硬石膏结核溶孔云岩是盆地东部地区最重要的有效储层类型,其分布受有利沉积相带、地层残存状况、次级古地貌单元的共同影响,主要出现在潮坪相硬石膏结核云岩较为发育、沉积地层保存相对完好、相对凸起的丘台、坡地等次级岩溶古地貌单元上。神木地区评价表明:马五$_1$亚段上部马五$_1^1$、马五$_1^2$小层地层严重缺失,有效储层不发育;马五$_1$亚段下部马五$_1^3$、马五$_1^4$小层地层保存较好,有利沉积或岩相发育,且主体位于丘台地貌背景,有效储层较为发育;马五$_2$亚段上部马五$_2^1$小层有利岩相不发育,有效储层不发育,下部马五$_2^2$小层有利岩相较为发育、地层保存较为完善,主体位于丘台地貌背景,有效储层非常发育;马五$_3$亚段整体以膏岩沉积为主,有利岩相不发育,虽然地层保存良好,但有效储层不发育;马五$_4$亚段有利岩相发育,地层保存相当完整,但由于岩溶改造作用偏弱,有效储层发育较差。

据钻井资料揭示,神木气田厚度大于 2m 的有效储层钻遇率为 51.81%,厚度主要分布于 2~8m(图9)。整个盆地东部地区有效储层发育状况应该同神木地区基本类似,总体而言,盆地东部地区仍具备相对较好的有效储层发育基础。

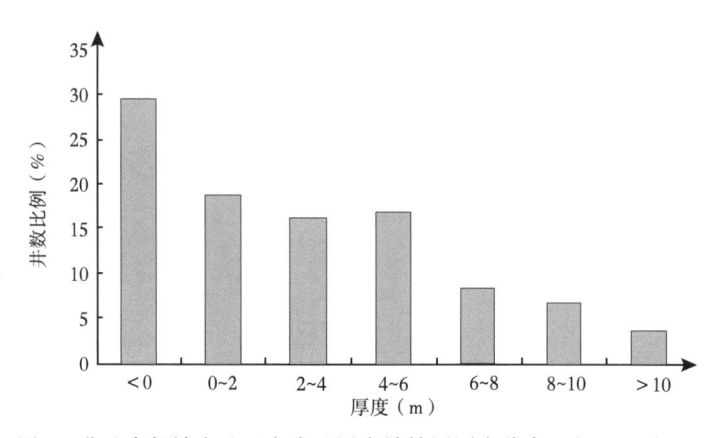

图9 盆地东部神木地区岩溶地层有效储层厚度分布(马五$_1$—马五$_4$)

#### 4.1.2 源—运—储配置关系

鄂尔多斯盆地上古生界煤系烃源岩有机质丰度高,煤岩平均有机碳为 67.3%,碳质泥岩平均有机碳为 2.93%,具有较强的生气能力,是奥陶系顶部气藏的主要气源岩。盆地东部地区上古生界源岩生烃强度大,生烃强度范围为 $(20~40)\times10^8 m^3/km^2$,平均为 $36\times10^8 m^3/km^2$,

高于盆地中西部地区,气源供给充分。在生排烃高峰期,天然气沿古沟槽与岩溶不整合面向下近距离运移,与风化壳岩溶储层构成良好上生下储组合关系[20-24]。

在气源岩、有效储层、运移通道都具备的条件下,三者良好的配置关系是盆地东部区域下古生界天然气能够有效富集的关键。盆地东部区域神木地区下古生界奥陶系顶部地层岩溶侵蚀差异明显,总体而言,神木地区的西部岩溶作用强于东部,西部地区岩溶沟槽多切割至马五$_3$,局部甚至可达马五$_4$,而东部普遍溶蚀切割至马五$_2$(图10)。发育于丘台、斜坡部位的马家沟组上部马五$_{1-2}$亚段的有效储层在侧向或垂向可与岩溶不整合面及岩溶古沟槽等优势运移通道紧密相邻,并且临近上古生界气源岩,源—运—储配置关系最佳,有利于天然气富集。马五$_4$亚段总体发育完整,虽然其上部有有效储层发育,但因被马五$_3$亚段厚层稳定泥云岩遮挡,优势运移通道难以与有效储层充分接触,源—运—储配置关系不佳,总体不利于天然气富集,仅在西部距离深切沟槽较近的区域有气层发育,东部区域则主要发育气水层、含气水层(图11)。此外,盆地东部的西部地区岩溶作用比东部地区相对强烈,有效储层发育程度高,岩溶不整合面及古沟槽切割深度大,且上覆本溪组厚度薄,天然气运送距离短,源—运—储配置关系良好,因此气层发育程度比东部地区大。

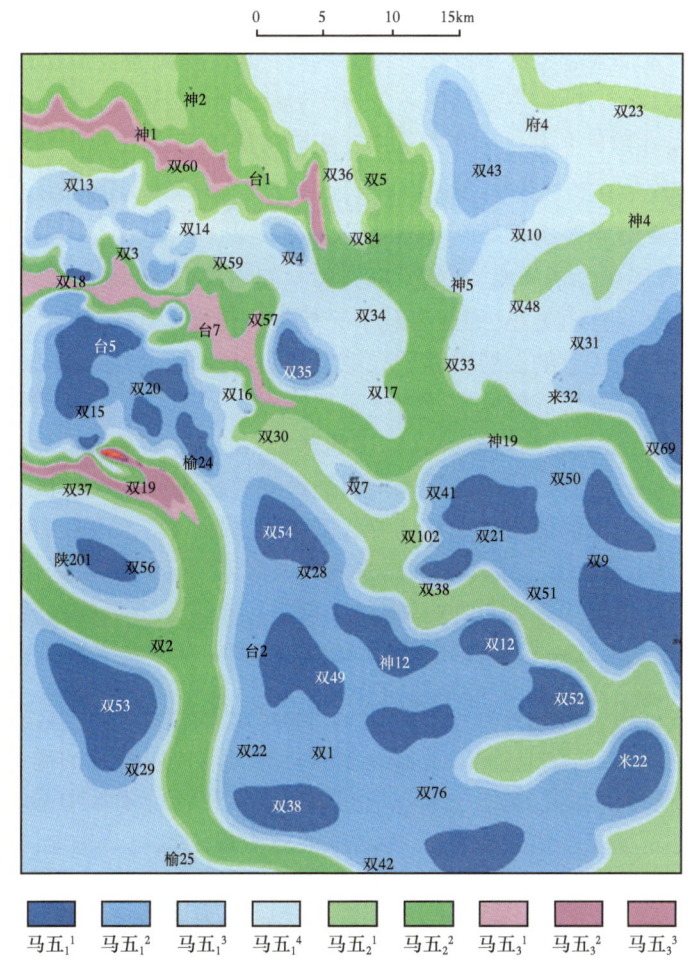

图 10　盆地东部神木地区岩溶地层残存图

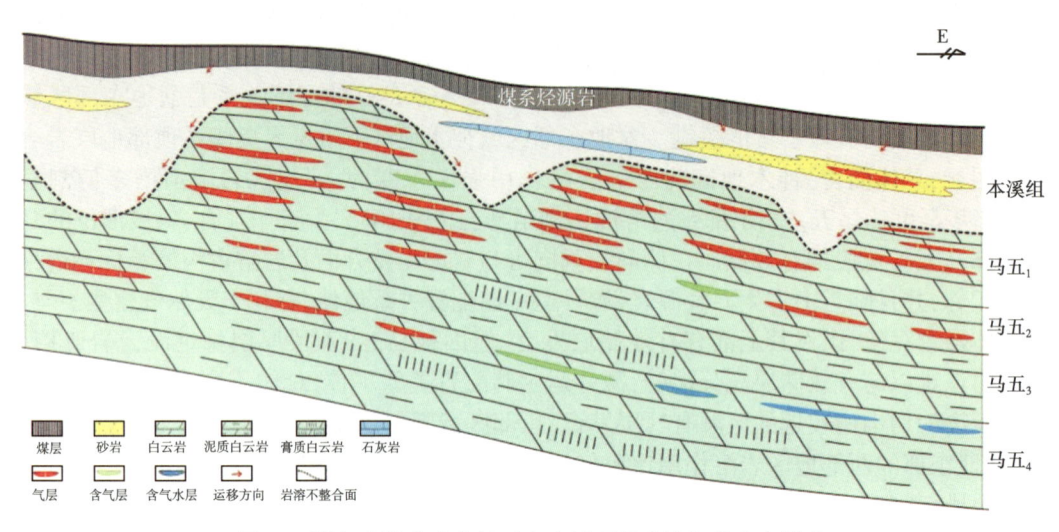

图 11 鄂尔多斯盆地东部下古生界顶部岩溶气藏富集模式

总体而言,盆地东部区域上部层位比下部层位源—运—储配置关系佳,西部地区比东部地区源—运—储配置关系佳,即盆地东部区域发育于岩溶丘台、斜坡部位的马五$_{1-2}$亚段比保存相对较为完整的马五$_4$亚段更有利于天然气富集,盆地东部区域靠近岩溶斜坡的西部地区比处于岩溶盆地的东部地区更利于天然气富集。

#### 4.1.3 岩溶储层物性的强非均质性

室内模拟实验及相关分析表明,有效储层孔隙空间相对较大、物性较好,在天然气富集过程中起始充注压力低、运移阻力小,因此有利于天然气完全驱替可动地层水而高效富集,往往形成纯气层,而物性相对较差的储层因储层孔隙空间相对狭窄,天然气难以完全驱替地层可动水而不利于高效富集,往往形成气水层及含气水层[25-29]。

受沉积、岩溶、成岩等复杂多因素综合作用的影响,盆地东部下古生界岩溶储层非均质性要明显强于盆地中部靖边地区,储层的强非均质性影响着盆地东部下古生界储层气、水分布格局。从宏观来看,平面上西部储层物性整体要优于东部,纵向上上部地层储层物性要整体优于下部,因此,西部地区上部马五$_{1-2}$亚段纯气层发育比例较高,而东部地区下部马五$_4^1$亚段气水层、含气水层比例较高。从局部来看,西部地区上部层位马五$_{1-2}$亚段物性普遍较好的储层中,也局部发育少量物性相对较差储层,因此表现为纯气层中夹有少量气水层的特征。

### 4.2 与盆地中部气田对比

靖边气田位于鄂尔多斯盆地中部,是盆地碳酸盐岩岩溶地层发现的唯一探明储量超千亿立方米的气田。对比评价表明,盆地东部有利沉积相或岩相组合不如中部靖边地区发育程度好,表生岩溶作用亦不如靖边地区高效,岩溶孔隙的填充程度也比靖边地区高,储层品质总体相对偏差(表1)。但盆地东部地区下古生界碳酸盐岩岩溶储层平均有效厚度为4.5m,局部地区可达8m以上(图9),多层叠置也可大面积连片分布,总体仍较为发育,且盆地东部平均生烃强度达 $36×10^8m^3/km^2$,远高于盆地中部地区,气源供给充分。盆地东部神木地区评价结果表明,马五$_{1-4}$亚段平均储量丰度为 $0.52×10^8m^3/km^2$,局部地区可与靖边地区平均值相当。鄂尔多斯盆地东部地区面积广阔,围绕神木—米脂—清涧—宜川一线,面积可达 $3\times10^8 km^2$,下古生界碳酸盐岩地层具备局部较大天然气勘探开发潜力。

表1 鄂尔多斯盆地中、东部下古碳酸盐岩岩溶储层参数对比表

| 对比指标 | 盆地中部 | 盆地东部 |
| --- | --- | --- |
| 平均孔隙度(%) | 6.0 | 3.80 |
| 平均渗透率(mD) | 2.63 | 0.82 |
| 平均储层厚度(m) | 6.6 | 4.5 |
| 平均生烃强度($10^8 m^3/km^2$) | 24 | 36 |
| 储量丰度($10^8 m^3/km^2$) | 0.72 | 0.52 |
| 孔喉特征 | 以溶蚀孔为主,晶间孔及膏模孔次之,属微米至毫米级孔隙,孔隙充填程度中等—偏高;可见微裂缝,可有效改善储层渗透性 | 以溶蚀孔为主,晶间孔及膏模孔次之,裂缝属微米至毫米级孔隙,孔隙充填程度普遍较高;微裂缝大多被充填,对储层渗透性的改善作用有限 |
| 有利沉积相或岩相组合 | 潮上含硬石膏结核云坪,分布广、连续性普遍较好,大面积连片分布 | 潮间含硬石膏结核云坪,分布局限、单层连续性较差,但多层叠合也可连片 |
| 岩溶古地貌背景 | 主体位于岩溶斜坡淋滤溶蚀区,存在垂直扩散渗滤、水平潜流及潜水面以下深部缓流等三种水流方式作用,表生岩溶作用较强,形成大量石膏结核溶蚀孔 | 主体位于岩溶盆地汇水区,表生岩溶作用偏弱,填充作用偏强,局部微构造高部位(丘台、坡地)存在强溶蚀区,有利于石膏结核溶蚀孔发育 |
| 储层分布特征 | 主力层位大面积连片分布,储层连续及连通性好;非主力层呈孤立井点状、孤岛状分布 | 主体以孤立井点状、孤岛状分布为主,多层叠置可连片,储层连通性总体偏差 |

## 5 结论

(1)鄂尔多斯盆地东部奥陶系顶部碳酸盐岩风化壳地层储层发育,以硬石膏结核云岩为主,储层孔隙直径分布于微米级至毫米级,喉道直径分布于几微米至几十微米,孔隙直径 $30\mu m$、喉道直径 $5\mu m$ 为储层与非储层的孔—喉界限标准。下古生界储层总体表现为低孔—特低孔、低渗—致密的特征。

(2)有效储层的发育受有利沉积相或岩相组合、高效岩溶作用、成岩综合作用及裂缝的综合影响。有利岩相发育是有效储层得以形成的基本条件,高效岩溶改造是有效储层形成的关键,而成岩综合作用决定有效储层的最终形成,半充填型硬石膏结核溶孔云岩为盆地东部地区最重要的有效储层类型,由于大多被充填,裂缝系统对有效储层的形成影响有限。

(3)天然气富集受有效储层发育与分布、源—运—储配置关系、岩溶储层强非均质性的共同影响。有效储层发育是天然气能够富集的基本条件,源—运—储配置关系是盆地东部区域下古生界天然气能够富集的关键,而储层物性强非均质性影响气、水分布格局。总体而言,盆地东部碳酸盐岩风化壳储层较为发育,气源供给充足,具备天然气大规模富集的条件,具有较大的天然气勘探开发潜力。

## 参 考 文 献

[1] 马永生,李启明,关德师,等.鄂尔多斯盆地中部气田奥陶系马五碳酸盐岩微相特征与储层不均质性研究[J].沉积学报,1996,14(1):22-32.
[2] 何江,方少仙,侯方浩,等.风化壳古岩溶垂向分带与储集层评价预测以鄂尔多斯盆地中部气田区马家沟组马五$_5$—马五$_1$亚段为例[J].石油勘探与开发.2013,40(5):534-539.
[3] 杨华,付金华,魏新善,等.鄂尔多斯盆地奥陶系海相碳酸盐岩天然气勘探领域[J].石油学报,2011,32(5):733-741.
[4] 杨华,刘新社,张道峰.鄂尔多斯盆地奥陶系海相碳酸盐岩天然气成藏主控因素及勘探进展[J].天然气工业,2013,33(5):1-11.
[5] 韩品龙,张月巧,冯乔,等.鄂尔多斯盆地祁连海域奥陶纪岩相古地理特征及演化[J].现代地质,2009,23(5):822-827.
[6] 杨华,付金华,包洪平.鄂尔多斯地区西部和南部奥陶纪海槽边缘沉积特征与天然气成藏潜力分析[J].海相油气地质,2010,15(2):1-13.
[7] 付金华,魏新善,等.鄂尔多斯盆地奥陶系顶面形成演化与储集层发育[J].石油勘探与开发,2012,39(2):154-161.
[8] 黄正良,包洪平,任军峰,等.鄂尔多斯盆地南部奥陶系马家沟组白云岩特征及成因机理分析[J].现代地质,2011,25(5):925-930.
[9] 付金华,白海峰,孙六一,等.鄂尔多斯盆地奥陶系碳酸盐岩储集体类型及特征.石油学报,2012,32(2):110-117.
[10] 冯增昭,鲍志东.鄂尔多斯奥陶纪马家沟期岩相古地理[J].沉积学报,1999,17(1):1-8.
[11] 王国亭,何东博,王少飞,等.苏里格致密砂岩储层岩石孔隙结构及储集性能特征[J].石油学报,2013,34(4):660-666.
[12] 吴永平,王允诚.鄂尔多斯盆地靖边气田高产富集因素[J].石油与天然气地质,2007,28(4):473-478.
[13] 夏日元,唐健生,关碧珠,等.鄂尔多斯盆地奥陶系岩溶地貌及天然气富集特征[J].石油与天然气地质,1999,20(2):133-136.
[14] 顾岱鸿,代金友,兰朝利,等.靖边气田沟槽高精度综合识别技术[J].石油勘探与开发,2007,34(1):60-64.
[15] 徐波,孙卫,宴宁平,等.鄂尔多斯盆地靖边气田沟槽与裂缝的配置关系对天然气富集程度的影响[J].现代地质,2009,23(2):299-304.
[16] 苏中堂,陈洪德,林良彪,等.靖边气田北部下奥陶统马五$_4^1$段古岩溶储层特征及其控制因素[J].矿物岩石,2011,31(1):89-96.
[17] 拜文华,吕锡敏,李小军,等.古岩溶盆地岩溶作用模式及古地貌精细刻画——以鄂尔多斯盆地东部奥陶系风化壳为例[J].现代地质,2002,16(3):292-298.
[18] 杨华,王宝清,孙六一,等.鄂尔多斯盆地南部奥陶系碳酸盐岩储层的胶结作用[J].沉积学报,2013,31(3):527-535.
[19] 夏新宇,赵林,李剑锋,等.长庆气田天然气地球化学特征及奥陶系气藏成因[J].科学通报,1999,44(10):1116-1119.
[20] 戴金星,李剑,罗霞,等.鄂尔多斯盆地大气田的烷烃气碳同位素组成特征及其气源对比[J].石油学报,2005,26(1):18-26.
[21] 陈安定.陕甘宁盆地中部奥陶系天然气的成因与运移[J].石油学报,1994,15(2):1-9.
[22] 林家善,周文,张宗林,等.靖边气田气田下古气藏相对富水区控制因素及气水分布模式研究[J].大庆

石油地质与开发,2007,26(5):72-74.
- [23] 程付启,金强,刘文汇,等.鄂尔多斯盆地中部气田奥陶系风化壳混源气成藏分析[J].石油学报,2007,28(1):38-42.
- [24] 肖晖,赵靖舟,王大兴,等.鄂尔多斯盆地奥陶系原生天然气地球化学特征及其对靖边气田气源的意义[J].石油与天然气地质,2013,34(5):601-609.
- [25] 姜福杰,庞雄奇,姜振学,等.致密砂岩气藏成藏过程的物理模拟实验[J].地质评论,2007,53(6):844-849.
- [26] 姜福杰,庞雄奇,武丽.致密砂岩气藏成藏过程的地质门限及其控气机理[J].石油学报,2010,31(1):49-54.
- [27] 邹才能,陶士振,张响响,等.中国低孔渗大气区地质特征、控制因素和成藏机制[J].中国科学D辑,2009,39(11):1607-1624.
- [28] 邹才能,陶士振,袁选俊,等.连续型油气藏形成条件与分布特征[J].石油学报,2009,30(3):324-331.
- [29] 邹才能,陶士振,朱如凯,等."连续型"气藏及其大气区形成机制与分布——以四川盆地上三叠统须家河组煤系大气区为例[J].石油勘探与开发,2009,36(3):307-319.

# 塔里木盆地库车坳陷深层大气田气水分布与开发对策

贾爱林 唐海发 韩永新 吕志凯 刘群明

张永忠 孙贺东 黄伟岗 王泽龙

(中国石油勘探开发研究院)

**摘要**：库车坳陷深层碎屑岩气田是近年来塔里木盆地天然气勘探开发的热点,该类气田具有储量规模大、埋藏深、高温高压、区块间裂缝发育差异大、边底水普遍存在等特点。气藏非均匀水侵导致气井产能快速下降,严重制约气田开发效果,是目前气田高效开发面临的普遍难题。从静态气水分布、微观水侵机理和动态水侵评价入手,系统建立了气水分布描述、水侵规律和控水开发技术对策为一体的静动态评价技术。研究认为,库车坳陷深层大气田气水分布受基质物性和缝网发育的共同控制,气水分布模式可划分为两类：薄气水过渡带型(含基质物性好型和裂缝特别发育型两个亚类)和厚气水过渡带型。气水分布和裂缝发育的差异性,直接导致了气田水侵部位与水侵动态特征的不同,表现为3种水侵类型：边底水整体抬升侵入型、边底水沿微细裂缝带锥进型和边底水沿大裂缝纵窜型。基于不同气水分布模式及水侵动态,提出了构造高部位布井避水、降速控压控水和边部水淹井强排等开发技术对策,为塔里木盆地库车深层大气田的高效开发和调整提供技术支撑,并为国内同类气田的开发提供方法借鉴。

**关键词**：塔里木盆地;库车坳陷;深层气田;裂缝;水侵;开发对策

近年来,随着中国对油气能源需求的增长以及中浅层油气勘探开发程度的日益成熟,中浅层新的勘探发现难度越来越大,深层、甚至超深层领域逐渐成为油气勘探开发的重点和热点[1-6]。特别是,伴随着深层油气勘探理论和工程技术的进步,更加速了深层油气资源向产能转化的进程。目前,在塔里木盆地库车坳陷深层碎屑岩和塔中碳酸盐岩[5]、四川盆地川中深层碳酸盐岩[6]等领域,均取得了重要的勘探突破和规模开发,成为盆地未来油气增储上产的主战场。

塔里木盆地库车坳陷深层—超深层碎屑岩领域是中国乃至世界上都罕见的大型陆相碎屑岩气田群,继克拉2气田(埋深3500~4100m)发现及成功开发后[7],近年来在库车超深层(埋深大于4500m)又相继发现和开发了大北、克深、博孜等系列气田[8],目前已有13个气田陆续投入开发和试采,累计探明天然气地质储量近万亿立方米[9],建成天然气产能规模近150×$10^8$m³,成为塔里木盆地天然气开发的主要领域,也是"西气东输工程"的主力气源。

库车坳陷深层气田群主要含气层系为白垩系巴什基奇克组,气藏埋深跨度大(除了克拉2气田外,埋深超过了4500m)、高温超高压、储层厚、区块间裂缝发育差异大[10,11]、边底水普遍存在、气水关系复杂[12]。气田投入开发2~3年后,构造低部位、高部位气井不同程度产水,气井见水后产能快速下降,导致储量动用不均衡,部分气田开发指标达不到方案设计要求,严重制约了气田开发效果,亟需开展水侵动态评价与控水开发对策研究。前人研究多集中在深层气田成藏机理、储层致密成因、裂缝分布及定量描述、气水分布影响因素等[10-18]地质研究方面,对气田开发研究的文章相对较少[19-23],且主要针对单个气田的产能和水侵动态研究,而对库车坳陷气田群水侵特征的差异性尚未开展系统研究。本文以塔里木库车坳陷深层气田群为主要研究对象,基于气田群内部不同气田储层、裂缝及气水分布的特点,通过静态气水分布描述、微观水侵机理与动态水侵评价研究相结合,针对性提出控水开发技术对策,为塔里木库车坳陷深层气田群的高效开发和调整提供技术支撑,并为国内同类气田的科学开发提供方法借鉴。

# 1 库车深层气田群地质特征

## 1.1 地质概况

库车坳陷位于塔里木盆地北部,南天山褶皱带的南缘,南为塔北隆起,东起阳霞凹陷,西至乌什凹陷,以中新生代沉积为主,整体呈 NEE 向展布,面积约为 $3.7\times10^4\mathrm{km}^2$。库车坳陷可进一步划分为克拉苏构造带、依奇克里克构造带和秋里塔格构造带,以及北部单斜带和南部斜坡带等次级构造单元[12]。近年来发现的库车坳陷深层气田群主要分布在克拉苏构造带,该带为天山南麓第一排冲断构造,依据主控断裂构造特征的不同,由北向南划分为克拉区带和克深区带。其中克深区带自西向东按构造特征又被细分阿瓦特段、博孜段、大北段及克深段(图1),由北向南被克拉苏断裂派生的多条次级逆断裂切割成6~7排构造,形成大范围的冲断叠瓦构造,加之上覆区域性膏盐盖层的有效封堵[13,14],形成了克拉、克深、大北、博孜等呈东西向排列的多个深层大气田(图1)。

## 1.2 基本地质特征

库车坳陷深层气田群主要含气层系为白垩系巴什基奇克组,该组自上而下发育3个岩性段,第一岩性段(巴一段)遭受不同程度的剥蚀,与上覆古近系库姆格列木群膏盐层呈角度不整合接触,第二(巴二段)、第三岩性段(巴三段)发育较好。巴什基奇克组沉积时期北部南天山存在多个物源出口,物源供应充分,湖盆宽缓,山前由北向南沉积相依次为冲积扇、扇三角洲或辫状和三角洲、滨浅湖[14]。巴三段古气候以干旱炎热,盆地高差较大,沉积物快速堆积入湖,在库车坳陷形成扇三角洲沉积,巴一段、巴二段构造互动减弱,地势变平坦,扇三角洲演化为辫状河三角洲。巴什基奇克组砂体垂向上多期切割叠置、横向连片,形成巨厚砂体(300~500m),高砂地比(40%~65%)。

储层岩性以中砂岩、细砂岩和粉砂岩为主,碎屑颗粒中岩屑、长石含量较高,岩石类型以岩屑砂岩、岩屑长石砂岩和长石岩屑砂岩为主,储集空间以原生粒间孔、粒间溶孔、粒内溶孔为主,微孔隙及微裂缝局部发育(图2)。受埋藏深度、沉积成岩作用及后期构造挤压运动改造程度的不同[10,17],库车坳陷深层气田群储集物性在空间上表现出一定的规律性,即沿克拉苏构造带自东向西、自北向南,随着埋深的增加,储层越致密,从克拉2、大北、克深到博孜气田,由常规储层演变为致密储层,6000m以深埋深基质孔隙度一般在1.5%~8%,空气渗透率0.01~0.2mD(图3)。

受南北两侧及垂向上强烈的构造挤压作用影响,库车坳陷深层气田群断裂(裂缝)普遍发育,平面上呈"南北分带、东西分段"的特征[10]。"南北分带"即在克拉苏断裂以北的山前带为

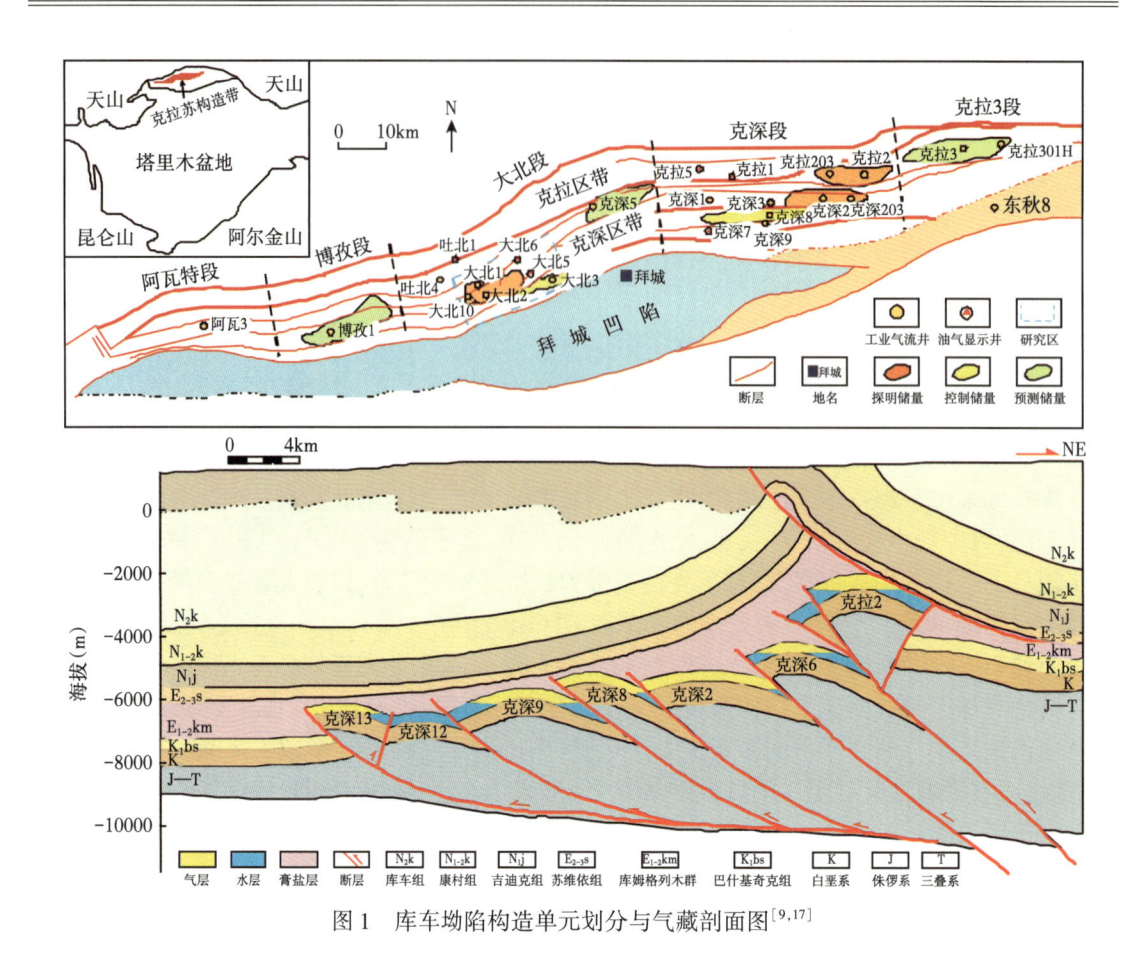

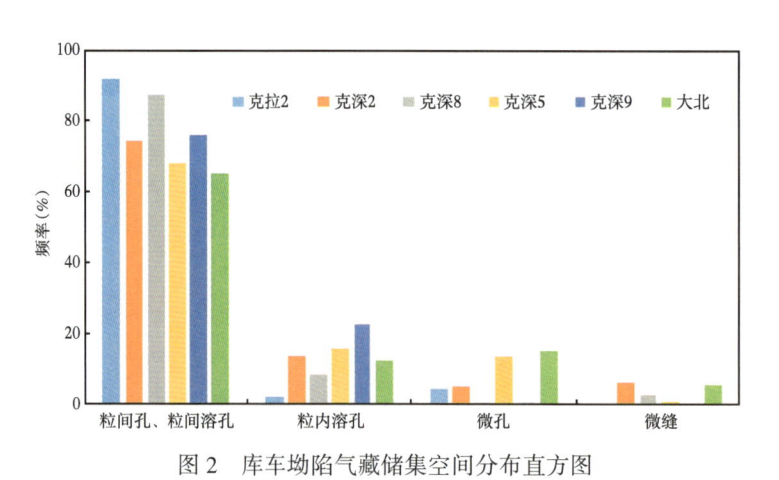

图 1 库车坳陷构造单元划分与气藏剖面图[9,17]

图 2 库车坳陷气藏储集空间分布直方图

垂向裂缝强发育带,克拉苏断裂与克深断裂之间为零星裂缝发育带,克深断裂与拜城断裂之间为网状裂缝和高角度裂缝强发育带,拜城断裂以南为零星裂缝发育带,"东西分段"表现为克深段以高角度垂向裂缝为主,大北—博孜段主要发育网状裂缝。相较而言,库车坳陷东部的克

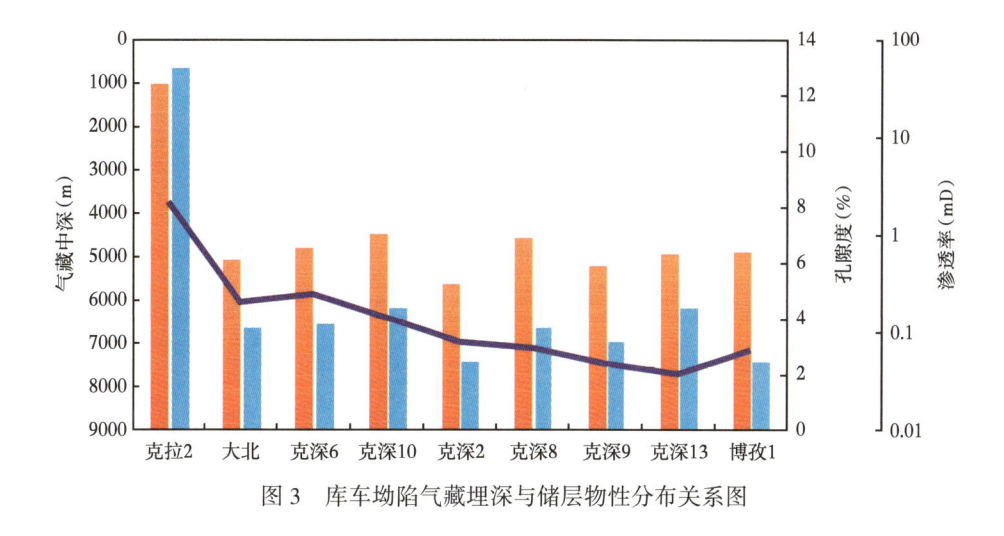

图 3 库车坳陷气藏埋深与储层物性分布关系图

拉 2 气田区内三级断裂更为发育[24],可识别 240 余条,成为气藏流体流动的"高速通道"。不同级次断裂(裂缝)及其空间组合,是深层气田气井高产的主控因素,同时也是开发过程中水体侵入的主要通道。

## 2 气水分布及其主控因素

### 2.1 气水分布主控因素

常规气藏因储层物性好,气藏成藏过程中气驱水较为彻底,气水关系简单,气水分布主要与构造有关,高部位产气低部位产水,试气测试不存在气水同出的现象,通常具有统一的气水界面。致密砂岩气藏由于储层物性致密,多级次孔喉发育不均一,成藏过程中当连续性气柱高度产生的浮力不足以克服毛细管阻力驱替孔隙内可动水时,往往在气顶纯气区下部形成一段含气饱和度向上逐渐升高的气水过渡带[25],在该带内射孔测试产气一般为气水同出,测井解释及试气结论从下到上依次为含气水层、气水同层、含水气层,气水过渡带下部为纯水区。

气水过渡带在苏里格气田西区[26]、广安须家河组气藏[27]、吐哈盆地巴喀气田[28]等国内多个致密砂岩气藏中均有发育,气水过渡带厚度主要受生烃强度、源储距离、构造幅度、基质物性及裂缝发育程度五大因素控制,其中前三者提供了成藏气柱浮力,后两者决定了成藏毛细管阻力。生烃强度越强、源储距离越近、构造闭合幅度越高,则成藏气柱浮力越大,在储层物性一致且裂缝不发育的情况下,气层相对越发育,气水过渡带厚度则越小甚至消失。相比之下,在成藏气柱浮力一定的前提下,储层物性越致密,孔喉条件越差,则毛细管阻力越大,气水过渡带厚度越厚,而裂缝缝网提供了气驱水的渗流通道,降低了毛细管阻力,缝网发育的地方气水分异一般相对彻底。

苏里格气田西区、须家河组气藏作为"源储共生性"气藏类型,生烃强度及源储距离控制了气水分布的总体宏观格局,如须家河组须六段气藏因烃源岩厚度较大、生烃强度强,气水分异程度明显好于下伏须四段气藏。塔里木盆地库车坳陷深层气田群作为"源储显著分离型"[12]构造气藏,烃源岩与储层被巨厚隔层分隔开距离较远,生烃强度、源储距离对气水分布影响可以忽略不计,且考虑构造幅度一般都为数百米甚至上千米,构造幅度基本一致,所以控制深层气田群气水分布的主控因素主要为基质物性与裂缝发育程度。

## 2.2 气水分布模式及特征

依据"储层基质物性、裂缝缝网发育程度、气水分布特征"的不同,将深层气田气水分布模式划分为薄气水过渡带型和厚气水过渡带型两大类,其中薄气水过渡带型可进一步细分为基质物性好型和裂缝特别发育型两个亚类(图4)。

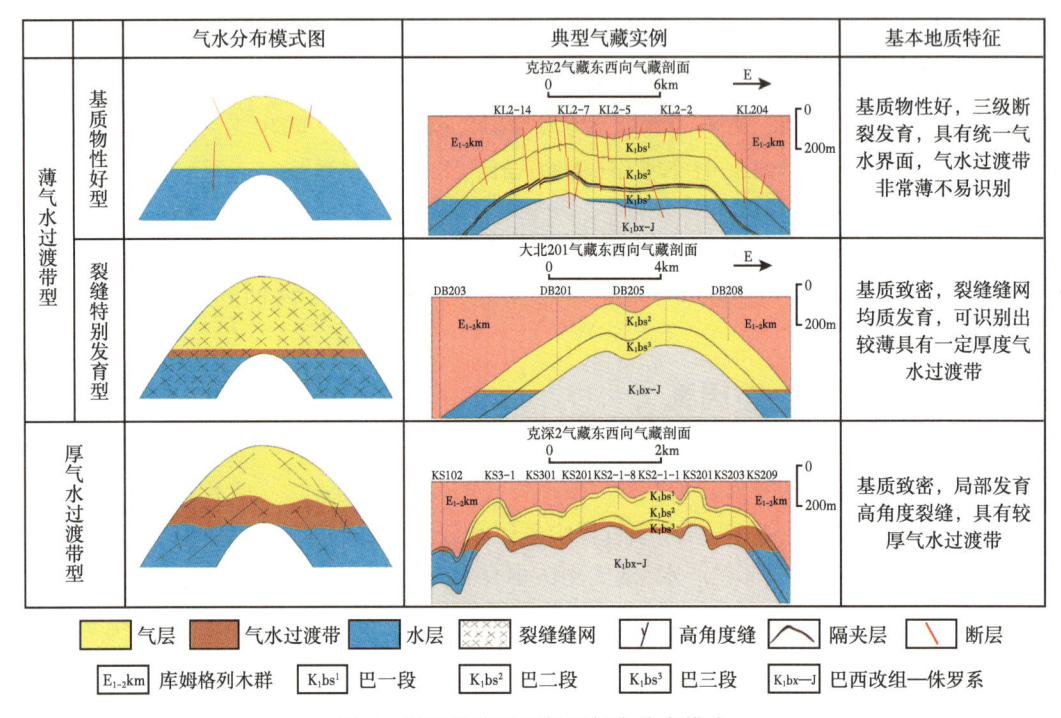

图4 库车坳陷深层气田气水分布模式

### 2.2.1 薄气水过渡带型

(1)基质物性好型:气藏储层物性好,裂缝基本不发育或呈零星分布状态,不构成裂缝网络体系,裂缝的存在主要起到改善局部储层物性的作用,基质孔隙仍然是流体渗流的主要通道,试井曲线表现为单孔单渗孔隙型均质储层特征,气藏类型属孔隙型气藏。典型代表气藏为克拉2气藏,气藏埋藏深度3500~4100m,平均孔隙度12.44%,平均渗透率49.42mD,为中孔中渗储层,储集空间以粒间溶孔和残余原生粒间孔为主,裂缝少量发育,岩心及成像测井统计裂缝密度小于0.05条/m,且裂缝多被膏泥质或云质充填—半充填。该类气藏成藏过程中因储层物性较好,气驱水较充分,气水正常分异,气水过渡带厚度一般较薄。在不同开发阶段评价的该类气藏动静态储量基本一致,气井稳产时间较长,见水时间较晚,后期水侵主要因为次级断裂沟通边底水所致。

(2)裂缝特别发育型:气藏储层基质致密,但裂缝比较发育,多级次裂缝组合形成裂缝缝网体系,空间上分布相对均匀,流体渗流具有基质与裂缝双重介质特征,气藏类型属裂缝—孔隙型气藏。典型代表气藏如大北201气藏,埋深5500~6500m,平均孔隙度7.3%,基质平均渗透率0.08mD,为致密储层,储集空间为多类型、多级次裂缝—孔隙复合空间,高角度构造缝裂缝缝网发育,岩心及成像测井统计裂缝密度为1.67~16.67条/m。该类气藏成藏过程中因均质裂缝系统的存在,水体驱替相对较为彻底,往往形成统一的气水界面或较薄的气水过渡带,厚度一般在米到十米级。该类气藏在不同开发阶段评价的动静态储量相差不大,误差基本在10%以内。该类型气藏生产过程中不存在高部位出水的情况,裂缝发育均质,一般不存在裂缝形成的水侵高速通道。

#### 2.2.2 厚气水过渡带型

气藏储层基质致密,裂缝发育但空间分布不均,局部可形成裂缝缝网,裂缝发育的非均质性较强,流体渗流同样具有基质与裂缝双重介质特征,气藏类型属裂缝—孔隙型气藏。典型代表气藏如克深2气藏,埋藏深度6500~8000m,平均孔隙度4.4%,基质平均渗透率0.05mD,孔隙类型以粒间溶孔和粒内溶孔为主,裂缝较为发育,岩心及成像测井统计裂缝密裂缝线密度为2~10条/m,裂缝类型主要为构造缝,且多为高角度半充填—未充填裂缝。平面上裂缝主要分布在构造高部位及边界断层附近[16],其中边界断层附近裂缝线密度最高,其次为次级断层控制区,背斜高点控制区相对较低,但发育程度仍高于其他部位。纵向上裂缝主要分布在目的层中上部,下部裂缝发育较差。下部裂缝及基质孔喉发育的非均质性导致过渡带界面起伏不平,厚度大小不一,过渡带厚度80~200m[9]。气水过渡带认识不清是导致气藏开发不同阶段,尤其是中后期评价的动静态储量差异仍较大的主要原因。克深2气藏动静储量比小于40%,并且该类气藏因裂缝发育不均造成局部位置纵向形成水侵高速通道,导致气井见水早,产量递减严重,稳产期较短的生产特征。

## 3 水侵动态特征

### 3.1 微观水侵特征

气藏微观水侵特征受基质物性和裂缝发育程度共同控制,基质致密时,裂缝既是气藏高产的主要因素,也是水侵的主要通道。水侵入裂缝后,边底水会沿断裂、裂缝快速突进,封堵基质中的气相渗流通道,产生"水封气"效应,降低气井产量和气藏采收率[29-32]。微观上天然气主要以绕流封闭气、卡断封闭气和水锁封闭气的形式存在,宏观上表现为水对气的封闭、封隔和水淹3种水侵现象[29]。裂缝性储层发生水窜后,边、底水沿裂缝水侵推进的速度受多种因素影响,水沿裂缝推进过程中与基质的作用机理差异明显。水在基质中呈活塞式推进,水侵前缘推进速度慢;在裂缝中呈快速非均匀突进,水侵速度随裂缝导流能力、水体大小、驱替压差的增加而增大[32]。

库车坳陷深层气田群埋藏深,裂缝不同程度发育,边底水发育,地层高温、超高压条件下渗流机理复杂,常规物理模拟实验方法很难揭示气水两相渗流规律,通过高温、超高压条件下全直径岩心的渗流实验装置,实现了模拟地层条件下(最高温度为160℃、最高压力为116MPa)的水驱气相渗模拟实验[30]。选取库车坳陷深层气田具有代表性的全直径致密岩心,经人工造缝处理,进行无裂缝、有微裂缝和有大裂缝的岩心相渗实验:基质无裂缝岩心(#1)、含有微裂缝岩心(#2)、含有大裂缝岩心(#3),其中#1、#2直接选取,#3需要进行人工造缝。实验结果表明,带裂缝的岩心在地层条件下驱替效率较低,见水后气相相对渗透率急剧下降,且含大裂缝的岩心这种特征更加明显(图5),说明裂缝性气藏快速水侵,气井在见水以后产气量会快速下降,从而使累产气量降低。

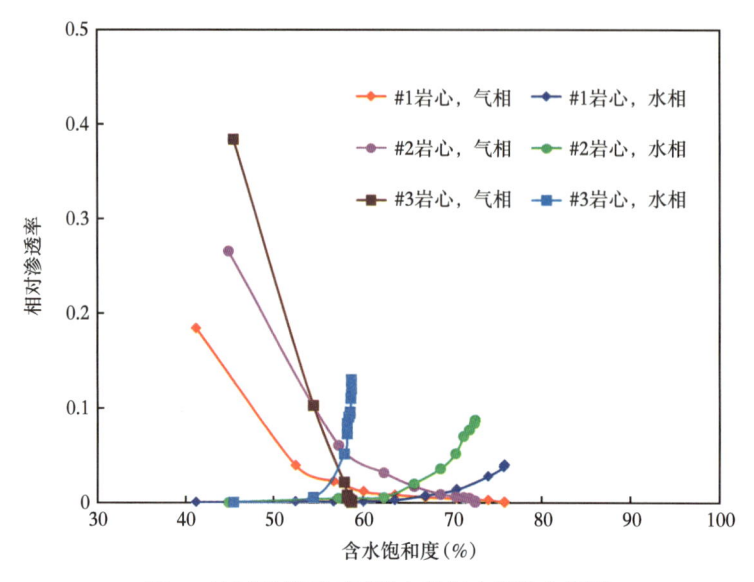

图5 地层条件下不同岩心的气水相渗曲线图

## 3.2 水体活跃程度

边、底水气藏在一定水体规模条件下,储层物性的差异决定了水从原始水区向原始气区侵入的难易程度,一般用水侵替换系数来表示,它的物理意义是在地层压降条件下,到某一生产时间,天然有效累计水侵量与气藏亏空体积之比,反映了地层水的活跃程度。水体倍数为与气藏连通水体体积与气区孔隙体积之比,反映了气藏周围水体规模的大小。

$$I = \frac{\omega}{R} = \frac{W_e W_p B_w}{G_p B_{gi}} \tag{1}$$

$$n = \frac{V_{pw}}{G_p B_{gi}/(1 - S_{wi})} \tag{2}$$

基于考虑水侵的高压气藏物质平衡方程[21],评价塔里木库车坳陷深层气田群水侵替换系数普遍小于0.3(图6),动态水体倍数为2~3倍(图7),属于次活跃水体。但与常规有水气藏不同的是,裂缝性气藏气井生产所形成的压降首先沿大裂缝传到远处,在大裂缝中形成低能带,如果大裂缝与水体连通则水沿大裂缝迅速到达井底,形成裂缝水窜,严重影响气井生产。因此,即使裂缝性气藏水侵替换系数不高,但不代表水体活跃程度低,边底水对气井生产的影响仍极其严重。

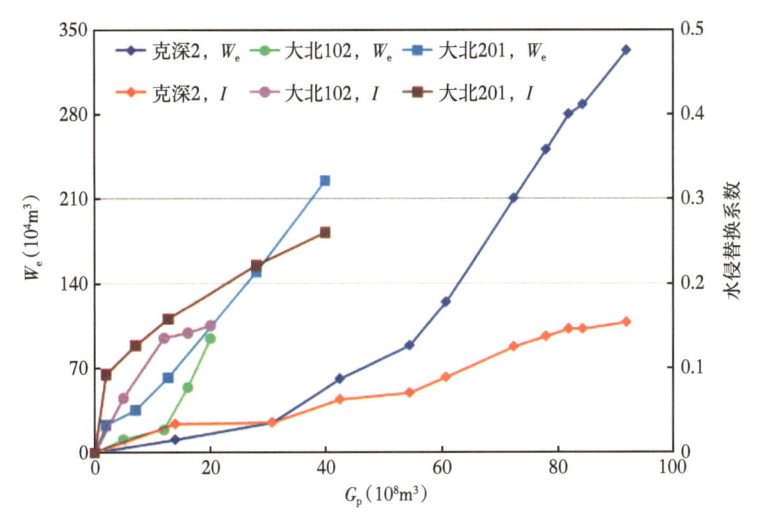

图 6　库车深层气田累计水侵量与水侵替换系数

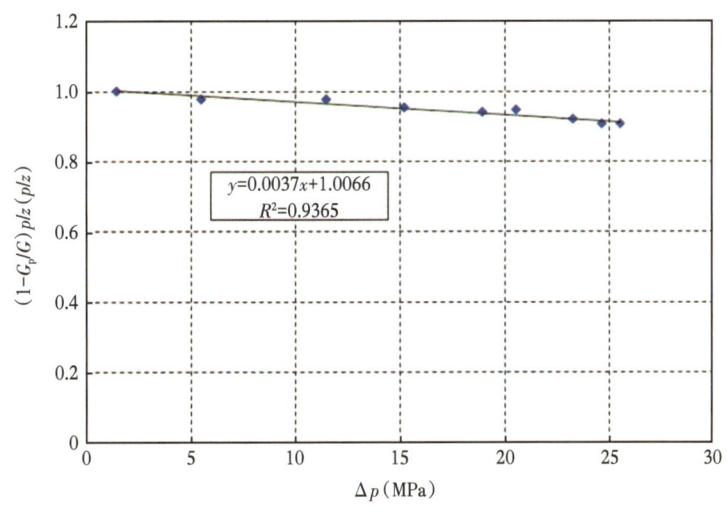

图 7　水体倍数计算拟合曲线图

## 3.3　气藏水侵规律

一般来说,由于储层类型、不同部位裂缝发育程度的差异,气藏水侵活动表现出多种多样的水侵模式,如水锥型、纵窜型、横侵型和复合型等[31]。裂缝性有水气藏大裂缝发育区水侵形式主要表现出"水窜"特征,即生产压差使底水很快沿高渗裂缝窜至局部气井,生产压差越大水窜越快。裂缝性水窜可导致很多气井投产短时间内出现地层水或气水同产,不久就被水淹。

水侵活跃程度取决于储层裂缝发育程度、边底水水体能量的大小以及气藏配产的高低。库车深层气田群由于储层基质物性、不同部位裂缝发育程度以及基质与裂缝组合方式的差异,导致气田生产动态与水侵特征明显不同。基于气水分布模式,库车坳陷深层气田群水侵表现为 3 种类型(表 1):边、底水整体抬升侵入型,边、底水沿微细裂缝带指进型和边、底水沿大裂缝锥进型。

表1 库车坳陷深层气田三种水侵类型

| 水侵类型 | 储层类型 | 典型试井曲线 | 产水图版 | 水气比 (m³/10⁴m³) | 产出液氯根含量 (10⁻²mg/L) |
|---|---|---|---|---|---|
| KL-A | 单孔单渗 | | | >1000 | >10 |
| DB-B | 双孔双渗 | | | <0.5 | 5~6 |
| KeS-C | 双孔单渗 | | | 2~2.5 | 8~10 |

— 254 —

(1)边、底水整体抬升侵入型:以克拉2气田为代表,储层物性较好,裂缝基本不发育,流体主要渗流通道仍是基质孔喉,试井曲线中径向流段导数曲线表现为0.5的水平直线,反映出单孔单渗储层特征。边、底水整体抬升侵入井筒,构造低部位气井见水后产出液氯根含量在$10×10^{-2}$mg/L以上,油压快速下降,出水量大,水气比急剧上升,气井暴性水淹,产能下降至0。

(2)边、底水沿微细裂缝带锥进型:以大北气田为代表,储层基质致密,但裂缝比较发育,多级次裂缝组合形成裂缝缝网,试井曲线表现为基质与裂缝双重介质特征。弹性储容比表示裂缝中天然气的储存比例,控制下凹深度;窜流系数表示基质部分采出难易程度,决定了下凹段出现的时间。气藏投入开发2年后,边、底水沿微细裂缝带渗入井筒,均匀推进,气井见水后油压基本稳定,气量逐渐下降,水气比较稳定,氯根含量$(5~6)×10^{-2}$mg/L,可长期带水生产,产能下降30%~50%。

(3)边、底水沿大裂缝纵窜型:以克深2气田为代表,储层基质致密,裂缝发育非均质性较强,试井压力曲线表现长期线性流特征,对应导数曲线斜率为1/2,晚期拟径向流不易观测到,反映出双孔单渗大裂缝储层特征。气藏投入开发2年后,边、底水沿大裂缝快速窜入井筒,构造低部位及高部位裂缝发育区气井很快见水,油压、气量快速下降,产能下降45%以上。水气比见水初期快速上升,生产一段时间后水气比趋于稳定,氯根含量$(8~10)×10^{-2}$mg/L,构造高部位气井可长期带水生产。

# 4 控水开发技术对策

气藏开发过程中如果气井与边底水之间存在裂缝沟通,边、底水会沿裂缝快速向井筒突进,同时水在裂缝运移过程中,储层基质会渗吸一部分水,基质渗吸水后减少或封堵气相渗流通道,从而增加储层基质气相渗流阻力,降低气藏稳产能力和最终采出程度[31,32]。因此,为了延长裂缝性有水气藏无水采气期、提高气藏采收率,需要在气藏开发全生命周期内采取一切防水、控水和排水措施。基于库车坳陷深层气田群不同的基质物性、裂缝发育、气水分布及水侵动态特征,从3个方面提出了深层气田群控水开发技术对策。

## 4.1 精细化地质研究,构造高部位布井避水,延长气藏无水采气期

库车坳陷深层气田群气藏类型均为边、底水构造气藏,埋藏深到超深,加之顶部巨厚膏盐屏蔽效应,地震资料信噪比低,构造落实难,裂缝分布、水体规模准确预测难度更大。为了有效规避见水风险,提高开发井成功率,在井位部署中逐渐形成了"沿轴线高部位集中布井"的部署思路,即在裂缝发育、远离边、底水的轴线部位集中布井,增加气井避水高度,优化打开程度,有效规避构造偏移风险和水侵风险。两者结合使库车坳陷深层气田的钻井成功率由早期的50%提高到100%,产能到位率由64%提高到100%,实现了高效布井。

## 4.2 差异化气井配产,气藏整体降速控压控水,实现水驱气藏均衡开发

边、底水发育对气藏开发具有双重影响,一方面可以弥补因采气过程中压力下降而损耗的地层能量;另一方面由于裂缝的沟通和输导作用,容易发生快速水侵,严重影响气田开发效果。因此,在气田开发过程中需要在两者之间建立一种动态平衡,合理优化气田采气速度,控制合理压降,延缓水侵速度,实现气田均衡开发。对于克拉2型孔隙性气藏,气水分布正常分异,气

水过渡带薄,边、底水近似于活塞式水侵,水侵速度缓慢,边、底水界面缓慢抬升,通过降低采气速度,回归合理开发制度后,气藏压降趋于平稳,气田生产指标转好(图8)。对于大北、克深2型裂缝—孔隙型气藏,基质致密,具有一定厚度的气水过渡带,裂缝非均匀水侵严重,边、底水沿裂缝快速水侵,开发过程中需要根据水侵前缘动态不断优化采气速度,同时考虑裂缝分布、水体能量等因素进行差异化气井配产,实现气井与气藏之间动态均衡开发。构造高部位见水气井按照临界携液流量配产带水生产,降低裂缝水窜风险;无水采气井兼顾气藏采气速度进行合理配产,延长气井无水采气期。

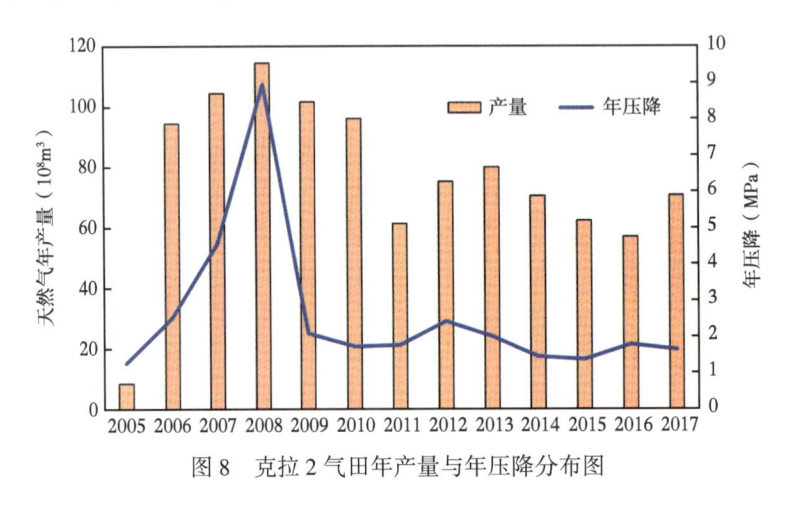

图 8 克拉 2 气田年产量与年压降分布图

## 4.3 系统化水侵监测,气藏边部水淹井强排,延缓递减提高气藏采收率

对于裂缝性边、底水气藏,水侵是其主要开发特征,因此,动态监测是贯穿气藏开发整个生命周期的一项重要工作[33]。加强动态监测,及早制定控排水对策,可以避免气藏大幅度水淹。苏联奥伦堡气田岩心一维和三维毛细管渗吸、径向水驱气及高压水驱气采收率实验表明[33]:封闭气须在发生膨胀且占据50%以上孔隙空间时才能流动。由此得出,气藏部分气井水淹后,继续降压开采,使被水封闭的天然气不断膨胀,冲破水封,进入生产井底。因此,提高有水气藏采收率的方法是从水淹井中强化排水采气,地层能量逐渐消耗后,借助压力差和水在基质孔隙的渗吸驱气作用,使"死气区"的天然气逐渐"复活"释放,能够提高采收率10%~20%。

库车坳陷深层克拉2、大北、克深2气田开发2~3年后,构造边部、低部位气井快速见水。为控制边底水侵入速度,一方面通过降低气田采气速度,减缓水侵;另一方面采取主动排水措施,利用边部、构造底部位水淹井强排水,降低水区与气区的压力差,从而达到降低水侵量、延缓见水时间、降低废弃压力、提高采收率的目的。克深2区块是目前水侵最严重的气田,产能下降幅度大。通过数值模拟优化设计了综合治水方案,与见水关井相比,通过利用边部位老井排水(合计日排水600m³),高部位井见水时间延缓明显(推迟2~3.6年),采收率由29.22%升至43.69%,提升14.47%(图9)。

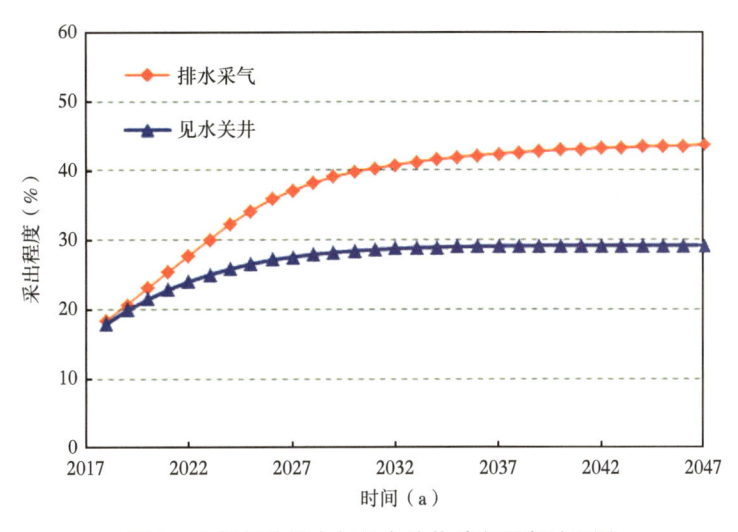

图 9 边部气井排水与见水关井采出程度对比图

## 5 结论

库车坳陷深层大气田埋藏深、区块间裂缝发育差异大,气水分布受储层基质物性和裂缝缝网发育程度共同控制,可划分为两类气水分布模式:薄气水过渡带型(基质物性好型、裂缝特别发育型)和厚气水过渡带型。基质物性好型薄气水过渡带主要分布在物性好的孔隙性储层,裂缝特别发育型薄气水过渡带型和厚气水过渡带型则主要分布在埋藏 6000m 以深、基质致密、裂缝发育的裂缝—孔隙性储层。

库车坳陷深层大气田边底水发育,水侵替换系数普遍小于 0.3,动态水体倍数为 2~3 倍,整体属于次活跃水体,但是由于裂缝的存在,气藏呈非均匀水侵,气井在见水后产气量快速下降。气水分布及缝网发育程度的差异性,导致深层气田表现为 3 种水侵类型:边、底水整体抬升侵入型,边、底水沿微细裂缝带锥进型,边、底水沿大裂缝纵窜型。

基于不同的气水分布及水侵特征,提出了 3 种控水开发技术对策:(1)精细化地质研究,构造高部位布井避水,延长气藏无水采气期;(2)差异化气井配产,气藏整体降速控压控水,实现水驱气藏均衡开发;(3)系统化水侵监测,气藏边部水淹井强排,延缓递减提高气藏采收率。从"构造高部位布井避水、降速控压控水和边部水淹井强排"等方面形成控水开发技术对策,可有效延长气井无水采气期,降低气藏水侵风险,减缓水侵前缘推进速度,同时研发与深层气田相配套的新型排水采气工艺,可进一步提升气田开发效益和最终采收率,实现深层大气田的科学高效均衡开发。

### 符号注释

$I$——水侵替换系数,无因次;$\omega$——水侵体积系数,无因次;$R$——地质储量的采出程度,小数;$W_e$——累计天然水侵量,$10^8 m^3$;$W_p$——累计产水量,$10^8 m^3$;$G_p$——累计产气量,$10^8 m^3$;$B_w$——地层压力下的地层水体积系数(一般可取为 1.0),$Rm^3/STm^3$;$B_{gi}$ 为原始地层压力下的气体体积系数,$Rm^3/SCm^3$;$V_{pw}$ 为水体在地层条件下的体积,$10^8 m^3$;$S_{wi}$ 为原始含水饱和度,小数。

## 参 考 文 献

[1] 贾承造,庞雄奇. 深层油气地质理论研究进展与主要发展方向[J]. 石油学报,2015,36(12):1457-1569.
[2] 张光亚,马锋,梁英波,等. 全球深层油气勘探领域及理论技术进展[J]. 石油学报,2015,36(9):1156-1166.
[3] 白国平,曹斌风. 全球深层油气藏及其分布规律[J]. 石油与天然气地质,2014,35(1):19-25.
[4] 冯佳睿,高志勇,崔京钢,等. 深层、超深层碎屑岩储层勘探现状与研究进展[J]. 地球科学进展,2016,31(7):718-736.
[5] 孙龙德,邹才能,朱如凯,等. 中国深层油气形成、分布与潜力分析[J]. 石油勘探与开发,2013,40(6):641-649.
[6] 邹才能,杜金虎,徐春春,等. 四川盆地震旦系-寒武系特大型气田形成分布、资源潜力及勘探发现[J]. 石油勘探与开发,2014,41(3):278-293.
[7] 李保柱,朱忠谦,夏静,等. 克拉2煤成大气田开发模式与开发关键技术[J]. 石油勘探与开发,2009,36(3):392-397.
[8] 杜金虎,王招明,胡素云,等. 库车前陆冲断带深层大气区形成条件与地质特征[J]. 石油勘探与开发,2012,39(4):385-393.
[9] 江同文,孙雄伟. 库车前陆盆地克深气田超深超高压气藏开发认识与技术对策[J]. 天然气工业,2018,38(6):1-8.
[10] 张惠良,张荣虎,杨海军,等. 超深层裂缝—孔隙型致密砂岩储集层表征与评价——以库车前陆盆地克拉苏构造带白垩系巴什基奇克组为例[J]. 石油勘探与开发,2014,41(2):158-167.
[11] 刘春,张荣虎,张惠良,等. 库车前陆冲断带多尺度裂缝成因及其储集意义[J]. 石油勘探与开发,2017,44(3):463-472.
[12] 赵力彬,张同辉,杨学君,等. 塔里木盆地库车坳陷克深区块深层致密砂岩气藏气水分布特征与成因机理[J]. 天然气地球科学,2018,29(4):500-509.
[13] 付晓飞,贾茹,王海学,等. 断层-盖层封闭性定量评价——以塔里木盆地库车坳陷大北—克拉苏构造带为例[J]. 石油勘探与开发,2015,42(3):300-309.
[14] 初广震,石石,邵龙义,等. 库车坳陷克深2气藏与克拉2气田白垩系巴什基奇克组储层地质特征对比研究[J]. 现代地址,2014,28(3):604-610.
[15] 朱光有,张水昌,陈玲,等. 天然气充注成藏与深部砂岩储集层的形成——以塔里木盆地库车坳陷为例[J]. 石油勘探与开发,2009,36(3):347-357.
[16] 韩登林,李忠,寿建峰. 背斜构造不同部位储集层物性差异——以库车坳陷克拉2气田为例[J]. 石油勘探与开发,2011,38(3):282-286.
[17] 张荣虎,杨海军,王俊鹏,等. 库车坳陷超深层低孔致密砂岩储层形成机制与油气勘探意义[J]. 石油学报,2014,35(6):1057-1069.
[18] 鲁雪松,赵孟军,刘可禹,等. 库车前陆盆地深层高效致密砂岩气藏形成条件与机理[J]. 石油学报,2018,39(4):365-378.
[19] 罗瑞兰,张永忠,刘敏,等. 超深层裂缝性致密砂岩气藏水侵动态特征分析——以库车坳陷克深2气田为例[J]. 浙江科技学院学报,2017,29(5):322-327.
[20] 李保柱,朱玉新,宋文杰,等. 克拉2气田产能预测方程的建立[J]. 石油勘探与开发,2004,31(2):107-111.
[21] 夏静,谢兴礼,冀光,等. 异常高压有水气藏物质平衡方程推导及应用[J]. 石油学报,2007,28(3):96-99.

[22] 李勇,张晶,李保柱,等．水驱气藏气井见水风险评价新方法[J]．天然气地球科学,2016,27(1):128-133.

[23] 李勇,李保柱,夏静,等．有水气藏单井水侵阶段划分新方法[J]．天然气地球科学,2015,26(10):1951-1955.

[24] 江同文,张辉,王海应,等．塔里木盆地克拉2气田断裂地质力学活动性对水侵的影响[J]．天然气地球科学,2017,28(11):1735-1744.

[25] 吴红烛,黄志龙,童传新,等．气水过渡带和天然气成藏圈闭闭合度下限问题讨论——以莺歌海盆地高温高压带气藏为例[J]．天然气地球科学,2015,26(12):2304-2314.

[26] 孟德伟,贾爱林,冀光,等．大型致密砂岩气田气水分布规律及控制因素——以鄂尔多斯盆地苏里格气田西区为例[J]．石油勘探与开发,2016,43(4):607-635.

[27] 陈涛涛,贾爱林,何东博,等．川中地区须家河组致密砂岩气藏气水分布规律[J]．地质科技情报,2014,33(4):66-71.

[28] 王国亭,何东博,程立华,等．吐哈盆地巴喀气田八道湾组致密砂岩气藏分布特征[J]．天然气地球科学,2014,33(4):2012,26(2):370-376.

[29] 樊怀才,钟兵,李晓平,等．裂缝型产水气藏水侵机理研究[J]．天然气地球科学,2012,23(6):1179-1184.

[30] 方建龙,郭平,肖香姣,等．高温高压致密砂岩储集层气水相渗曲线测试方法[J]．石油勘探与开发,2015,42(1):84-87.

[31] 冯昇勇,贺胜宁．裂缝性底水气藏气井水侵动态研究[J]．天然气工业,1998,18(3):40-44.

[32] 胡勇,李熙喆,万玉金,等.裂缝气藏水侵机理及对开发影响实验研究[J].天然气地球科学,2016,27(5):910-917.

[33] 贾爱林,闫海军,郭建林,等．全球不同类型大型气藏的开发特征及经验[J]．天然气工业,2014,34(10):33-46.

# Control of Fault Related Folds on Fracture Development in Kuqa Depression, Tarim Basin

Yongzhong Zhang[1]　Jianwei Feng[2]　Baohua Chang[1]
Zhaolong Liu[1]　Zhenhua Guo[1]

(1. PetroChina Research Institute of Petroleum Exploration & Development, Beijing, China;
2. School of Geosciences, China University of Petroleum, Qingdao, China

**Abstract**: In the multi-stage evolution process, the fault-related folds in the Kuqa depression are derived or associated with a large number of structural fractures, and the spatial distribution is very complicated. It is of great significance for gas field development to know the distribution law and main control factors of fractures. Taking Dabei, Keshen, Dina2 and Kela2 gas fields as examples, the fracture development characteristics of different fold types are analyzed by using core, imaging logging and seismic data, and the control effect of fold style evolution on fracture development is revealed by using finite element simulation method. Fracture development has obvious correlation with reservoir quality and fold style. The density of fractures in Dabei and Keshen gas fields with low porosity are much higher than that in Dina and Kela gas fields with relatively high porosity. Dabei and Keshen anticlines, are dominated by shear network fractures and exist less tension fractures due to their low amplitude (relatively large interlimb angle), while the tension fracture proportion increases within Dina and Kela anticlines due to their high amplitude (relatively small interlimb angle). The high-value areas of tension fracture is mainly located near the cores of anticlines and the faults, while that of shear fracture is obviously correlation with faults. Further study shows that open folds with interlimb in range of 70°~120° have significant influence and controlling on effectiveness of fractures, but they are not apparent in the gentle folds with interlimb angle in range of 120°~180°. The fracture development heterogeneity of the imbricated double thrust structure with very different scale fracture is stronger than that of the back thrust structure with relatively small scale difference fracture. Finite element simulation showing the development and distribution of structural fractures are obviously controlled by fold uplifting amplitude, distance from fault plane and fault sliding displacement, especially when the fold uplifting amplitude is in the range of 410~450m, the fracture density value is the largest. This study has effective guidance for development plan of newly discovered gas fields and comprehensive management of developed gas fields of the piedmont thrust belt in Kuqa depression.

**Keywords**: Kuqa depression; Fault related fold; Fracture; Finite element simulation.

# 1　Introduction

Due to the southward multi-stage nappe compression of the Tianshan orogenic belt in the late Cenozoic, various fault related folds are widely developed in Kuqa depression, which is rich in natural gas resources. A series of gas reservoirs, such as Kela2, Dina2, Dabei and Keshen, have been successively discovered, with the depth of Kela2 Gas Reservoir being 3500~4000m and other gas reservoirs being 4700~8000m belonging to deep/ultra-deep gas reservoirs. Under the influence of several tectonic movements[1,2], structural fractures are developed in these reservoirs to different degrees. This extensively developed structural fracture system is not only an important seepage channel and space of the reservoirs, but also seriously affects the productivity of a single well for those deep sandstone reservoirs[3,4]. Therefore, understanding where the subsurface structural fractures develop and when they are formed, including their development degree, physical properties and seepage characteristics, is very important for the effective exploration and fine development of fractured tight reservoirs. In this paper, the characteristics of fracture development in Keshen, Dabei, Dina and Kela gas field of Kuqa depression are analyzed comparatively based on core description, thin slice, CT and FMI data. At the same time, the distributions of natural fracture within varied folds in Keshen structural belt were predicted by using finite element (FE) simulation method. Finally, the relationship between the parameters of structural fracture and the distance to the fault and the amplitude of structural uplift is revealed. This study has reference value for exploration and deployment of new gas reservoirs and comprehensive treatment of developed gas reservoirs.

## 2 Geological setting

As an enrichment area for natural gas, Kuqa depression is developed in the joint of Tianshan fold belt and the northern margin of Tarim plate, and a series of large-scale thrust faults and fault-related folds are widely developed in the Kuqa depression, mainly formed in the Cenozoic Himalayan movement. Kuqa depression can be divided into 3 secondary structural units (Fig.1a), which are rich in natural gas resources, including Kelasu thrust belt, Qiulitage thrust belt and Baicheng sag[5]. Based on the difference of fault structures, the Kelasu belt is further divided into Kela zone and Keshen zone from north to south. The southern boundary of Keshen zone is bounded by Baicheng fault and the northern boundary is bounded by Northern Kela fault. The general trend is nearly E-W, and it reverse to NEE-SWW in Dabei area. The Keshen zone displays a series of imbricated thrust structures, with pop-up structure and fault-related fold in each imbricate fan (Fig.1b), where the important commercial discoveries including the Dabei, Keshen, and Bozi gas reservoir groups have been found. The Qiulitage thrust belt, where the Dina gas field have been found is an arc-shaped structural belt, which is distributed nearly in the east-west direction. The fold thrust uplift in the middle part is higher, and gradually leans to the East. In the east section of Qiulitage structural belt, there are two North dipping thrust faults, one is Northern Dina fault, which developed and shaped in Himalayan period, the other is East Qiulitage fault, which began to develop in Yanshan period, continued to move in Himalayan period, and shaped in late Himalayan period. These two faults not only control the structural pattern of the East Qiulitage structural belt, but also control the formation of Dina structure.

Kuqa Depression is a Meso-Cenozoic sedimentary foreland basin[6]. Neogene, Paleogene and Cretaceous are developed from top downwards in the drilled strata of Kela, Dabei, Keshen and Dina gas fields. The gas bearing rocks are mainly the Cretaceous Bashijiqike ($K_1bs$) in Kela, Dabei and

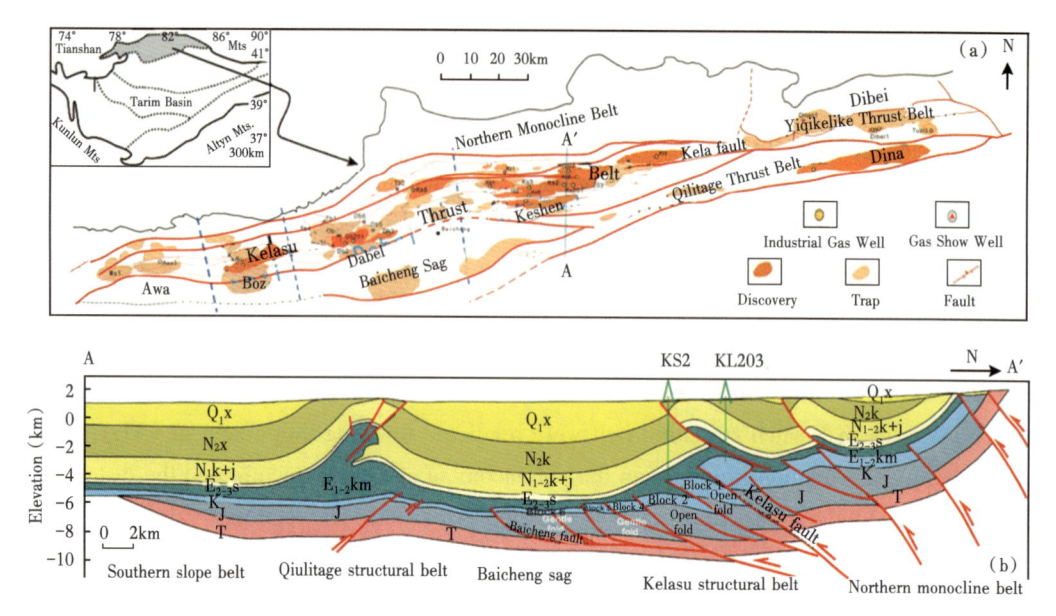

Fig. 1 Maps showing the elementary structural features of the Kuqa Depression. (a) Location of the study area and subdivision of the Kuqa Depression; (b) Tectonic cross-section of the location shown in (a). T: Triassic; J: Jurassic; K: Cretaceous; $E_{1-2}km$: Paleogene Kumugeliemu Formation; $E_{2-3}s$: Paleogene Suweiyizu Formation; $N_1k$: Neogene Kangcun Formation; $N_2k$: Neogene Kuqa Formation; $Q_1x$: Quaternary Xiyu Formation

Keshen which were deposited in a braided river delta front-fan delta front environment, while those are mainly the Paleogene Suweiyi ($E_{2-3}s$) and the Kumugeliemu ($E_{1-2}km$) Formations in Dina gas field, which were deposited in a fan delta front environment. The depth of the reservoirs in Kela 2 gas field is from 3500m to 4000m, while in Dina is from 4700m to 5400m, and in Dabei and Keshen is from 5500m to 8000m. The property of reservoirs becomes poor with the increase of burial depth (Table 1). The porosity of the Kela 2 sandstones mainly ranges from 10% to 18%, and the horizontal permeability mainly varies from 1mD to 100mD. Although Dina, Dabei and Keshen are belong to deep/ultra-deep reservoirs, some primary pores are remained due to overpressure. The porosity of the Dabei sandstones mainly lies in the range of 3% to 8%, and the horizontal permeability vary from 0.05 mD to 0.5mD; while the porosity of the Keshen sandstones mainly ranges from 2% to 8%, and the horizontal permeability mainly varies from 0.01 mD to 0.1 mD. The physical properties of Dina gas reservoir are obviously poorer than Kela 2, but better than Dabei and Keshen.

Table 1 Geological characteristics of the main gas fields in Kuqa depression

| Gas field | Depth (m) | Pressure coefficient | Average pore throat radius (μm) | Porosity (%) | Permeability (mD) |
| --- | --- | --- | --- | --- | --- |
| Kela | 3500~4000 | 2.2 | 1.5~2.6 | 10~18 | 1~100 |
| Dina | 4700~5400 | 2.1 | 0.02~1 | 5~12 | 0.05~1 |
| Dabei | 5500~7100 | 1.53~1.73 | 0.01~0.5 | 3~8 | 0.05~0.5 |
| Keshen | 6000~8000 | 1.65~1.85 | 0.01~0.05 | 2~8 | 0.01~0.1 |

## 3  Characteristics of the structural fracture

There are lots of drilling core and formation micro image logging (FMI) data which can help to research the characteristics of the structural fracture in Kelasu and Qiulitage structural belts. Drilling core description including core observation, thin section and ICT provides the most direct and intuitive method to evaluate fractures. It can finely measure the filling composition of fractures and the change in aperture[7]. However, due to the high cost and high process requirements, coring is generally carried out in the exploration wells and appraisal wells, and only a few development wells are coring. Meanwhile the drilling core is discontinuous in the well, which limits the understanding of the spatial distribution characteristics of fractures in gas field. At present, logging methods are widely used to identify wellbore fractures, which compensates for this deficiency and enables continuous imaging of wellbore formations and fractures to some extent. However, practical work showed that whether conventional logging or unconventional logging curves are not sensitive to identify the micro fracture in the boreholes. Furthermore, oil-based mud is often used as drilling fluid in many wells, thus makes logging curves less sensitive to identifying fractures. Therefore, the number of fractures identified by logging is often lower than that identified by core description.

Based on core observation, the structural fractures in the study area mainly include tension fracture formed under the condition of tensile stress, shear fracture formed under the condition of tensile stress, and tension shear fracture formed under the action of tension stress and shear stress. The tension fractures with irregular and rough surfaces usually develop in wells at the top of anticline, extending long and having large opening. Shear fractures with smooth surface can extend up to several tens of meters on the cores. These shear fractures can form conjugate patterns under strong compression condition in some positions. The characteristics of the tension shear fracture are between the shear fracture and the tension fracture. It is shown in the core that the fracture surface is bent locally, but it is still flat on the whole. Under the microscope, the tension fractures often have dendritic structure, passing around the sedimentary particles; while shear fractures often pass through sedimentary particles. Most fractures were filled or half-filled, with dolomite, calcite and siliceous mineral as the main filling minerals. Core observation and FMI analysis show that the fracture dip angle is mainly in range of 75°~90°, followed by ranges 45°~75° and 0°~45°. The fracture distribution is related to the structural position. The vertical fracture is mostly distributed in the high part of the structure, the high angle oblique fracture is mostly distributed in the structural wing, and the network shear fracture is mostly found near the fault.

There are some differences in the development characteristics of structural fractures between different gas fields in Kelasu and Qiulitage structural belt due to the reservoir property and Tectonic stress (Table 2). In Dabei gas field, shear fracture is the main fractures. The fractures are filled with calcite and dolomite, mainly with calcite. Among them, fully filled fractures account for 68.3%, half-filled fractures account for 13.7%, and unfilled fractures account for 18%. Core observation and FMI analysis show that the fracture dip angle is mainly in range of 50°~90°, and those with a dip angle of less than 30° are rare. Fracture strike is NW-SE, followed by E-W strike, and the N-S strike fracture is rare. The fracture linear density of single well varies from 2.27 to 16.67m$^{-1}$ based on the core description and 0.26 to 3.38m$^{-1}$ based on the FMI analysis.

Compared with Dabei gas field, the fractures in Keshen gas field are mainly filled with dolomite and siliceous minerals, and fully filled fractures account for about 60%. Shear fracture takes dominate place, for example, in Keshen 2 block, which accounting for 79.5%[8]. The proportion of tension shear fracture and tension fracture is small (i.e. 18.7% and 1.8% respectively), mainly distributed at the high point of anticline and the boundary of mudstone interlayer. The fractures strike is mainly E-W above the neutral plane, and NW-SE and NE-SW below the neutral plane. From the angle of inclination, the fractures are mainly vertical fractures (>75°), followed by high angle fractures (45°~75°) and low angle fractures (15°~45°). The average fracture linear density is 2.98m$^{-1}$ based on the core description, which is obviously lower than the value of 7.78 in Dabei gas field. In contrast, the fracture density appears to be relatively low in the limbs and saddles of anticlines. Fracture density tends to decrease from north to south (Fig.2). For example, within Keshen 2 block in the north area it reaches 1m$^{-1}$ based on FMI analysis, while it is only 0.06 m$^{-1}$ within Keshen 2 block in the south area.

Table 2 Geological characteristics of the main gas fields in Kuqa depression

| Fracture characteristics | | Dabei | Keshen | Dina | Kela |
|---|---|---|---|---|---|
| Fracture types | | Dominated by shear fracture | Shear fracture (79.5%), tension shear fracture (18.7%), tension fracture (1.8%) | Shear fracture (54.2%), tension shear fracture (27.7%), tension fracture (18.1%) | Shear fracture (72%), tension shear fracture (21%), tension fracture (7%) |
| Dip angle | | In range of 50°~90° | Mainly in range of 75°~90°, followed by 30°~75°. | In range of 60°~90° | In range of 60°~90° |
| Strike | | NW-SE, followed by E-W | Mainly E-W above the neutral plane, and NW-SE and NE-SW below the neutral plane | Mainly NEE-SWW and NE-SW at top, NEE-SWW and NE-SW at flank | Mainly E-W |
| Linear density, (min-max)/avg. | From core | (2.27~16.67)/7.78 | (0.4~8.5)/2.98 | (0.23~1.83)/0.56 | (0.14~1.84)0.54 |
| | From FMI | (0.26~3.38)/1.5 | Northern block: (0.26~1.83)/0.77 | (0.1~0.8)/0.43 | (0.02~0.47)/0.11 |

In Dina and Kela gas fields, the proportion of tension shear fracture and tension fracture increase owing to their high structural amplitude. The top of the anticline not only has large curvature value, but also develops tensile stress environment, which provides necessary conditions for the

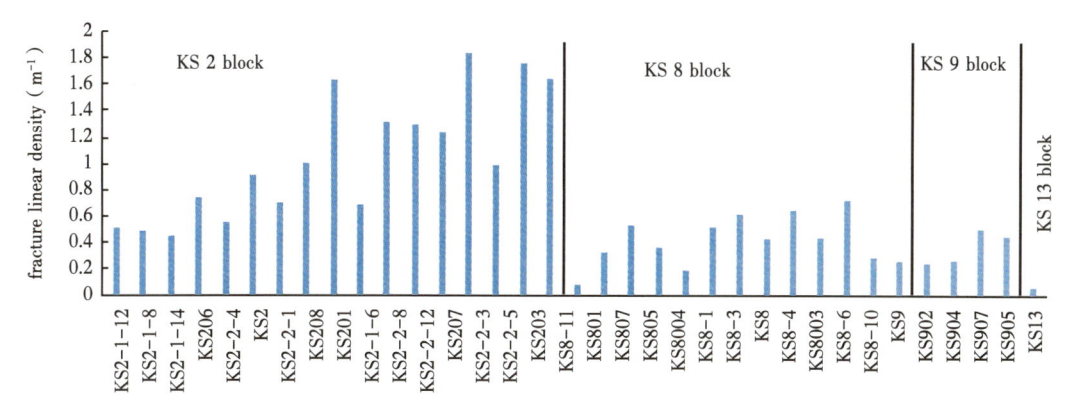

Fig. 2 Comparison of fracture linear density in different blocks of Keshen gas field interpreted by FMI. Data are partly cited from literature[9]

development of tensile faults and fractures. Except for individual well points, the fracture linear density is generally low, most of which is lower than $1m^{-1}$, especially in Kela 2 gasfield, linear density from FMI is less than $0.1\ m^{-1}$.

Fracture density is a simple, effective, and intuitive method for characterizing intensity of fracture development[10], which has obvious correlation with reservoir physical properties (Fig.3). The FMI interpretation shows that the Fracture density and permeability become lower with the porosity decreasing. Thin section analysis of Dina gas field shows that the fractures are mainly developed in the rocks with porosity less than 6%, and only a few fractures are developed in the rocks with porosity higher than 6%. The sandstone with low porosity and permeability is brittle and easy to crack, which is the reason why the fracture development of Dabei and Keshen gas fields is much higher than that of Dina and Kela gas fields.

There structural deformation mechanisms are varied in different areas of Kelasu structural belt[11], which lead to different fracture development. Due to the influence of regional plate movement, Kuqa depression has a strong near north-south tectonic compression, forming a strong tectonic compression environment in the Kelasu structural belt, and there are different stress field environments and thrust nappes between different areas, resulting in varied tectonic deformation in different areas. In Keshen area, a series of fault anticline assemblages and pop-up structures are developed, which are mainly the structural deformation mechanism of forward compression, back thrust uplift and front detachment contraction. Dabei area is located in the stress transformation area, which is mainly the structural deformation mechanism of oblique compression and torsion, paleo uplift control and thrust superposition, and develops a series of thrust and imbrication structures with strike slip property; the compression and torsion type fault anticline is subject to compression and torsion, with high degree of structural transformation. Therefore, the fracture development degree of Dabei gas field is higher than that of Keshen.

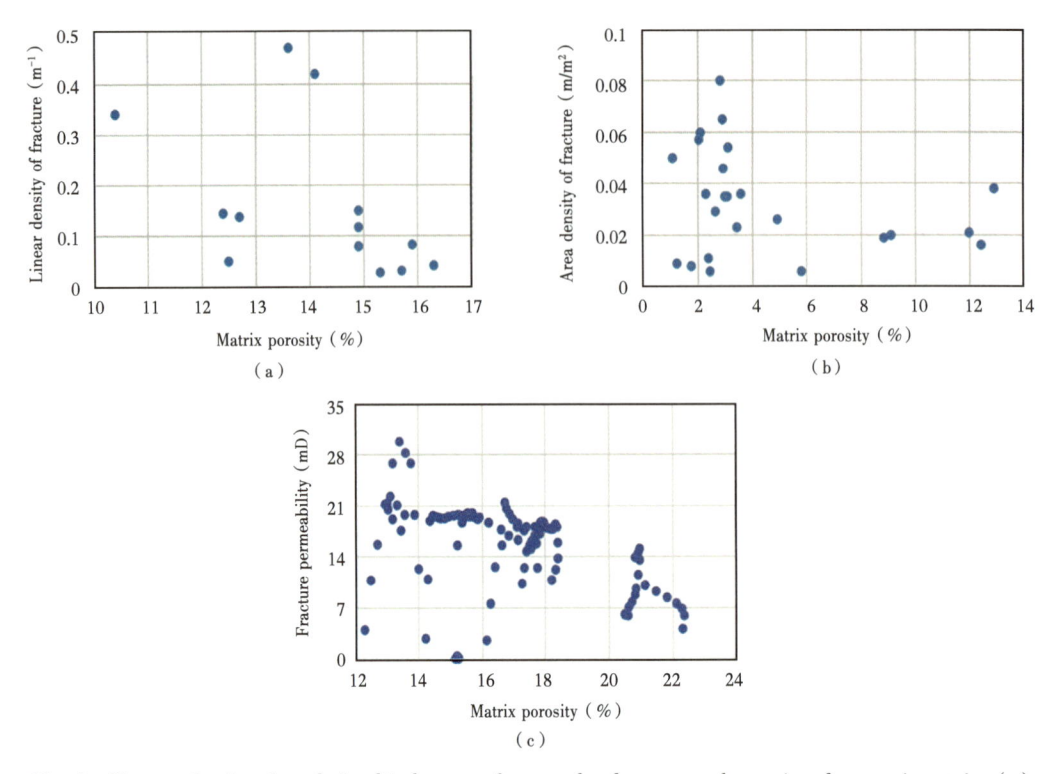

Fig. 3 Figures showing the relationship between fracture development and porosity of reservoir matrix. (a) Relationship between fracture linear density within single well from FMI interpretation and matrix porosity of reservoir in Kela 2 gas field; (b) Relationship between fracture area density from thin sections statistics and matrix porosity of reservoir in Dina 2 gas field. (c) Relationship between permeability from FMI interpretation and matrix porosity of reservoir within Kela 201 well

## 4  The control of fold style on fracture development

The fracture development is closely related to the fold style. According to the division scheme of the traditional fault related fold, Kelasu structure is dominated by fault propagation folds, while Dina 2 structure is fault bending fold. Many geologists have studied the correlation between fractures and faults, but the scheme of fault related fold division mainly based on the control of fault action limits the study of fracture spatial distribution. In recent years, many geologists believe that the main hydrocarbon-bearing formations in the folds are likely to be closely related to the location of the neutral surface of the folds, suggesting that folding can significantly influence/control the development and distribution of structural fractures[12,13]. Degree of fold deformation can be reflected by angle between two limbs of fold (i.e., interlimb angle $\theta$)[14], as well by the subdivision of the combined shape of the faults with the size of the angle. According to the geometric classification scheme of fold interlimb angle[15], fold can be divided into gentle fold ($120°<\theta<180°$), open fold ($70°<\theta<120°$) normal fold ($30°<\theta<70°$; closed fold ($5°<\theta<30°$) and isoclinal fold, ($0°<\theta<$

5°). Kuqa Piedmont structure is mainly composed of gentle fold and open fold. According to the fault combination configuration, the anticline in the study area is further divided into double-thrust and back-thrust style. double-thrust structure refers to the imbricated structure held by two boundary thrust faults with the same dip, and the back thrust structure refers to the structure enclosed by two boundary thrust faults with opposite dip and upward divergence (i.e. pop-up structure).

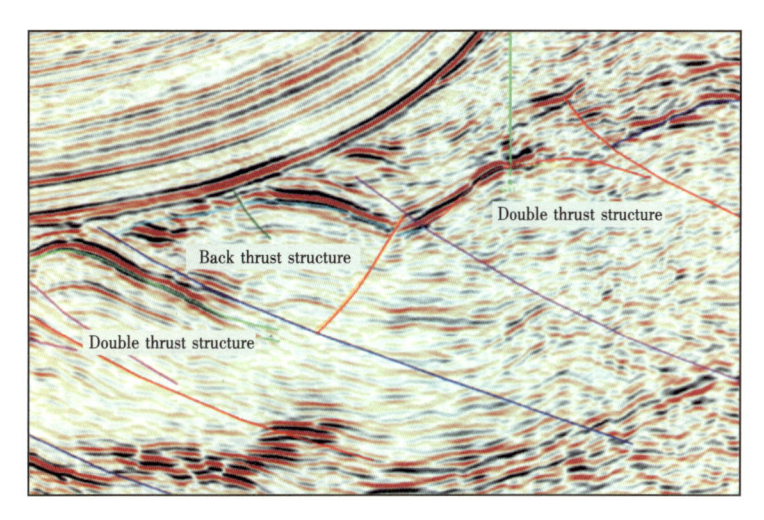

Fig. 4　Seismic section showing typical structure style in Kelasu

Dabei anticline belongs to the typical gentle type (>120°), and Kela2 and Dina2 gas fields are open type, while The fold type of each fold of Keshen gas field changes from open type to gentle type with the interlimb angle increasing from north to south. The strain distribution in the fold is closely related to the interlimb angle. The thickness of the tension strain increases with the decrease of the interlimb angle. Therefore the thickness of the tension fracture zone increases with the decrease of the interlimb angle, and the distribution range of the tension fracture zone expands from the hinge to the wing. The structural style differences lead to the difference of structural fracture distribution. Dabei and Keshen anticlines are dominated by shear network fractures, with a small amount of tension fractures at the top of the anticlines due to their low amplitude (relatively large interlimb angle), while the tension fracture proportion increases within Dina and Kela anticlines due to their high amplitude (relatively small interlimb angle).

The distribution of fracture parameters was analyzed by the imaging logging data and by statistics of finite element numerical simulation in Keshen area (Fig.5). The high-value areas of tension fracture is mainly located near the cores of anticlines and the faults (Fig.5a, b), while that of shear fracture has better correlation with faults than with cores of anticlines (Fig.5c, d). The tension fracture density of Keshen 2 block is slightly low in a small range of the core of the anticline (Fig.5a), but it is dominated by large fractures, with large fracture aperture, while it is larger within a certain distance to the structural top, but the fracture aperture is obviously smaller. It has a

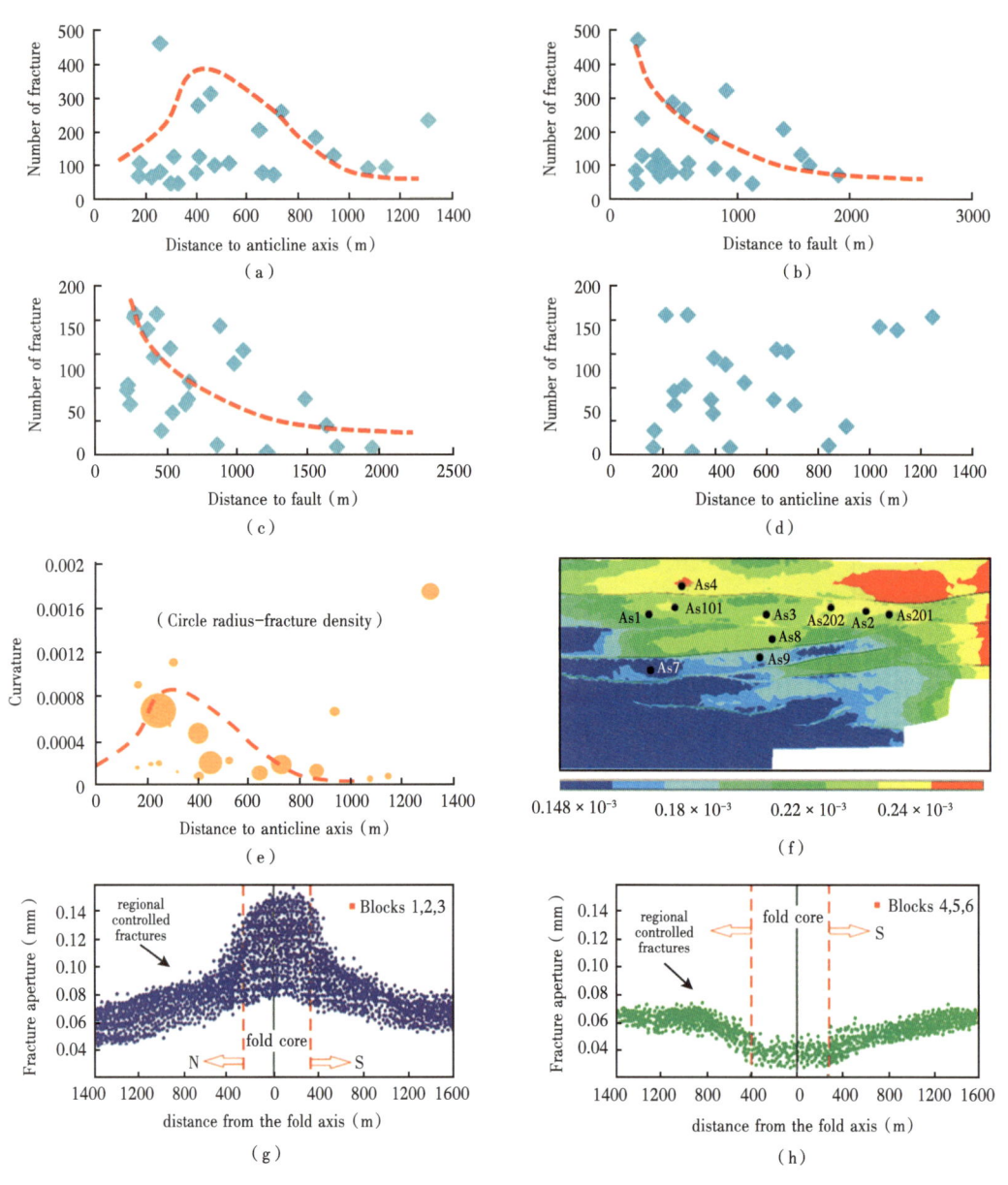

Fig. 5 Analysis of influencing factors of fracture parameters (Fracture data in fig. a, b, c, d are from FMI interpretation of the gas well in Keshen 2 block and fracture data in fig. f, g, h are from finite element simulation). (a) Relationship between distance from fold axis and tension fracture number; (b) Relationship between distance from faults and tension fracture number; (c) Relationship between distance from faults and sheer fracture number; (d) Relationship between distance from fold axis and sheer fracture number; (e) Relationship between distance from fold axis and structure curvature; (f) The distribution maps of simulated fracture parameters of Keshen area; (g) Relationship between distance from fold axis and fracture aperture in Blocks 1, 2 and 3; (h) Relationship between distance from fold axis and fracture aperture in Blocks 4, 5 and 6

good correlation with the rock strain at the top of the structure (Fig.5e). Finite element numerical simulation shows that fracture aperture within in single anticline tend to become smaller gradually from north to south in Keshen area due to the change of interlimb angle (Fig.5f). As described above, the high-value areas of fracture aperture is mainly located near the cores of anticlines. However, due to the difference in shape and amplitude of folds, the control effect on the development of fractures is not greatly different. Folding of open anticlines in northern part of the Keshen area (Blocks 1, 2 and 3 in Fig.1b) have significant influence and controlling on effectiveness of fractures(Fig.5g), but are not apparent in the gentle anticlines in Blocks 4, 5, and 6 (Fig.5h). In detail, the positive effect of folding on fracture apertures is about 0~370m to the axis. Similarly, the maximum influence distance of faulting on fracture apertures in Keshen area reach about 400m, and the maximum distance on fracture porosity is about 380m (Fig.5g, h).

Fracture development is also different in folds with different fault combination configuration due to the stress distribution[15]. The fracture of the imbricated double thrust structure is mainly developed near the core and the hanging wall of the boundary fault (south of the anticline), and that of the back thrust structure (i.e. pop-up structure) is mainly concentrated in the core of the anticline (Fig.6). Moreover, tension fracture zone of the former is smaller than that of the latter, but there exist very large tension fracture in the small tension zone and near the faults. In addition, structural neutral plane of the imbricated double thrust structure changes more complicated with the different structural areas and the stage of tectonic evolution, which results in the fracture heterogeneity of the imbricated double thrust structure is stronger than that of the back thrust structure. In contrast, the heterogeneity of back thrust structural fractures is relatively moderate. For example, in Keshen 2 block, it rise from west to East in turn, the highest one is located in the south wing, moving from top to bottom from ancient times to the present, with local fluctuations, which results in heterogeneity of tension fracture distribution. It can be proved by well test analysis. The 1/2 slope of straight line section of the derivative curve of well test pressure illustrates that the pressure presents obvious single direction linear propagation, indicating that there exist very large fractures in the reservoir; while the slope is close to indicates that the multi-directional natural fractures connect with each other, which bring about the pressure diffuses in multi directions, and soon reaches the flow quasi steady state of the fracture system, and then the matrix flows into the fracture; The slope of the curve is between 1/2 and 1 (Fig.6a), indicating the anisotropy of fracture development. In Keshen 2 block, the slope of the curves are mainly 1/2 or between 1/2 and 1, indicating that large fractures with unidirectional seepage characteristics developed in Keshen 2 reservoir. However, the slope of straight line section of well test curve in Keshen 8 block is close to 1 (Fig.6b), indicating that the effective fracture system in Keshen 8 block is more developed than that in Keshen 2 block.

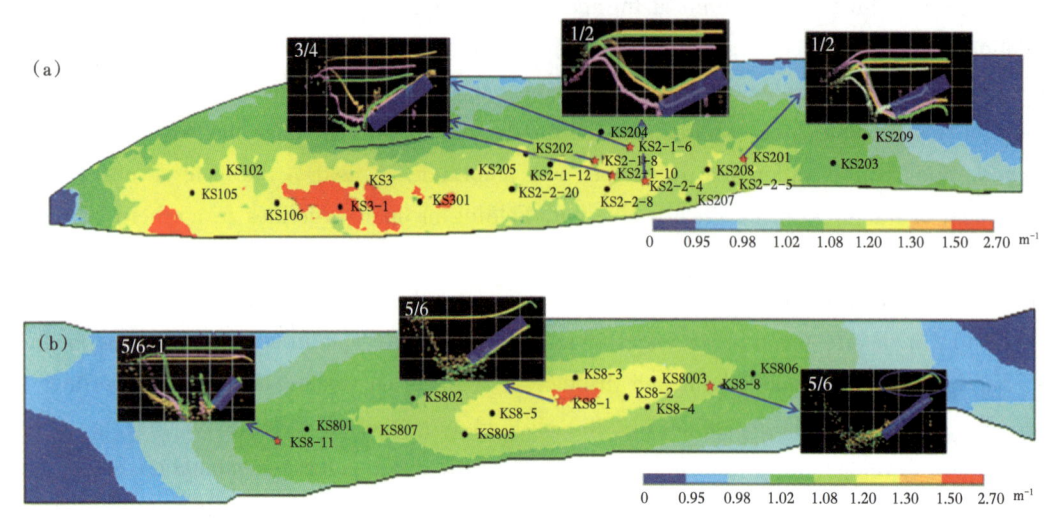

Fig. 6 The distribution of simulated fracture density and well test curves
of Keshen 2 (a) and Keshen 8 blocks (b)

## 5 Effects of fold revolution on fracture development

Since the late Cretaceous, the tectonic stress field of Kuqa depression has undergone a gradual transformation from weak extension to compression under the influence of Tianshan uplift[16], the extrusion direction experienced a transition from NNW-SSE to NW-SE[17]. According to the rock acoustic emission test and brittle structure sequence, the lower Cretaceous strata in Kuqa depression have memorized four major tectonic movements, namely, the late Yanshanian, early Himalayan, middle Himalayan and late Himalayan tectonic movements, while the Paleogene and Neogene strata have memorized three and two major paleotectonic movements respectively[18]. It can be seen that the activity intensity of Himalayan movement in the early and Middle Cretaceous strata is higher than that in the Paleogene strata However, the intensity of movement in the later period was significantly lower than that in the Paleogene[19].

In Kuqa depression, Dabei, Keshen, Dina and Kela anticlines also experienced several key stages of regional horizontal compression, including initial uplift, strong uplift and final finalization. The intensity of tectonic activity experienced a process of migration from deep to shallow from Cretaceous to Paleogene. Because the stress and strain energy in the foreland thrust belt have the characteristics of forward and shallow transfer and reduction[19], the deep Dabei and Keshen anticlines are strongly affected by regional tectonic stress in the first and second development stages, which is conducive to the development of large-scale imbricated faults. At the same time, in the deep III stress environment[20], a large number of near vertical conjugate regional fractures are developed, mainly distributed in the faults. Due to the small uplift amplitude and fold curvature, the core of the anticline has not yet formed a local extensional stress environment, but has been basically separated from the regional tectonic environment. Therfore only a small number of tensile fractures appear. In contrast, although the stress intensity of the first two stages of the shallow Dina and Kela anticlines is lower than that of the Dabei and Keshen anticlines, in the third stage, i.e. the strong uplift and stereotype stage, the tectonic activity is obviously strengthened, which leads to the appearance of the obvious local extension stress field at the top of the anticline, thus controlling the formation of a series of secondary normal faults and the development of derived and derived tensile fractures.

Based on the theory of rock mechanics and energy conservation, there is a certain upper limit of the releasable strain energy stored in the tight reservoir rock, and there is a positive correlation between the strain energy density and the fracture bulk density[19]. Both faults and fractures are the result of strain energy release. Dabei and Keshen anticlines are characterized by the deep burial depth, high principal stress value, large confining pressure, strong brittleness and high releasable strain energy of, while Dina and Kela anticlines are characterized by the shallow burial, relatively low stress value, relatively small confining pressure, relatively low brittleness and relatively low releasable strain energy, which lead to low fracture density but high fracture density in Dabei and Keshen areas, while high fracture density but dense fracture in Dina and Kela areas Low degree.

In order to further quantify the influence and control of faults and folds on fracture development in the process of fault related fold evolution, taking the Dina anticlines with large fold amplitude as an example, a comprehensive geological model is established. Based on the establishment of geological model, the fault plane is set as the mechanical detachment plane, and the stress field simulation and fracture quantitative prediction in different structural evolution stages are carried out by using the elastic – plastic finite element simulation method and the rock composite fracture criteria, and the fracture density, dip angle and other parameters are extracted. Based on the data statistics of a series of finite element simulation, it is found that the development and distribution of fractures in fault-related folds represented by Dina anticline are not only directly related to lithology and neutral plane, but also obviously controlled by factors such as fold uplift amplitude, distance from fault plane and fault sliding displacement (Fig. 7). It can be seen from Fig. 7a that with the continuous increase of fold uplift amplitude, the fracture density generally shows an increasing trend, but when the fold amplitude exceeds about 450m, the fracture density value gradually becomes stable or even has a decreasing trend. It can be seen from Fig. 7b that after removing the abnormal data points, the distance from the fault plane to the fracture density shows a certain power exponential relationship, and the correlation reaches 0.775. With the increase of the distance from the fault plane, the fracture density value gradually decreases. After the distance value exceeds 65m, the fracture density value becomes stable. It can be considered that the maximum effective control distance of general faults to fractures is about 65m. In comparison, with the continuous fault activity or the increase of fault displacement, the fracture density value increases gradually, and the highest value of fracture density in the whole fault related fold is generally located in the upper stratum or above the middle plane, but when the fault displacement is more than 630m, the fracture density values in the upper and lower strata become similar again (Fig. 7c). It can be seen from Fig. 6d that the change of fracture dip angle is directly related to the fold uplifting amplitude. For example, the fracture dip angle in the shallow Suweiyi formation (EI) increases almost linearly with the increase of fold uplifting amplitude, while the fracture dip angle in the deep Kumgeliemu group (EII) increases first and then decreases with the increase of fold amplitude, which is in the process of fault related fold evolution. In contrast, due to the gradual expansion of the local extension stress field in the shallow part of the fold, the high angle tension fractures become more developed. From the above analysis, no matter in the structural evolution mode of multi fault related folds or single fault related folds, the quantitative simulation or prediction results of fractures are in good agreement with the actual fracture measurement results.

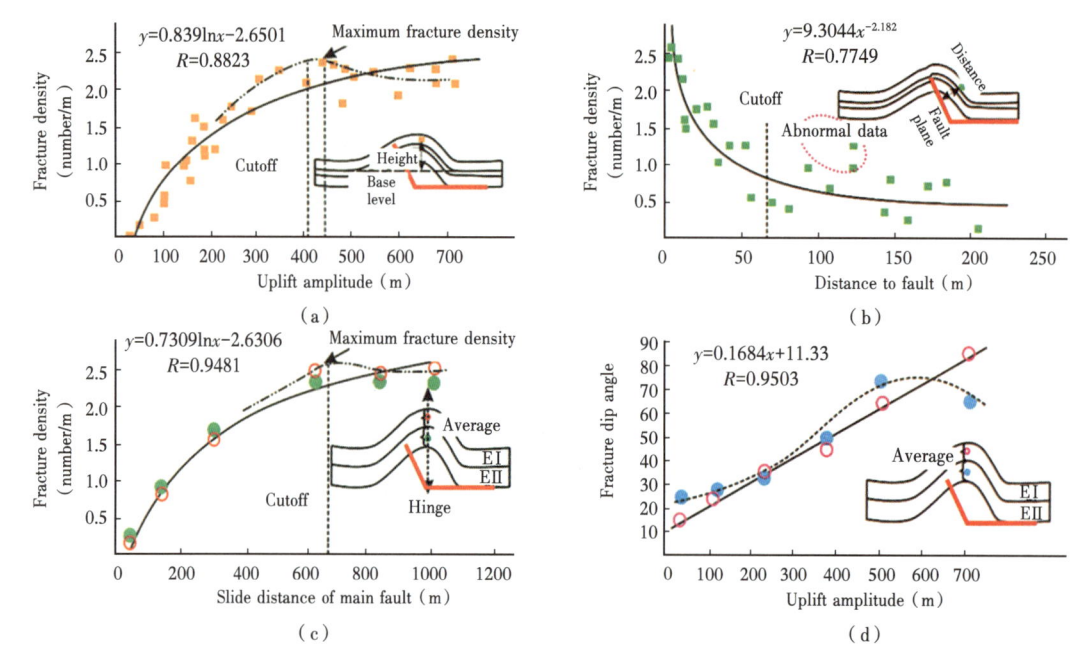

Fig. 7 Statistical results of the relationship between fault-related fold and fracture development in Kuqa depression. (a) Fold amplitude vs fracture density; (b) Distance to fault vs fracture density; (c) major fault slip vs fracture density; (d) Fold amplitude vs fracture dip angle

# 6 Conclusions

In view of the differences of fracture distribution within the anticlines in Kuqa depression, under the study of the structural characteristics, reservoir characteristics, physical characteristics and fracture statistics of single well, combined with the finite element numerical simulation, the spatial distribution characteristics of fractures are analyzed, and the main control factors of fracture development are summarized as flows: (1) Fracture density has obvious correlation with reservoir physical properties. The density of fractures in Dabei and Keshen gas fields with poor physical properties are much higher than that in Dina and Kela gas fields with relatively good physical properties. (2) The fracture development mode is controlled by the structural style. Dabei and Keshen anticlines are dominated by shear network fractures due to their low amplitude (relatively large interlimb angle), while the tension fracture proportion increases within Dina and Kela anticlines due to their high amplitude (relatively small interlimb angle). (3) The high-value areas of tension fracture is mainly located near the cores of anticlines and the faults, while that of shear fracture is obviously correlation with faults but not apparent with cores of anticlines. Further study shows that open anticlines in northern part of the Keshen area have significant influence and controlling on effectiveness of fractures, but are not apparent in the gentle anticlines in the southern area. (4) Fracture development is different in folds with different fault combination configuration. The fracture development heterogeneity of the imbricated double thrust structure with very different scale fracture is stronger than that of the back thrust structure with relatively small scale difference fracture. (5) The structural evolution process and fracture simulation results of single fault related folds show that the development and distribution of structural fractures are not only directly related to lithology and neutral plane, but also obviously controlled by factors such as fold uplifting amplitude, distance from fault plane and fault sliding displacement, especially when the fold uplifting amplitude is in the range of 410~450m, the fracture density value is the largest. The structural styles of the main elements can be used to judge the favorable structures for further oil and gas exploration and development. The open fold in the study area has better exploration prospects than the gentle fold. As to the imbricated double thrust structure, Measures should be taken as early as possible in the development process to prevent fissured water channeling.

## References

[1] Wang Z M. Formation mechanism and enrichment regularities of the Kelasu subsalt deep large gas field in Kuga Depression, Tarim Basin. Natural Gas Geoscience. 2014,25(2):153-166.

[2] Zeng L B, Tan C X, Zhang M L. Tectonic stress field and its effect on hydrocarbon migration and accumulation in Mesozoic and Cenozoic in Kuqa depression, Tarim basin. Science in China (Series D: Earth Sciences). 2004,S2: 114-124.

[3] Cumella S P, Shanley K W, Camp, W K.The influence of stratigraphy and rock mechanics on Mesaverde gas distribution, Piceance Basin, Colorado. AAPG Hedberg Series 3,2008. 137-155.

[4] Ju W, Wu C F, Wang K, et al.Prediction of tectonic fractures in low permeability sandstone reservoirs: A case study of the Es3m reservoir in the Block Shishen 100 and adjacent regions, Dongying Depression. Journal of Petroleum Science and Engineering. 2017,156: 884-895.

[5] Xu Z P, Li Y, Ma Y J, et al. Future gas exploration orientation based on a new scheme for the division of structural units in the central Kuqa Depression, Tarim Basin. Geology and exploration. 2011,31(3):31-36.

[6] Tian Z J, Song J G. Tertiary structure characteristics and evolution of Kuqa foreland basin. Acta Petrolei sinica. 1999,20(4):7-13.

[7] Zeng L B, Li Y G. Tectonic Fractures in Tight Gas Sandstones of the Upper Triassic Xujiahe Formation in the Western Sichuan Basin, China. Acta Geologica Sinica (English Edition),2010,84(5):1229-1238.

[8] Wang, K, Wang G W, Xu B, et al. Fracture classification and structural fractures in Keshen 2 well area.

Progress in Geophysics. 2015,30(3): 1251-1256.

[9] Wang K, Yang H J, Zhang H L, et al.Characteristics and effectiveness of structural fractures in ultra-deep tight sandstone reservoir: A case study of Keshen-8 gas pool in Kuqa Depression, Tarim Basin. Oil & Gas Geology. 2018,39(4):719-729.

[10] Ju W, Hou G T, Zhang B. Insights into the damage zones in fault-bend folds from geomechanical models and field data. Tectonophysics. 2014,610: 182-194.

[11] Wei G Q, Wang J P, Zeng L B, et al.Structural reworking effects and new exploration discoveries of subsalt ultra-deep reservoirs in the Kelasu tectonic zone. Natural Gas Industry. 2020,40(1):20-30.

[12] Qian L, Yu H Z, Ding C H, et al. The relationship between fold neutral planes and hydrocarbon accumulation in Tarim basin. Marine Origin Petroleum Geology. 2011,16(4), 62-65.

[13] Zhou X G, Cao C J, Yuan J Y.The status and prospects of quantitative structural joint prediction of reservoirs and research of oil and gas seepage law. Advances in Earth Science. 2003,18: 398-404.

[14] Sun S, Hou G T, Zheng C F.Fracture zones constrained by neutral surfaces in a fault-related fold: Insights from the Kelasu tectonic zone, Kuqa Depression. Journal of Structural Geology. 2017,104, 112-124.

[15] Hou G T, Sun S, Zheng C F, et al.Subsalt Structural Styles of Keshen Section in Kelasu Tectonic Belt. Xinjiang Petroleum Geology. 2019,40(1):21-26.

[16] He G Y, Lu H F, Wang L S, et al.Evidence for Paleogene Extensive Kuqa Basin, Tarim. Journal of Nanjing University (Natural Science). 2003,39(1):40-45.

[17] Zhang Z P, Wang Q C. Development of joints and shear fractures in the Kuqa depression and its implication to regional stress field switching. Science in China(Series D:Earth Sciences). 2004,S2, 74-85.

[18] Zheng C F, Hou G T, Zhan Y, et al.An analysis of Cenozoic tectonic stress fields in the Kuqa depression. Geological Bulletin of China. 2016,35(1):130-139.

[19] Zhang M L, Tan C X, Tang L J, et al. An Analysis of the Mesozoic-Cenozoic Tectonic Stress Field in Kuqa Depression, Tarim Basin. Acta Geoscientica Sinica. 2004,25(6):615-619.

[20] Dai J S, Feng J W, Li M, et al.Discussion on the extension law of structural fracture in sand-mud interbed formation. Earth Science Frontiers. 2011, 18(2): 277-283.

# 靖边气田低效储量评价与可动用性分析

贾爱林　付宁海　程立华　郭建林　闫海军

(中国石油勘探开发研究院)

**摘要**：靖边气田是风化壳型碳酸盐岩气藏,目前处于开发中期,保持气田稳产是靖边气田面临的核心问题,而低效储量是靖边气田储量接替的重要领域。在低效储量分布特征的基础上,将靖边气田低效储量分布分为三种类型：低效储量与优质储量垂向叠加型；低效储量与优质储量侧向连通型；低效储量孤立分布型。研究结果表明,孤立型低效储量规模小,垂向叠加型和侧向连通型低效储量是靖边气田储量接替的重要类型。针对垂向叠加型和侧向连通型低效储量分布特征,采用数值模拟方法对低效储量可动用性进行分析和评价。研究结果表明,低效储层渗透率、低效储层布井方式(井距)、低效储层气井配产是影响低效储量动用程度的最主要因素。

**关键词**：碳酸盐岩气藏；储量接替；低效储量；分布特征；可动用性分析；靖边气田

靖边气田发现于 1989 年,先后经历了前期评价和开发试验、探井试采、规模开发以及气田稳产 4 个阶段。靖边气田产能建设始于 1997 年,从 1999 年开始进入规模开发阶段,至 2003 年底具备 $55 \times 10^8 m^3/a$ 的生产能力,目前一直稳产在 $55 \times 10^8 m^3/a$ 左右。气层位于奥陶系马家沟组碳酸盐岩顶部,具有层薄、低渗透、非均质性强、低丰度、顶部裂缝发育、中下部溶蚀孔洞发育的特点,属于低孔、低丰度、无边(底)水、深层大型定容碳酸盐岩气藏[1-5]。

储层的非均质性严重制约气藏继续稳产和高效开发[6,7],目前气田内部大量低效储量采出程度较低,随着优质资源的减少,低效储量所占的分量日益突出,这部分储量将是气田接替稳产的重要资源,但是该部分储量的有效动用对策不明确。针对保持气田稳产面临的新形势和靖边气田目前开采阶段的特点,通过对低效储量分布和可动用性分析,对低效储量可动用性进行评价,为气田的稳产接替提供技术支持,发展和完善了风化壳型碳酸盐岩气田稳产技术。

## 1 靖边气田低效储量特点与分布特征

选取靖边气田南部为研究的典型区块,包括南区、南二区、陕 227 井区、陕 106 井区、陕 230 井区和陕 100 井区。其中,南区和南二区 1998 年投产,代表了气田本部开发多年老区特征；陕 227、陕 106 和陕 230 井区为 2003 年以后投产区块,相对投产时间较晚,且储层条件略差,代表了靖边气田后期投产、采出程度偏低的区块特征；陕 100 井区提交了探明储量,尚未整体投入开发,代表了气田扩边建产区块。因此,典型区内的 6 个区块均具有代表性,同时在平面上又具有衔接性,便于作整体研究。

### 1.1 低效储量特点

储量分类评价的目的是优选出优质储量以指导气田开发。Ⅰ+Ⅱ类储量属于易动用储量,Ⅲ类储量属于在现有技术经济条件下,因储层条件差而较难动用的储量。因此,整体上Ⅲ类储量属于低效储量。参考长庆油田的划分标准[8]以及动、静态参数,以储层分类和气井分类为基础,对低效储量进行了划分。储量分类计算表明,低效储量达到总储量的26%,挖潜空间较大。

从储层分布统计分析,Ⅰ、Ⅱ类层在马五$_1^3$最发育,构成靖边气田的主力气层,低效储量分布的Ⅲ类层在马五$_1^2$、马五$_1^3$、马五$_2^2$和马五$_4^{1a}$均有发育。从气层厚度分布比例看,气田本部Ⅲ类气层厚度占29.8%,Ⅰ+Ⅱ类气层厚度占70.2%;气田东侧Ⅲ类气层厚度占40.9%,Ⅰ+Ⅱ类气层厚度占59.1%。总体上,Ⅲ类低效层占有较大比例,是气田接替稳产的潜力资源。从储层特征分析,低效储量储层相对致密,孔隙度小于5%,基质渗透率小于0.04mD,含气饱和度为65%~75%,总渗透率为0.5~1.5mD,声波时差为160~155μs/m。从储集空间类型分析,低效储量储层主要为充填残余溶孔、针孔及晶间微孔,发育微裂缝。压汞曲线表现排驱压力高,偏细歪度;反映孔喉半径小,退汞效率低。

## 1.2 低效储量分布特征

在储层分类的基础上,对单井进行有效储层分类解释,勾画有效储层连井对比剖面,参照地质建模程序方法[9-12],建立地质模型,分析有效储层的空间分布规律。

从低效储量分布分析,垂向上,低效储量在各层都有分布,与优质储量相间发育。平面上,低效储量存在3种分布形式(图1):(1)分布在优质储量内部,被优质储量包围或半包围;(2)分布在优质储量边部,与优质储量相通,呈条带状;(3)孤立分布,发育较少。

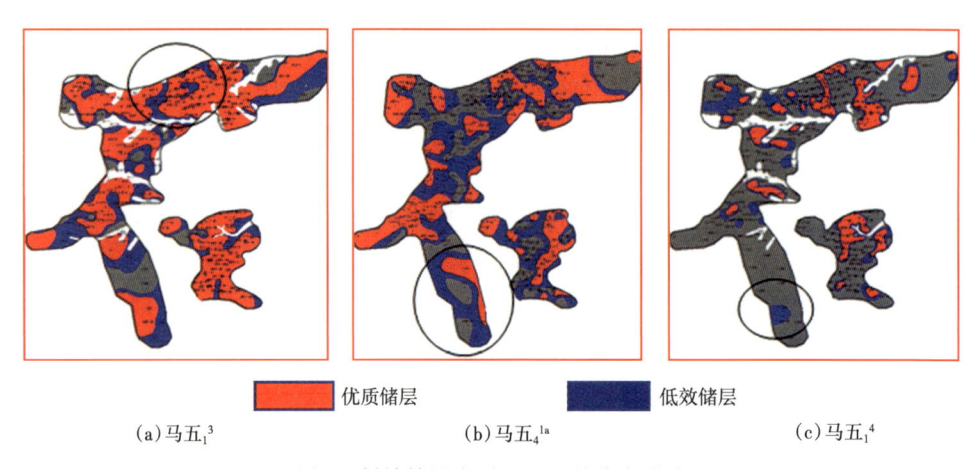

(a)马五$_1^3$  (b)马五$_4^{1a}$  (c)马五$_1^4$

图1 低效储量在平面上三种分布形式

## 1.3 低效储量分布模式

针对储量在剖面上和平面上分布的非均质性,划分为如下3种基本分布形式:(1)低效储量与优质储量垂向叠加型;(2)低效储量与优质储量侧向连通型;(3)低效储量孤立分布型。以上3个基本形式的组合可以构建出靖边气田低效储量分布的复杂情况。

孤立分布的低效储量分布少,大部分低效储量与优质储量并存,因此主要针对前两种分布形式,建立数值模拟模型,模拟开采过程中低效储量的动用情况。

## 2 低效储量可动用性分析模型设计

### 2.1 与优质储量垂向叠加型

#### 2.1.1 模型参数设计

对于与优质储量垂向叠加型低效储量的动用性模拟的设计是基于无边底水均质气藏模型,单井生产。垂向上发育两个有效储层:上部有效层段物性好,相当于Ⅰ、Ⅱ类储层;下部有效层段物性差,相当于Ⅲ类储层,对应低效储量,中间发育隔层。目的是分析优质储层、低效储层合采以及低效储层单独开采时低效储量的动用情况。

模型为均质模型,网格横纵步长为50m;网格数为41×41×3=5043个;模拟面积为2.05×2.05=4.2km²。模型计算中的物性参数见表1。

表1 模型参数表

| 储层 | 有效厚度(m) | 孔隙度(%) | 含气饱和度(%) | 渗透率(mD) | 配产($10^4$m³) |
|---|---|---|---|---|---|
| 优质储层 | 5 | 6.8 | 77 | 2 | 3.2 |
| 低效储层 | 3 | 3.5 | 70 | 0.1~0.4 | 0.28~0.8 |

#### 2.1.2 模拟过程

设计3组不同渗透率级差的数模模型,每组共包括3个模型,即优质储层单采模型、低效储层单采模型以及优质储层低效储层合采模型(表2)。

表2 模型渗透率与配产参数表

| 模拟组次 | 渗透率级差 | 优质储层渗透率(mD) | 低效储层渗透率(mD) | 优质储层配产($10^4$m³/d) | 低效储层配产($10^4$m³/d) |
|---|---|---|---|---|---|
| 1 | 5 | 2 | 0.4 | 3.2 | 0.6、0.65、0.7、0.75、0.8 |
| 2 | 10 | 2 | 0.2 | 3.2 | 0.4、0.45、0.5、0.55、0.6 |
| 3 | 20 | 2 | 0.1 | 3.2 | 0.28、0.31、0.34、0.37、0.4 |

### 2.2 与优质储量侧向连通型

对于与优质储量侧向连通的低效储量的动用性模拟,设计的基础模型是基于平面均质气藏模型,中部为优质储层,有6口气井生产,井距为3km,优质储层外围6km范围为低效储层,对应低效储量(图2a)。在基础模型基础上,通过改变低效储层渗透率、含气饱和度、有效厚度、布井方式、低效储层气井配产以及优质储层渗透率、优质储层配产等参数模拟了优质储层气井生产时,不同参数变化对外围低效储量的采出程度的影响。

基础模型设计为,网格横纵步长为50m;网格数361×361×1=130321个;模拟面积为18.05×18.05=325.8km²。模型参数见表3。

表 3 基础模型参数设计表

| 储层 | 有效厚度（m） | 孔隙度（%） | 含气饱和度（%） | 渗透率（mD） | 配产（$10^4 m^3/d$） |
| --- | --- | --- | --- | --- | --- |
| 优质储层 | 5 | 6.8 | 77 | 2 | 6 |
| 低效储层 | 4 | 4.6 | 70 | 0.01~1.5 | 1 |

## 3 低效储量可动用性评价

根据模型设计,对于与优质储量侧向连通型低效储量,模拟了以下 7 种不同因素对低效储量采出程度的影响:(1)低效储层布井方式(井距);(2)低效储层渗透率;(3)低效储层含气饱和度;(4)低效储层有效厚度;(5)低效储层气井配产;(6)优质储层渗透率;(7)优质储层气井配产。

### 3.1 低效储层布井方式

考虑到气藏的非均质性和开采中的不均衡性,在生产中,低渗透区的气可以补给高渗透区,高渗透区的气井不仅可以采出本井区的储量,而且可以采出邻区的部分储量,因此,在研究低效储层布井方式对低效储量采出程度的影响时,设计了 3 种布井方式(图 2)。

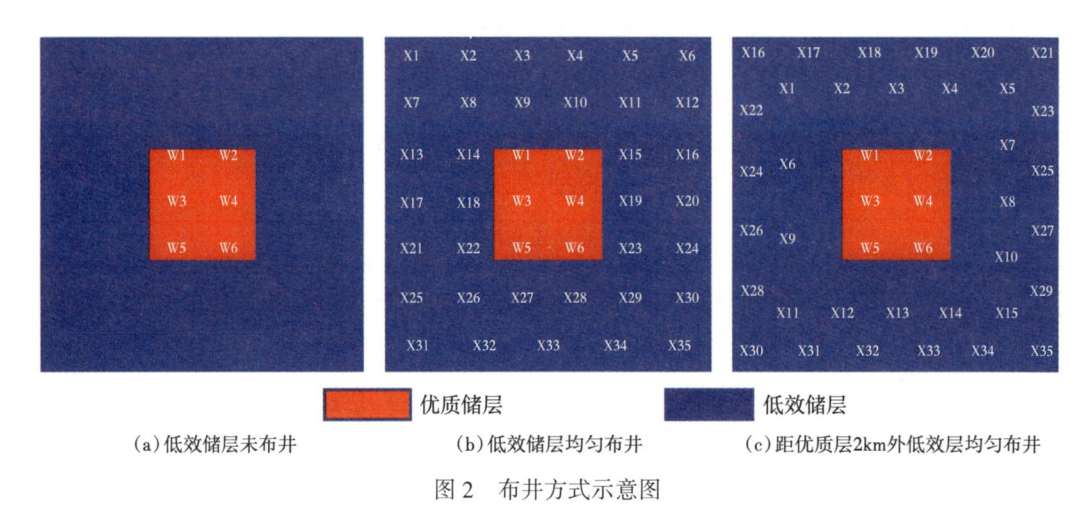

图 2 布井方式示意图

(a)低效储层未布井　　(b)低效储层均匀布井　　(c)距优质储层2km外低效层均匀布井

(1)低效储层未布井:中部优质储层井距为3km,有6口气井生产,外围低效储量区无生产井。(2)低效储层均匀布井:中部优质储层井距为3km,有6口气井生产,外围低效储量区均匀布 35 口井。(3)距优质储层 2km 外均匀布井:中部优质储层井距为3km,有6口气井生产,在距离优质储层 2km 外的低效储量区均匀布 35 口井。

当低效层不布井时,随着与优质储层距离的增加,低效储量采出程度急剧下降,但在距离优质储层 2km 范围内,低效储量采出程度能达到 20% 以上,2km 以外储量难以动用,须进一步钻井才能有效动用;低效层布井时,在距离优质储层 2km 范围外均匀布井的效果好于整个低效储层均匀布井,低效储层储量采出程度要提高 1.3% 左右(图 3)。

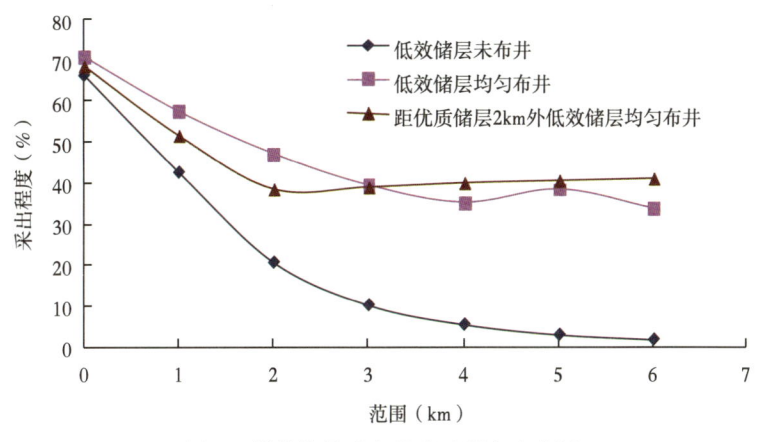

图 3 低效储量采出程度随距离变化图

在距优质储层 2km 外均匀布井,井数在 4 口、8 口、14 口、21 口、28 口、35 口、56 口变化,对应井距在 7.50km、5.30 km、4.01 km、3.27 km、2.83 km、2.53 km、2.00 km 变化,最大井距与最小井距对应的优质储量采出程度变化范围不到 3%,而低效储量采出程度从 14.23% 变化到 52.29%,变化范围近 40%(图 4)。

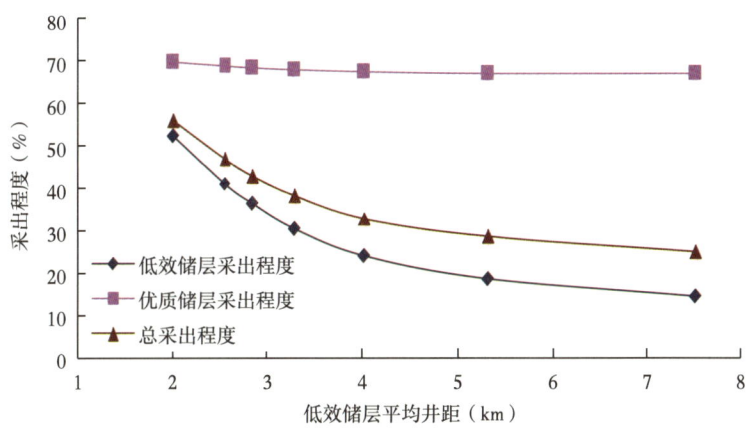

图 4 井距对低效储层采出程度的影响图

## 3.2 低效储层渗透率与含气饱和度

渗透率是影响低效储层动用能力的主要因素,渗透率在 0.01~1.5mD 变化时,低效储量采出程度从 2.72% 变化到 53.48%,变化范围达 50%。渗透率在 0.3~1.5mD 变化时,对低效储量采出程度影响较小,低效储量采出程度变化在 7.5% 左右,渗透率小于 0.3mD 时,低效储量采出程度急剧降低,低效储量采出程度变化达到 43%(图 5)。

低效储层含气饱和度在 45%~77% 变化时,低效储量采出程度从 22.75% 变化到 42.58%,变化范围在 20% 左右(图 6)。

— 279 —

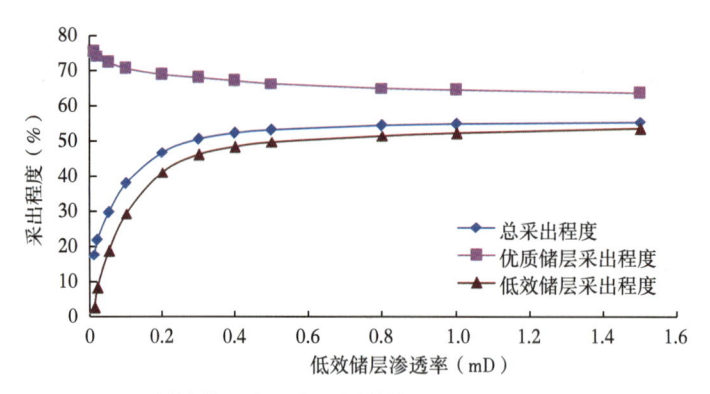

图 5　低效储层渗透率对低效储层采出程度的影响图

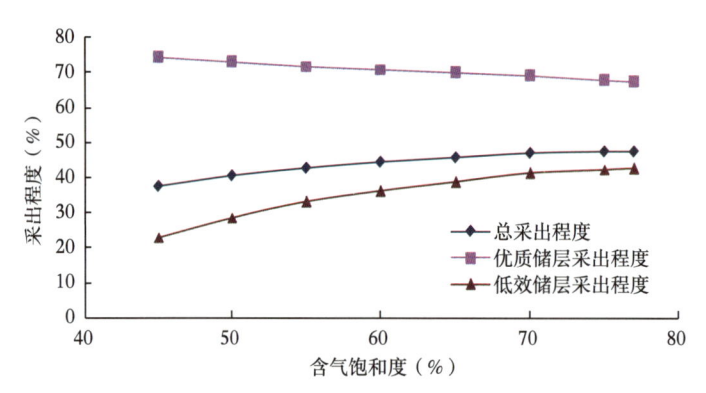

图 6　低效储层含气饱和度对低效储层采出程度的影响图

## 3.3　低效储层有效厚度与气井配产

当低效储层厚度在 2~8m 变化时,低效储量采出程度从 43.45% 变化到 27.20%,变化范围在 16% 左右(图 7)。

低效储层配产主要影响气井的稳产期,保持优质储层配产不变,低效储层配产在 (0.8~2) ×10⁴m³/d 之间变化时,稳产期从 20 年减小到 0.33 年,对低效储量采出程度影响较小,影响程

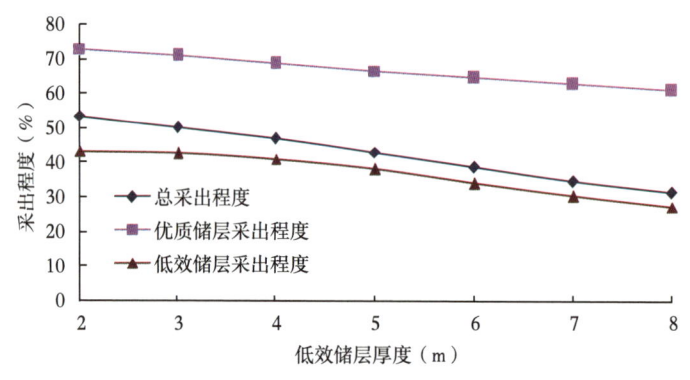

图 7　低效储层有效厚度对低效储层采出程度的影响图

度在 4.3% 以内(图 8)。

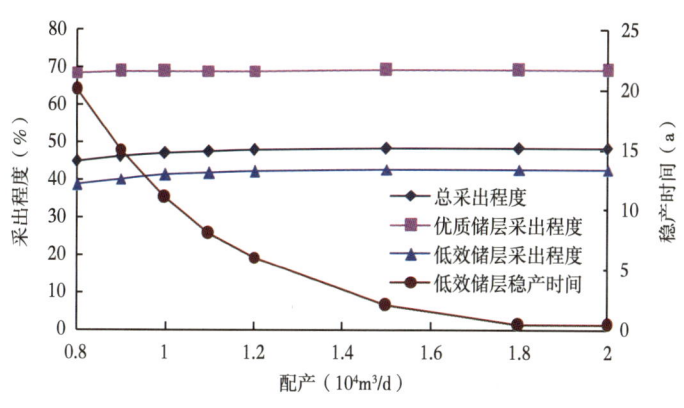

图 8 低效储层气井配产对低效储层采出程度的影响图

## 3.4 优质储层渗透率与配产

优质储层渗透率主要影响优质储层采出程度,优质储层渗透率在 1~10mD 变化时,优质储层采出程度变化范围在 20% 左右,对低效储层采出程度影响很小,低效储层采出程度变化在 2.3% 左右(图 9)。

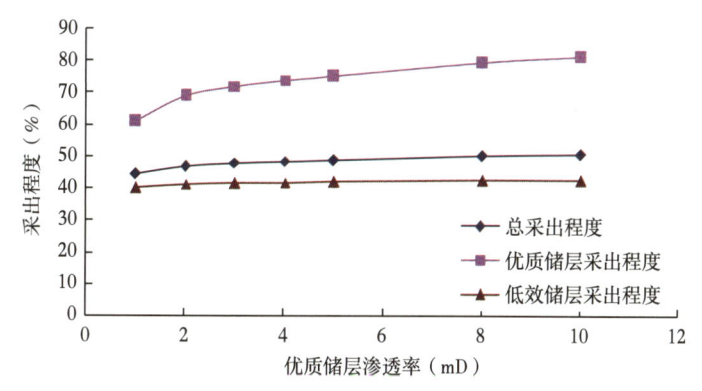

图 9 优质储层渗透率对低效储层采出程度的影响图

优质储层配产主要影响优质储层稳产期,优质储层配产从 $4×10^4m^3/d$ 到 $20×10^4m^3/d$ 变化,优质储层稳产期从 20 年降至 0.04 年,对低效储层采出程度影响不大,影响范围在 1.8% 以内(图 10)。

通过以上分析,对于与优质储量侧向连通型低效储量,其采出程度的影响因素由强到弱依次为:(1)低效储层渗透率;(2)低效储层布井方式(井距);(3)低效储层含气饱和度;(4)低效储层有效厚度;(5)低效储层气井配产;(6)优质储层渗透率;(7)优质储层气井配产。

## 3.5 渗透率级差与低效储层配产

对于与优质储量垂向叠加型低效储量,模拟了不同渗透率级差与低效储层配产对低效储量采出程度的影响(图 11、图 12)。

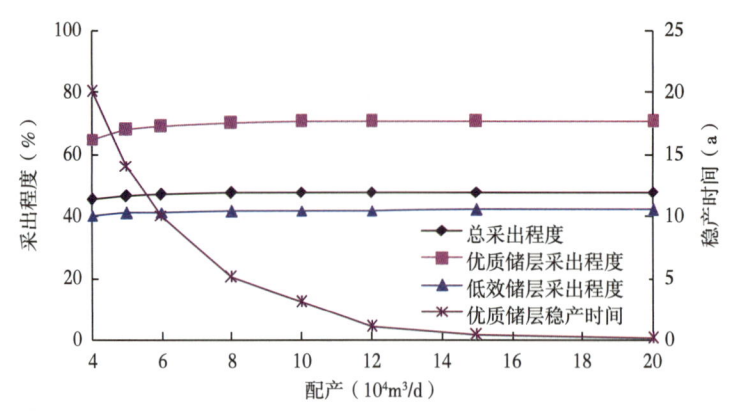

图 10 优质储层配产对低效储层采出程度的影响图

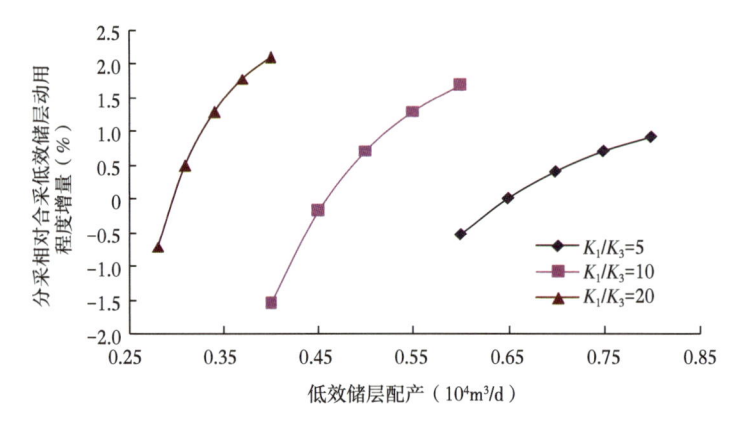

图 11 渗透率与配产对低效储层采出程度的影响图

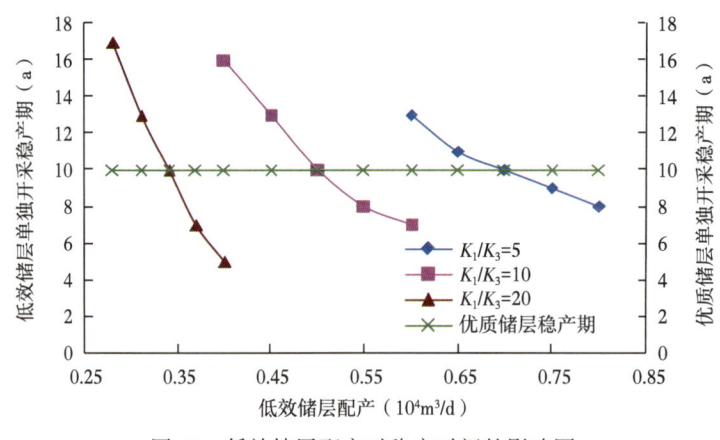

图 12 低效储层配产对稳产时间的影响图

结果表明,低效储量采出程度受气井配产影响大。优质储层与低效储层分采与合采时低效储量采出程度的差异主要受气井配产影响,低效储层单独生产配产的稳产期不高于优质储层稳产期时,分采采出程度高。渗透率级差越大,优质储层与低效储层分采与合采时储量采出程度的差异越大。因此,选择与低效层相匹配的产量可提高低效储层采出程度。

## 4 结论

(1)低效储量在气田普遍发育,是气田稳产接替的潜力资源。根据储量在剖面上和平面上的分布特征,将低效储量分布划分为与优质储量垂向叠加型、与优质储量侧向连通型和孤立分布型三种形式。

(2)根据与优质储量垂向叠加型、与优质储量侧向连通型两种情况下低效储量的动用情况,低效储层渗透率、低效储层布井方式(井距)以及低效储层气井配产是影响低效储量采出程度的最主要因素。因此,通过储层改造措施提高低效层渗透率,优化低效储层布井与配产是提高低效储量采出程度的关键。

### 参 考 文 献

[1] 何自新,郑聪斌,王彩丽,等.中国海相油气田勘探实例之二 鄂尔多斯盆地靖边气田的发现与勘探[J].海相油气地质,2005,10(2):37-44.
[2] 孙来喜,李允,陈明强,等.靖边气藏开发特征及中后期稳产技术对策研究[J].天然气工业,2006,26(7):82-84.
[3] 吴永平,王允诚.鄂尔多斯盆地靖边气田高产富集因素[J].石油与天然气地质,2007,28(4):473-478.
[4] 付金华,魏新善,任军峰,等.鄂尔多斯盆地天然气勘探形势与发展前景[J].石油学报,2006,27(6):1-4.
[5] 张歧,黄文科,刘茂果,等.靖边气田碳酸盐岩储层精细描述[J].石油化工应用,2011,30(11):47-50.
[6] 田冷,何顺利.长庆靖边下古生界气藏储层非均质性研究[J].山东科技大学学报:自然科学版,2009,28(3):13-16.
[7] 晏宁平,张宗林,何亚宁,等.靖边气田马五$_{1+2}$气藏储层非均质性评价[J].天然气工业,2007,27(5):102-103.
[8] 王东旭,王鸿章,李跃刚.长庆气田难采储量动用程度评价[J].天然气工业,2000,20(5):64-66.
[9] 贾爱林,程立华.数字化精细油藏描述程序方法[J].石油勘探与开发,2010,37(6):709-715.
[10] 贾爱林.中国储层地质模型20年[J].石油学报,2011,32(1):181-188.
[11] 李元觉,冯军祥.长庆气田靖边区气藏地质建模[J].天然气工业,2000,20(6):50-54.
[12] 黄文科,吴正,晏宁平,等.储层建模优化技术在靖边气田下古气藏中的应用[J].石油化工应用,2010,29(4):60-63.

# 苏里格致密砂岩气田水平井开发地质目标优选

刘群明[1] 唐海发[1] 冀 光[1] 孟德伟[1] 王 键[2]

(1. 中国石油勘探开发研究院;2. 振华石油控股有限公司)

**摘要**:水平井技术是提高致密气单井产量、实现致密气经济有效开发的关键技术之一,与国外致密砂岩气田稳定分布的储层条件相比,国内致密砂岩气田一般具有储层规模小、纵向多层、整体分散、局部相对富集等特点,水平井开发地质目标优选是实现国内致密砂岩气田水平井规模化应用的关键技术问题。以国内典型致密砂岩气田苏里格气田为例,通过实钻水平井地质综合分析和密井网区精细地质解剖,应用储层构型层次分析方法,根据砂体及有效砂体叠置样式的不同,将苏里格气田水平井划分为3大类6小类水平井钻遇储层地质模型:$A_1$垂向切割叠置型、$A_2$侧向切割叠置型、$B_1$夹层堆积叠置型、$B_2$隔层堆积叠置型、$C_1$单层孤立型、$C_2$横向串联型。其中分布在辫状河体系叠置带内的$A_1$垂向切割叠置型和$B_1$夹层堆积叠置型是水平井开发的主要地质目标。依据储层地质、生产动态、储量丰度、井网密度等关键参数,建立了水平井整体开发和"甜点"式开发两种开发模式下的井位优选标准,并成功应用于苏中X区块,取得了较好的应用效果,同时该地质目标优选方法与井位优选标准对我国同类气藏的开发具有很好的借鉴作用。

**关键词**:致密砂岩气;苏里格气田;水平井;地质目标优选;叠置带

苏里格气田作为国内致密砂岩气田的典型代表[1,2],随着水平井钻井提速及多段压裂技术的提高,气田开发方式已经实现直井开发向水平井开发的迅速转变[3,4]。但与国外致密气稳定分布的储层条件相比[5,6],苏里格气田有效储层规模小,连通性差,纵向多层且分散,但部分井区仍然存在砂岩和气层集中分布段,气田储层强非均质性的特点[7]要求水平井部署时要优选地质目标。前人[8-10]在该气田水平井开发部署选区方面的研究较少且不系统,研究过程也限于水平井资料有限多通过解剖直井来确定水平井选区参数。本文基于大量的已完钻水平井资料,重点从完钻水平井精细地质解剖和动静态参数综合分析的角度出发,并结合直井密井网区储层构型研究成果,来最终确立水平井开发有利目标区定量识别标准,从而为苏里格气田下一步规模化应用水平井提高单井产量和采收率提供技术支撑,并为国内同类型气藏水平井开发提供指导借鉴。

## 1 水平井分类评价

参考前人[11]在致密气水平井分类评价方面的研究成果,并考虑到评价参数要方便现场录取,最终优选出7个静态指标和3个动态指标建立了苏里格气田水平井分类评价标准(表1),为下一步准确评价完钻水平井的开发效果提供了标准和依据。该标准将气田水平井分为高产Ⅰ类井、普通Ⅰ类井、Ⅱ类井、Ⅲ类井4类水平井。7个静指标态具体为"水平井钻遇气层长度""水平井气层钻遇率""相邻直井测井相""井间储层连续性""临井主力层累计气层厚度"

及"临井目标气层单层厚度",3个动态指标具体为"无阻流量""初期产量""稳产3年配产"。其中Ⅰ类井经济效益明显,Ⅱ类井为有一定经济效益,Ⅲ类井经济效益差。

表1 水平井分类评价标准

| 动静态参数 | | | Ⅰ类井 | | Ⅱ类井 | Ⅲ类井 |
|---|---|---|---|---|---|---|
| | | | 高产Ⅰ类井 | 普通Ⅰ类井 | | |
| 静态 | 水平井 | 钻遇气层长度(m) | >800 | 600~800 | 400~600 | <400 |
| | | 气层钻遇率(%) | 80 | 60~80 | 40~60 | 40 |
| | 相邻直井 | 测井相 | 平滑箱形 | 微齿化箱形 | 齿化箱形 | 齿化钟形 |
| | | 井间储层连续性 | 好 | 较好 | 一般 | 差 |
| | | 主力层累计气层厚度(m) | >18 | 15~18 | 10~15 | <10 |
| | | 目标气层厚度(m) | >10 | 8~10 | 6~8 | <6 |
| 动态 | | 无阻流量($10^4 m^3/d$) | >80 | 60~80 | 20~60 | <20 |
| | | 初期产量($10^4 m^3/d$) | >12 | 8~12 | 4~8 | <4 |
| | | 稳产三年配产($10^4 m^3/d$) | >9 | 9~6 | 3~6 | <3 |

按上述标准对苏里格地区生产时间超过1年,且静态数据录取资料较为齐全的208口水平井进行了分类评价,统计得到目前苏里格气田Ⅰ+Ⅱ类井比例平均为65.72%。其中按静态指标分类比例为65.61%,动态指标为65.83%,两者差别不大,说明静态储层地质特征与单井动态表现相关性较好。

## 2 水平井钻遇储层地质模型分析

### 2.1 水平井钻遇储层地质模型分类

通过对苏里格气田208口实钻水平井的精细地质解剖,根据水平井钻遇砂体及有效砂体叠置样式的不同,将水平井钻遇储层地质模型按形成水动力条件由强到弱依次划分为A切割叠置型、B堆积叠置型、C分散孤立型3大类,每一大类又细分为2类共6小类(图1)。其中切割叠置型按有效砂体切割叠置位置的不同分为$A_1$垂向切割叠置型和$A_2$侧向切割叠置型,堆积叠置型按有效砂体间隔夹层种类的不同分为$B_1$夹层堆积叠置型和$B_2$隔层堆积叠置型,分散孤立型按水平井钻遇1个或2个有效砂体细分为$C_1$单层孤立型和$C_2$横向串联型。

### 2.2 水平井钻遇储层地质模型优选

对实钻水平井按上述6类水平井模型分别统计静动态参数,并依据水平井分类评价标准分模型对钻遇水平井进行归类,统计结果显示(表2)目前已完钻的水平井模型以$A_1$垂向切割叠置型和$B_1$夹层堆积叠置型为主,占总井数的58%。$A_1$垂向切割叠置型为强水动力条件下的河道心滩砂体垂向多期次切割叠置形成,有效砂体空间呈块状富集;$B_1$夹层堆积叠置型为中等水动力条件下的2期辫状河道心滩粗砂岩相垂向堆积叠置形成,有效砂体中间沉积物性或泥岩夹层,压裂改造可沟通上下气层。2模型气层横向分布稳定,水平段气层钻遇率高,通常大于70%,临井揭示有效厚度一般大于8m,水平井类型基本为Ⅰ类和Ⅱ类水平井,平均日产

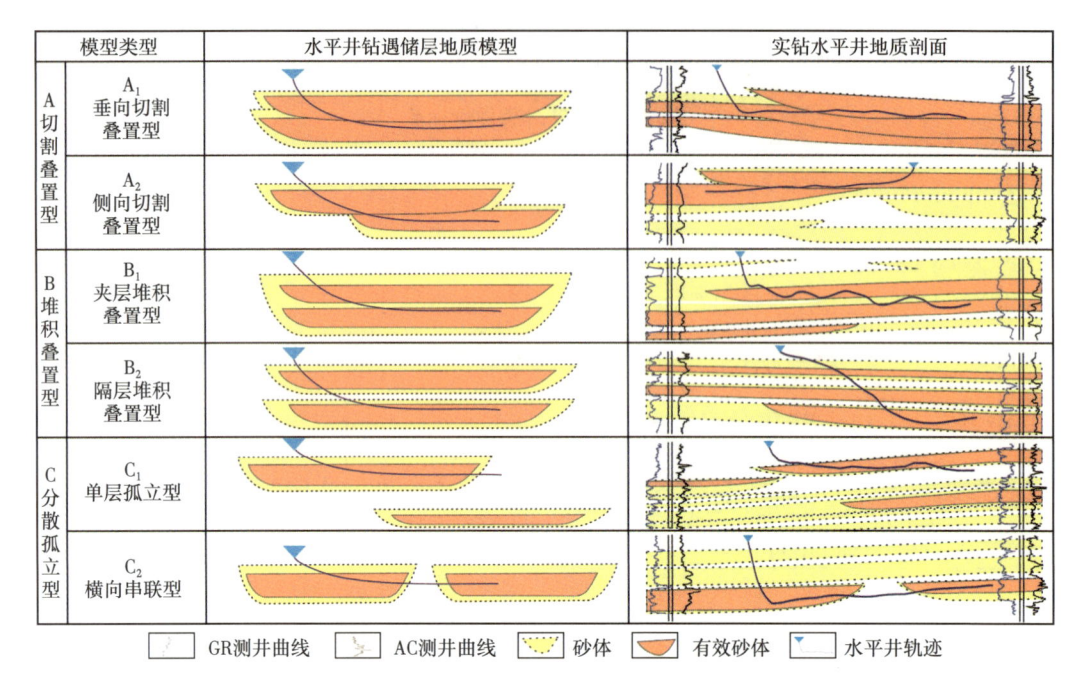

图 1 苏里格气田水平井钻遇储层地质模型

气量 $A_1$ 型 $7.1×10^4 m^3$ 和 $B_1$ 型 $6.2×10^4 m^3$，属高产模型，这两个模型可作为下一步水平井开发的主要地质目标类型。

表 2 水平井模型静动态参数分析

| 水平井模型 | 模型钻遇比例（%） | 钻遇气层长度（m） | 钻遇气层厚度（m） | 气层钻遇率（%） | 无阻流量（$10^4 m^3/d$） | 初期产量（$10^4 m^3/d$） | 平均日产气（$10^4 m^3$） | Ⅰ类井比例（%） | Ⅱ类井比例（%） | Ⅲ类井比例（%） |
|---|---|---|---|---|---|---|---|---|---|---|
| $A_1$ 型 | 34 | 826.6 | 9.1 | 82.4 | 71.2 | 8.9 | 7.1 | 62 | 38 | 0 |
| $A_2$ 型 | 9 | 682.2 | 6.7 | 67.1 | 35.6 | 3.8 | 3.1 | 25 | 20 | 55 |
| $B_1$ 型 | 24 | 723.5 | 8.5 | 74.5 | 66.9 | 7.6 | 6.2 | 48 | 52 | 0 |
| $B_2$ 型 | 15 | 567.3 | 6.6 | 60.2 | 21.9 | 3.2 | 2.5 | 0 | 21 | 79 |
| $C_1$ 型 | 11 | 471.9 | 5.2 | 49.3 | 16.2 | 2.8 | 2.2 | 0 | 0 | 100 |
| $C_2$ 型 | 7 | 431.2 | 4.3 | 48.5 | 13.3 | 2.6 | 2.1 | 0 | 0 | 100 |

相比之下 $A_2$ 侧向切割叠置型因侧向切割程度的不同导致气层累计厚度分布区间范围较大，为 3~10m，相对较高的气层钻遇率和不确定性的储层厚度导致水平井钻遇类型多样，该模型开发具有一定的风险，属风险模型。$B_2$ 隔层堆积叠置型为中等偏弱水动力条件下 2 期河道心滩砂体垂向堆积叠置而成，其间在洪水间歇期沉积了一套泥质隔层，隔层厚度分布不均，导致水平井纵向压裂无法完全沟通上下两套气层而常出现Ⅲ类水平井，该模型属低产模型。$C_1$

单层孤立型和 $C_2$ 横向串联型形成于弱水动力条件,储层连续性较差,有效砂体厚度较薄且多孤立,即使横向钻穿两个有效砂体,中间较长的洪泛平原泥岩段也导致气层钻遇率较低而多见Ⅲ类井,属低产模型,水平井开发过程中应该尽量避免钻遇该两种模型。

## 3 水平井开发有利模型所处相带优选

### 3.1 水平井开发辫状河体系相带划分

苏里格气田储层沉积类型为大型辫状河复合砂体,前人[12]从储层分级构型的角度确定二级构型河道复合体为井型井网部署研究基本构型单元,相当于地层单位砂组级,苏里格气田含气层段包含盒 8 上亚段、盒 8 下亚段、山 1 段 3 个砂组。应用储层构型层次分析方法[13],将苏里格辫状河道复合体按砂组内砂岩百分含量及储层叠置样式的不同划分为叠置带、过渡带和体系间洼地 3 个主要岩相带(图 2)。

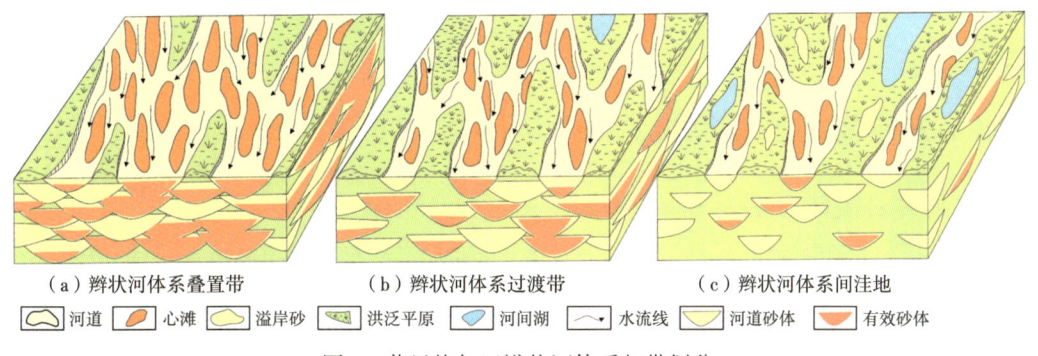

图 2 苏里格气田辫状河体系相带划分

各岩相带具体储层地质特征为:辫状河体系叠置带砂岩百分含量大于 70%,形成于古地貌地势最低洼处,为持续性河流发育部位,河道砂体纵横向不断切割叠置导致储层空间分布较为集中,累计气层厚度常大于 10m。过渡带砂岩百分含量介于 30%~70%,位于古地貌中等低洼处,河流间歇性发育导致河道砂岩与洪泛平原间泥岩交互出现,纵向叠置气层间隔夹层发育。体系间洼地砂岩百分含量小于 30%,洪水期有河流发育沉积砂岩,空间呈"泥包砂"状态,有效砂体呈孤立分散状分布于泥岩中。

### 3.2 水平井开发辫状河体系相带优选

苏里格密井网区砂体及有效砂体解剖(图 3)显示高产模型 $A_1$ 垂向切割叠置型和 $B_1$ 夹层堆积叠置型及风险模型 $A_2$ 侧向切割叠置型主要分布在辫状河体系叠置带内,优选辫状河体系叠置带作为水平井开发优势相带类型。低产模型 $B_2$ 隔层堆积叠置型、$C_1$ 单层孤立型、$C_2$ 横向串联型主要分布于辫状河体系过渡带,少部分在体系间洼地,所以过渡带和体系间洼地不适宜水平井开发,建议采用丛式井和直井开发。苏里格气田叠置带主要发育的地层层位为盒 8 下亚段砂组,确定盒 8 下亚段砂组作为气田下一步水平井开发的主要目标层位。

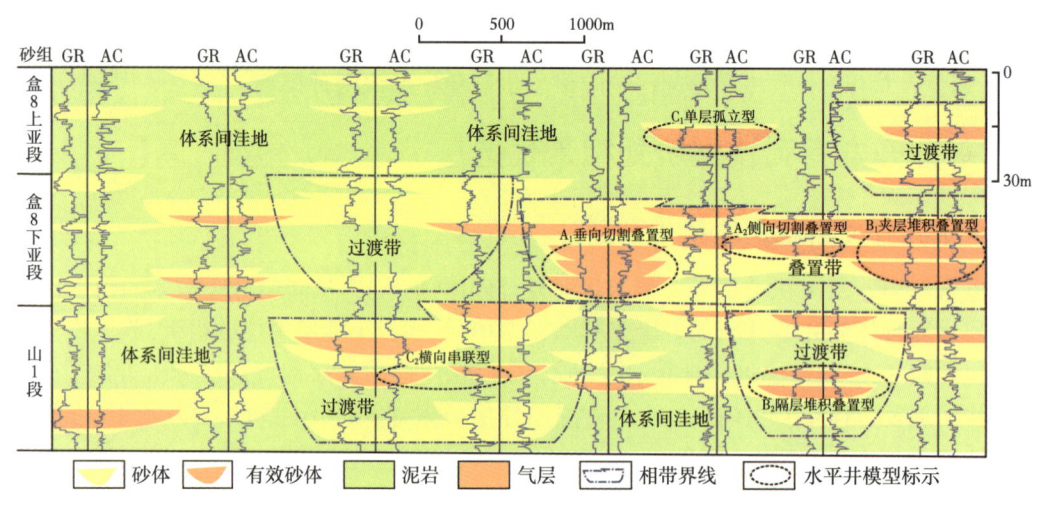

图 3 苏里格密井网区砂体及有效砂体解剖

## 4 水平井开发地质目标优选

### 4.1 水平井开发地质目标优选标准

苏里格气田目前水平井开发部署方式主要有两种：一是在产能建设投入较晚，井网密度较低的区域优选富集区采用水平井整体部署，二是对直井井网相对较为完善的已建产区采用"甜点"式优选水平井井位进行加密部署。在前述水平井分类评价标准建立的基础上，综合高产水平井钻遇储层地质模型静动态参数和水平井开发相带优选结果，选取储层地质、生产动态、储量丰度、井网密度等关键参数，从整体开发和"甜点"式开发两个思路分别建立了苏里格气田水平井整体开发区块优选标准（表3）和水平井"甜点"式开发井位优选标准（表4），从而为气田下一步规模化应用水平井进行开发提供指导。

表 3 水平井整体开发区块优选标准

| 类型 | 参数 | | 指标 |
|---|---|---|---|
| 区块储层地质 | 地震含气性检测（如异常振幅属性、AVO 等） | | 含气有利区 |
| | 主力层砂岩 | 平均钻遇砂岩厚度 | >20m |
| | | 平均钻遇泥岩隔层单层厚度 | <3m |
| | 主力层有效砂岩 | 平均有效砂岩单层厚度 | >6m |
| | | 平均累计气层厚度 | >10m |
| | | 纵向分布 | 集中分布 |
| 生产动态 | 区块内直井 I+II 类井比例 | | >75% |
| 储量丰度 | 主力层储量控制程度 | | >60% |

**表 4　水平井甜点式开发井位优选标准**

| 类型 | 参数 |  | 指标 |
|---|---|---|---|
| 相邻直井储层地质 | 三维地震含气性检测(AVO,异常振幅属性等) |  | 含气有利区 |
|  | 二维地震远近道叠加测线含气响应 |  | 较好 |
|  | 邻井主力层测井曲线形态 |  | 箱形 |
|  | 主力层砂岩 | 平均砂岩厚度 | >20m,且横向分布稳定 |
|  |  | 平均钻遇泥岩隔层单层厚度 | <3m |
|  | 主力层有效砂岩 | 平均有效砂岩单层厚度 | >6m |
|  |  | 平均累计气层厚度 | >10m,且纵向分布集中,井间储层可对比性较好 |
| 井网密度 | 井距 |  | >600m |
|  | 排距 |  | >1600m |
| 生产动态 | 无阻流量 |  | 相对较高,试采效果较好 |
|  | 临井平均日产气量 |  | $>2×10^4 m^3/d$,且生产相对稳定 |

其中储层地质参数标准值的确定主要参考了Ⅰ+Ⅱ类水平井相邻直井静态参数和水平井开发优势相带叠置带地质特征参数的统计结果,按该标准实施首先确保所选区块和井位位于辫状河体系叠置带内,进而来保证水平井具有一定的砂岩及有效储层钻遇率,提高钻遇高产水平井比例。生产动态参数标准值的选取主要考虑了Ⅰ+Ⅱ类水平井相邻直井动态参数和叠置带内钻遇Ⅰ+Ⅱ类直井比例的统计结果,来确保水平井单井动态指标良好,并具有一定的单井控制储量,从而尽可能减少低效及无效井钻遇比例。井网密度参数标准值主要参考了前人有关苏里格井网井距研究成果[14],来确保水平段延伸方向及长度满足目前井网井距。储量丰度取值主要为了约束目标气层的垂向集中程度,通过尽可能地提高储量平面动用程度和减少纵向储量地损失来保证区块整体采收率水平。

## 4.2　水平井开发地质目标优选应用实例

按上述优选标准对盒8下亚段储量集中度较高的苏中X区块开展了水平井部署选区,研究过程首选以"盒8下亚段平面平均砂岩厚度大于20m,有效厚度大于10m,泥岩隔层厚度小于3m"为标准识别出Ⅰ—Ⅵ等6块63.22km² 叠置带区域(图4),叠置带内考虑到已完钻井井网密度情况确定Ⅰ、Ⅱ2区为水平井整体开发区,区域内按600m×1600m的井网[15]顺有效砂体走向均匀部署水平井39口,Ⅲ—Ⅳ区适合"甜点"式开发,按邻井动态情况部署7口水平井。截至目前该区已完钻并投产水平井24口,统计Ⅰ+Ⅱ类井比例比气田平均值提高20.10%为85.82%,其中Ⅰ类井平均产气量$9.3×10^4 m^3/d$,Ⅱ类井平均产气量$4.5×10^4 m^3/d$,达到临近直井的4~5倍,水平井地质目标优选部署实施效果良好。

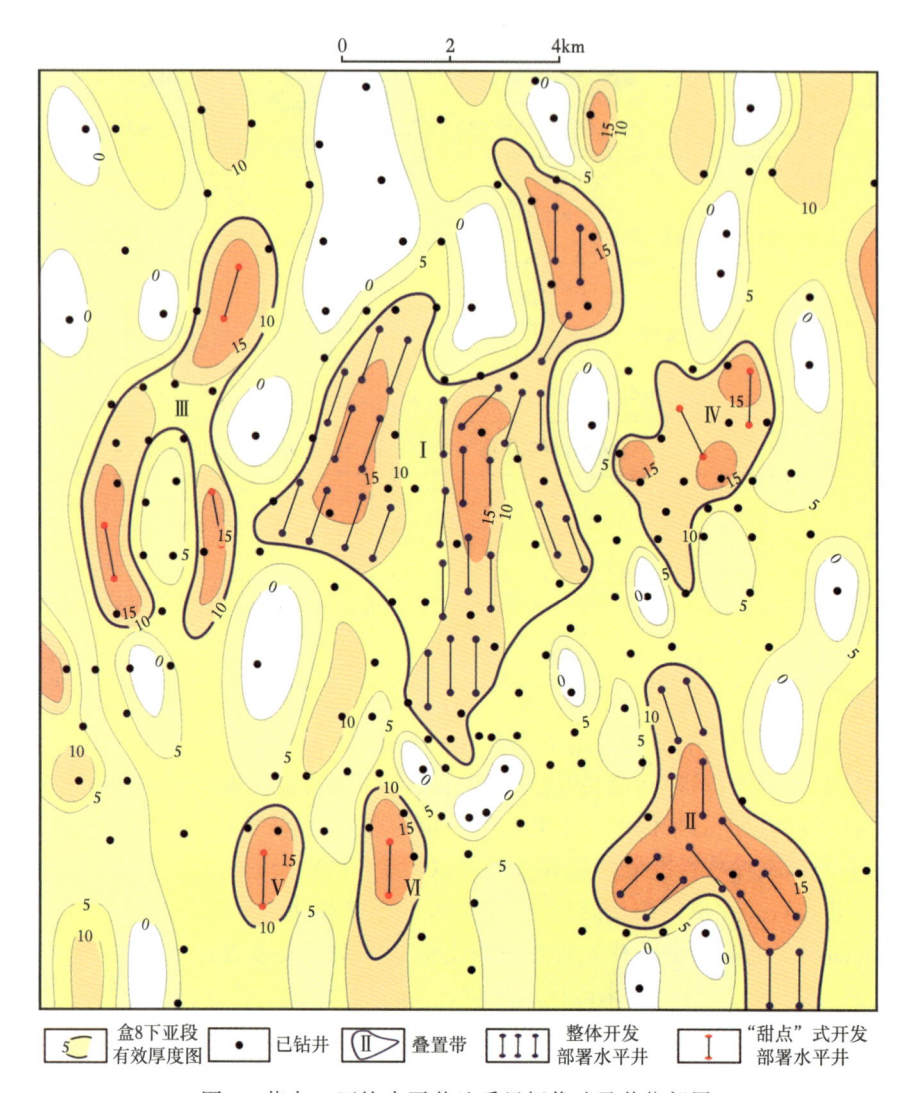

图 4 苏中 X 区块水平井地质目标优选及井位部署

## 5 结论

(1) 建立了水平井分类评价标准，分类指标包括有"水平井钻遇气层长度""临井主力层累计气层厚度""无阻流量""初期产量"等 10 项静动态参数，并按上述标准统计气田目前完钻 I+II 类水平井比例为 65.72%。

(2) 根据水平井钻遇砂体及有效砂体叠置样式的不同，建立了 3 大类 6 小类水平井钻遇储层地质模型，具体包括 $A_1$ 垂向切割叠置型、$A_2$ 侧向切割叠置型、$B_1$ 夹层堆积叠置型、$B_2$ 隔层堆积叠置型、$C_1$ 单层孤立型、$C_2$ 横向串联型，各模型静态动参数统计结果显示 $A_1$ 垂向切割叠置型和 $B_1$ 夹层堆积叠置型为高产模型，可作为下一步水平井开发的主要地质目标类型。

(3) 苏里格辫状河道复合体按砂组内砂岩百分含量及储层叠置样式的不同可细分为叠置带、过渡带和体系间洼地三个主要岩相带，其中叠置带是高产水平井模型的主要分布相带，是水平井开发的优势相带类型。

（4）从储层地质、生产动态、储量丰度、井网密度等方面提出了水平井整体开发区块优选标准和水平井"甜点"式开发井位优选标准，并按上述标准在苏中X区块优选了63.22km² 水平井开发有利目标区，并部署了39口整体开发井和7口"甜点"式开发井，目前完钻水平井显示水平井部署实施效果较好。

## 参 考 文 献

[1] 贾承造,郑民,张永峰.中国非常规油气资源与勘探开发前景[J].石油勘探与开发,2012,39(2):129-136.

[2] 杨华,付金华,刘新社,等.苏里格大型致密砂岩气藏形成条件及勘探技术[J].石油学报,2012,33(增刊1):27-35.

[3] 何光怀,李进步,王继平,等.苏里格气田开发技术新进展及展望[J].天然气工业,2011,31(2):12-16.

[4] 张明禄,樊友宏,何光怀,等.长庆气区低渗透气藏开发技术新进展及攻关方向[J].天然气工业,2013,33(8):1-7.

[5] Baihly J,Grant D,Fan L,et al. Horizontal wells in tight gas sands:A methodology for risk management to maximize success[R]. SPE110067,2007.

[6] Bagherian B,Sarmadivaleh M,Ghalambor A,et al. Optimization of multiple-fractured horizontal tight gas well[R]. SPE 127899,2010.

[7] 李易隆,贾爱林,何东博.致密砂岩有效储层形成的控制因素[J].石油学报,2013,34(1):71-82.

[8] 余淑明,刘艳侠,武力超,等.低渗透气藏水平井开发技术难点及攻关建议——以鄂尔多斯盆地为例[J].天然气工业,2013,33(1):54-60.

[9] 卢涛,张吉,李跃刚,等.苏里格气田致密砂岩气藏水平井开发技术及展望[J].天然气工业,2013,33(8):38-43.

[10] 费世祥,王东旭,林刚,等.致密砂岩气藏水平井整体开发关键地质技术——以苏里格气田苏东南区为例[J].天然气地球科学,2014,25(10):1620-1629.

[11] 位云生,贾爱林,何东博,等.苏里格气田致密气藏水平井指标分类评价及思考[J].天然气工业,2013,33(7):47-51.

[12] 何东博,贾爱林,冀光,等.苏里格大型致密砂岩气田开发井型井网技术[J].石油勘探与开发,2013,40(1):79-89.

[13] Miall A D. Reservoir heterogeneities in fluvial sandstone:Lessons from outcrop studies[J]. AAPG,1988,72(6):682-697.

[14] 何东博,王丽娟,冀光,等.苏里格致密砂岩气田开发井距优化[J].石油勘探与开发,2012,39(4):458-464.

[15] 位云生,何东博,冀光,等.苏里格型致密砂岩气藏水平井长度优化[J].天然气地球科学,2012,23(4):775-779.

# 苏里格大型致密砂岩气田储层结构与水平井提高采收率对策

唐海发 吕志凯 刘群明 位云生 王国亭

(中国石油勘探开发研究院)

**摘要**：苏里格气田是中国储量和产量规模最大的致密砂岩气田，具有储层致密、非均质性强、储量丰度低、单井产能低等特点，水平井开发技术因其可以显著提高单井产量而得到了规模化应用，实现了单井产量是直井产量3倍的开发效果，但是在提高气藏采收率方面，仍存在诸多争议。基于此，文章在辫状河体系及储层结构特征研究基础上，提出了剖面储量集中度的概念，建立了单期厚层块状型、多期垂向叠置泛连通型、多期分散局部连通型三种砂体分布模式，探讨了不同储层结构下的水平井采出程度，提出了水平井提高采收率技术对策。研究结果表明，辫状河沉积体系复合有效砂体由于"阻流带"的存在，直井动用不完善，水平井能克服"阻流带"的影响，提高层内储量动用程度；但由于砂体多层状分散分布，水平井开发会导致纵向含气层系储量动用不充分，影响层间采出程度。对于剖面储量集中度大于60%的单期厚层块状型、多期垂向叠置泛连通型储层，采用水平井整体开发，Ⅰ+Ⅱ类井比例达70%以上，可显著提高储量动用程度和采收率。对于剖面储量集中度小于60%的多期分散局部连通型储层，采用直井井网开发后进行甜点式优选水平井井位加密部署，可提高采收率10%以上。

**关键词**：苏里格气田；致密砂岩；储层结构；水平井；采收率

鄂尔多斯盆地苏里格气田具有储层致密、非均质性强、低压、低产等特征，是目前中国发现并投入开发的储量和产能规模最大的天然气田，也是中国致密砂岩气田的典型代表[1-5]。2016年，苏里格致密砂岩气产量已达$230×10^8 m^3$，占同期中国天然气总产量的近1/5，对缓解中国天然气供需紧张局面作出了重要贡献。经过近15年的探索和实践，苏里格气田从直井发展到丛式井、水平井等多种井型，形成了直井分层压裂和水平井多段压裂工艺技术，在油气资源品位日趋劣质化的现状下，有力地支撑了苏里格气田的规模上产。目前气田已进入稳产阶段[6]，提高气藏采收率、最大限度地延长气田稳产时间成为现阶段气田开发的重点和难点。

利用水平井开发致密砂岩气是有效解放储层、提高单井产量的重要手段[7]。目前苏里格已投产水平井1000余口，其产量占气田总产量30%以上，水平井开发成效显著。但随着气田继续深入开发，储量品质越来越差，自2011年以来新投产水平井平均无阻流量和初期产量逐年下降，低产、低效、产水井数增多，水平井部署风险越来越大[8]。同时，水平井开发究竟提高了气藏采收率，还是采气速度，目前尚无定论。因此，总结苏里格气田储层结构特征，分析水平井提高气藏采收率机理，建立水平井适应的砂体分布模式，提出水平井优化部署建议，对苏里格气田水平井开发部署、提高采收率和长期稳产具有重要指导意义。

# 1 苏里格辫状河体系沉积特征

## 1.1 气田地质概况

苏里格气田主体位于鄂尔多斯市乌审旗境内,区域构造属于鄂尔多斯盆地伊陕斜坡,勘探面积约 $5×10^4 km^2$,埋藏深度主要为 3000~3600m,主要产层为二叠系盒 8 段—山 1 段。主力储层盒 8 属于宽缓背景下的辫状河沉积,单期河道 0.2~1.0km。沉积盆地北部物源供给充足,砂体延伸远,横向展布宽,砂体展布面积超过 $4×10^4 km^2$,这些特征构成了大型岩性气藏的形成基础[2]。目前钻井揭示,数万平方千米气藏范围内具有整体含气的特征,气藏主体不含水,没有明显的气藏边界,具有"连续型油气聚集"的气藏分布特征[9]。但气层厚度较薄,砂岩厚度30~50m,主力气层厚度约 10m;地质储量丰度一般为 $(0.5~2.0)×10^8 m^3/km^2$,局部发育"甜点"。

苏里格气田井网加密区储层精细研究表明,气田主力含气砂体主要为辫状河心滩微相沉积砂体,单个砂体规模较小,厚度主要为 2~5m,宽度主要为 200~400m,长度 600~800m;同一小层内,心滩钻遇率为 10%~40%,心滩砂体占总面积的 10%~40%。虽然主力含气砂体呈多层状分散分布,但将多层主力含气砂体投影叠置后,可覆盖近100%的气田面积,所以苏里格气田具有整体含气的特征[10]。

## 1.2 辫状河体系沉积特征

苏里格辫状河沉积体系的形成是地质历史时期物源、水动力、古地形、可容纳空间及沉积物供给等地质因素共同作用的结果[10]。按照其空间演化所表现出的区域性差异,可分为辫状河体系叠置带、辫状河体系过渡带和辫状河体系间三个相带。不同辫状河体系带成因和储层特征差异很大,从辫状河体系叠置带、过渡带到辫状河体系间,沉积水动力由强到弱,可容纳空间由大到小,沉积物岩性由粗到细,砂体叠置期次由多到少,砂体连通性和连续性由好到差[11]。主力储层辫状河体系过渡带、体系间水动力条件相对较弱,有效砂体较薄甚至不发育。叠置带内砂体通过多期叠置形成规模较大的泛连通体,储层岩石颗粒分选好,岩性纯,物性好,心滩较发育,有效砂体分布相对集中,为水平井部署提供了较有利的地质条件。不同相带在平面上的演化具有一定的规律性。整体上,辫状河体系主砂体由南向北呈条带性展布;中区砂体较东区相对更为发育;盒 8 上亚段辫状河体系叠置带发育较差,盒 8 下亚段发育最好(尤其是中区北部的苏 10 区块),山 1 段在中区苏 6、苏 14 区块局部发育较好(图1)。

垂向上,辫状河体系叠置带的发育程度由北向南逐渐变差(图2)。西区与中区的发育程度整体上高于东区,表现在叠置砂体厚度大,水动力强度大,砂体相对均匀等。不同开发区块的砂体结构特征具有明显的差异,苏 10、苏 11 区块以厚层块状叠置砂体发育为主。苏里格气田不同期次的辫状河砂体叠加形成多套砂体,控制着有效储层纵向上的多层分布格局。

(a) 盒8上亚段砂体厚度图　(b) 盒8下亚段砂体厚度图　(c) 山1段砂体厚度图

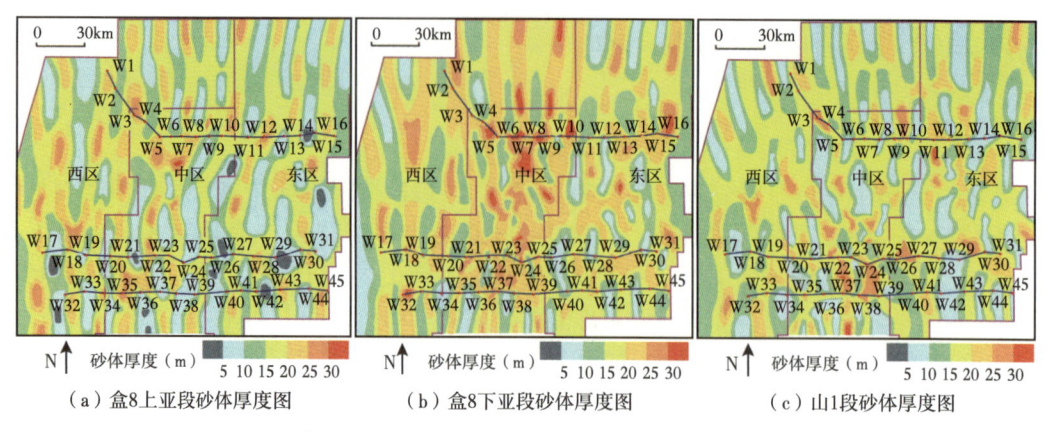

图1　苏里格盒8上亚段、盒8下亚段、山1段砂体厚度图

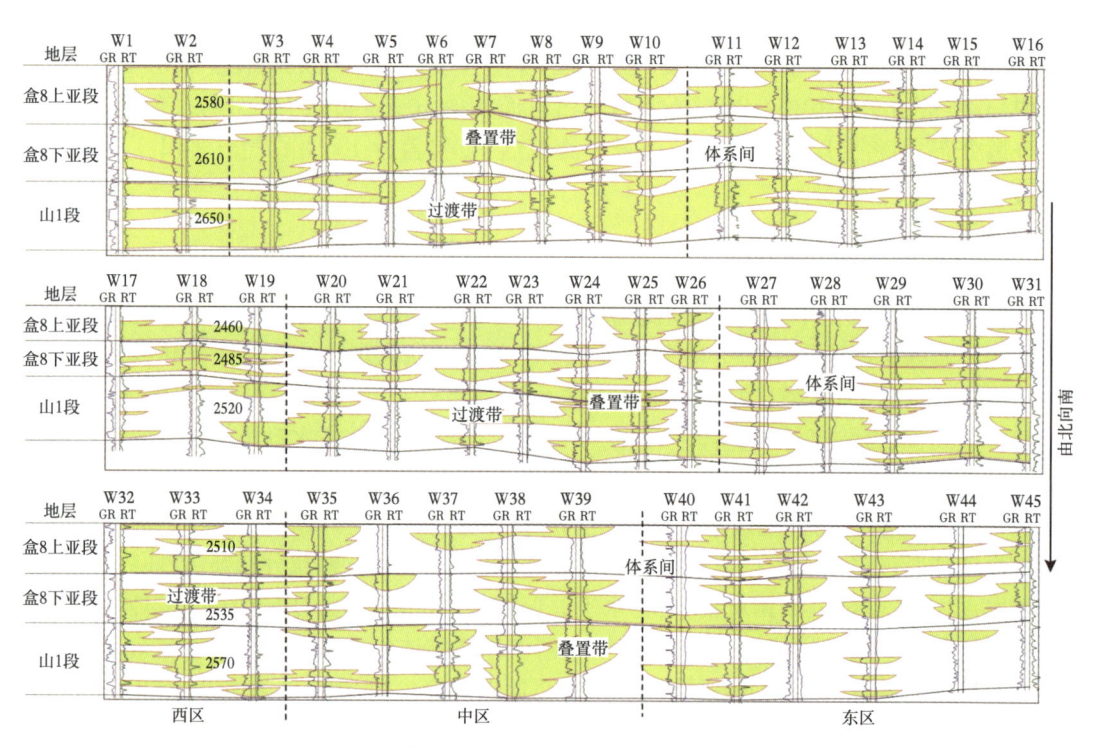

图2　苏里格气田由北向南砂体结构特征

## 2　砂体分布模式及地质特征

对于大型致密砂岩气藏,在集群化水平井井位部署之前,首先从地质背景、体系类型和沉积相带识别入手进行层段复合砂体描述;其次在沉积微相,砂体叠置类型和主河道识别的基础上刻画小层复合砂体;进而进行单砂体划分、单砂体规模表征来定性描述单砂体规模,最后查明剩余储量分布,为水平井井位部署提供科学依据[6]。苏里格气田目的层为大型辫状河沉

积,多期次辫状河河道的频繁迁移与切割叠置作用,使得含气砂体多以小规模孤立状分布在垂向多个层段中。为了研究不同储层结构下的水平井开发效果,基于密井网区地质解剖,定义剖面储量集中度的概念,即最大砂层泛连通体有效厚度与剖面有效储层总厚度的比值。

$$剖面储量集中度 = \frac{最大砂层泛连通体有效厚度}{剖面有效储层总厚度}$$

根据有效储层垂向剖面的集中程度及砂体分布特点,建立了覆盖全区的三种砂体分布模式,按形成时水动力条件由强到弱分别为单期厚层块状型、多期垂向叠置泛连通型、多期分散局部连通型(图3)。

(a)单期厚层块状型　　(b)多期垂向叠置泛连通型　　(c)多期分散局部连通型

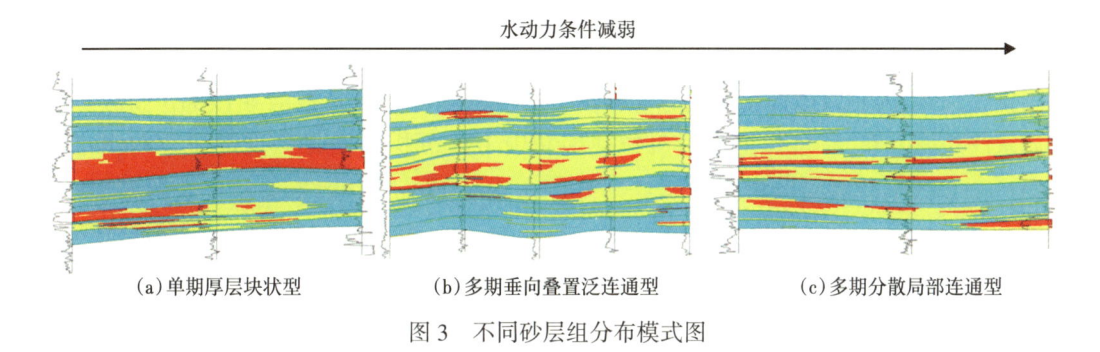

图3　不同砂层组分布模式图

## 2.1　单期厚层块状型

剖面储量集中度大于75%,主力层系有效砂岩主要集中在某一个砂层组内(图3a),有效砂岩纵向切割叠置,累计厚度一般超过8m,中间无或少有物性和泥质夹层,有效砂岩横向可对比性较好。通过密井网井组JZ1精细解剖发现,有效砂岩纵向主要集中在盒8下亚段砂层组2小层内,盒8上亚段砂层组不含气,仅在山西组2小层内可见较薄气层存在,为典型的单期厚层孤立型(图4)。该井组地质储量分布高度集中,有效砂体单层厚度大,主力层绝对突出,储

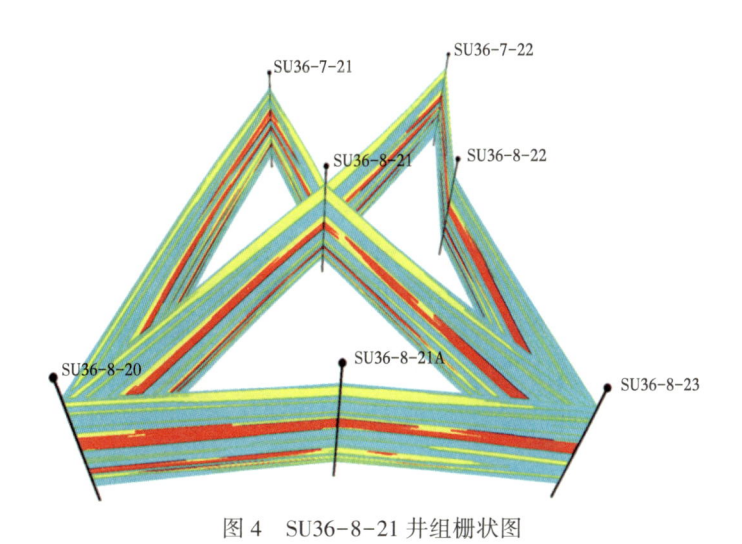

图4　SU36-8-21井组栅状图

量占比80.09%。推测形成原因为持续强水动力条件仅在盒8下亚段2小层砂体沉积时出现，其他时期水动力条件都相对较弱不足以形成粗粒岩相。

## 2.2 多期垂向叠置泛连通型

剖面储量集中度在60%~75%，主力层系有效砂岩集中在两个或多个砂层组内（图3b），主力层系砂层组间砂岩纵横向相互切割叠置形成叠置泛连通体砂岩。有效砂岩在泛连通体内呈多层分布，叠置方式多呈堆积叠置和切割叠置出现，单层或累计厚度一般在5~8m，中间多存在物性夹层，有效砂岩横向可对比性较差。通过密井网井组JZ2精细解剖发现，有效砂岩主要发育在盒8下亚段砂层组内，其中2小层更为发育，有效砂岩累计厚度平均可达8m，但在盒8上亚段砂层组1小层和山1段砂层组1小层内也发育有效砂体，二者累计厚度可达6m（图5）。该井组地质储量分布集中，主要分布在盒8下亚段两个小层，储量占比61.63%，两个砂组切割叠置，砂组内有效砂体呈多层分布特征。推测除在盒8下亚段时期该部位持续处在强水动力条件环境下外，在盒8上亚段沉积晚期和山1段沉积晚期也出现过短暂的强水动力条件。

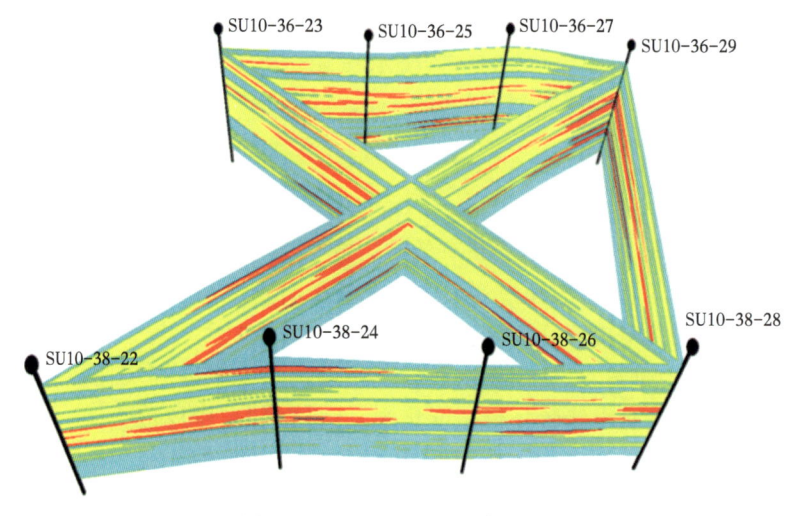

图5 SU10-36-25井组栅状图

## 2.3 多期分散局部连通型

剖面储量集中度小于60%，即纵向不发育主力层系，砂岩及有效砂岩纵向多层分布，砂岩横向局部连通，有效砂岩多为孤立状，单层厚度一般在3~5m（图3c），中间多存在泥质夹层，夹层厚度一般大于3m。通过密井网井组JZ3精细解剖发现，有效砂岩在盒8下亚段砂层组内1、2小层和山1段砂层组2、3小层都有发育，且厚度较为平均，无主力层（图6）。该井组地质储量分布比较分散，剖面上分布三套砂组，最大砂组连通体储量占比37.77%。推测该部位水动力条件变化频繁，时强时弱，无持续强水动力条件出现。

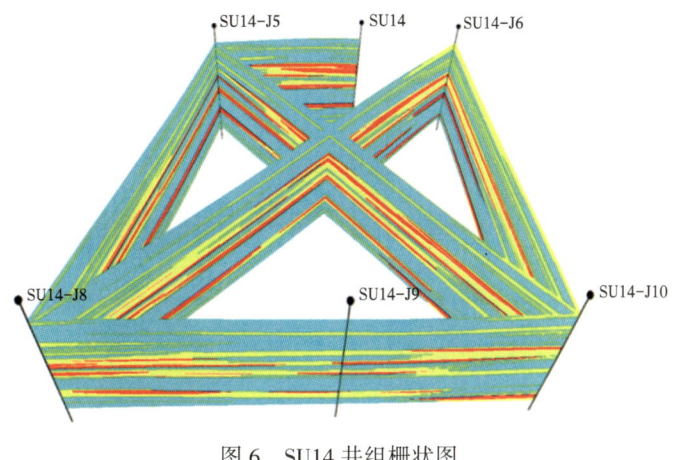

图 6　SU14 井组栅状图

## 3　不同储层结构下水平井采出程度

对于连通性好的气藏,假设一口气井可以控制整个气藏,那么井型的差别,仅仅决定了采气速度,对最终采收率影响不大。苏里格气田有效单砂体呈孤立状、多层分散分布,采用直井多层合采能充分动用纵向上多个有效单砂体;但是对于复合砂体,由于砂体内部存在着阻碍连通的"阻流带"[12-14],限制了直井的控制范围,而采用水平井开发通过分段压裂技术可以有效克服阻流带的影响,从而提高层内储量动用程度。另一方面,苏里格气田储层非均质性强,在目前技术条件下,水平井以钻遇 1 套有效砂体(单砂体或复合砂体)为主[10],使多层系含气储层储量纵向上动用不充分,层间采出程度存在进一步提高的潜力。

### 3.1　水平井层内采出程度分析

苏里格气田储层厚度较薄,且有效砂体规模小,当直井开发采收水平较低时,通过井网加密可以一定程度提高采收率,但难以确保单井经济极限累计产量。根据密井网解剖的结果,复合有效砂体规模尺度大,厚度一般在 5~10m,宽 500~1000m,长 800~1500m,地质储量一般大于 $5000 \times 10^4 m^3$。气藏工程论证结果表明[15],直井的实际泄流面积一般小于 $0.48km^2$,泄流半径一般在 200~400m,这反映出复合有效砂体内部是不连通的,存在阻流带,限制了直井的控制范围。在直井开发方式下,压裂缝东西向展布,难以克服南北两侧阻流带的影响,而使储量的动用程度不充分;水平井则可以钻穿东西向展布、南北向排列的阻流带,提高储量的动用程度(图 7)。经数值模拟计算,水平井的动态储量可以达到直井动态储量的 2~3 倍甚至更高,水平井有效控制层段的采收率可达 80%以上[16]。另外,水平井压裂可沟通垂向上未钻遇的有效砂体,提高单井控制储量和采收率。

### 3.2　水平井层间采出程度分析

水平井与分段压裂技术的组合应用,可以一定程度上提高钻遇有效砂体及邻近井筒有效砂体的动用程度,但在垂向剖面上仍会有剩余部分被较厚泥岩隔开的气层,导致垂向储量动用不充分,气藏采收率存在进一步提高的潜力。数值模拟研究表明,剖面储量集中度高(>60%)

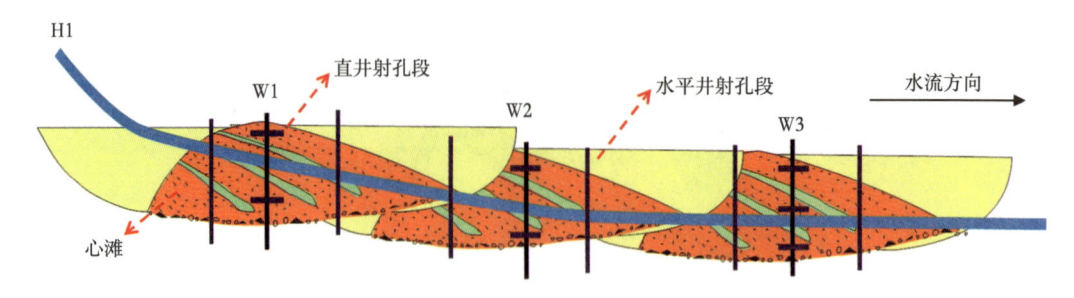

图 7　直井与水平井钻遇心滩内"阻流带"射孔示意图

的井组,由于主力层比较突出,采用水平井开发,水平井控制层段采出程度可达65%以上,层间采出程度在40%以上;剖面储量集中度低(<60%)的井组,由于剖面储量分布比较分散,采用水平井开发,平井控制层段采出程度小于60%,层间采出程度小于25%,此时采用直井密井网开发,纵向上多层合采,可获得较高的采出程度(表1)。因此,对于单期厚层块状型、多期垂向叠置泛连通型储层,剖面储量集中度高,采用水平井整体开发可大幅提高采收率;而对于多期分散局部连通型储层,剖面储量分布分散,不建议采用水平井整体开发。

表 1　不同砂体分布模式下水平井采出程度数值模拟评价结果表

| 砂组组合模式 | 模拟井组 | 剖面储量集中度(%) | 地质储量($10^8 m^3$) | 累计产气量($10^8 m^3$) | 水平井控制层段采出程度(%) | 采出程度(%) |
| --- | --- | --- | --- | --- | --- | --- |
| 单期厚层块状型 | SU36-8-21 | 80.09 | 9.48 | 5.7293 | 75.47 | 60.44 |
| 多期垂向叠置泛连通型 | SU10-36-25 | 61.63 | 8.06 | 3.3333 | 67.11 | 41.36 |
| 多期分散局部连通型 | SU14 | 37.77 | 5.56 | 1.043 | 49.67 | 18.76 |

# 4　水平井提高采收率技术对策

井型井网是致密强非均质砂岩气田采收率的主要影响因素之一[18,19],目前苏里格气田投入开发的井型主要有三种:直井、直井丛式井组、水平井。鉴于苏里格气田含气砂体小而分散、多层分布的地质特征,主要在主力气层发育好的区块应用水平井,其他区块主要采用直井或直井丛式井组开发。基于提高单井控制储量与气藏采出程度,苏里格气田水平井开发部署方式主要有两种:一是在产能建设投入较晚,井网密度较低的区域优选剖面储量集中度较高的富集区(单期厚层块状型、多期垂向叠置泛连通型)采用水平井整体部署;二是剖面储量集中度低(多期分散局部连通型)但局部储量丰度大的区域或直井井网相对较为完善的已建产区采用"甜点式"优选水平井井位进行加密部署。

## 4.1　水平井整体开发提高采收率

按照上述水平井整体开发部署原则,并参照水平井地质目标优选条件[15],对盒8下亚段储量集中度较高的苏X区块开展了水平井整体开发。建产区整体区块面积为157.17km²,含气面积145.98km²,整体区块储量175.23×$10^8 m^3$,盒8下亚段砂体厚度大,横向连续性、连通性好,有效砂体以物性夹层叠置型和切割叠置型为主(图8)。区内已钻71口直井,其中在产的投产井63口,平均单井日产气1.27×10⁴m³,平均单井控制储量2309×10⁴m³。直井已开始进入递减期,新建产能和弥补递减产能全部采用水平井。除直井动用面积外,按照1600m×600m的井排距顺着有效砂体走向均匀部署水平井82口。截至目前该区已完钻并投产水平井24口,统计Ⅰ+Ⅱ类井比例为86%,平均日产气量6.2×10⁴m³,平均单井控制储量9547×10⁴m³,达到临近直井的5倍左右,预计最终采出程度46%,较目前苏里格气田600m×800m井网条件下35%的采收率有显著提高。

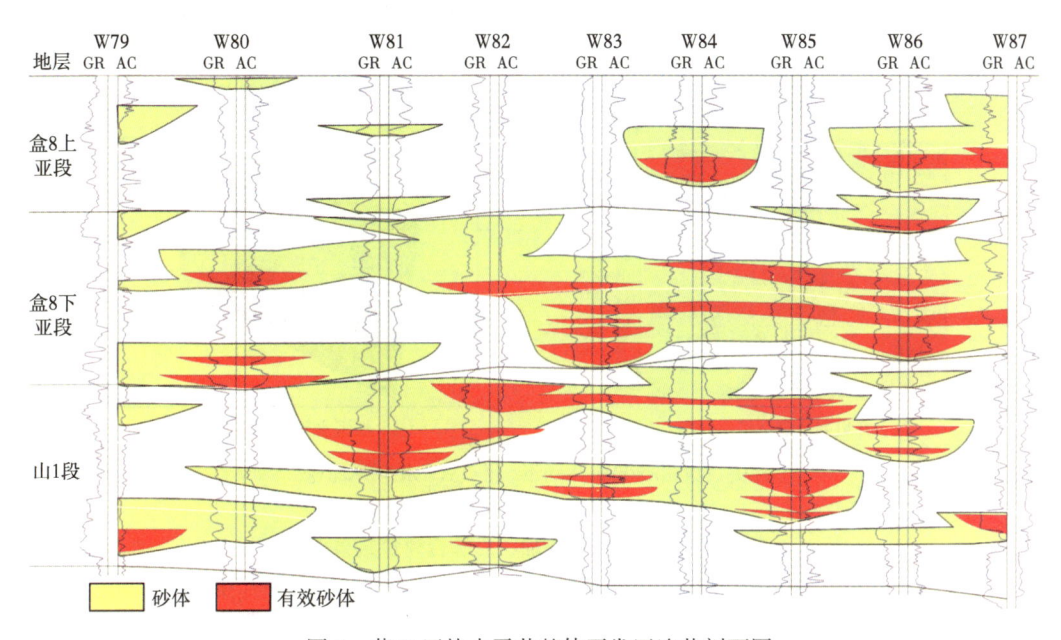

图8 苏X区块水平井整体开发区连井剖面图

## 4.2 加密水平井提高采收率

苏里格气田垂向上发育盒8上亚段、盒8下亚段和山1段3套层系,单井多具有2~6个气层,目前水平井动用的主要为盒8段储层内部单层有效厚度大于5m的单个气层,水平井开发后,非主力层和主力层内仍剩余部分储量且分布高度分散,后续开发难度更大。对于剖面储量集中度低的多期分散局部连通型新开发区块,受限于目前常规水平井井型和气藏非均质性,水平井整体部署后造成井间储量动用盲区。因此,针对该型储层仍采用直井平行四边形600m×800m井网开发,同时探索缩小排距为500m×650m以进一步提高采收率[20]。现有开发井网不宜规模采用水平井加密,但在局部剩余储量较高的区域在储层精细描述的基础上,可选水平井加密方式提高采收率。

苏里格气田水平井水平段长度一般设计为1000m,为了提高有效储层钻遇率,水平段方位为近南北向,或顺着有效砂体的展布方向,相当于部署在直井600m×800m井网的长轴对角线上。针对不同井组,通过精细地质解剖,选择局部剩余储量较高的区域加密部署水平井(图9)。数值模拟表明,通过"甜点式"水平井加密,三个井组的最终采出程度分别提高了14.1%、17.6%和19.9%(表2)。

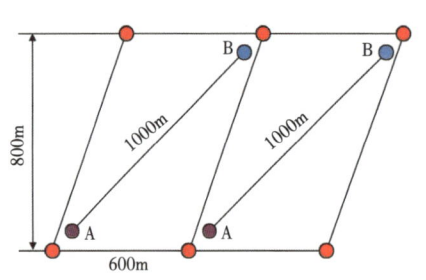

（a）600m×800m直井井网与1000m水平段的水平井位示意图

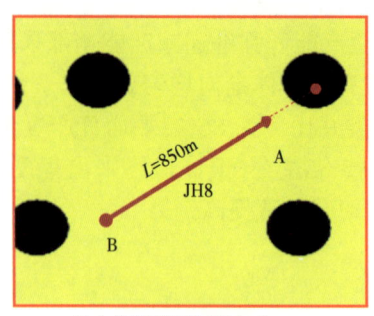

（b）实际井组加密方式

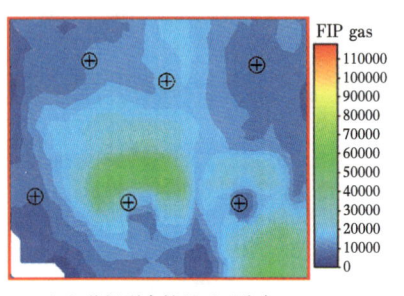

（c）井组剩余储量平面分布

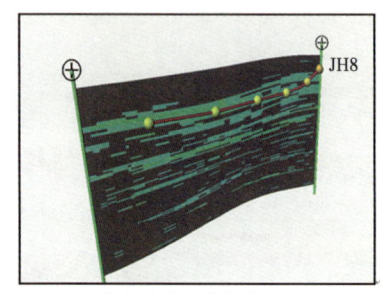

（d）加密水平井轨迹设计

图9　600m×800m井网下"甜点式"水平井加密

表2　三个井组水平井加密后综合评价结果表

| 井组 | 储量丰度($10^8m^3/km^2$) 原始 | 储量丰度($10^8m^3/km^2$) 剩余 | 加密水平井 井号 | 加密水平井 开发层位 | 加密水平井 水平段长度(m) | 加密水平井 累计产量($10^4m^3$) | 最终采出程度(%) 加密前 | 最终采出程度(%) 加密后 | 采出程度增加幅度(%) |
|---|---|---|---|---|---|---|---|---|---|
| G1 | 1.724 | 0.879 | JH1 | S12 | 920 | 1853 | 29.7 | 43.8 | 14.1 |
| G2 | 2.808 | 1.295 | JH8 | H821 | 1097 | 5361 | 28 | 45.6 | 17.6 |
| G3 | 2.099 | 1.141 | JH6 | S11 | 765 | 1692 | 25.4 | 45.3 | 19.9 |

## 5　结论

(1)大型致密砂岩气藏辫状河沉积体系,含气砂体多层状分散分布,剖面储量集中度的提出有效地解决了覆盖苏里格全区的砂体分布模式的定量描述问题,为水平井提高采收率研究奠定了基础。

(2)苏里格辫状河体系复合有效砂体由于"阻流带"的存在,直井动用不完善,水平井能克服"阻流带"的影响,提高层内储量动用程度;但对于剖面储量集中度低的多期分散局部连通型储层,采用水平井开发,会导致纵向含气层系储量动用不充分,影响层间采出程度。

(3)深入不同储层结构下水平井采出程度研究,提出提高采收率技术对策:储层剖面储量集中度大于60%,采用水平井整体开发可显著提高储量动用程度和采收率;储层剖面储量集中度小于60%,在现有开发井网下局部剩余储量较高的区域,进行储层精细描述,采用"甜点式"水平井加密方式提高采收率。

（4）为进一步提高苏里格气田采收率，应继续开展储层分布模式精细化研究，在井网加密提高采收率开发试验的基础上，加强直井、丛式井、水平井优化配置论证；同时，继续攻关老井侧钻、重复改造、增压开采、排水采气等低产低效井挖潜工艺技术。

## 参 考 文 献

[1] 杨华,付金华,刘新社,等. 苏里格大型致密砂岩气藏形成条件及勘探技术[J]. 石油学报,2012,33（增刊1）：27-35.

[2] 张明禄,樊友宏,何光怀,等. 长庆气区低渗透气藏开发技术新进展及攻关方向[J]. 天然气工业,2013,33(8)：1-7.

[3] 李易隆,贾爱林,何东博. 致密砂岩有效储层形成的控制因素[J]. 石油学报,2013,34(1)：71-82.

[4] 马新华,贾爱林,谭健,等. 中国致密砂岩气开发工程技术与实践[J]. 石油勘探与开发,2012,39(5)：572-579.

[5] 卢涛,张吉,李跃刚,等. 苏里格气田致密砂岩气藏水平井开发技术及展望[J]. 天然气工业,2013,33(8)：38-43.

[6] 谭中国,卢涛,刘艳侠,等. 苏里格气田"十三五"期间提高采收率技术思路[J]. 天然气工业：2016,36(3)：30-37.

[7] 李建奇,杨志伦,陈启文,等. 苏里格气田水平井开发技术[J]. 天然气工业,2011,31(8)：60-64.

[8] 李波,贾爱林,何东博,等. 苏里格强非均质性致密气藏水平井产能评价[J]. 天然气地球科学,2015,26(3)：539-548.

[9] 邹才能,杨智,陶士振,等. 纳米油气与源储共生型油气聚集[J]. 石油勘探与开发,2012,39(1)：13-26.

[10] 何东博,贾爱林,冀光,等. 苏里格大型致密砂岩气田开发井型井网技术[J]. 石油勘探与开发,2013,40(1)：79-89.

[11] 郭智,贾爱林,何东博,等. 鄂尔多斯盆地苏里格气田辫状河体系带特征[J]. 石油与天然气地质,2016,37(2)：197-203.

[12] Miall A D. Architectural-element analysis: a new method of facies analysis applied to fluvial deposits[J]. Earth Science Reviews, 1985, 22(2):261-308.

[13] Miall A D. Reservoir heterogeneities in fluvial sandsatone: lessons from outcrop studies[J]. AAPG, 1988, 72(6):682-697.

[14] 廖保方,张为民,李列,等. 辫状河现代沉积研究与相模式[J]. 沉积学报,1998,16(1)：34-39.

[15] 罗瑞兰,雷群,范继武,等. 低渗透致密气藏压裂气井动态储量预测新方法——以苏里格气田为例,天然气工业, 2010, 30(7)：28-31.

[16] 位云生,贾爱林,何东博,等. 苏里格气田致密气藏水平井指标分类评价及思考[J]. 天然气工业,2013,33(7)：47-51.

[17] 刘群明,唐海发,冀光,等. 苏里格致密砂岩气田水平井开发地质目标优选[J]. 天然气地球科学,2016,27(7)：1360-1365.

[18] 卢涛,刘艳侠,武力超,等. 鄂尔多斯盆地苏里格气田致密砂岩气藏稳产难点与对策[J]. 天然气工业,2015, 35(6)：43-52.

[19] 余淑明,刘艳侠,武力超,等. 低渗透气藏水平井开发技术难点及攻关建议——以鄂尔多斯盆地为例[J].天然气工业,2013,33(1)：54-60.

[20] 李跃刚,徐文,肖峰,等. 基于动态特征的开发井网优化——以苏里格致密强非均质砂岩气田为例[J]. 天然气工业,2014,34(11)：56-61.

# 苏里格致密砂岩气田潜力储层特征及可动用性评价

王国亭 贾爱林 闫海军 郭 智 孟德伟 程立华

(中国石油勘探开发研究院)

**摘要**:苏里格气田开发已全面进入稳产阶段,后备优质储量资源相对不足,对低于物性下限标准的储层进行潜力储层筛选并评价其开发可动用性,具有重要意义。面对气田致密砂岩储层分类不尽完善的问题,继承性开评展了储层系统性综合评价,将储层划分成5大类6种类型,并明确Ⅲ$_2$类是未来最具有开发潜力的储层类型,其内蕴含约 $2.90 \times 10^{12} m^3$ 天然气资源。潜力储层形成主控于同砂岩粒度密切相关的储层物性,即发育于高能心滩—辫状河道微相中上部、中—低能心滩—辫状河道微相主体及中下部物性变差的中—细粒岩相部分。总结了潜力储层的6种发育模式,提出了间接动用、组合动用和直接动用等3种开发方式。可动用性评价表明,直接和间接动用方式开发效果相对较好,潜力储层具备较好的开发前景。潜力储层特征及可动用性评价可为未来此类储层开发提供有力支撑。

**关键词**:苏里格气田;致密砂岩;储层评价;潜力储层;可动用性评价

中国致密砂岩气具有巨大的资源潜力和可观的储量规模,分布于鄂尔多斯盆地、四川盆地、松辽盆地、塔里木盆地等沉积盆地。位于鄂尔多斯盆地的苏里格气田是中国目前储量和产量规模皆最大的致密砂岩气田[1,2],累计探明储量达 $4.21 \times 10^8 m^3$,年产量超过 $230 \times 10^8 m^3$。"十一五"、"十二五"期间,气田探明储量的提交以孔隙度5%、渗透率0.1mD、含气饱和度50%为物性下限,当前开发动用的重点目标也是针对高于下限标准的有效砂体,即"甜点"进行的[3]。

经过10余年的开发建设,苏里格气田经历了开发早期评价、规模效益开发和快速上产等3个阶段后,目前已全面进入稳产期。致密气田主体采用衰竭式开发,由于气井产量递减快,稳产期间每年需新动大量储量以弥补递减,受地层水、储层品质变差和地面保护区等因素的影响,剩余可开发动用的优质储量不断减少,后备优质储量相对不足是气田保持长期稳产面临的关键问题[4-7]。随着"甜点"类储层的不断减少,"非甜点"类储层中蕴含的致密气资源将会逐渐受到重视,开发工艺的进步也将助推此类资源的开发。本文通过对苏里格气田低于下限标准的"非甜点"储层开展评价,优选未来有开发潜力的储层并进行可动用性分析,以期为气田可动资源基础的扩大和长期稳产提供支撑。

## 1 基本地质特征

气田主体位于鄂尔多斯市乌审旗境内,区域构造属于伊陕斜坡,勘探面积约 $5 \times 10^4 km^2$,主要目的层段为二叠系下石河子组盒8段和山西组山1段。随着开发的持续深入,针对气田的储层地质认识逐渐清晰明确[8-16],总体可概括为以下3个方面:

## 1.1 砂岩致密且普遍含气

储层总体致密,砂岩厚度大且表现为普遍含气特征,优质"甜点类"有效砂体呈"透镜状"包裹于厚层的低于下限标准的"非甜点"基质储层内。实验分析表明,85%以上样品的覆压渗透率小于0.1mD,按致密气国家评价标准(GB/T 30501—2014),苏里格气田整体属致密砂岩气田。目的层段沉积期,盆地北部因发育多个大型辫状河水系而形成广泛分布的厚层辫状河复合砂体,受普遍发育的煤系烃源岩影响砂体具有普遍含气特征,砂体富气程度与其物性呈正相关关系。有效砂体为辫状河心滩及河道沉积微相中物性较好的薄层砂岩部分,其含气饱和度高于厚层围岩砂体。

## 1.2 有效砂体规模小且空间多层叠置分布

有效砂体规模较小,横向连通性差,多层发育的有效砂体空间上呈非连通性叠置,平面上表现出广泛分布的特征。密井网精细解剖、野外露头及干扰试井等综合分析表明,苏里格气田有效砂体规模相对较小,有效单砂体厚、宽、长的主要范围分别为 1~5m、200~500m、400~700m,其中长小于700m、宽小于500m的比例达70%。单井可钻遇2~4个有效单砂体,井间多不连通,呈多层孤立状分散于三维空间,平面上有效砂体多期叠合分布,表现出大面积连片分布特征,展布面积可达上万平方千米。

## 1.3 储量丰度低且平面差异性大

优质储量集中分布于气田中部区域,东部与南部区因储层物性变差储量品质降低,西部及北部局部区储量受水影响严重。气田探明储量丰度范围为 $(0.5~2.6)×10^8m^3/km^2$,储量品质平面分布不均。目前主要选取丰度大于 $1.4×10^8m^3/km^2$ 的富集区进行效益开发,富集区呈连片状分布于气田中部区,东区与南区呈分散状分布,而西区约50%的气井受水影响严重,总体为富水储量区,难以有效开发。

# 2 储层系统划分与潜力储层筛选

## 2.1 储层系统划分

先前苏里格气田致密砂岩研究的重点侧重于有效砂体,对低于物性下限的储层未开展相关研究,因此储层分类并不完善,难以满足气田未来深入开发需求。基于有效砂体研究取得的成果认识,继承性开展苏里格气田致密砂岩储层系统性评价,重点剖析低于下限标准的基质类储层。

基于10789块样品的物性实验数据,对气田储层的常压孔、渗关系进行了系统分析。分析发现,常压孔隙度2.5%和5%两处对应的常压渗透率为0.01mD和0.1mD,在孔渗半对数坐标上(2.5%,0.01mD)和(5.0%,0.1mD)处表现出两个明显转折拐点,孔渗关系呈现3段式特征(图1)。孔渗关系的阶段性差异变化是储层储集、渗流性能差异的综合体现,可作为储层系统分类的基础。基于此,将上述两拐点作为储层类型划分的关键点,(5%,0.1mD)恰是目前界定有效砂岩的下限标准点,(2.5%,0.01mD)可作为含气基质储层进一步细分的界限点。而针对高于下限标准的有效砂岩,紧密结合生产特征,将其在(10%,0.5mD)和(12%,1mD)两处进行进一步细分。综上所述,依据孔渗关系将苏里格气田储层划分为I~V等5种基本类型(表1)。

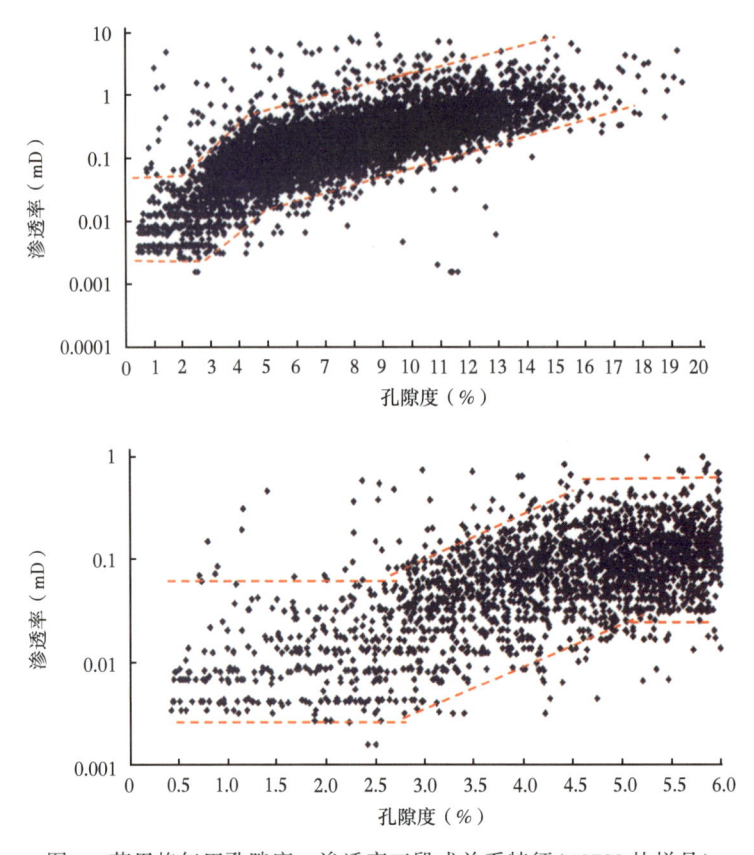

图1 苏里格气田孔隙度—渗透率三段式关系特征(10789块样品)

核磁共振和气驱水实验等综合分析表明,含气饱和度同常压渗透率具有较好正相关关系,渗透率越大含气饱和度越大,束缚水饱和度则越小。同特定渗透率值对应的含气饱和度并非为固定值,而是幅度约为20%~30%区间范围(图2)。对常压渗透率<0.01mD、0.01~0.1mD、0.1~0.5mD、0.5~1mD、>1mD的储层而言,相对应的含气饱和度平均范围分别为<25%、25%~35%、35%~60%、60%~70%、>70%。目前含气饱和度50%是提交探明储量的气饱下限标准,I、II类储层气饱和度都大于此标准,而III类储层含气饱和度范围为35%~60%,因此以气饱50%为界将III类储层进一步细划分为III$_1$和III$_2$两类(表1)。III$_1$类为富气有效砂岩部分,III$_2$类为含气基质储层部分。

高分辨率场发射扫描电镜和Naco-CT等方法分析表明,苏里格气田储层纳米级孔隙约占总孔隙空间的主体部分[17],区分苏里格气田致密砂岩储层微、纳米级孔隙的临界喉道半径为0.5μm[3]。以压汞实验分析为重要手段,对苏里格气田不同类型储层的微、纳米级孔隙的发育程度进行了探讨性分析。对常压渗透率<0.01mD、0.01~0.1mD、0.1~0.5mD、0.5~1mD、>1mD的储层而言,纳米级孔隙体积占比依次约为>98%、80%~98%、75%~80%、65%~75%、>65%(图3)。对基质储层而言,纳米级孔隙空间占总孔隙空间的绝大部分,成为储层储集性能的主要贡献者。

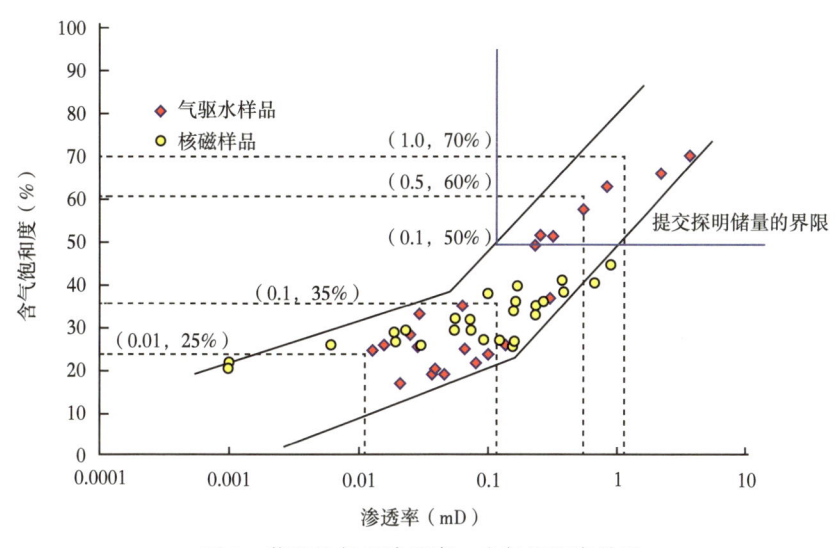

图 2 苏里格气田渗透率—含气饱和度关系

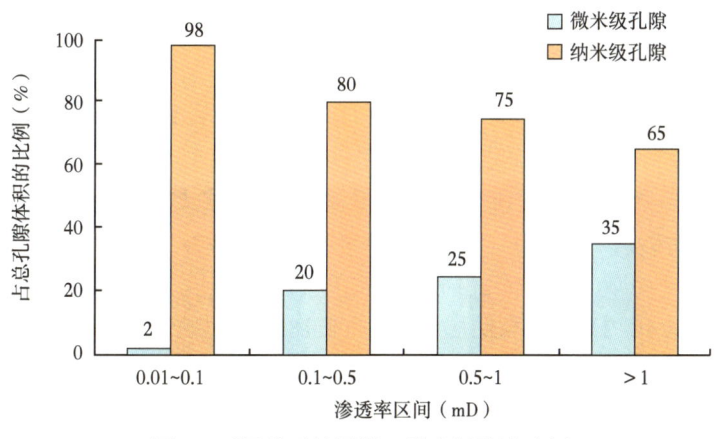

图 3 不同类型储层微—纳米级孔隙比例

在孔渗、含气饱和度、微—纳级孔隙比例等关键要素分析的基础上,进一步结合地层覆压渗透率、密度、孔隙结构和启动压力梯度等参数,最终将苏里格致密砂岩储层类型系统划分成 5 个大类 6 种类型,其中,Ⅰ、Ⅱ、Ⅲ₁ 类储层是富气有效砂体,即"甜点",Ⅲ₂、Ⅳ、Ⅴ 类储层是含气基质储层(表 1)。

表 1 苏里格气田致密砂岩储层综合系统划分

| 类型 | Ⅰ | Ⅱ | Ⅲ₁ | Ⅲ₂ | Ⅳ | Ⅴ |
|---|---|---|---|---|---|---|
| 岩石类型 | 含砾粗、粗砂岩 | 粗砂岩 | 粗及中细砂岩 | | 细、中及少量粗砂岩 | 细、粉细砂岩 |
| 常压孔隙度(%) | >12 | 12~10 | 10~5 | | 5~2.5 | <2.5 |
| 常压渗透率(mD) | >1 | 1~0.5 | 0.5~0.1 | | 0.1~0.01 | <0.01 |

续表

| 类型 | I | II | III₁ | III₂ | IV | V |
|---|---|---|---|---|---|---|
| 地层渗透率(mD) | >0.5 | 0.5~0.1 | 0.1~0.005 | 0.005~0.0001 | <0.0001 |  |
| 含气饱和度(%) | >70 | 70~60 | 60~50 | 50~35 | 35~25 | <25 |
| 密度(g/cm³) | <2.34 | 2.34~2.40 | 2.40~2.56 | 2.56~2.64 | >2.64 |  |
| 纳米级孔隙比例(%) | <65 | 65~75 | 75~80 | 80~98 | >98 |  |
| 主要喉道半径(μm) | >0.16 | 0.16~0.13 | 0.13~0.10 | 0.10~0.06 | <0.06 |  |
| 排驱压力(MPa) | <0.5 | 0.5~1 | 1~1.5 | 1.5~2.8 | >2.8 |  |
| 最大进汞饱和度(%) | >85 | 85~75 | 75~60 | 60~25 | <25 |  |
| 启动平方压力梯度(MPa²/cm) | <0.3 | 0.3~0.6 | 0.6~2.4 | 2.4~18 | >18 |  |
| 开发动用情况 | 有效储层、"甜点"目前开发动用的主体目标 |  |  | 基质类储层还未开发动用 |  |  |

## 2.2 潜力储层筛选

含气基质类储层厚度大、连通性好且分布广泛,同时蕴含丰富的天然气资源,从其中筛选出具有未来开发动用潜力的储层对气田可动储量的扩大和长期稳产具有重要意义。以苏里格气田东部4450km²区域为重点解剖对象,在储层厚度、储量丰度及储量规模等统计分析的基础上开展了潜力储层筛选。

统计表明,I+II+III₁类有效砂体均厚为8.22m,占砂岩总均厚的22.33%,储量丰度为1.11×10⁸m³/km²,占天然气总资源量为51.63%。有效储层厚度比例虽低却蕴含了大部分天然气资源。III₂、IV、V类储层目前还未开发动用,潜力储层主要从此3类储层中筛选。III₂储层均厚为10.29m,占砂岩总均厚的27.95%,储量丰度为0.77×10⁸m³/km²,占天然气总资源量的35.81%;IV类储层均厚为7.12m,占砂岩总均厚的19.34%,储量丰度为0.22×10⁸m³/km²,占天然气总资源量的10.23%;V类储层均厚为11.18m,占砂岩总均厚的30.37%,储量丰度为0.05×10⁸m³/km²,占天然气总资源量的2.32%(图4)。

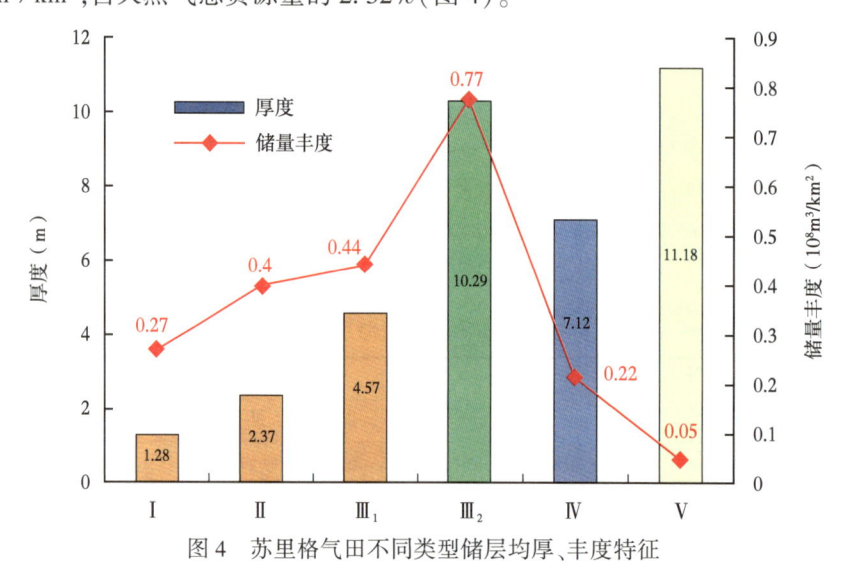

图4 苏里格气田不同类型储层均厚、丰度特征

分析表明，Ⅲ₂类储层具有厚度大、物性好、储量丰度相对较高和天然气资源相对集中的特点。目前，苏里格气田有效储层内探明储量规模为 $4.7×10^{12}m^3$，据此推算Ⅲ₂类储层内天然气资源基础巨大，蕴含约 $2.90×10^{12}m^3$ 天然气，是未来最具有开发潜力的储层类型，Ⅳ、Ⅴ类储层厚度虽大，但其内所蕴含的天然气资源量有限，难以成为未来开发潜力储层。

## 3 潜力储层形成主控因素、发育模式与动用方式

### 3.1 潜力储层形成主控因素

鄂尔多斯盆地山1段、盒8段沉积期古气候开始向热带、亚热带干旱气候转变，干旱气候下河流补给以间歇性大气降水为主，在洪水期辫状河水系搬运能力强，沉积物载荷量大，而枯水期水系搬运能力有限[18]。根据辫状河水体能量的强弱与碎屑组分构成的差异，将苏里格气田目的层段普遍发育的辫状河划分高能辫状河和中—低能辫状河：高能辫状河集中发育于洪水期，由于水体能量较强粗粒、中—细粒碎屑沉积都发育，粗粒碎屑组分主要沉积于高能心滩、高能河道微相中—下部和底部，而中—细粒组分则主要发育于上述微相中—上部和上部；中—低能辫状河主要发育于枯水期，由于水体能量偏弱，粗粒碎屑组分不发育，中—细粒沉积组分主要发育于中低能心滩、中—低能河道微相主体部位(图5)。

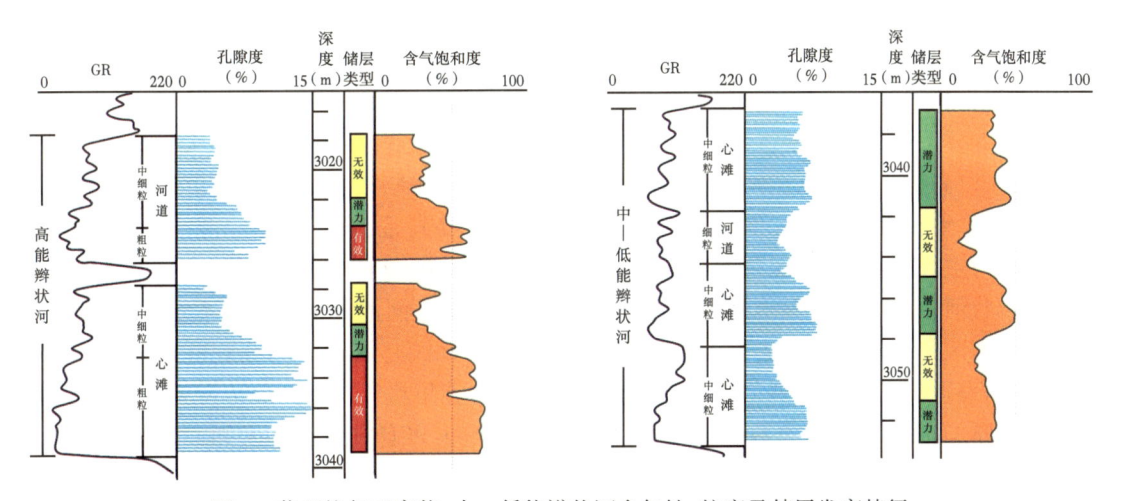

图 5 苏里格气田高能、中—低能辫状河含气性、粒度及储层发育特征

苏里格气田天然气富集受生烃强度、储层物性和微构造幅度等多种因素的综合影响[5-7,19]，目前气田西部主体及北部部分区受水影响严重，在宏观生烃强度一定的条件下，储层物性对天然气富集起着决定性影响。研究表明，苏里格气田储层物性同砂岩粒度有较好的对应关系，随着砂岩粒度的由粗变细，物性较好储层的发育比例逐渐降低，储层物性逐渐变差(图6)。有效砂体主要受控于粗粒岩相，主要发育于高能心滩中下部及高能河道底部；潜力储层发育主要受控于中细粒岩相，主要发育于高能心滩及高能河道的中上部、中—低能心滩及中低能河道的主体部位及中下部(图5)。

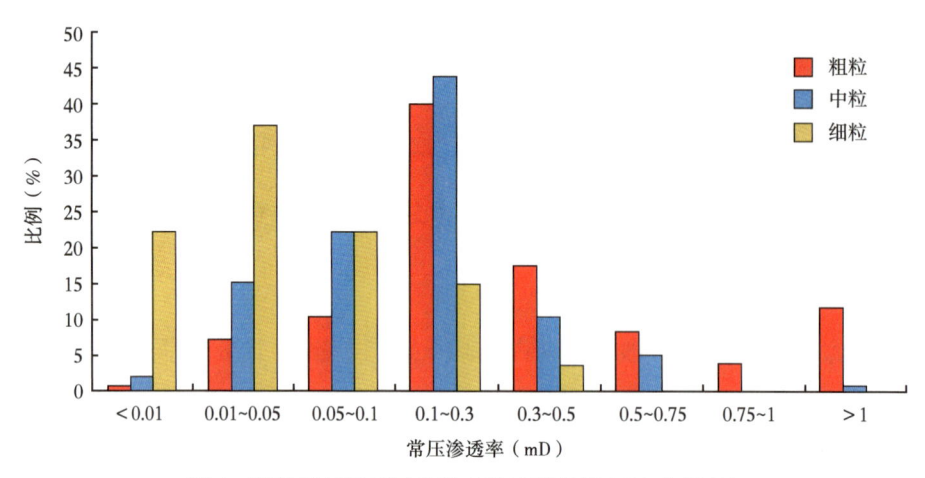

图 6　苏里格气田储层物性与粒度相关性(622 个样品)

## 3.2　潜力储层发育模式

### 3.2.1　有效砂体发育模式

受沉积物源、水动力及古地形等条件控制,苏里格气田辫状河沉积体系可分为辫状河体系叠置带、辫状河体系过渡带和辫状河体系间 3 个宏观相带单元[20]。辫状河体系根据水系能量强弱、沉积物构成、沉积微相类型等可划分为高能辫状河体系和中—低能辫状河体系(图 7)。高能辫状河体系叠置带处于古地形最低洼处,为古河道持续发育部位,砂地比大且高能辫状河集中发育,以高能心滩和高能河道等微相为主。中—低能辫状河体系叠置带也处于古地形相对低洼处,砂地比大且以中—低能辫状河集中发育为主要特征,微相以中—低能心滩、中—低能河道等为主,高能微相不发育;其过渡带砂地比变小,微相以中—低能心滩、中—低能河道等为主。辫状河体系间则以泥质沉积为主,砂地比最低,心滩、河道等富砂微相不发育。

密井网区解剖及野外露头等综合分析表明,苏里格地区有效砂体的叠置模式可划分为孤立型、垂向叠置型和侧向切割型等 3 种类型。孤立型是苏里格气田有效砂体发育的主要模式,有效砂厚 1~5m,横向展布数百米;垂向叠置型指纵向上多套有效砂体紧邻或切割叠置发育形成厚层砂体单元,累计厚度可达 5~10m,横向展布近千米;横向切割型指有效砂体侧向上切割叠置连通,有效砂体厚 3~8m,连通规模可达千米(图 7)。

### 3.2.2　潜力储层发育模式

在高能辫状河体系发育带,潜力储层多与有效砂体相伴发育,因此其发育模式同有效砂体发育紧密相关。以密井网解剖为重要手段,同时紧密结合有效砂体发育模式,总结了苏里格气田潜力储层发育的 6 种主要模式(图 7):

(1)高能辫状河孤立型:单期高能心滩或河道孤立发育,有效砂体垂向上和横向上不与其他有效砂体接触。剖面上潜力储层发育于有效砂体中上部及两侧,空间上将单个有效砂体包裹于其中。潜力储层厚较小。

(2)高能辫状河垂向叠置型:高能心滩或河道垂向上多期发育,多个有效砂体垂向切割叠置或紧密相邻。剖面上潜力储层填充于多套有效砂体之间及两侧,空间上潜力储层将多套有

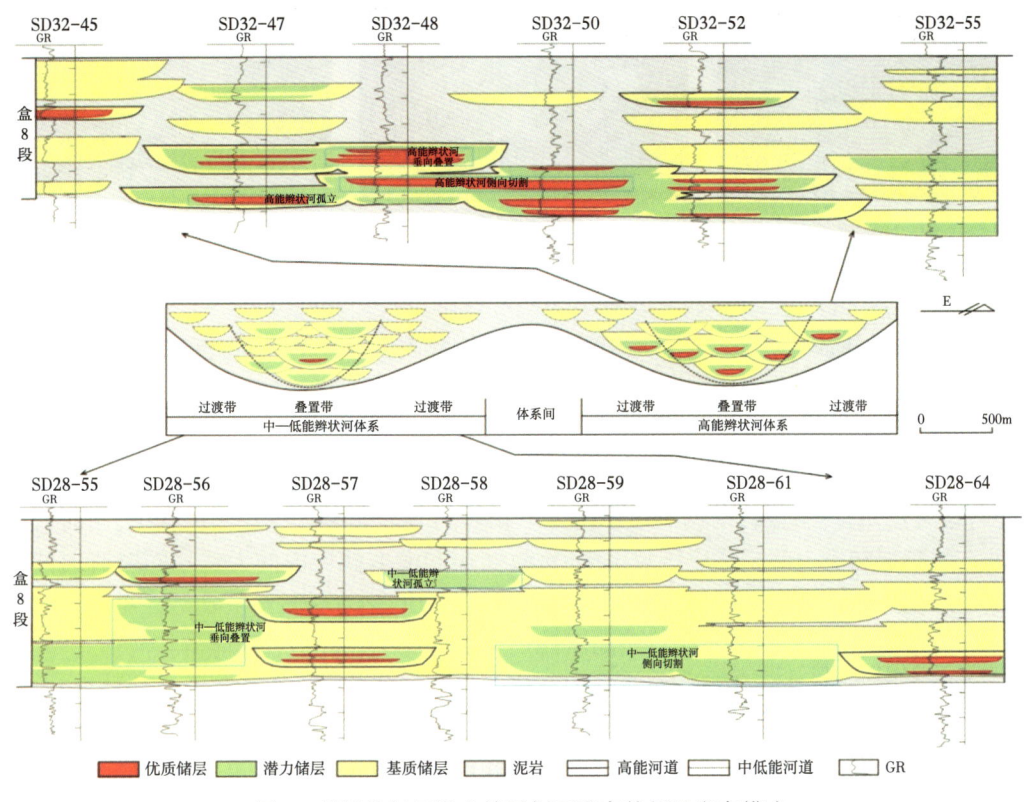

图 7 苏里格气田潜力储层剖面发育特征及发育模式

效砂体整体包裹于其中。潜力储层累计厚度大。

(3)高能辫状河侧向叠加型:高能心滩或河道侧向上多期切割,多个有效砂体侧向切割连通。剖面上潜力储层分布于侧向连通有效砂体两侧,空间上潜力储层与将有效砂体整体包裹。潜力储层厚度也较大。

(4)中—低能辫状河孤立型:中—低能心滩或河道孤立发育,无有效砂体发育,潜力储层垂、横向上不与其他潜力储层接触。潜力储层规模较小。

(5)中—低能辫状河垂向叠置型:多期中—低能心滩或河道垂向上切割叠置,无有效砂体发育,潜力储层垂向上多期切割叠置,可形成厚层连通储渗单元。

(6)中—低能辫状河侧向叠加型:中—低能心滩或河道侧向上多期切割叠加,无有效砂体发育,潜力储层侧向可形成较大规模连通储渗单元。

## 3.3 潜力气层动用方式

目前,苏里格气田开发主体目标是有效砂体,主要采取直井多层和水平井多段压裂的开发动用方式,潜力储层的开发动用还处于探索阶段。各类垂向叠置、侧向叠加型潜力储层发育厚度大、延伸范围广,是开发动用的重点类型。在前述潜力储层发育模式分析的基础上,总结3种开发动用方式(图8)。

(1)间接动用型:主要针对于分布于有效砂体外围的潜力储层。在有效砂体持续开采过

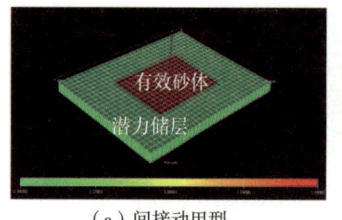

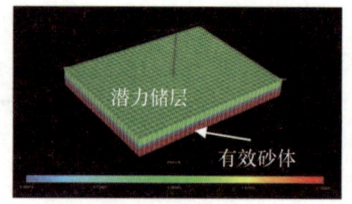

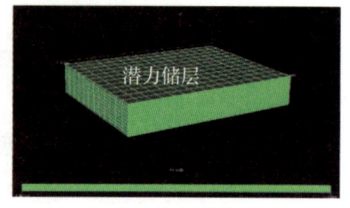

（a）间接动用型　　　　　　（b）组合动用型　　　　　　（c）组合动用型

图8　苏里格气田潜力储层开发动用模式

程中,渗流边界范围不断扩大,当扩大至潜力储层时,其内天然气便开始向有效砂体区供给,并最终通过井筒开采出来。这种动用方式也可称之为"高渗透采低渗透",即通过高渗透有效砂体的开发带动外围相对低渗透潜力储层的动用。适用于高能辫状河孤立型、高能辫状河垂向叠置型、高能辫状河侧向叠加型等模式中有效砂体外围潜力储层的开发。

（2）组合动用型:主要针对与有效砂体上下相邻的潜力储层。在开采过程中,同时对上下相邻的有效储层和潜力储层采取压裂措施,有效储层和潜力储层内的天然气同时流向井筒并开采出来。同"高渗透采低渗透"的间接动用方式相比,组合动用方式主要是针对与有效储层垂向相邻的潜力储层,而非分布于有效储层四周的潜力储层。适用于高能辫状河孤立型、高能辫状河垂向叠置型、高能辫状河侧向叠加型等模式中与有效砂体上下相邻的潜力储层开发。

（3）直接动用型:主要针对于无有效砂体发育的类型。可选择叠置层厚度较大的潜力储层,采取工艺措施进行直接动用。与前两种潜力储层的动用方式相比,直接动用模式中潜力储层的开发动用不受有效储层的影响。适用于中—低能辫状河孤立型、中—低能辫状河垂向叠置型、中—低能辫状河侧向叠加型等模式中潜力储层的开发。

## 4　潜力储层可动用性分析

### 4.1　潜力储层开发现场试验

目前潜力储层开发动用的方式主要为间接动用型和直接动用型。间接动用重点分析生产时间较久且后期表现为间歇式生产特征的气井。苏38-16-5井是气田中区苏6区块的典型气井,于2003年10月投产,气井生产历史可划分为初始生产(2003—2005年)、拟稳态生产(2005—2006年)、间歇生产(2006—2015年)3个阶段,前两个阶段天然气产自有效砂体,间歇生产中后期天然气应主要产自潜力储层。分析表明,目前有效储层累计产出天然气$4316\times10^4 m^3$,潜力储层累计产出约$455.8\times10^4 m^3$,目前产量贡献率为10.56%,表现出较好的开发效果。直接动用重点针对有效砂体不发育的潜力储层,压裂试气表明,潜力储层自身也具有较好的产气能力,如苏314井,山1段潜力储层厚度为11m、孔隙度6.26%、渗透率0.044mD、含气饱和度40.46%,试气产量为$1.32\times10^4 m^3/d$;苏364井,山1段潜力储层厚度为4m、孔隙度5.43%、渗透率0.047mD、含气饱和度38.46%,试气产量为$0.57\times10^4 m^3/d$。

### 4.2　潜力储层可动用性评价

为了系统评价3种潜力储层开发动用方式的效果,设计了系列组合条件下的单井机理预测模型。间接动用模式下有效砂体分别与5种不同物性非有效储层横向等厚度组合；组合动用模式下有效砂体分别与5种不同物性非有效储层垂向等厚度组合；直接动用模式下分别评价5种不同物性非有效储层各自的开发效果(表2)。配产方式为先定产生产,后定压生产,配产范围为$(0.1\sim1.5)\times10^4m^3/d$。以地层压力2.3MPa、经济极限产量$0.1\times10^4m^3/d$为废弃条件。

表2 3种动用模式下单井理论模型储层组合参数表

| 储层 | 孔隙度(%) | 渗透率(mD) | 气饱和度(%) | 设计厚度(m) |
| --- | --- | --- | --- | --- |
| 非有效储层 | 7.5 | 0.25 | 50 | 10 |
|  | 6 | 0.125 | 40 | 10 |
|  | 5 | 0.1 | 35 | 10 |
|  | 4 | 0.05 | 30 | 10 |
|  | 2.5 | 0.01 | 25 | 10 |
| 有效砂体 | 10 | 0.75 | 65 | 10 |

分析表明,3种动用方式下潜力储层天然气采出程度主要与其储层物性相关,物性越好采出程度越高。间接动用与组合动用相比,前者潜力储层开发效果相对较好,天然气采出程度、产量贡献率都相对较高(图9、图10)。主要因为在组合动用开发过程中,有效砂体和潜力储层自始至终都同时向井筒供给天然气,在开发过程及达到废弃条件时,受有效砂体产气能力较强的影响,潜力储层压力释放和天然气产出不彻底；而间接动用开发过程中,前期为有效砂体单独供气,后期产量则主要为潜力储层贡献,达到废弃条件时,压力释放和天然气的产出都相对彻底。直接动用方式下,产量都由潜力储层贡献,当储层物性低于0.05mD时,由于难以达到经济极限产量难以开发动用,随着储层物性变好,采出程度快速增加并迅速超过间接动用型和组合动用型。

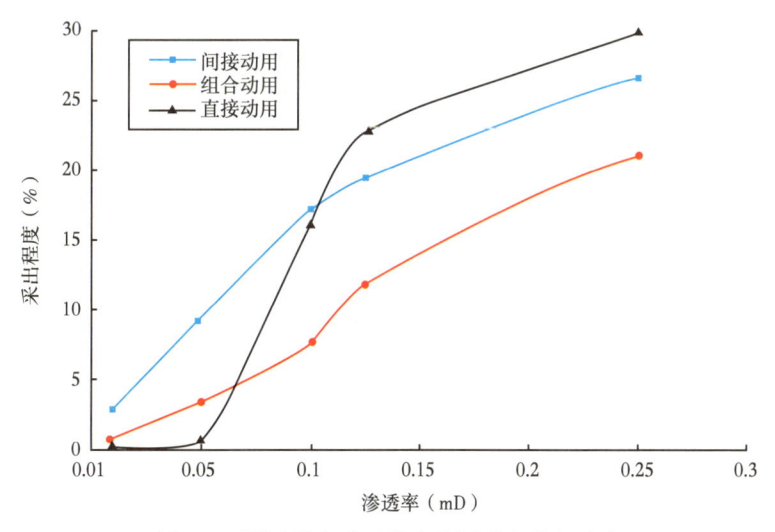

图9 不同动用方式下潜力储层采出程度对比

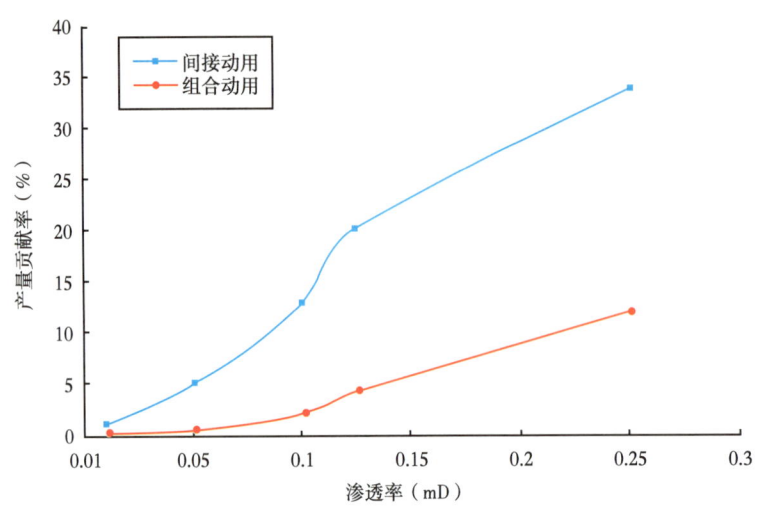

图10 不同动用方式下潜力储层产量贡献对比

比较而言,直接动用型开发效果最好,间接动用型次之,组合动用效果较差。潜力储层具备采出20%天然气蕴含量的开发前景,可有效扩大苏里格气田长期稳产的资源基础。

## 5 结论

经过多年的开发建设,苏里格气田已全面进入稳产阶段,后备优质储量资源相对不足是实现气田长期稳产面临的关键问题,需进行"非甜点"类储层中潜力储层的筛选并评价其开发可动用性。

通过系统开展苏里格气田致密砂岩储层评价,将"非有效"类储层进行细分,并结合探明储量物性下限标准,将储层划分成5大类6种类型,其中Ⅰ、Ⅱ、Ⅲ$_1$类储层为富气有效砂体,Ⅲ$_2$、Ⅳ、Ⅴ为含气基质储层。通过重点区解剖,评价了各类储层蕴含的天然气资源,分析表明Ⅲ$_2$类储层是未来最具有开发前景的潜力储层类型。

潜力储层形成主控因素为储层物性,发育于高能心滩和高能辫状河道微相中上部、中低能心滩和中—低能辫状河道微相主体及中下部位物性偏差的中细粒岩相部分。总结了潜力储层的6种发育模式,高能、中—低能辫状垂向叠置及侧向叠加型潜力储层是开发动用的重点。

针对潜力储层的地质发育特征,提出间接动用、组合动用、直接动用等3种开发方式。采用现场试验与理论模型分析相结合的方法,开展了潜力储层开发动用方式的效果评价,分析表明,直接动用型、间接动用型开发效果优于组合动用型,潜力储层具有较好的开发前景。

### 参 考 文 献

[1] 杨华,刘新社.鄂尔多斯盆地古生界煤层气勘探进展[J].石油勘探与开发,2014,41(2):129-138.
[2] 杨华,付金华,刘新社,等.苏里格大型致密砂岩气藏形成条件及勘探技术[J].石油学报,2012,31(S1):27-36.
[3] 王国亭,何东博,王少飞,等.苏里格致密砂岩储层岩石孔隙结构及储集性能特征[J].石油学报,2013,3(44):660-666.

[4] 卢涛,刘艳侠,武力超,等.鄂尔多斯盆地苏里格气田致密砂岩气藏稳产难点与对策[J].天然气工业,2014,35(6):43-52.

[5] 窦伟坦,刘新社,王涛.鄂尔多斯盆地苏里格气田地层水成因及气水分布规律[J].石油学报,2010,31(5):767-773.

[6] 王泽明,鲁保菊,段传丽,等.苏里格气田苏20区块气水分布规律[J].天然气工业,2010,30(12):37-40.

[7] 代金友,李建霆,王宝刚,等.苏里格气田西区气水分布规律及其形成机理[J].石油勘探与开发,2012,35(5):524-529.

[8] 贾爱林,唐俊伟,何东博,等.苏里格气田强非均质致密砂岩储层的地质建模[J].中国石油勘探,2007,12(1):12-16.

[9] 何东博,贾爱林,田昌炳,等.苏里格气田储集层成岩作用及有效储集层成因[J].石油勘探与开发,2004,31(3):69-71.

[10] 何东博,贾爱林,冀光,等.苏里格大型致密砂岩气田开发井型井网技术[J].石油勘探与开发,2013,40(1):79-89.

[11] 何东博,王丽娟,冀光,等.苏里格致密砂岩气田开发井距优化[J].石油勘探与开发,2012,39(4):458-464.

[12] 唐俊伟,贾爱林,何东博,等.苏里格低渗透强非均质性气田开发技术对策探讨[J].石油勘探与开发,2007,33(1):107-110.

[13] 李跃刚,徐文,肖峰,等.基于动态特征的开发井网优化——以苏里格致密强非均质砂岩气田为例[J].天然气工业,2014,34(11):56-61.

[14] 李波,贾爱林,何东博,等.苏里格气田强非均质性致密气藏水平井产能评价[J].天然气地球科学,2015,26(3):539-549.

[15] 马旭,郝瑞芬,来轩昂,等.苏里格气田致密砂岩气藏水平井体积压裂矿场试验[J].石油勘探与开发,2014,41(6):742-747.

[16] 费世祥,王东旭,林刚,等.致密砂岩气藏水平井整体开发关键地质技术——以苏里格气田苏东南区为例[J].天然气地球科学,2014,25(10):1620-1629.

[17] 邹才能,朱如凯,白斌,等.中国油气储层中纳米孔首次发现及其科学价值[J].岩石学报,2011,27(6):1857-1864.

[18] 刘锐娥,肖红平,范立勇,等.鄂尔多斯盆地二叠系"洪水成因型"辫状河三角洲沉积模式[J].石油学报,2013,34(S1):660-666.

[19] 王国亭,冀光,程立华,等.鄂尔多斯盆地苏里格气田西区气水分布主控因素[J].新疆石油地质,2012,33(6):657-659.

[20] 郭智,孙龙德,贾爱林,等.辫状河相致密砂岩气藏三维地质建模[J].石油勘探与开发,2015,42(1):76-83.

# 气藏应用篇

# Evaluation of Dynamic in Ultra-deep Naturally Fractured Tight Sandstone Gas Reservoirs

Ruilian Luo[1]　Jichen Yu[1]　Yujin Wan[1]　Xiaohua Liu[1]
Lin Zhang[1]　Qingyan Mei[2]　Yi Zhao[2]　Yingli Chen[2]

(1. RIPED, PetroChina; 2. PetroChina Southwest Oil&Gas Company)

## Abstract

Ultra-deep naturally fractured tight sandstone gas reservoirs have the characteristics of tight matrix, natural fractures development, strong heterogeneity and complex gas-water relations. There is strong uncertainty of gas reserves estimation in the early stage for such reservoirs, which brings big challenge to the development design of gas fields. Taking Keshen gas field in Tarim basin as example, during the early development stage, the dynamic reserves were much less than those of proven geologic reserves. As results, the actual production performances are obviously different from those of conceptual design. What are the reasons? How to adjust the development program of gas field? Based on special core analysis, production performance analysis, gas reservoir engineering method, and numerical simulations, influencing factors on evaluation of dynamic reserves for ultra-deep fractured tight sanstone gas reservoirs are analyzed. The results show that rock pore compressibility, recovery percent of gas reserves, gas supply capacity of matrix rock, water invasion are the major factors affecting the evaluation of dynamic reserves. On the basis of above analysis, some suggestions are given for the evaluation of dynamic reserves in Ultra-deep fractured tight sandstone gas reservoirs. For this kind of reservoirs, it is reasonable to determine the gas production scale based on dynamic reserves instead of proven geological reserves.

## 1 Introduction

With the development of petroleum exploration, deep formation and ultra-deep formation are attracting more and more attention and some new progresses have been made[1-5]. In the 21st century, a series of major breakthroughs of petroleum exploration have been made in deep formation in China, especially in the mid-western basins.

There are abundant natural gas resources in the deep formation of Kuqa depression in Tarim Basin, which is enriched by gas reservoirs group[6-9], such as Dabei, Keshen, and Bozi gas reservoirs. Among these resvervoirs, Keshen gas reservoirs have the largest proven reserves, with depth of

6500~8000m; formation pressure of 116~128MPa, formation temperature of 165~175℃, formation pressure coefficient of 1.70~1.80, matrix porosity of 2%~6%; matrix permeability of 0.01~0.1mD, and accompanied with high-angle tectonic fractures. They are typical ultra-deep fractured tight sandstone gas reservoirs. At present, this kind of gas reservoirs have no development precedent, and no experience can be used for reference.

Unlike conventional gas reservoirs, ultra-deep fractured tight sandstone gas reservoirs are characterized by large burial depth, complex structure, tight matrix, strong heterogeneity, and complicated gas-water relationship, so the early evaluation of recoverable reserves is highly uncertain. It is difficult to accurately determine the production scale of gas field based on proven geological reserves. For Keshen gas field, after Keshen-2, Keshen-8 and Keshen-9 blocks being put into production testing, the early evaluation of dynamic reserves of each block is less than half of the proven geological reserves, which lead that gas field can not be developed smoothly according to the development plan. Therefore, for this kind of gas reservoir, it is crucial to evaluate of recoverable reserves accurately in the early stage.

## 2　Influencing factors of dynamic reserves

### 2.1　Rock pore compressibility

During production, the driving energy of gas reservoir is provided by gas, formation water and rock together. The total compressibility is as follows:

$$C_t = C_p + C_g S_g + C_w S_w \tag{1}$$

Ultra-deep fractured tight sandstone gas reservoir is characterized by well-developed natural fracture and abnormal high formation pressure. For such gas reservoirs, because of undercompaction of reservoir and strong stress sensitivity of fracture, the pore compressibility $C_p$ ranges from $10\times10^{-4}$ MPa$^{-1}$ to $50\times10^{-4}$ MPa$^{-1}$, which is several times higher than that of the conventional gas reservoir. At the same time, owing to the high formation pressure in the early development stage, gas compressibility $C_g$ is in the same order of magnitude as rock pore compressibility, so the driving energy provided by rock compression plays an important role and cannot be ignored[10-11].

The original formation pressure of the Keshen gas field is up to 116~128MPa. Fig. 1 shows the curve of the ratio of pore compressibility to total compressibility under different formation pressures. It can be seen that the higher the formation pressure and the larger the pore compressibility, the higher the proportion of driving energy provided by the rock is. When formation pressure is greater than 100MPa and rock pore compressibility is more than $30\times10^{-4}$ MPa$^{-1}$, the driving energy provided by the rock is more than 50%. Therefore, for ultra-deep fractured tight sandstone gas reservoir, pore compressibility is an important parameter affecting the evaluation of dynamic reserves. Pore compressibility is influenced by the lithology, physical property, clay content, cementation degree and pore-throat structure of the reservoir[12], which has high heterogeneity.

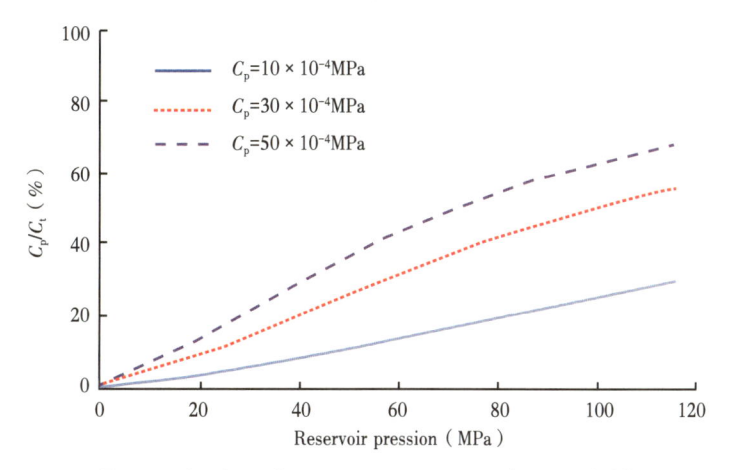

Fig. 1  $C_p/C_t$ vs. formation pression in Keshen gas field

Pore compressibility can be measured directly by experiments, but there are some uncertainties due to the limitation of the number and and representativeness of test samples. For fractured reservoirs, it is recommended to use whole core samples for testing.

The test results of 29 whole core samples in Keshen-2 block show that under the original reservoir conditions, the initial pore compressibility $C_{pi}$ mainly ranges from $15\times10^{-4}\,\text{MPa}^{-1}$ to $25\times10^{-4}\,\text{MPa}^{-1}$, with an average value of $20\times10^{-4}\,\text{MPa}^{-1}$ (Fig. 2).

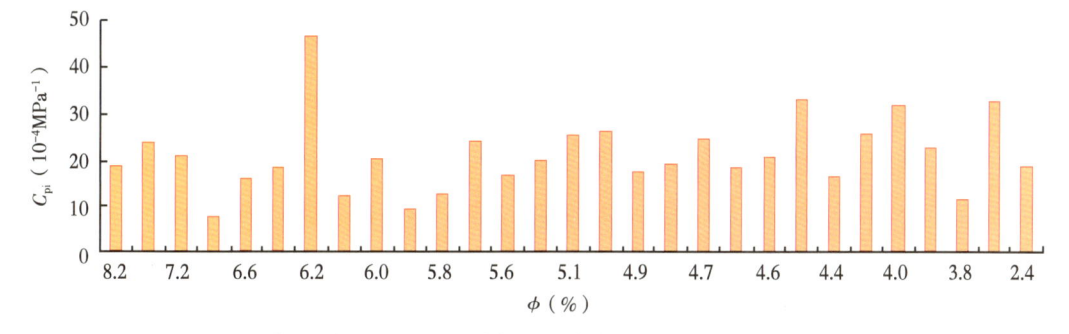

Fig. 2  Bar chart of pore compressibility of whole core samples in Keshen-2 block

With $C_{pi}$ setting as $15\times10^{-4}\,\text{MPa}^{-1}$, $20\times10^{-4}\,\text{MPa}^{-1}$ and $25\times10^{-4}\,\text{MPa}^{-1}$ respectively, based on three-year production data, the dynamic reserves of Keshen-2 block are calculated by the pressure decline method (based on material balance theory), and the results range from $440\times10^{8}\,\text{m}^{3}$ to $530\times10^{8}\,\text{m}^{3}$ (Fig. 3). In order to reduce the influence of uncertainty of pore compressibility on dynamic reserves, it is necessary to track the production performance of gas reservoir for a long time to obtain more accurate value of pore compressibility, thus to provide reference for similar gas reservoir.

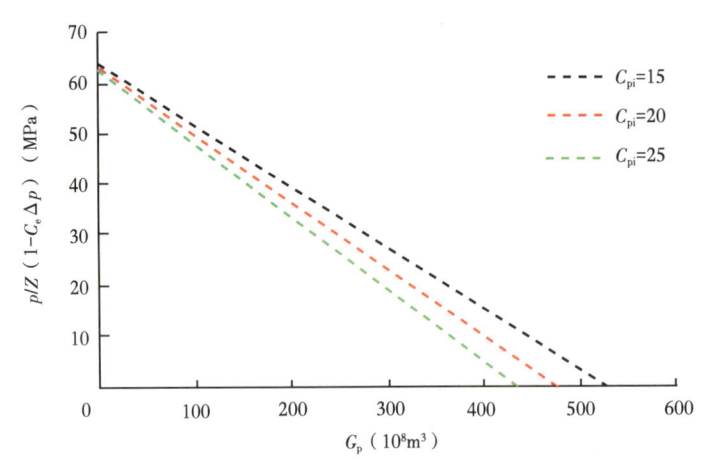

Fig. 3  *p/Z* vs. cumulative gas production in Keshen-2 block

## 2.2 Recovery percent of gas reserves

In early stage of development, the recovery degree is relatively low and the pressure drop has not yet fully affected the whole reservoir. Under such conditions, the dynamic reserves will be underestimated, in particularly for low permeability and tight gas reservoirs[13-14]. For ultra-deep fractured tight sandstone gas reservoir, matrix is tight and fluid flow rate is low, but because of fracture communication, the internal connectivity of gas reservoir is good, and its production performances are different from both conventional gas reservoir with good connectivity and tight gas reservoir with poor connectivity.

Table. 1  Single well model of fractured tight sandstone gas reservoir

| Reservoir Parameters | Value |
| --- | --- |
| Matrix permeability(km/mD) | 0.001 |
| Fracture permeability(kf/mD) | 1 |
| Matrix porosity(%) | 6 |
| Fracture porosity(%) | 0.06 |
| Reservoir length(m) | 1500 |
| Reservoir width(m) | 1500 |
| Reservoir thickness(m) | 100 |
| Gas saturation(%) | 65 |
| Interporosity flow coefficient/(1/m$^2$) | 0.05 |
| Gas reserves of model(10$^8$m$^3$) | 39.6 |

In order to investigate the influence of recovery percent on dynamic reserves of such gas reservoirs, a dual-porosity dual-permeability model with a single well is established, model parameters are shown in Table 1. The annual rate of gas production 2.88%.

Using the simulated gas production and pressure data, the dynamic reserves of gas reservoir in different production stages are calculated by pressure decline method. The result shows that when the recovery degree of gas reservoir is less than 10%, the dynamic reserves of early evaluation is 10% ~ 20% less than the volumetrical reserves of model (Fig. 4).

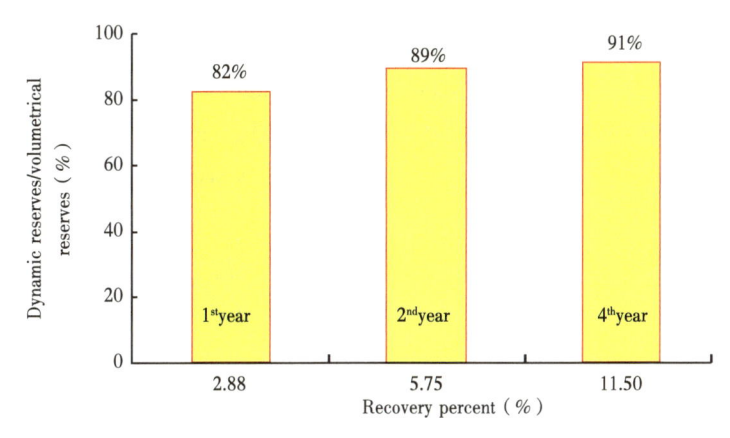

Fig. 4  Bar chart of dynamic reserves/ model reserves and recovery percent

Tracking the production performance of Keshen-2 block, calculating dynamic reserves at different stage by pressure drop method. The correlation between cumulative gas production and dynamic reserves is illustrated in Fig. 5, which indicats a good logarithmic relationship. It can be seen that the result is similar to that of numerical simulation. During the first year of gas production (cumulative gas production is less than $20 \times 10^8 m^3$), the variation of dynamic reserves is large, and then gradually slows down and stablilizes. The reasonable range of final dynamic reserves of gas reservoir can be predicted according to the fitting relationship of the figure.

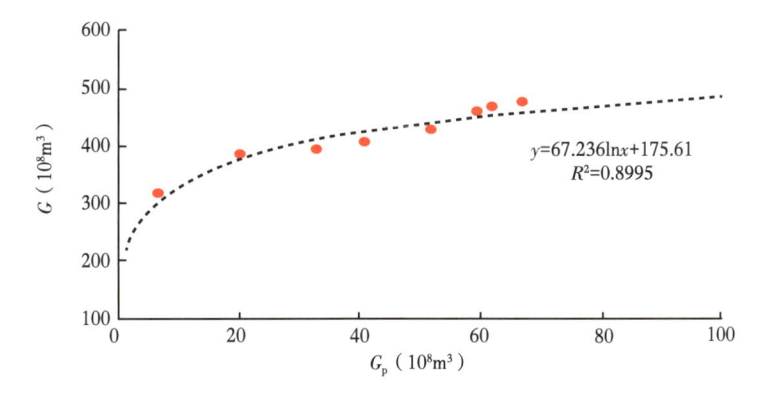

Fig. 5  Dynamic reserves vs. cumulative gas production in Keshen-2 block

## 2.3  Gas supply capacity of matrix rock

Ultra-deep naturally fractured tight sandstone gas reservoir is characterized by tight matrix and developed fractures. Gas is mainly stored in matrix pore. During production, gas flows mainly form matrix to frature and then from fracture to wellbore. The shorter the time of gas flows from matrix to fracture, the stronger the gas supply capacity of matrix.

For one-dimensional flow, the time of gas flowing from matrix to fracture is:

$$t = \frac{zp_aT_f}{T_f} \cdot \frac{2\mu_g\phi L_m^2}{K_m\Delta p^2} \cdot \tau \quad (2)$$

According to Eq. 2, the correlations between gas flow time and the matrix physical property/dimension are plotted, shown as Fig. 6. It can be seen that when matrix permeability $K_m$ is less than 0.001mD and matrix dimension $L_m$ is more than 30m, it will take tens of day for gas to flow from matrix to fracture.

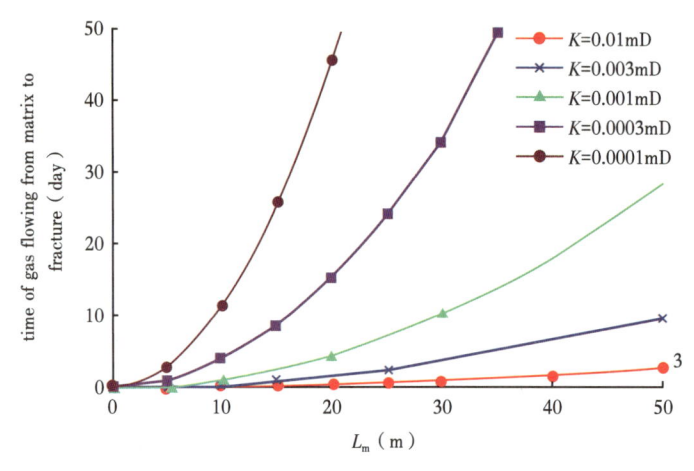

Fig. 6 Relationship of seepage time and matrix dimension

Gas supply capacity of matrix directly affects the pressure drop during gas production, and the pressure build-up during well shut-in. When gas flow rate is too high, the limitation of the gas supply capacity of the matrix will lead to rapid decline of bottom hole pressure and long time of build-up after shut-in. Well test results of KS2-2-4 show that the the bottom hole pressure keeps rising for a long time after shut-in, the pressure build-up rate is 27.2kPa/d(0.82MPa/month) by the 8th day, indicating that the matrix is continuously supplying gas.

For Keshen gas filed, because of high formation pressure and temperature, it is difficult to put pressure gauge into bottom hole. Most of shut-in formation pressure in Keshen-2 block are obtained by converting static wellhead pressures by temporary shut-in. Because the temporary shut-in time of gas wells is relatively short, which is about 1~2 hours, the recovery of formation pressure buildup is not enough. As a result, the dynamic reserves will be underestimated, which is more obvious in the early stage of gas reservoir development.

## 2.4 Water invasion through fractures

Water invasion will greatly reduce the productivity and dynamic reserves of individual well. It is an important factor affecting the stable production of gas fields.

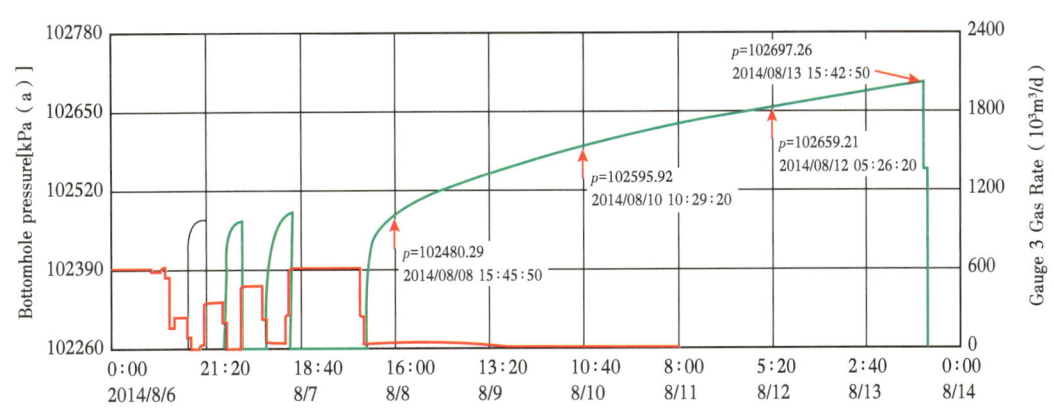

Fig. 7  Pressure build-up test of well KS2-2-4 in Keshen-2 block

In the early stage of production, water invasion occurred in some gas wells located at the edge of Keshen-2 block. Three gas wells located at southern limb of Keshen-2 block produced water after being put into production for three months, with water production of 10m$^3$/d to 140m$^3$/d. The curves of dynamic gas deliverablility of these wells show that water invasion leads to a rapid deline of open-flow capacity of 60%~90%. Soon after water breakthrouth, the three gas wells were shut down, with cumulative gas production less than 5000×10$^4$m$^3$. By using of software RTA, the production performances of these wells are analyzed stage-by-stage, and dynamic reserves are listed in Table 2.

Table. 2  Dynamic reserves of individual well at different stages in Keshen-2 block

| Well No. | matching Stage | Dynamic reserves of individual well($10^8$m$^3$) |
| --- | --- | --- |
| KS2-2-3 | Jul. 2013~Oct. 2013 | 2.64 |
|  | Oct. 2013~Jan. 2014 | 1.16 |
|  | Jan. 2014~Feb. 2014 | 0.55 |
| KS2-2-5 | Jun. 2013~Sep. 2013 | 4.67 |
|  | Sep. 2013~Jan. 2014 | 1.08 |
|  | Jan. 2014~Mar. 2014 | 0.91 |

In the early stage of gas reservoir development, there were not many water-producing wells, and water invasion has little influence on the dynamic reserves of gas reservoir. However, with water invasion advancing into the gas reservoir, the influence of water invasion on dynamic reserves of gas reservoir will gradually increase.

For fractured tight sandstone gas reservoirs, the risk of water invasion should be fully considered in the early stage of development. Gas production should not be too high, so as to avoid rapid water invasion affecting the overall recovery of gas reservoirs.

## 3　Requirements for production testing

Production testing is a critical step for obtaining dynamic data, investigating development features of gas reservoir and determining development scale during early appraisal stage.

For ultra-deep fractured tight sandstone gas reservoirs, because of tight matrix, natural fractures development, strong heterogeneity and complex gas-water relations, there is strong uncertainty of gas reserves estimation in the early appraisal stage. If development plan is based on proven geological reserves, the production scale is generally large, resulting in waste of surface construction. It is suggested that large-scale production test should be conducted to confirm producible reserves, so as to determine the appropriate production scale of gas reservoir.

Based on the comprehensive analysis of the factors affecting the dynamic reserves evaluation of ultra-deep fractured tight sandstone gas reservoirs, and combined with the production performance of Keshen gas field, the requirements for production testing and dynamic reserves evaluation of this kind of gas reservoirs are put forward as follows:

(1) Using whole core samples to test the rock pore compressibility under the original reservoir condition as an important parameter for the evaluation of dynamic reserves.

(2) Gas wells for production test should be located in different parts of the gas reservoir, and the number of test wells is more than 30% of the total number of wells, to obtain a comprehensive understanding of production performance of gas reservoir

(3) During production testing, the shut-in pressure is monitored once a month by high precision pressure gauge at wellhead for each test well, with shut-in time from 1 hour to 2 hours. In the calculation of formation pressure, the effects of gas supply capacity of matrix rock on the pressure buildup should be considered.

(4) Production testing should last 10~12 months, with apparent formation pressure ($p/Z$) drop of 3% to 5%. The relationship curve between dynamic reserves and cumulative gas production is established to predict ultimate dynamic reserves.

(5) For gas wells in edge and bottom water areas, chlorine is monitored once a half month to observe the water invasion performance.

## 4　Conclusion

(1) For ultra-deep fractured tight sandstone gas reservoirs, there is strong uncertainty of gas reserves estimation in the early appraisal stage. It is suggested that large-scale production testing should be conducted to confirm producible reserves, to determine the appropriate produciotn scale of gas reservoir.

(2) Rock pore compressibility, recovery percent of gas reserves, gas supply capacity of matrix rock, water invasion are the major factors affecting evaluation of dynamic reserves of ultra-deep fractured tight sandstone gas reservoirs. The influence degree of each factor varies with the development stage.

(3) For ultra-deep fractured tight sandstone gas reservoirs, when production test has been conducted for 10 to 12 months, and the apparent formation pressure decreased by 3%~5%, the final producible reserves of gas reservoir can be evaluated according to the relationship curve between early dynamic reserves and cumulative gas production, which provids a reliable basis for the gas reservoir development plan.

## 5 Nomenclature

$G_p$—Cumulative gas production, $10^8 m^3$;

$G$—Dynamic reserves, $10^8 m^3$;

$p$—Formation pressure, MPa;

$Z$—Gas deviation factor;

$p/Z$—Apparent formation pressure;

$C_p$—Pore compressibility, $MPa^{-1}$;

$C_w$—Formation water compressibility, $MPa^{-1}$;

$C_g$—Gas compressibility, $10^{-4} MPa^{-1}$;

$C_t$—Total compressibility, $C_t = C_p + C_g + C_w S_w$, $MPa^{-1}$;

$C_e$—Effective compressibility, $C_e = (C_w S_w + C_p)/(1 - S_w)$, $MPa^{-1}$;

$C_{pi}$—Reservoir rock pore compressibility under original reservoir conditions, $MPa^{-1}$;

$S_w$—Water saturation, %;

$S_g$—Gas saturation, %;

$K_m$—Matrix permeability, md;

$\phi$—Porosity, %;

$\tau$—Tortuosity;

$\Delta p$—producing pressure dropdown, MPa;

$L_m$—Matrix rock dimension, m;

$T_a$—Temperature at the standard condition, K;

$T_f$—Formation temperature, K;

$p_a$—Pressure at the standard condition, MPa;

$\mu_g$—Gas viscosity, mPa·s

## References

[1] Sun L, Fang C, Sa M, et al. Innovation and prospect of geophysical technology in the exploration of deep oil and gas[J]. Petroleum Exploration and Development, 2015, 42(4): 414-424.

[2] Sun L, Zou C, Zhu R, et al. Formation, distribution and potential of deep hydrocarbon resources in China [J]. Petroleum Exploration and Development, 2013, 40(6): 641-650.

[3] Bai G, Cao B. Characteristics and distribution patterns of deep petroleum accumulations in the world [J]. Oil & Gas Geology, 2014, 35(1): 19-25.

[4] Wang Y, Su J, Wang K, et al. Distribution and accumulation of global deep oil and gas [J]. Nat Gas Geosci,

2012, 23(3): 526-533.

[5] Zhang G, Ma F, Liang Y, et al. Domain and theory-technology progress of global deep oil & gas exploration [J].Acta Petrolei Sinica, 2015, 36(9): 1156-1166.

[6] Zhang R, Yang H, Wang J, et al. The formation mechanism and exploration significance of ultra-deep, low-porosity and tight sandstone reservoirs in Kuqa Depression, Tarim Basin [J]. Acta Petrolei Sinica, 2014, 35(6): 1057-1069.

[7] Zhang R, Wang J, Ma Y, et al. The sedimentary microfacies, palaeo-geomorphology and their controls on gas accumulation of deep-burried cretaceous in Kuqa Depression, Tarim Basin, China[J]. Natural Gas Geoscience, 2015, 26(4),667-678.

[8] Wang J, Zhang R, Zhao J, et al. Characteristics and evaluation of fractures in ultra-deep tight sandstone reservoir: Taking Keshen gas field in Tarim Basin, NW China[J]. Natural Gas Geoscience, 2014, 25(11),1735-1745.

[9] Xu Z, Li Y, Ma Y, et al. Future gas exploration orientation based on a new scheme for the division of structural units in the central Kuqa Depression, Tarim Basin[J]. Natural Gas Industry, 2011, 31(3): 31.

[10] Liu X. A discussion on several key parameters of gas reservoir dynamic reserves calculation [J]. Natural Gas Industry, 2009, 29(9): 71-74.

[11] Wei J, Zheng R. The influence of formation compressibility and edge and bottom water size of abnormally pressured gas reservoir on recovery performance [J]. Petroleum Exploration and Development, 2002, 29(5): 56-58.

[12] Jiang Y, Li M, Xue X, et al. Compressibility derivation and experimental verification of two-component rock [J]. Rock and Soil Mechanics, 2013, 34(5): 1279-1286.

[13] Feng X, He W, Xu Q. Discussion on calculating dynamic reserves in the early stage of heterogeneous gas reservoir development [J]. Natural Gas Industry, 2002, 22(Supplement): 87-90.

[14] Zhou S, Zhou J, Shen D, et al. One calculation method of dynamic reserve of low-permeability reservoir single well [J]. Journal of Chongqing University of Science and Technology (Natural Science edition), 2013, 15(2): 10-13.

# Optimization of Managed Drawdown for A Well with Stress-Sensitive Conductivity Fractures: Workflow and Case Study

Yunsheng Wei[1]  Junlei Wang[1]  Ailin Jia[1]  Cheng Liu[2]  Chao Luo[3]  Yadong Qi[1]

(1.PetroChina Research Institute of Petroleum Exploration and Development, Beijing, China;
2.PetroChina Zhejiang Oilfield Company, Hangzhou, Zhejiang, China;
3.Chongqing University of Science & Technology, Chongqing, China)

**Abstract:** The effect of bottomhole-pressure (BHP) drawdown schedule on the well performance is generally attributed to the stress sensitivity in propped finite-conductivity fractures. The purpose of this work is to develop a detailed workflow of optimizing BHP drawdown schedule to improve long-term performance by finding a tradeoff between delaying conductivity degradation and maintaining drawdown. First, according to experimental data of propped fracture, an alternative relationship between conductivity and pressure drawdown is developed to mimic the change of fracture conductivity with effective stress. Second, based on the dimension-transformation technique, the coupled fracture-reservoir model is semi-analytically solved, and seamlessly generates the time-dependent equation (i.e., transient inflow performance relationship, IPR) which provides the production-rate response to any BHP variation. Next, the value of BHP on the reversal behavior of rate is defined as the optimum BHP on the specified time-dependent IPR, and then the optimum profile of BHP-drawdown over time is achieved. Finally, we corroborate the effectiveness of this workflow with a field case from Zhaotong shale in China. Field case substantiates that: (1) the well with restricted drawdown has more advantage of improving the performance than that with unrestricted drawdown; (2) after inputting the optimum BHP drawdown into the history-unrestricted case, the long-term cumulative-gas production could indeed be increased.

**Key words:** multiple fractured horizontal well; pressure sensitivity; dynamic conductivity; transient IPR; optimization workflow

# 1 Introduction

In the last decades, the applications of horizontal drilling and fracturing-stimulation treatment have gained great success in economic development of unconventional resources. A large volume of fluids is pumped to create huge contact area between fractures and matrix in a tight formation[1]. Although large amount of proppants are pumped to prevent fracture closure to some extent, the increases in effective stress caused by the extraction of fluids inevitably results in fracture closure. Fracture closure is often referred to as "geomechanics effect"[2,3]. Given that fractures are more deformable than the matrix and the conductivity of fractures dominates the production[4-6], productivity loss is often attributed to geomechanics-related factors in conductivity of propped fractures, which is caused by the proppant embedment, crushing or fracture-face creep, and the flowback of proppant.

Numerous studies demonstrate that stress sensitivity of fractures has a direct relationship with the production performance. Production might benefit or damage from the stress-drawdown management for fractured wells in shale formations[7-9]. High initial production rate of the wells is accomplished by a very-high pressure drawdown, however resulting in a very-steep productivity decline. As a contrast, a proper managed BHP drawdown schedule might improve cumulative-gas-production by reducing the magnitude of effective stress on the fractures. Production wells applying restricted drawdown in the Haynesville shale have an average performance with a first-year decline of only 38%, lower than 83% for these wells on unrestricted drawdown[8]. Inversely, an unrestricted drawdown strategy may cause abnormal productivity declines generally observed within a short-term period, with the substantial cumulative-gas-production reduction of up to 20% in the Vaca Muerta shale[10]. It is almost certainty that using restricted stress-drawdown to the wells indeed yields higher long-term cumulative production compared with the unrestricted. Building on a relationship between drawdown and productivity degradation, Rojas and Lerza defined an optimum drawdown management for the Vaca Muerta wells by integrating production performance analysis of a hundred of fractured horizontal wells with the economic evaluation, and proposed a workflow to accomplish optimum pressure decline with the right chock selection over time to optimize well performance[10]. Karantinos et al. presented a choke-management strategy that aims to minimizing wellbore pressure gradients along the fracture, and proposed the optimal chock-management strategy with beanup operations achieving a relatively higher reduction in pressure gradients for the case of low values of dimensionless fracture conductivity[11]. Kumar et al. provided a set of drawdown management scenarios that assumed constant pressure decline rate until a stabilized bottomhole flowing pressure has been reached, and used a net present value (NPV) maximization approach to determine an optimum drawdown strategy[12]. From the viewpoint of field applications, Mirani et al. suggested that the optimum long-term drawdown strategy should be determined over imposing an economic optimization[13].

To the authors' knowledge, for the high BHP drawdown, the change of reservoir pressure might lead to higher effective stresses in the formation/matrix. To find the optimum drawdown, the first step is to identify the relationship between conductivity degradation and in-situ stress changing. The most of experimental studies directly investigated the values of fracture conductivity under changing stress conditions, rather than pore pressure. The results found that the relationship between the effective stress and fracture conductivity is nonlinear. According to the poroelastic theory that the effective normal stress is the difference between the total normal stress and the pore pressure, when Biot coefficient is assumed to be 1, the change in effective stress is typically about 0.4-0.7 times the change in formation pore pressure[9]. For the fracture, the change in closure stress can be also related to the drop in pore pressure by the relationship[14]:

$$\Delta\sigma_{closure} = \alpha \frac{1-2v}{1-v}\Delta p \qquad (1)$$

where $\alpha$ is the Biot coefficient, $v$ is the Poisson's ratio. Note that the relationship is not a straightforward concept. Traditional reservoir simulation approaches incorporate compaction tables (pore pressure and fracture conductivity), relating pore pressure to effective stress, to mimic the fundamental physics of fracture closure. This contributes to the nonlinearity of the governing equations. Jiang et al. used the Pedrosa's transform formulation based on a perturbation method to couple the reservoir flow with stress-dependent fractures with constant value of permeability modulus in hydraulic fractures and natural-fracture subsystems[15]. Whereafter, they provided a more accurate linearization scheme and an iteration method were adopted to improve the nonlinearity algorithm[16]. These approaches assume that the change in pore pressure has a predefined relationship with effective stress, which ignores the path-dependency of the stress changing caused by different depletion schedules. For example, Okouma et al. considered different permeability modulus for different drawdown cases (i.e., a higher value of permeability modulus for unrestricted case and a lower value for the restricted case)[8]. Familiar field-observed trends with lower long-term cumulative production corresponding to higher drawdown were reproduced. To capture the process of stress path dependency that governs the effective stress on the fractures, more accurate fully coupled-geomechanics reservoir simulation models were constructed[17,18]. The stress changes can be calculated during reservoir depletion, but the coupled-geomechanic simulations are very specialized and time-consuming.

Without understanding the tradeoff between minimizing fracture degradation and maintaining a large enough driving force, drawdown-strategy optimization would remain solely a trail-and-error process. In this work, a model of dynamic permeability-decay modulus was developed to equivalently estimate the direct relation between pore pressure and fracture conductivity. Next, an efficient and accurate method using semi-analytical model was provided to predict the relationship between fracture closure and the productivity. The approach is able to reproduce the coupled-geomechanical results across a wide range of drawdown management strategies with an acceptable error. Then, by use of transient inflow performance relation (IPR), a workflow is calibrated to find the optimal path of drawdown strategy. A field case study from a shale gas reservoir in China is used to demonstrate the effectiveness of the workflow.

# 2 Model Development

## 2.1 Permeability-decay Coefficient

The permeability in the fracture is related with effective stress. There are various models to calculate the permeability value. Considered that the simple hydrostatic compaction testing can be done to replicate effective stress changes, the Yilmaz model is selected[19], which is given by:

$$K_f = K_{fi}\exp[-d_f(\sigma_{eff} - \sigma_{eff,i})] \qquad (2)$$

where $K_\text{fi}$ is the initial permeability, $\sigma_\text{eff}$ is effective stress and $d_\text{f}$ is the permeability modulus. A larger $d_\text{f}$-value indicates that the fracture is more stress sensitive. The exponential relationship is found in Devonian shale, and the Marcellus shale. Jia et al. summarized the fitting results of $d_\text{f}$-value in the Marcellus shale, and the value ranges from $3.6 \times 10^{-4}$ to $1.49 \times 10^{-3}$ psi$^{-1}$ for propped fractures[20].

The changing of effective stress determines fracture aperture and the resulting conductivity. Note that due to the mismatch of asperities as shown in Fig. 1, fractures can still retain part of their conductivity even if fracture walls have come into contact. The residual fracture permeability is denoted as $K_{\text{f min}}$, which is still considerably larger than matrix permeability.

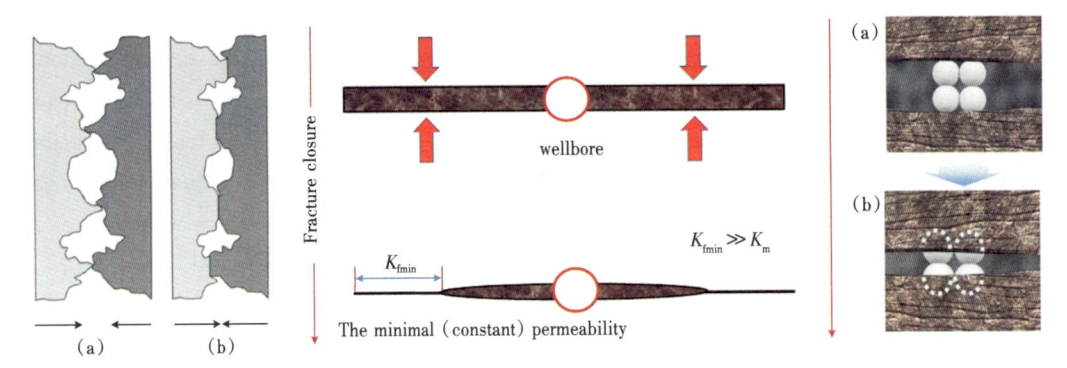

Fig.1 Schematic of fracture closure in the (a) initial condition and (b) final condition

After further using the relation given in Eq. (1), Eq. (2) is rewritten by

$$\frac{K_\text{f} - K_{\text{f min}}}{K_\text{fi} - K_{\text{f min}}} = \exp[-\gamma_\text{f}(p_\text{i} - p)] \tag{3}$$

where permeability-decay coefficient $\gamma_\text{f}$ satisfies the relationship: $\gamma_\text{f} = d_\text{f} v \alpha / (1-v)$.

According to the laboratory measurement data for shale samples in Sichuan Basin China[21], Fig. 2a presents the relationship between permeability and closure pressure ($\Delta p$) rather than effective stress in the unpropped fracture. For the stiff sample, the $\gamma_\text{f}$-value is a constant, and the relationship satisfies the exponential expression given by Eq. (3). However, for the medium and soft sample, the $\gamma_\text{f}$-value is not a constant. Fig. 2b achieved the data of propped fracture, and provided a binomial fitting. It is found that the binomial could achieve a better matching than the linear fitting. Therefore, it is acceptable to regard the permeability-decay coefficient to be a function of closure pressure. The permeability-decay coefficient is a line function of closure pressure (pressure drawdown) with non-zero intercept as follows[22]:

$$\gamma_\text{f} = a\Delta p_\text{f} + b \tag{4}$$

where $a$ and $b$ are the characteristic parameters determined by experimental data.

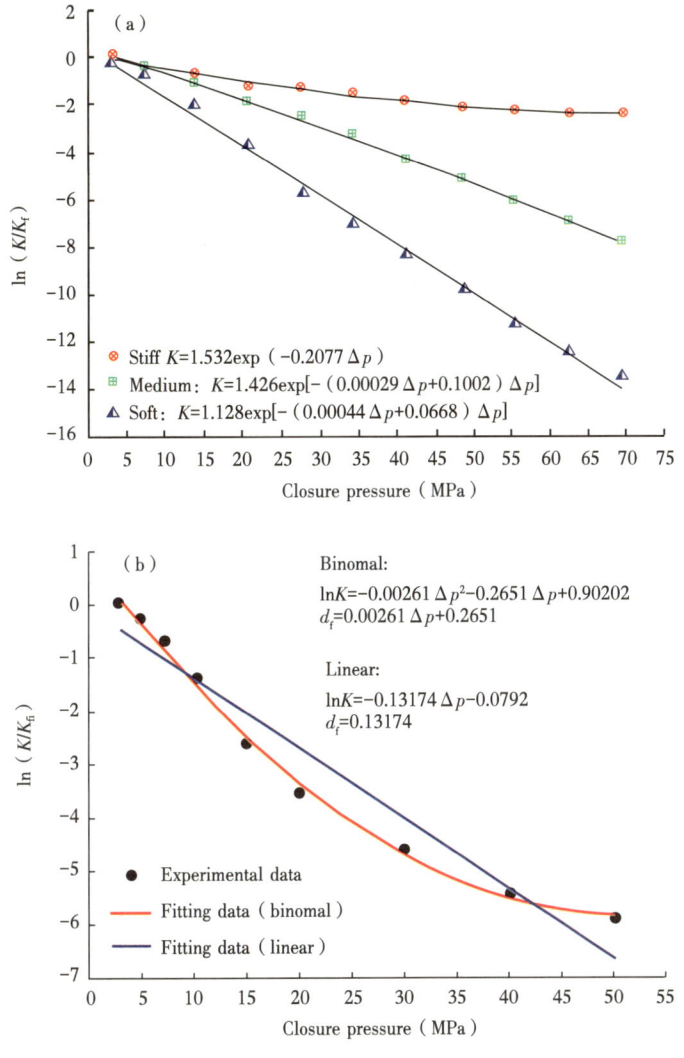

Fig.2 The relationship between permeability and closure pressure for (a) the unpropped and (b) the propped fracture

## 2.2 Semi-Analytical Model

For fractures with stress-sensitivity conductivity in a box-shaped reservoir (Fig. 3), the assumptions are made as follows: (1) the matrix is homogeneous, having box shape with no flow outer boundaries; (2) the transverse hydraulic fractures have a finite conductivity with pressure sensitivity; (3) the properties of fractures are assumed to be identical, all penetrating the formation; (4) the single-phase fluid obeys Darcy-flow law.

According to our previous study[22], in the $n$-th fracture for multiple-fracture system ($N_f$), the diffusivity equation was revisited to consider pressure/stress dependency, which is expressed as the following dimensionless form (Appendix A):

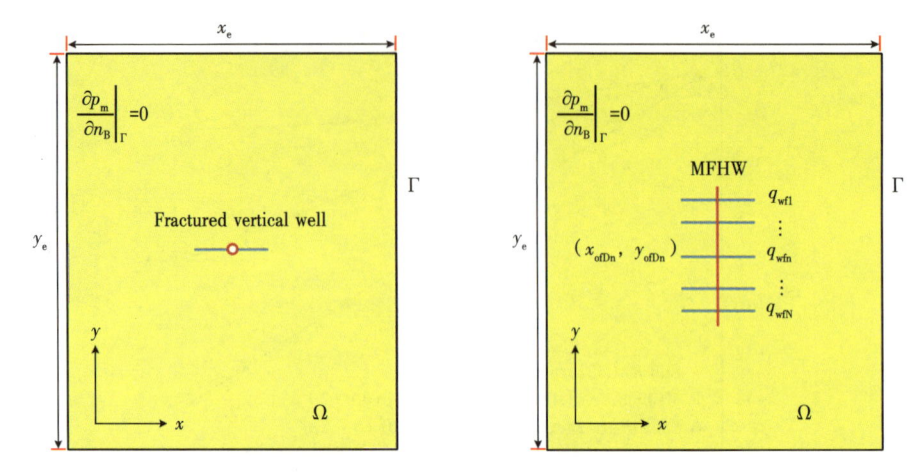

Fig.3 Schematic of (a) a vertical fractured well and (b) a horizontal well with multi-stage fractures in a box-shaped reservoir

$$\begin{cases} \dfrac{\partial}{\partial x_{Dn}}\left(C_{fDn}(p_{fDn})\dfrac{\partial p_{fDn}}{\partial x_{Dn}}\right) - 2\pi q_{fDn}(x_{Dn}) + 2\pi q_{wfDn}(t_D)\delta(x_{Dn},x_{wfDn}) = 0 \\ q_{cDn}(x_{Dn}) = \int_{x_{Dn}}^{L_{fDn}} q_{fDn}(\zeta)\,d\zeta,\ \left.\dfrac{\partial p_{fDn}}{\partial x_{Dn}}\right|_{x_{Dn}=0} = \left.\dfrac{\partial p_{fDn}}{\partial x_{Dn}}\right|_{x_{Dn}=L_{fDn}} = 0 \end{cases} \quad (5)$$

where the dynamic conductivity is given by

$$\frac{C_{fDn}(p_{fDn})}{C_{fDi}} = \left(1 - \frac{C_{fDmin}}{C_{fDi}}\right)\exp[-\gamma_{fD}p_{fD}] + \frac{C_{fDmin}}{C_{fDi}} \quad (6)$$

Integrating Eq. (5) with respect to $x_{Dn}$ twice yields the following expression:

$$p_{wfDn} - p_{fDn}(x_{Dn}) = 2\pi q_{wfDn}G(x_{Dn},x_{wfDn}) - 2\pi I(x_{Dn},x_{wfDn}) + q_{wfDn}S_{cn} \quad (7)$$

where $G(\ )$ is the integral of Heaviside unit step function [i.e., $H(\ )$], and $I(\ )$ is the Fredholm integral function. Note that these two function are modified with varying conductivity, which are given by respectively,

$$G(x_{Dn},x_{wfDn}) = \int_0^{x_{Dn}} \frac{H(\zeta,x_{wfDn})}{C_{fDn}(\zeta)}d\zeta,\ \text{and}\ I(x_{Dn},x_{wfDn}) = \int_{x_{wfDn}}^{x_{Dn}} \frac{1}{C_{fDn}(\zeta)}d\zeta\int_0^{\zeta} q_{fDn}(\zeta)d\zeta \quad (8)$$

Since the flow regime of transverse fracture is more complex due to the convergence of fluid into the horizontal wellbore, a convergence flow skin $S_c$ is incorporated. For vertical fracture, the value of $S_c$ equals to 0.

In the reservoir, the pressure depletion in the region $\Omega$ is caused by the flow from fracture, and the boundary condition satisfies $(\partial p_m/\partial n_B)|_\Gamma = 0$. By use of instantaneous point solution, the pressure solution in the matrix is presented as follows:

$$p_{\mathrm{mD}}(x_{\mathrm{D}},y_{\mathrm{D}},t_{\mathrm{D}}) = \sum_{n=1}^{N_{\mathrm{f}}} \int_{0}^{t_{\mathrm{D}}} \mathrm{d}\tau \int_{0}^{L_{\mathrm{fDn}}} q_{\mathrm{fDn}}(u,\tau) \frac{\partial p_{\mathrm{uD}}(x_{\mathrm{D}},y_{\mathrm{D}},t_{\mathrm{D}}-\tau,x'-u,y')}{\partial t_{\mathrm{D}}} \mathrm{d}u. \tag{9}$$

where $p_{\mathrm{uD}}$ is the stress-drop solution of unit-rate point source[15], which is given by

$$p_{\mathrm{uD}} = \frac{2\pi}{x_{\mathrm{eD}} y_{\mathrm{eD}}} [t_{\mathrm{D}} + 2F_{\mathrm{pss}}(x_{\mathrm{D}},y_{\mathrm{D}},x',y') - 2F_{\mathrm{trans}}(x_{\mathrm{D}},y_{\mathrm{D}},t_{\mathrm{D}},x',y')] \tag{10}$$

and

$$F_{\mathrm{pss}} = \sum_{v=1}^{\infty} \frac{\cos(\gamma_v y_{\mathrm{D}})\cos(\gamma_v y')}{\gamma_v^2} + \sum_{v=1}^{\infty} \frac{\cos(\beta_v x_{\mathrm{D}})\cos(\beta_v x')}{\beta_v^2} \\ + 2\sum_{v=1}^{\infty} \cos(\beta_v x_{\mathrm{D}})\cos(\beta_v x') \sum_{v=1}^{\infty} \frac{\cos(\gamma_v y_{\mathrm{D}})\cos(\gamma_v y')}{\beta_v^2 + \gamma_v^2} \tag{11a}$$

$$F_{\mathrm{trans}} = \sum_{v=1}^{\infty} \frac{\exp(-\gamma_v^2 t_{\mathrm{D}})}{\gamma_v^2}\cos(\gamma_v y_{\mathrm{D}})\cos(\gamma_v y') + \sum_{v=1}^{\infty} \frac{\exp(-\beta_v^2 t_{\mathrm{D}})}{\beta_v^2}\cos(\beta_v x_{\mathrm{D}})\cos(\beta_v x') \\ + 2\sum_{v=1}^{\infty} \cos(\beta_v x_{\mathrm{D}})\cos(\beta_v x')\exp(-\beta_v^2 t_{\mathrm{D}}) \sum_{v=1}^{\infty} \frac{\exp(-\gamma_v^2 t_{\mathrm{D}})}{\gamma_v^2 + \beta_v^2}\cos(\gamma_v y_{\mathrm{D}})\cos(\gamma_v y') \tag{11b}$$

The fracture model [i.e. Eq. (7)] and matrix model [i.e. Eq. (9)] are able to couple together by using the stress-continuity condition along fracture, which is given by

$$p_{\mathrm{fDn}}(x_{\mathrm{Dn}}) = p_{\mathrm{mD}}(x_{\mathrm{ofD}} + x_{\mathrm{Dn}}, y_{\mathrm{ofD}}, t_{\mathrm{D}}) \tag{12}$$

Note that the pressure drop within wellbore is ignored. This approach is known as the boundary element method[23,24]. It is noted that the dependence of dynamic conductivity on pressure makes the diffusion equation [i.e. Eq. (7)] nonlinear. The technique of dimension transformation renders the nonlinear equation amenable to linear analytical treatment, which is given by

$$\xi_{\mathrm{Dn}}(x_{\mathrm{Dn}}) = \hat{C}_{\mathrm{fD}} \cdot \int_{0}^{x_{\mathrm{Dn}}} \frac{\mathrm{d}x_{\mathrm{D}}}{C_{\mathrm{fDn}}(x_{\mathrm{D}})} \mathrm{and} \hat{C}_{\mathrm{fD}} = L_{\mathrm{fDn}} / \int_{0}^{L_{\mathrm{fDn}}} \frac{\mathrm{d}x_{\mathrm{D}}}{C_{\mathrm{fDn}}(x_{\mathrm{D}})} \tag{13}$$

Once the operation condition ($q_{\mathrm{wfD}}$ or $p_{\mathrm{wD}}$) is incorporated by modifying inner boundary condition, the production performance is achieved. It is emphasized that the solution is generated in the real-time domain rather than Laplace domain as our previous work[22]. Discontinuities in rate or BHP schedule can be directly incorporated into the model according to Duhamel's theorem.

After discretizing coupled model in temporal and spatial domains, the unknown variables include the rate and pressure of fracture segments. In each time step, a closed-form iterative algorithm will be used and iterative equation is formulated. Taking constant-BHP condition for example (i.e. $p_{\mathrm{wfDn}}$ is the known), Eq. (7) is written as

$$p_{\mathrm{wfDn}} - p_{\mathrm{fDn}}^{\langle \kappa+1 \rangle}(\xi_{\mathrm{Dn}}^{\langle \kappa \rangle}) = \frac{2\pi}{\hat{C}_{\mathrm{fDn}}} q_{\mathrm{wfD}}^{\langle \kappa+1 \rangle} G(\xi_{\mathrm{Dn}}^{\langle \kappa \rangle}, \xi_{\mathrm{wfDn}}) - \frac{2\pi}{\hat{C}_{\mathrm{fDn}}} \int_{\xi_{\mathrm{wfDn}}}^{\xi_{\mathrm{Dn}}^{\langle \kappa \rangle}} \mathrm{d}\zeta \int_{0}^{\zeta} q_{\mathrm{fDn}}^{\langle \kappa+1 \rangle}(\zeta) \mathrm{d}\zeta + q_{\mathrm{wfD}}^{\langle \kappa+1 \rangle} S_{\mathrm{cn}} \tag{14}$$

The iterative calculation is illustrated as follows:

①With $\kappa=0$, $C_{fDn}^{\langle\kappa\rangle}$ is assumed to be a constant of initial conductivity. Calculating $q_{fDn}^{\langle\kappa\rangle}$ and $p_{fDn}^{\langle\kappa\rangle}$.

②Calculating dynamic conductivity of $C_{fDn}^{\langle\kappa\rangle}$, and transforming $x_{Dn}$ into $\xi_{Dn}^{\langle\kappa\rangle}$ according to Eq. (13).

③Solving the explicit linear equations of Eq. (14) by Gaussian elimination method, and obtaining the updated $q_{fDn}^{\langle\kappa+1\rangle}$, $p_{fDn}^{\langle\kappa+1\rangle}$ and $q_{wfD}^{\langle\kappa+1\rangle}$.

④If $|q_{wfD}^{\langle\kappa+1\rangle}-q_{wfD}^{\langle\kappa\rangle}|<\varepsilon(=10^{-5})$ then terminate the iterative process; otherwise, updating $p_{fDn}^{\langle\kappa\rangle} = p_{fDn}^{\langle\kappa+1\rangle}$ with $\kappa=\kappa+1$, and return step b) until convergence.

## 2.3 Model Verification

To verify the model proposed in this study, a synthetic case for a fully penetrating multi-fractured horizontal gas well in a rectangular reservoir was generated by a commercial simulator (KAPPA Workstation 5.20). According to the field experiences, basic reservoir and fluid properties are list in Table 1. Stress sensitivity parameters used in Eq. (4) are given as: $a=0.001\text{MPa}^{-2}$, $b=0.01\text{MPa}^{-1}$.

**Table 1  Rock and fractured gas well properties**

| property | Value | Unit | property | Value | Unit |
|---|---|---|---|---|---|
| Formation permeability, $K_m$ | 1 | mD | Fracture spacing, $L_s$ | 40 | m |
| Formation thickness, $h$ | 10 | m | Fracture length, $L_f$ | 200 | m |
| Formation porosity, $\varphi_m$ | 0.1 | | Number of fractures, $N_f$ | 15 | |
| Initial pressure, $p_i$ | 40 | MPa | Fracture conductivity, $F_c$ | 1000 | mD·m |
| Drainage area, $x_e \times y_e$ | 2000×2000 | m² | Min facture conductivity, $F_{c,min}$ | 10 | mD·m |

When the model is applied to gas flow equation, the concepts of pseudo pressure ($m$) and pseudo time ($t_a$) are used to render the equation of gas flow amenable to the linearized liquid-flow equation, which are defined as follows:

$$m(p) = \frac{\mu_{gi}Z_{gi}}{p_i}\int_{p_{sc}}^{p}\frac{p'}{\mu_g(p')Z_g(p')}dp', \text{ and } t_a(t) = \int_0^t \frac{\mu_{gi}c_{gi}}{\mu_g[p_{avg}(t')]c_g[p_{avg}(t')]}dt' \quad (15)$$

Gas properties are shown in Fig. 4a, and pseudo pressure vs. pressure plot is presented in Fig. 4b. Details on the calculation of pseudotime could found in the work of Ye and Ayala[25].

The comparisons between numerical and semi-analytical solutions are presented in Fig. 5 under constant BHP and variable BHP conditions. It can be seen that there are excellent agreements between the results computed by the solution developed in this work compared to those obtained by the numerical model.

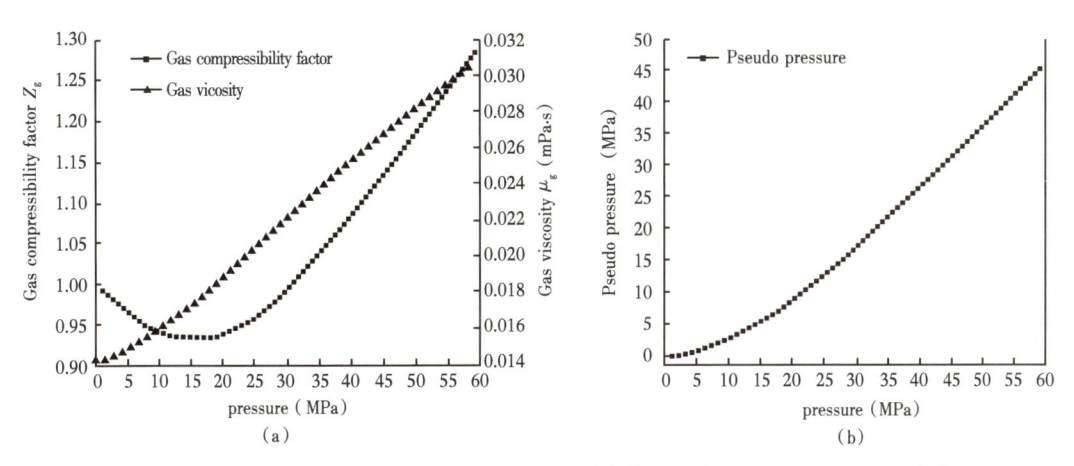

Fig.4 (a) gas properties at reservoir temperature and (b) pseudopressure vs. pressure behavior

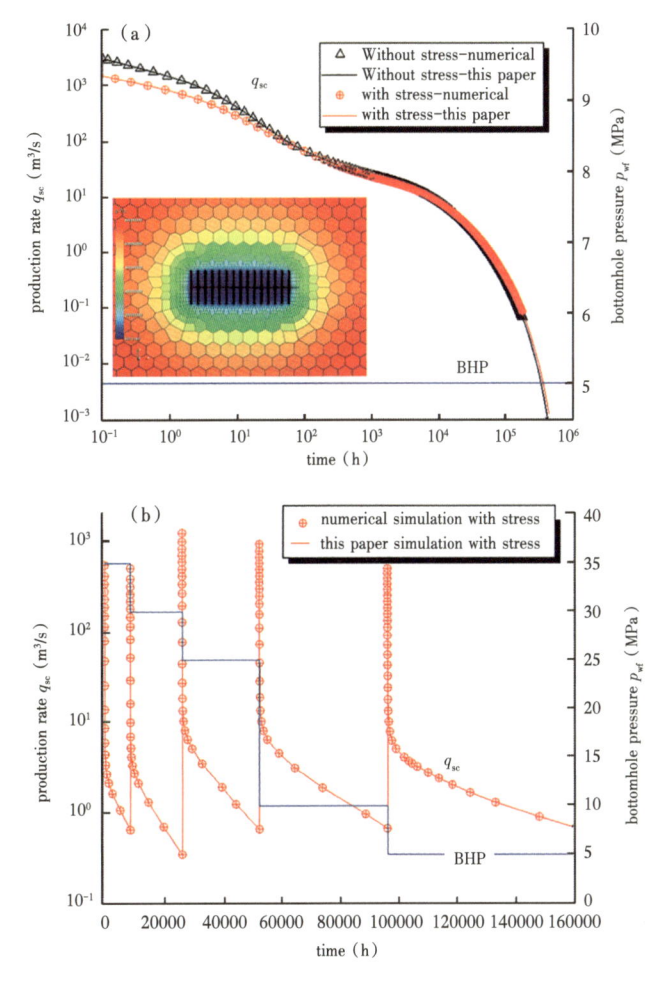

Fig.5 Validation against numerical simulation under the (a) constant-BHP condition, and (b) variable-BHP condition

## 3  An Integrated Approach

### 3.1  Transient IPR Based on Semi-Analytical Modeling

On the basis of semi-analytical modeling, the production performances across a wide range of pressure drawdowns are simulated. Afterwards, the production performance can be represented using IPR. IPR curve stands for the relationship between production rate and BHP. Conventional IPR is based on the knowledge of the average reservoir pressure and an assumption of stabilized-flow condition. For a slightly-compressible fluid under single-phase condition, the fluid flowing obeys simple Darcy's law, and IPR curve behaves a straight-line relationship between production rate and BHP. For stress-sensitive fractured gas well, the governing equation is nonlinear because of pressure dependence of gas properties and stress-sensitive permeability. As shown in Fig. 6, when BHP is in the near range of initial pressure, IPR behaves a linear approximation. When BHP further decreases, IPR exhibits a downward curvature.

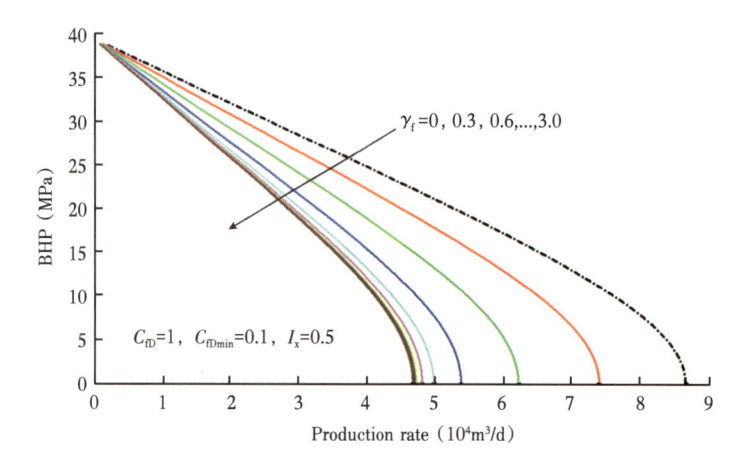

Fig.6  Conventional IPR curve for stress-dependent fractured gas well ( $I_x = L_f/x_e$ )

The average pressure varies with time during transient flow period. Transient IPR accommodates the time-dependent average pressure[26]. Semi-analytical model favors the generation of transient IPRs readily. First, a representative fracture-reservoir model is selected. Next, the model is used to forecast the production rate for a specified time under different BHP conditions. Fig. 7 illustrates the transient IPR curves and the normalized effective conductivity of fracture when $\gamma_f$ = constant. Here, based on Eq. (6), the normalized effective conductivity of fracture is expressed as the ratio of the integral average over fracture length to the initial conductivity, which is given by

$$\frac{C_{fDavg}(t)}{C_{fDi}} = \frac{1}{C_{fDi}L_f}\int_0^{L_f} C_{fD}[p_f(\xi,t)]\mathrm{d}\xi \tag{16}$$

The increase of BHP drawdown leads to a larger driving force, and the normalized effective conductivity continually decreases due to the stress sensitivity. The increase in the driving force always outweights the reduction in the effective conductivity. As a result, although the production gain is lower than predicted straight-line behavior, the production rate still increases. At each given time, the pairs of rate and BHP generate transient IPR. Meanwhile, both the production rate and the normalized effective conductivity decrease over time.

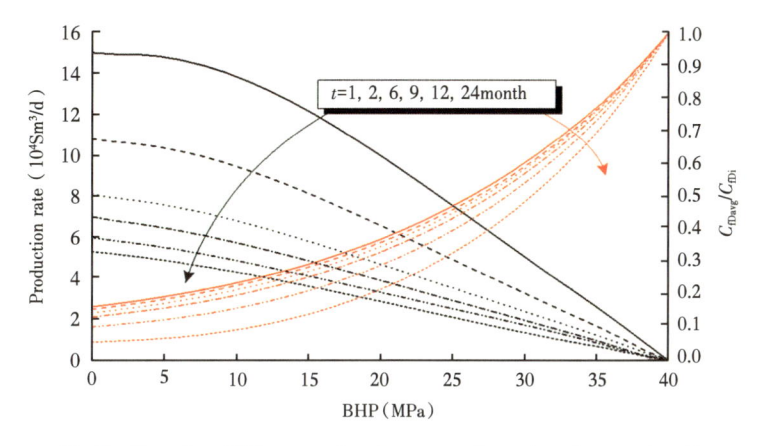

Fig.7 Transient IPR curves for the stress-dependent fractured well

Next, the dynamic permeability-decay coefficient is again incorporated into semi-analytical model. The normalized effective conductivity continually still decreases with the increase of BHP drawdown, but the change range of conductivity decline gradually becomes small. When a tradeoff is achieved by minimizing fracture degradation and maintaining a large enough driving force, the production rate reaches the maximum. As shown in Fig. 8, at a given time, the decreasing BHP leads to the production rate to increase up to a certain value denoted by red point. Once beyond the point, a reverse behavior appears and the production rate decreases with the BHP drawdown increasing. The same feature is also observed in the work of Tabatabaie et al. that a steady-state productivity equation accounting for the pressure dependent permeability in the fracture and the formation was established[27]. They concluded that when the constant permeability modulus is

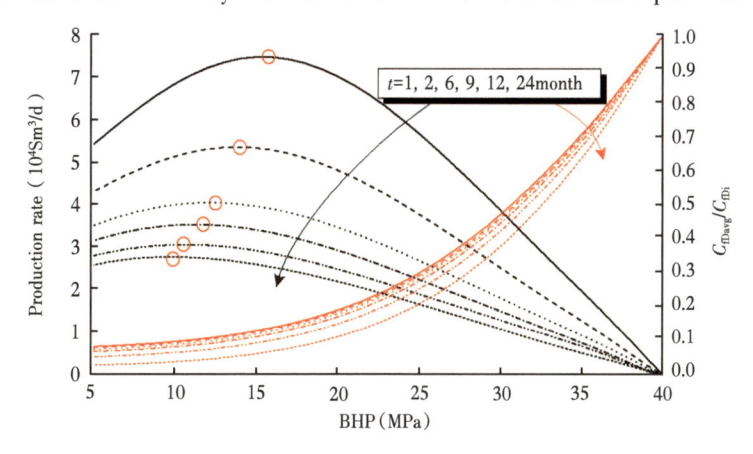

Fig.8 Reversal of productivity index on transient IPR

higher, there is an optimal BHP at which the flow rate is the maximum. However, it is noticed that their mechanism is completely different from our work.

## 3.2 Optimization Workflow

A workflow of optimizing BHP drawdown strategy is presented as shown in Fig. 9. First, based on experimental data, we construct custom compaction table and get the stress-dependent permeability decay coefficient by fitting experimental data. Second, the key inputs including reservoir & fracture & fluid properties are collected. Then, for each specified time, these variables are put into semi-analytical model to simulate production rate under different BHP-drawdown conditions.

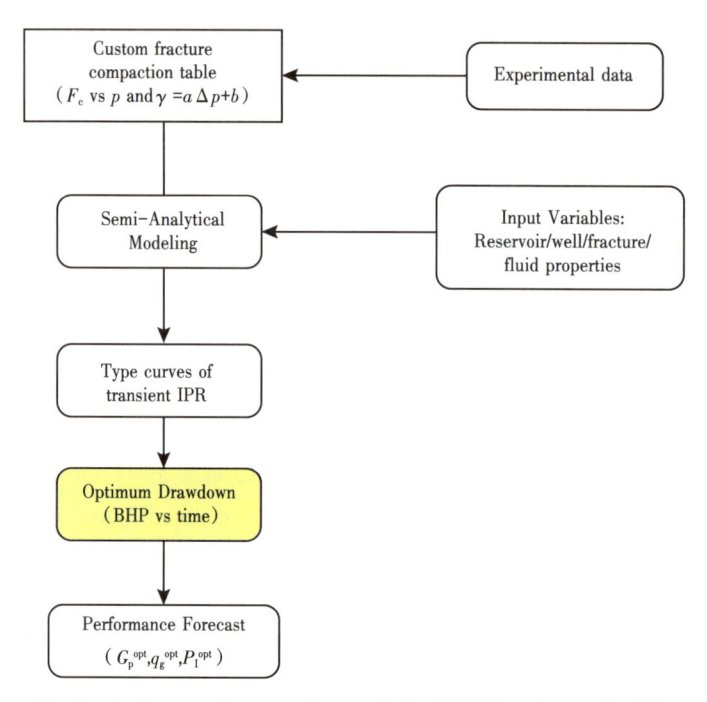

Fig.9　An integrated approach to optimize BHP drawdown schedule

As shown in Fig. 8, the operating points over time are defined as the optimum schedule of BHP-drawdown. The corresponding reversal behavior is attributed to the serious degradation of productivity, which depends on the BHP drawdown. Finally, the optimum profile of BHP-drawdown can be achieved by integrating operating points on transient IPR curves corresponding to different times. Note that the optimum BHP drawdown is increasing with time. Fig. 10 illustrates the effect of the order of stress-dependency magnitude on optimum drawdown profile, which indicates that the optimal drawdown depends mainly on the time and the stress sensitivity. The larger the permeability-decay coefficient, the smaller the drawdown at the same time. A more intense stress effect requires a more conservative drawdown to maximize the cumulative production rate. Alternatively, the optimum schedule has a BHP profile that declined very gradually until line BHP was reached, while the optimum schedule with weak stress sensitivity has a rapid BHP-decline profile. This observation is also consistent with the observation of Nguyen et al[28].

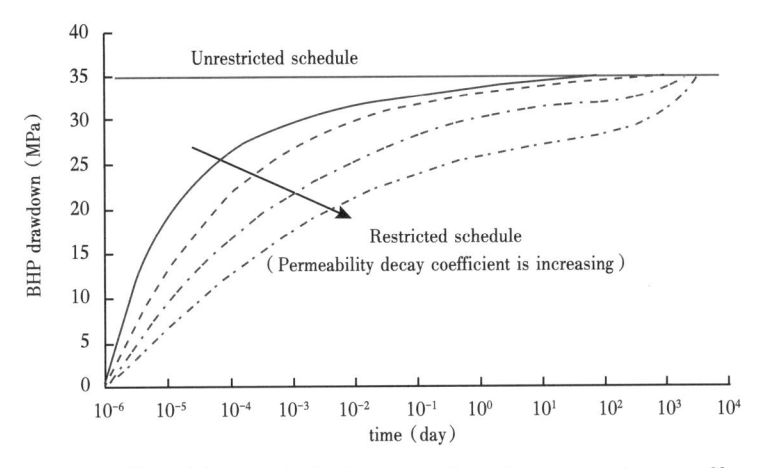

Fig.10  Effect of the magnitude of pressure dependency on optimum profile

It is emphasized that according to the principle of superposition, the current performance is influenced by the previous production history. The BHP drawdown should be calibrated by capturing the time-lapse behavior, considering the effect of production history on transient IPRs. In the practical drawdown-optimization process, the variable-BHP condition is regarded as a set of stepwise constant-BHP conditions which lasts very-short time scope. After taking the optimum BHP condition previously, the next time-scope transient IPRs are then generated. It is worth noting that even after considering the history the trend of optimum BHP-decline profile presented in Fig. 10b would still not be changed.

## 4  Field Case Study

Well performance is affected by many parameters, such as formation petrophysical properties, fluid properties, fracturing parameters, and operational condition. Numerous field practices demonstrate that when petrophysical and fluid properties are given, the performance is consistent with engineering specifications (i.e., horizontal length, number of fractures and volume of proppant).

Two horizontal wells in Zhaotong shale, China are selected to perform the analysis. Here, one well is controlled by unrestricted BHP drawdown, and the other is controlled by the (optimum) restricted BHP drawdown. Other geology and engineering parameters are list in Table 2. As analyzed from Table 2, horizontal length and number of fractures in the unrestricted well are smaller than the restricted well, but the volume of proppant is higher than restricted well by 50%. Therefore, without taking the effect of operational condition into account, the productivity of the unrestricted well approximates to the restricted well.

**Table 2  Geology and engineering parameters for unrestricted and restricted wells**

| Well name | Operational condition | Horizontal length (m) | Well spacing (m) | Horizontal fractured length (m) | Number of fractures | Volume of proppant (m³) |
|---|---|---|---|---|---|---|
| Unrestricted well | unrestricted | 1510 | 400 | 1318 | 18 | 950.5 |
| Restricted well | restricted | 1605 | 400 | 1568 | 22 | 625.6 |

\* The density of proppant is 1760kg/m³.

## 4.1  Unrestricted vs. Restricted Drawdown

By analyzing history-performance data given in Fig. 11, the decline rate of BHP for the unrestricted well is 0.278MPa/d, but the value for the restricted well is only 0.13MPa/d.

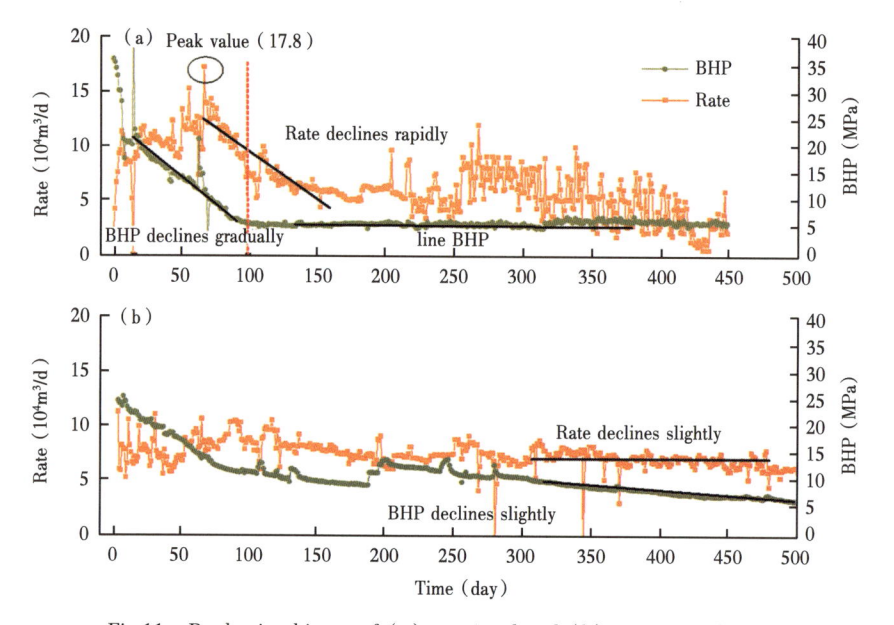

Fig.11  Production history of (a) restricted and (b) unrestricted wells

The well performances are further evaluated by quantitating the change normalized gas cumulative production with time, shown in Fig. 12. It indicates that the performance of the restricted well is better than the unrestricted well over the whole time range. During the first 3 months the value for the restricted well is about $1.52 \times 10^4 m^3/MPa$, being higher than $1.22 \times 10^4 m^3/MPa$ for the unrestricted well. The ratio of the restricted to the unrestricted is about 1.24. During the first 12 month, the ratio is further increased to 1.33. The difference of normalized gas cumulative production is attributed to the effect of operational condition.

Based on the parameters presented in Table 2, the conceptual reservoir model is established by use of semi-analytical modeling given in section 2. The initial pressure is 38 MPa, and the finial line BHP is set to be 5MPa. Characteristic parameter $a$ is $0.1108MPa^{-2}$, and characteristic parameter $b$ is $0.00167MPa^{-1}$. These parameters are all input into the semi-analytical model to perform history

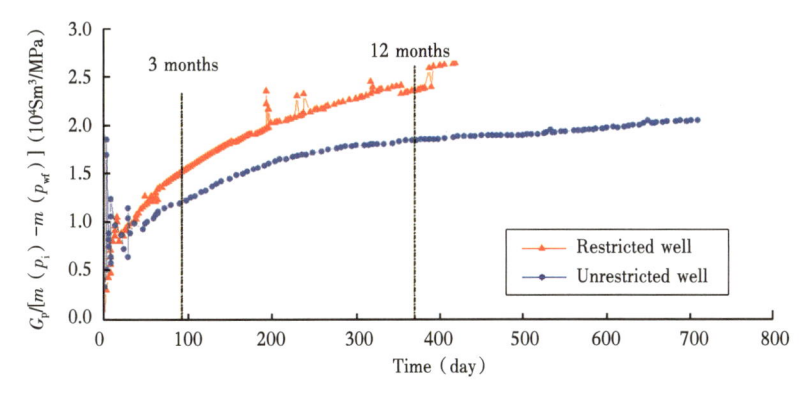

Fig.12 Comparison of the effect of restricted and unrestricted drawdowns on the performance

matching, and well matching is achieved as shown in Fig. 13a and Fig. 13b. By history matching[29], some parameters could be achieved: initial fracture conductivity and length respectively equal to 36mD·m and 90m; matrix porosity is 8%, and permeability equals to 0.0002mD.

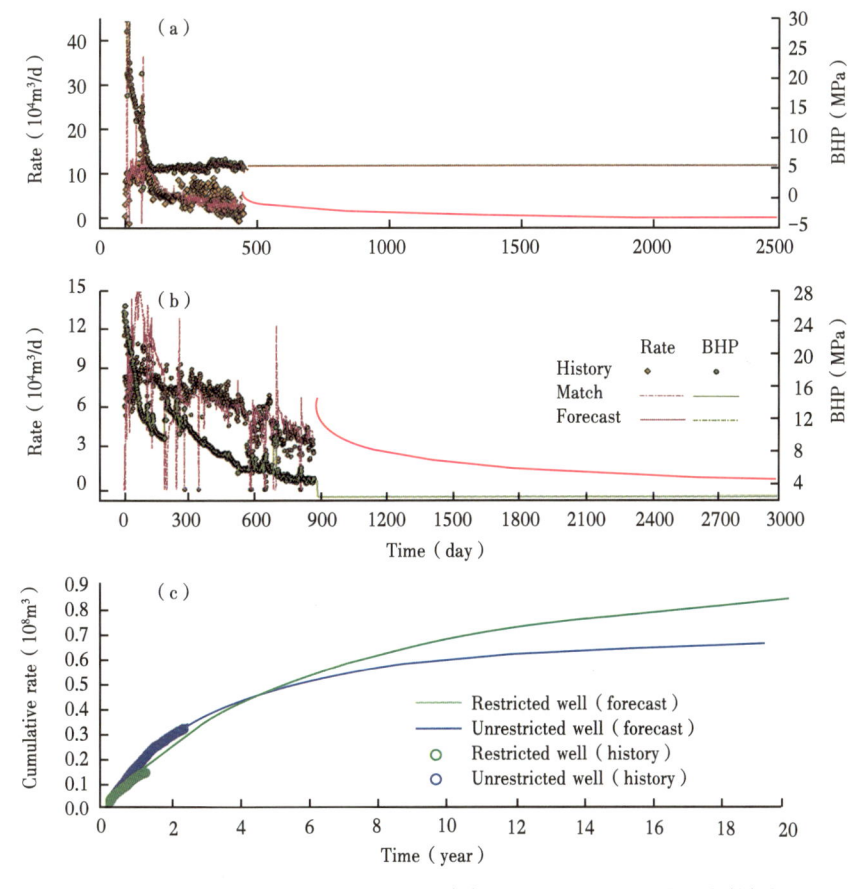

Fig.13 Production-history matching for the (a) history-unrestricted and (b) history-restricted wells; (c) Performance forecast

The performance is further predicted as shown in Fig. 14. It well reproduces the familiar trends from the fully coupled-geomechanics model, with lower long-term cumulative production and higher short-term cumulative production corresponding to the unrestricted drawdown. There is a transition time. Before the transition time, the cumulative-gas production of unrestricted case is higher than restricted case. In this work, cumulative-gas production of 20 years is especially defined as estimated ultimate recovery (EUR) according to the industrial standard in China[30]. EUR for the restricted well is about $0.836 \times 10^8 m^3$; while EUR for the unrestricted well is about $0.661 \times 10^8 m^3$. Compared with unrestricted schedule, the restricted drawdown has the ability of increasing EUR of up to 25%.

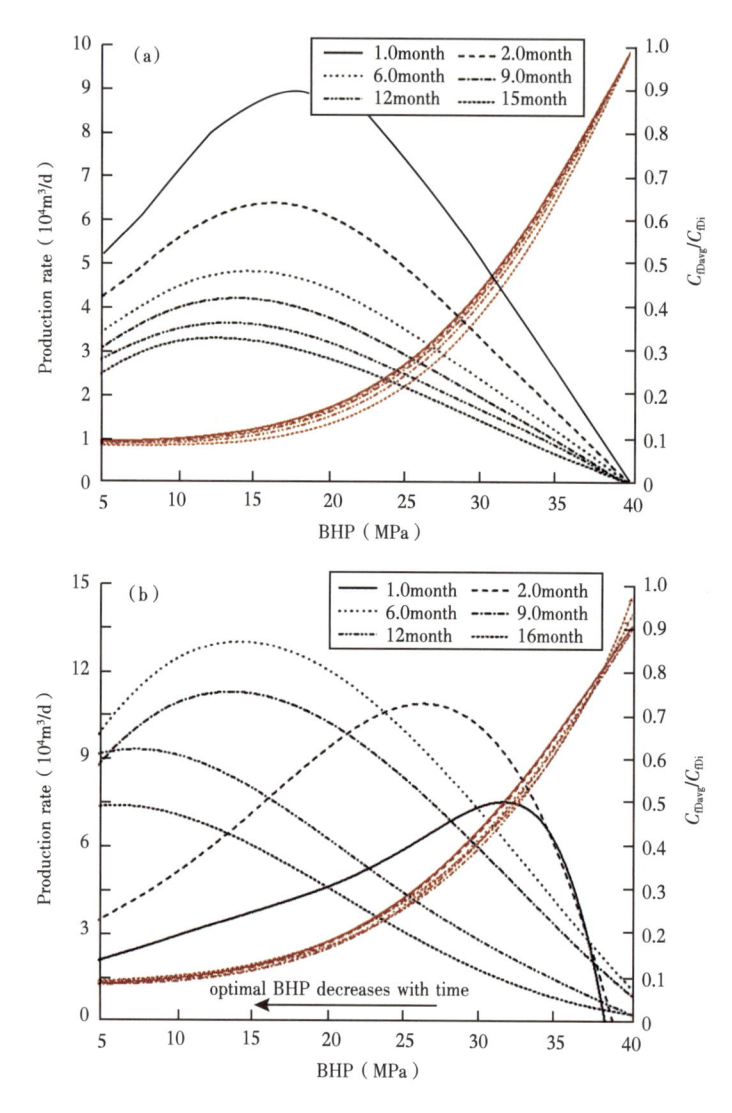

Fig.14 Optimum BHP drawdown for restricted well (a) without history and (b) with history

## 4.2 Increasing EUR of Unrestricted Well by Optimum Drawdown

According to the workflow, the optimum BHP schedule is achieved as shown in Fig. 14a. Taking the effect of production history into account, Fig. 14b generates a more practical transient IPR, which results in the optimum BHP after 447 day (forecast duration). Then, the optimum drawdown is put into the semi-analytical model for the history-unrestricted well to predict production performance. Constant BHP-drawdown schedule, 2.5 MPa, is put into the history-restricted well.

Fig. 15 shows the production forecast for history-unrestricted and history-restricted wells. As seen in Fig. 15a, although the drawdown of the history-unrestricted well is not managed in the initial time range (1~447 day), the drawdown of this well is strictly managed under the optimum strategy after 447 day. Note that the finial stabilized BHP of history-unrestricted well is set to be 2.5 MPa. As a comparison, seen in Fig. 15b, the history-restricted well always keep constant BHP ($p_{wf}$=2.5 MPa) during the forecast. Fig. 15c shows that the EURs for both wells would be increased due to the

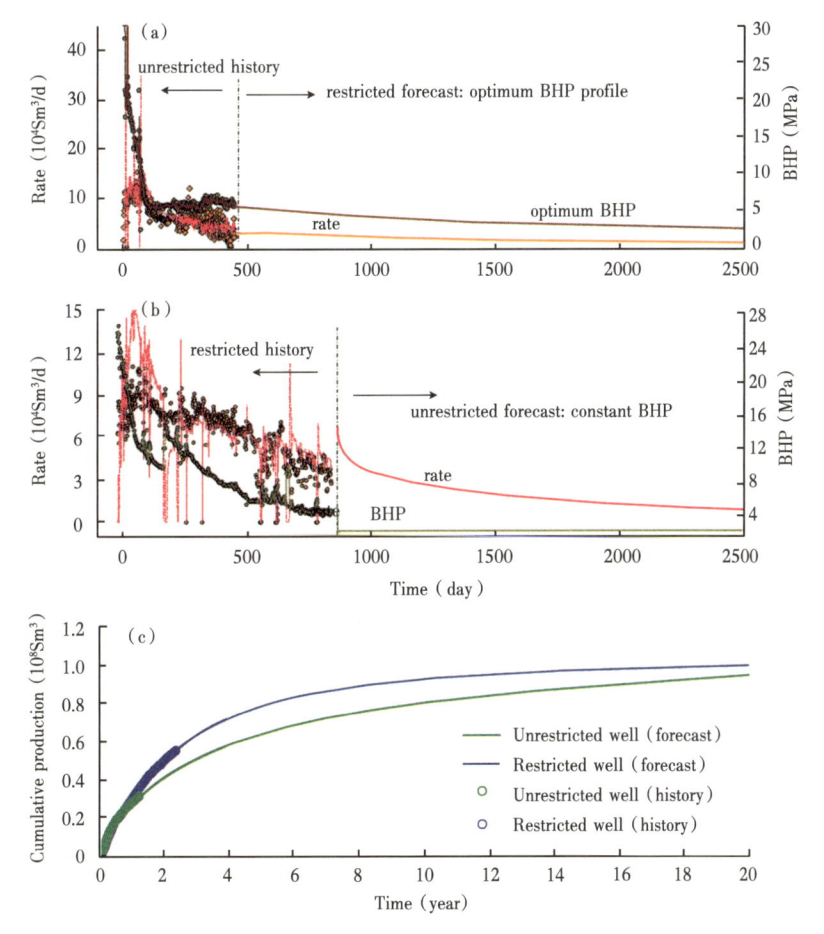

Fig.15 History-match and forecast for the history-unrestricted well with optimization and history-restricted well without optimization

increasing of finial BHP drawdown (i.e. BHP is decreased from 5 MPa to 2.5 MPa).

Results of EUR calculation are summarized in Table 3. It indicates that the EUR of the history-restricted well is increased from $0.836 \times 10^8 m^3$ to $1.001 \times 10^8 m^3$, but the EUR of the history-unrestricted well is increased from $0.661 \times 10^8 m^3$ to $0.949 \times 10^8 m^3$. The reason is explained that the degree of the EUR increase is determined by the BHP schedule during the forecast. The performance of the history-unrestricted well is greatly improved by using the optimum BHP drawdown, while the advantage of the history-restricted well at early stage is offset by the unrestricted management during forecasting.

Table 3　Results of EUR with and without BHP optimization

| Well name | History period | Without optimization | | With optimization | |
| --- | --- | --- | --- | --- | --- |
| | | Forecast period | EUR($10^8 m^3$) | Forecast period | EUR($10^8 m^3$) |
| Unrestricted well | High drawdown | Constant BHP ($p_{wf}$ = 5MPa) | 0.661 | Optimum drawdown (finial BHP = 2.5MPa) | 0.949 |
| Restricted well | Managed drawdown | Constant BHP ($p_{wf}$ = 5MPa) | 0.836 | Constant BHP ($p_{wf}$ = 2.5MPa) | 1.001 |

## 5　Summary and Conclusions

This study investigated the effect of stress-dependent conductivity of fractures on the performance of reservoir under different operational conditions. The relationship between production rate and BHP drawdown is quantified in the form of transient IPR. Afterward, an integrated approach was developed to give an efficient workflow of selecting an optimum BHP-drawdown schedule. Some important insights are summarized as follows:

(1) Different from conventional IPR, the productivity index deteriorates with time in the transient IPR. Due to pressure dependency of fracture conductivity, productivity index would deteriorate, which contributes to the production rate gain is lower than the predicted straight-line behavior.

(2) When the permeability-decay coefficient is a function of pressure drawdown, there exists a maximum value on transient IPR.

(3) Optimum BHP-drawdown schedule finds a reasonable tradeoff in which the fracture mains considerable conductive while maintaining a high enough drawdown to maximize the long-term cumulative gas-production.

## Nomenclature

### Filed variables

$a$ = the characteristic parameter, $Pa^{-2}$

$b$ = the characteristic parameter, $Pa^{-1}$

$c$ = compressibility, $Pa^{-1}$

$C_{fD}$ = dimensionless conductivity

$d_f$ = permeability modulus, $Pa^{-1}$

$F_c$ = fracture conductivity, mD·m

$h$ = formation thickness, m

$I_x$ = fracture penetration ratio, dimensionless

$k$ = permeability, mD

$k_{min}$ = residual permeability, mD

$L$ = length, m

$m$ = pseudo pressure, Pa

$p$ = pressure, Pa

$p_u$ = pressure drop under unit-rate condition, Pa

$q_f$ = flux density along fracture, $m^2/s$

$q_{wf}$ = production rate of fracture, $m^3/s$

$q_c$ = cross-section rate within fracture, $Sm^3/s$

$S_c$ = convergence flow skin, dimensionless

$t$ = time, s

$t_a$ = pseudo time, s

$x$ = x-direction coordinate, m

$w_f$ = fracture width, m

$x_e$ = the length of boxed reservoir in $x$ direction, m

$y$ = y-direction coordinate, m

$y_e$ = the length of boxed reservoir in $y$ direction, m

$Z$ = gas deviation factor, dimensionless

$\varphi$ = porosity, dimensionless

$\alpha$ = the Biot coefficient, dimensionless

$\sigma$ = closure stress, Pa

$\upsilon$ = Poisson's ratio, dimensionless

$\gamma$ = permeability-decay coefficient, $Pa^{-1}$

$\xi$ = transformed dimension, m

$\kappa$ = iterative number

$\mu$ = viscosity, Pa·s

**Subscripts**

D = dimensionless

f = fracture

m = matrix

w = well

ref = reference

i = initial

min = minimal

g = gas

m, n = count number

υ, ν = count number

## Appendix A: Dimensionless Definitions

For simplicity, the mathematical model is expressed in the dimensionless form, and the corresponding variables are defined as follows:

$$t_D = \frac{K_m t_a}{\phi_m \mu_{gi} c_{gi} L_{ref}^2}, L_D = \frac{L}{L_{ref}}, C_{fD} = \frac{K_f w_f}{K_m L_{ref}} \tag{A1}$$

The dimensionless permeability-decay coefficient is defined as

$$\gamma_{fD} = a_D p_{fD} + b_D \tag{A2}$$

In the condition of constant and variable-BHP condition, the dimensionless definitions are respectively given by

$$p_{fD} = \frac{m(p_i) - m(p_f)}{m(p_i) - m(p_w)} \tag{A3}$$

$$p_{mD} = \frac{m(p_i) - m(p_m)}{m(p_i) - m(p_w)} \tag{A4}$$

$$\gamma_{fD} = (p_i - p_w)\gamma_f \tag{A5}$$

$$q_{wfD} = \frac{q_{wf}\mu_{gi}B_{gi}}{2\pi K_m h[m(p_i) - m(p_w)]} \tag{A6}$$

$$q_{fD} = \frac{(2q_f L_{ref}) \cdot \mu_{gi} B_{gi}}{2\pi K_m h[m(p_i) - m(p_w)]} \tag{A7}$$

## References

[1] Cipolla C L, Williams M J, Weng X, Mack M, and Maxwell S. Hydraulic fracture monitoring to reservoir simulation: maximizing value. Proceedings of the SPE Annual Technical Conference and Exhibition, Florence, Italy, 2015, September 19-22, SPE 133877.

[2] Aybar U, Yu W, Eshkalak M O. Evaluation of production losses from unconventional shale reservoirs. Journal of Natural Gas Science and Engineering. 2015, 23:509-516.

[3] Britt L K, Smith M B, Klein H H, and Deng J Y. Production benefits from complexity - effects of rock fabric, managed drawdown, and propped fracture conductivity. Proceeding of the SPE Hydraulic Fracturing Technology Conference, Woodlands, Texas, 2016, 9-11 February, SPE-179159-MS.

[4] Guerra J, Zhu D, and Hill A D. Impairment of fracture conductivity in the Eagle Ford shale formation. SPE Production & Operations. 2018,33（4）:637-653.

[5] Tang Y, Ranjith P G, Perera M S A, and Rathnaweera T D. Influences of proppant concentration and fracturing fluids on proppant-embedment behavior for inhomogeneous rock medium: an experimental and numerical study. SPE Production & Operations. 2018,33（4）:666-678.

[6] Guo X Y, Song H Q, Wu K, and Killough, J. Pressure characteristics and performance of multi-stage fractured horizontal well in shale gas reservoirs with coupled flow and geomechanics. Journal of Petroleum Science and Engineering. 2018,163, 1-15.

[7] Huang X L, Guo X, Zhou X, Lu X Q, Shen C, Qi Z L, and Li J Q. Productivity model for water-producing gas well in a dipping gas reservoir with an aquifer considering stress-sensitive effect. ASEM Journal of Energy Resources Technology. 2019,141（2）:022903.

[8] Okouma M V, Guillot V F, Sarfare M, et al. Estimate ultimate recovery as a function of production practices in the Haynesville shale. Proceeding of the SPE Annual Technical Conference and Exhibition, Denver, 2011, 30 October-2 November, SPE-147623-MS.

[9] Seales M B, Ertekin T, and Wang Y L. Recovery efficiency in hydraulically fractured shale gas reservoirs. ASEM Journal of Energy Resources Technology: 2017,139（4）:042901.

[10] Rojas D, and Lerza, A. Horizontal well productivity enhancement through drawdown management approach in Vaca Muerta shale. Proceedings of the SPE Canada Unconventional Resources Conference, Calgary, Alberta, Canada, 2018,13-14 March, SPE-189822-MS.

[11] Karantinos E, Sharma M M, Ayoub J A, Parlar M, Chanpura R A. A general method for the selection of an optimum chock-management strategy. SPE Production & Operations. 2017,5,137-147.

[12] Kumar A, Seth P, Shrivastava K, Sharma M M. Optimizing drawdown strategies in wells producing from complex fracture networks. Proceeding of the SPE International Hydraulic Fracturing Technology Conference and Exhibition, Muscat, Oman, 2017,16-18 October, SPE-191419-18IHFT-MS.

[13] Mirani A, Marongiu-Porcu M, Wang H Y, and Enkababian P. Production-pressure-drawdown management for fractured horizontal wells in shale-gas formations. SPE Reservoir Evaluation & Engineering. 2018,21（3）:550-565.

[14] Wilson K. Efficient stress characterization for real-time drawdown management. Proceeding of the Unconventional Resources Technology Conference, Austin, Texas, 2017,24-26 July 2015, URTeC-221192-MS.

[15] Jiang L W, Liu T J, Yang D Y. A semianalytical model for predicting transient pressure behavior of a hydraulically fractured horizontal well in a naturally fractured reservoir with non-Darcy flow and stress-sensitive permeability effects. SPE Journal, 2019,24（3）:1322-1341.

[16] Jiang L W, Liu T J, Yang D Y. Effect of stress-sensitive fracture conductivity on transient pressure behavior for a horizontal well with multistage fractures. SPE Journal. 2019,24（3）:1342-1363.

[17] Wilson K. Analysis of drawdown sensitivity in shale reservoirs using coupled-geomechanics models. Proceeding of the SPE Annual Technical Conference and Exhibition, Houston, Texas, 2015,28-29 September, SPE-175029-MS.

[18] Gudala M, and Govindarajan S K. Numerical Modeling of Coupled Fluid Flow and Geomechanical Stresses in a Petroleum Reservoir. ASEM Journal of Energy Resources Technology. 2020,142（6）:063006.

[19] Yilmaz O, Nolen-Hoeksema R C, Nur A. Pore pressure profiles in fractured and compliant rocks. Geophysical Prospecting. 1994,42. 693-714.

[20] Jia B, Tsau J S, Ghahfarokhi R B.Investigation of shale-gas-production behavior: evaluation of the effects of multiple physics on the matrix". SPE Reservoir Evaluation & Engineering. 2020,23（1）:68-80.

[21] Guo W, Xiong W, Shu G. Experimental study on stress sensitivity of shale gas reservoirs. Special Oil and Gas Reservoirs. 2012, 19(1):95-98.

[22] Wang J L, Luo W J, Chen Z M, An integrated approach to optimize bottomhole-pressure-drawdown management for a hydraulically fractured well using a transient inflow performance relationship. SPE Reservoir Evaluation & Engineering. 2020,23（1）:95-111.

[23] Zhang F, and Yang D Y. Effects of non-Darcy flow and penetrating ratio on performance of horizontal wells with multiple fractures in tight formation". ASEM Journal of Energy Resources Technology. 2018,140（6）:032903.

[24] Chen Z M, Liao X W, Zhao X L, A practical methodology for production-data analysis of single-phase unconventional wells with complex fracture geometry", SPE Journal, 2019,22（2）:458-476.

[25] Ye P, and Ayala H L F. A density-diffusivity approach for the unsteady state analysis of natural gas reservoirs. Journal of Natural Gas Science and Engineering, 2012,7:22-34.

[26] Yuan B, Moghanloo G, and Shariff E. Integrated Investigation of Dynamic Drainage Volume and Inflow Performance Relationship (Transient IPR) to Optimize Multistage Fractured Horizontal Wells in Tight/Shale Formations. ASEM Journal of Energy Resources Technology. 2016,138（5）:052901.

[27] Tabatabaie S H, Pooladi-Darvish, and Mattar L. Draw-down management leads to better productivity in reservoirs with pressure-dependent permeability-or does it? . Proceeding of the SPE/CSUR Unconventional Resources Conference, Calgary, Alberta, Canada, 2015,20-22 October, SPE-175938-MS.

[28] Nguyen T C, Pande S, Bui D, et al. Pressure dependent permeability: unconventional approach on well performance. Journal of Petroleum Science and Engineering. 2015,193:107358.

[29] Jia A L, Wei Y S, Liu C, et al. A dynamic prediction model of pressure-control production performance of shale gas fractured horizontal wells and its appli. Natural Gas Industry B. 2020, 7:71-81.

[30] Bi H B, Hao M, Gao R L, et al, Evaluation method of recoverable reserves of single well in undeveloped area of shale gas. Acta Petrolei Sinia. 2020, 41（5）:565-573.

# Production Behavior Evaluation on Multilayer Commingled Stress-Sensitive Carbonate Gas Reservoir

Jianlin Guo[1]    Fankun Meng[2]    Ailin Jia[1]    Shuo Dong[3]    Haijun Yan[1]

(1. PetroChina Research Institute of Petroleum Exploration & Development,
Beijing, People's Republic of China;
2. Shengli College, China University of Petroleum, Dongying, People's Republic of China;
3. Department of Petroleum Engineering, Texas Tech University, Texas, USA)

**Abstract**: Influenced by complex sedimentary environment, a well always penetrates multiple layers with different properties. In practical, it is hard to analyze the production behavior for each layer. This paper proposed a semi-analytical model to evaluate the production performance of each layer in stress-sensitive multilayer carbonated gas reservoir. The flow of fluids in layers composed of matrix, fractures and vugs can be described by triple-porosity/single permeability model, and the other layers could be characterized by single porosity media. The stress sensitive exponents for different layers are determined by laboratory experiments and curve fitting, which are considered in pseudo-pressure and pseudo-time factor. Laplace transformation, Duhamel convolution, Stehfest inversion algorithm are used to calculate the well bottom-hole pressure and the production rate for each layer. Through the comparison with the classical solution, and the match with real bottom-hole pressure data collected from Gaoshiti-Moxi carbonate gas reservoir, the accuracy of the presented model is verified. A synthetic case which has two layers that the first one is tight and the second one is full of fractures and vugs, is utilized to study the effect of stress-sensitive exponents, skin factors, formation radius and permeability for these two layers on production performance. The results demonstrate that initially the well production is mainly derived from high permeable layer, which causes that in the early stage the bottom-hole pressure and the second layer production rate increase with the rise of formation permeability and radius, and the decrease of stress-sensitive exponents and skin factors. While the first layer contributes a lot to the total production in the later period, and the well bottom-hole pressure are influenced by the variation of formation permeability and radius, stress-sensitive exponent and skin factor at later stage. Compared with the second layer, the scales of formation permeability and skin factor for first layer have significant impacts on production behaviors. Hence, to enhance the well performance, it is meaningful to reduce the formation damage and improve the flow capacity for tight layer.

**Key words**: Production behavior; multi-layered formations; carbonate gas reservoir, stress sensitivity; commingled vertical well.

# 1 Introduction

Gaoshiti-Moxi carbonate gas reservoir is located in Sichuan Basin and mainly developed in formation $Z_2dn_4$[1,2]. According to the characteristics of lithology, resistance and sedimentary cycle, the formation $Z_2dn_4$ can be divided into two stratums, $Z_2dn_4^1$ and $Z_2dn_4^2$. Well logging and core analysis show that there are great differences in physical properties between these two stratums. $Z_2dn_4^1$ has abundant natural fractures and vugs, which has large porosity, permeability and stress-sensitivity. While $Z_2dn_4^2$ is tight and has few fractures and cavities, the porosity and permeability are much lower than $Z_2dn_4^1$. To improve the well performance and obtain better economic benefits, these two stratums are always seen as one layer and penetrated by vertical wells to produce. However, currently, it is no clear that the contribution of each layers, which maybe lead to the unreasonable plan for production schemes. It is imperative to propose an appropriate mathematical model to simulate the production process in multilayer carbonate gas reservoir with serious vertical heterogeneity and stress-sensitivity.

In terms of the modelling of multilayer commingled reservoirs with no-crossflow, a substantial number of investigators have presented various analytical or semi-analytical models to study the pressure transient behavior or production performance. For pressure transient analysis in multi-layered reservoirs, Lefkovits et al. made a rigorous study on the pressure behavior for the reservoirs composed of stratified layers with uniform pressure initially[3]. Since Lefkovits et al. pointed that it was hard to determine the properties of individual layers, the methods of estimating the formation properties were extended through the utilization of single layer plotting forms[4,5]. To solve the analytical model in Laplace space for layered reservoirs, Tariq and Ramey firstly introduced Stehfest numerical inversion method[6], and Spath et al. presented an efficient algorithm to compute pressure response through the application of single layer solution and Duhamel's theorem[7]. Additionally, Sun et al. used Crump numerical inversion method to give exact solution and typical curves for commingled reservoir[8]. Due to the difference of depth for each layer, it is unrealistic for the assumption of layers with uniform pressure. investigated the pressure and flow-rate transients caused by unequal initial pressure and various boundary conditions of individual zones in commingled multilayer reservoir[9,10]. Subsequently, numerous researchers presented a lot of mathematical models to study the pressure response in commingled reservoirs with different layer pressures, pressure-dependent or independent fluids and formation properties, and boundary conditions[11-16]. Furthermore, some of studies also proposed approaches, such as derivative extreme method (DEM), single-layer Mathews-Brown-Hazebroek function and layered reservoir test (LRT), for the interpretation of individual layer permeability, porosity and skin factor, and average reservoir pressure[17-22]. Recently, Sui and Zhu presented a wellbore/reservoir coupled model for multilayer commingled gas reservoir, which considered the non-Darcy and damage skin effect for each layer, and the variation of wellbore pressure with strategic locations[23]. Shi et al. proposed an analytical model to characterize the bottom-hole pressure behavior for acid fracturing stimulated wells in multi-layered carbonate gas reservoir[24]. Vieira Bela et al. extended the existing the falloff formulation to multilayer reservoirs[25].

With regard to the production performance analysis in multi-layer reservoirs, presented the method coupling material balance equation and stabilized non-Darcy gas flow equation with varied or constant bottom-hole flowing pressure, to match production data and then estimate OGIP and productivity of individual zone[26,27]. Arevalo-Villagran et al. improved the model developed by El-Banbi and Wattenbarger[28], to allow modelling of recompletions and different initial pressure in layers for multi-layer-commingled systems. Onwunyili and Onyekonwu coupled wellbore hydraulics and pay-zone fluid behavior with consideration of pressure difference between producing zones, and estimated the production contribution of various zones in multilayer reservoirs[29]. Hydraulically fractured vertical wells are always used in tight gas reservoirs. Ali et al. proposed a robust methodology to estimate fracture properties, permeability and drainage area of individual layers in multi-layer commingled tight gas reservoir produced with fractured vertical gas wells[30]. Wang et al. presented a case study of multilayered tight gas sands in eastern Ordos Basin, and investigated the influence of interference between different layers on production[31].

Through the above reviews on the pressure transient analysis and production performance evaluation in multilayered commingle reservoirs, it can be seen that the presented mathematical models do not consider the great differences of porous media types for individual layer. Moreover, the studies on stress sensitivity in naturally fractured vuggy carbonate gas reservoir indicate that there are strong stress sensitive effects which severely affects the production performance[32-35]. Therefore, in this paper, we propose a semi-analytical model that considers the differences of porous characteristics and stress sensitivity for each layer in multi-layer commingled reservoir, in which some layers are naturally fractured vuggy formations represented by triple porous medium, and the others are tight formations characterized by single porous medium. To solve the model, Laplace transformation, Stehfest numerical inversion and Duhamel's principle are employed. With the proposed model, the effects of prevailing factors on wellbore pressure and production performance of individual zone are studied.

## 2 Laboratory experiments for formation stress sensitivity

Two typical core samples collected from formations $Z_2dn_4^1$ and $Z_2dn_4^2$ in Gaoshiti-Moxi carbonate gas reservoir are used to conduct stress-sensitive experiments by applying core flooding apparatus. In these two experiments, the flowing pressure in core samples is decreased by 5 MPa to simulate reservoir depletion while the confining pressure keeps constant, then the core permeability is measured at each point. Fig. 1 shows the results of two experiments and curve fitting with two mathematical functions (power function and exponential function). Since for fractured core sample the natural fractures act as main channels, hence the measured core permeability can be seen as fracture permeability.

In Fig. 1, the net confining stress in horizontal axis is defined as the difference between overburden pressure and formation fluid pressure, and the dimensionless permeability in vertical axis represents the ratio of measured permeability to initial values. As Fig.1 shows, for these two kinds of formations, the experimental data could be matched with power function suitably, which is in

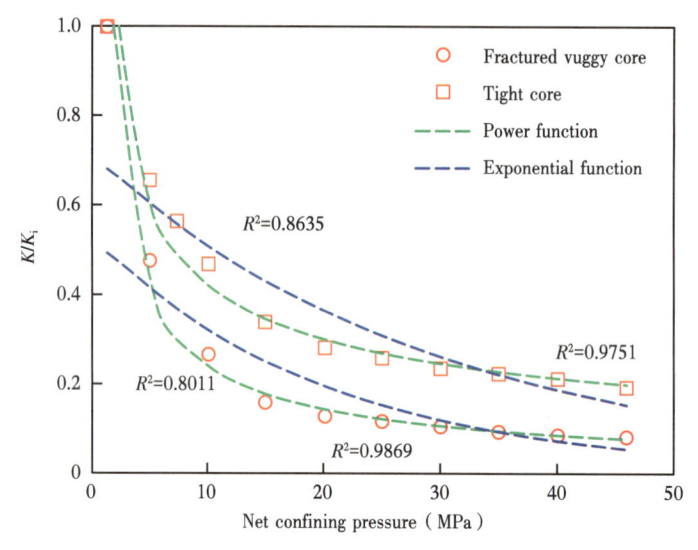

Fig. 1  The permeability stress sensitivity for naturally fractured vuggy and tight formations

compliance with the previous studies[34]. Thus the relation between net confining pressure and dimensionless permeability can be determined, as shown in Eq. 1.

$$\frac{K}{K_i} = \left(\frac{\sigma_s - p}{\sigma_s - p_i}\right)^{-\alpha} \quad (1)$$

Where α is named as stress-sensitive exponent, in this case, it equals to 0.738 and 0.493 for naturally fractured vuggy and tight formations, respectively. In addition, it also can be seen as the natural fractures in fractured vuggy formation are more sensitive to pressure variation compared with tight formation, thus the permeability stress sensitivity for fractured vuggy formation is more serious than tight formation, as shown in Fig.1.

## 3  Physical model and mathematical model

### 3.1  Physical model

Fig. 2 shows a schematic of multilayered commingled carbonate gas reservoir under study in this paper. In this reservoir, some upper layers are naturally fractured vuggy formations, and the other lower layers are tight formations with few fractures and cavities.

The assumptions of this model are listed as follows: (1) the reservoir is composed of $n$ cylindrical, homogenous and isotropic layers, and any of these layers are bounded. The formation radius and initial pressure prior to production for layer $j$ are $R_{ej}$ and $p_{ij}(j=1, 2, 3\cdots, n)$. (2) In fractured vuggy formation, matrix and vugs are the storage spaces of fluids, and fractures act as main channels to the wellbore. Provided that the transfer flow between factures and matrix or vugs is pseudo-steady state (PSS). (3) Permeability stress sensitivity for naturally fractured vuggy and tight formations can be described with Eq. 1, which stress-sensitive exponents are 0.738 and 0.493, respectively. (4) The gas is compressible and has pressure-dependent properties, and the compressibility of rock and connate water is ignored. Gas flow in porous media follows Darcy's law, and the gravity and capillary force are neglected. (5) The well is located in the center of the reservoir and produces with a constant production rate $q_g$, and the wellbore radius is $r_w$. All layers are penetrated completely by the vertical well, and the skin effect is considered. There is no interlayer flow and the crossflow between layers merely occurs through the wellbore.

In addition, it should be noted that because the triple porous medium with matrix, naturally fractures and vuggs can be simplified into single medium with matrix only under some certain circumstances, therefore, for simplicity, all layers are assumed to be triple porous medium during the derivation of the mathematical model.

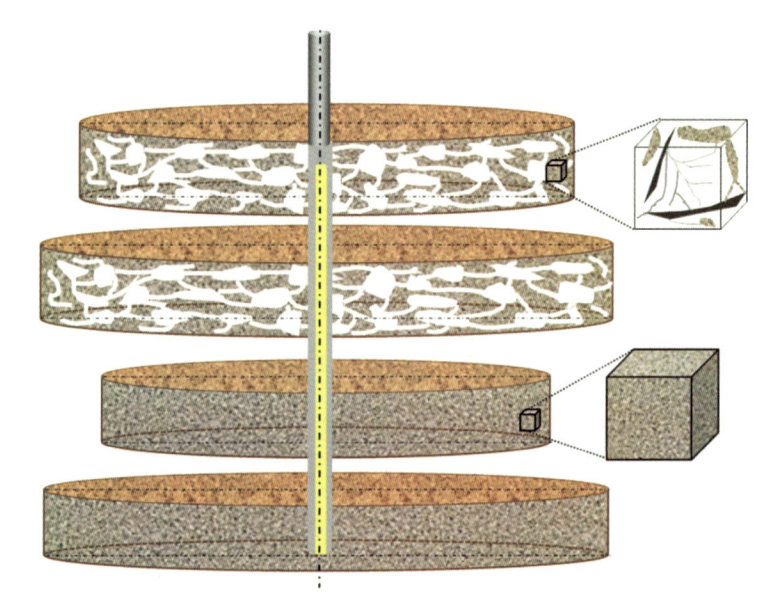

Fig. 2 Schematic of multilayer commingled carbonate gas reservoir

## 3.2 Mathematical model

According to the governing equations for layer $j$, the solution of dimensionless pseudo-pressure in Laplace domain can be obtained and expressed by (Appendix A):

$$\overline{m}_{fjD} = \frac{m_{ijD}}{s} + AI_0(\sqrt{D_j(s)}\, r_D) + BK_0(\sqrt{D_j(s)}\, r_D) \qquad (2)$$

In Eq. 2, $m_{ijD}/s$ is a characteristic solution for the governing equation, and the other parts are the general solution when $m_{ijD}$ is set to be zero. The production rate for any of layers varies with time, which can be deal with Duhamel's principle[7], and the general equation can be given by:

$$m_{wfj} = m_{ij} + \int_0^t q_{gj} B_{gij} \frac{\partial}{\partial \tau} m_{wfjr}(t-\tau)\, d\tau \qquad (3)$$

The dimensionless expression of Eq. 3 is:

$$m_{wfjD} = m_{ijD} + \int_0^{t_D} q_{jD} \frac{\partial}{\partial \tau} m_{wfjrD}(t_D - \tau) d\tau \tag{4}$$

Where

$$m_{wfjr} = \left(\frac{\mu_{gi} z_i}{p_i K_{fi}}\right)_j \int_{p_0}^{p_{wfjr}} \frac{K_{fj} p_j}{\mu_{gj} z_j} dp, \quad m_{wfjD} = \frac{(K_{fi} h/\mu_{gi})_r (m_{i,j} - m_{wfjr})}{a_p (q_g B_{gi})_r}$$

To obtain the solution of $m_{fjD}$ in Laplace space, on the basis of Duhamel's principle, $q_{jD}$ should be equal to 1, then the Eq (A.16) can be simplified and written as:

$$M_j r_D \frac{\partial \overline{m}_{fjD}}{\partial r_D} \bigg|_{r_D = 1} = -\frac{1}{S}, \quad \left(\overline{m}_{fjD} - S_j r_D \frac{\partial \overline{m}_{fjD}}{\partial r_D}\right)\bigg|_{r_D = 1} = \overline{m}_{wfjrD} \tag{5}$$

Substituting Eq. 2 into Eq. 5 and Eq. (A.9), then the parameters $A$ and $B$ in Eq. 2 can be given by:

$$A = \frac{K_1(\sqrt{D_j(s)} R_{ejD})}{sM_j \sqrt{D_j(s)} [I_1(\sqrt{D_j(s)} R_{ejD}) K_1(\sqrt{D_j(s)}) - I_1(\sqrt{D_j(s)}) K_1(\sqrt{D_j(s)} R_{ejD})]}$$

$$B = \frac{I_1(\sqrt{D_j(s)} R_{ejD})}{sM_j \sqrt{D_j(s)} [I_1(\sqrt{D_j(s)} R_{ejD}) K_1(\sqrt{D_j(s)}) - I_1(\sqrt{D_j(s)}) K_1(\sqrt{D_j(s)} R_{ejD})]} \tag{6}$$

Thus the solution of $m_{wfjrD}$ in Laplace domain with consideration of damage skin effects can be expressed by:

$$\overline{m}_{wfjrD} = \frac{m_{ijD}}{s} + \frac{K_1(\sqrt{D_j(s)} R_{ejD}) I_0(\sqrt{D_j(s)}) + I_1(\sqrt{D_j(s)} R_{ejD}) K_0(\sqrt{D_j(s)})}{sM_j \sqrt{D_j(s)} [I_1(\sqrt{D_j(s)} R_{ejD}) K_1(\sqrt{D_j(s)}) - I_1(\sqrt{D_j(s)}) K_1(\sqrt{D_j(s)} R_{ejD})]}$$

$$- S_j \sqrt{D_j(s)} \frac{K_1(\sqrt{D_j(s)} R_{ejD}) I_1(\sqrt{D_j(s)}) - I_1(\sqrt{D_j(s)} R_{ejD}) K_1(\sqrt{D_j(s)})}{sM_j \sqrt{D_j(s)} [I_1(\sqrt{D_j(s)} R_{ejD}) K_1(\sqrt{D_j(s)}) - I_1(\sqrt{D_j(s)}) K_1(\sqrt{D_j(s)} R_{ejD})]}$$

$$\tag{7}$$

In Eq. 7, $\overline{m}_{wfjrD}$ is the solution with unit rate. To calculate the well bottom-hole pseudo-pressure with real production rate for layer $j$, Eq. 4 is transformed into Laplace domain:

$$\overline{m}_{wfjD} = \frac{m_{ijD}}{s} + \overline{q}_{jD}(s\overline{m}_{wfjrD} - m_{ijD}) \tag{8}$$

For simplicity, the well bottom-hole pseudo-pressure for different layers is assumed to be equal:

$$\overline{m}_{wfjD} = \overline{m}_{wD} \tag{9}$$

After simple transformation of Eq. 8, the dimensionless production rate in Laplace domain, $\overline{q}_{jD}$ can be written as:

$$\bar{q}_{jD} = \frac{\bar{m}_{wfjD} - \dfrac{m_{ijD}}{s}}{s\bar{m}_{wfjrD} - m_{ijD}} = \frac{\bar{m}_{wD} - \dfrac{m_{ijD}}{s}}{s\bar{m}_{wfjrD} - m_{ijD}} \qquad (10)$$

Substituting Eq. 10 into Eq. (A. 20), and taking appropriate transformation, the dimensionless well bottom-hole pseudo-pressure in Laplace space can be given by:

$$\bar{m}_{wD} = \frac{1 + \sum_{j=1}^{n} \dfrac{m_{ijD}}{s\bar{m}_{wfjrD} - m_{ijD}}}{\sum_{j=1}^{n} \dfrac{s}{s\bar{m}_{wfjrD} - m_{ijD}}} \qquad (11)$$

Furthermore, $\bar{q}_{jD}$ can be obtained by substituting Eq. 11 into Eq. 10. Applying Stehfest numerical inversion proposed by Schmittroth[36] in Eq. (10) and (11), the dimensionless well bottom-hole pseudo-pressure or dimensionless production rate of layer $j$ in real domain can be obtained. Finally, the well bottom-hole pseudo-pressure and production rate of layer $j$ can be computed as:

$$m_{wfj} = m_{i,J} - \frac{a_p(q_{sc}B_{gi})_r}{(K_{fi}h/\mu_{gi})_r} m_{wfjD} \qquad (12)$$

$$q_{scj} = \frac{(q_{sc}B_{gi})_r}{B_{gij}} q_{jD} \qquad (13)$$

The detailed solution process for this model is similar with the method proposed by Meng et al.[2], and can be divided into five steps: (1) Calculate the cumulative production of individual layer (see Appendix B), and determine the formation average pressure of each layer with MBE (material balance equation). (2) Compute the pseudo-time factor and dimensionless time after obtaining the values of stress-sensitive permeability, pressure-dependent gas viscosity and compressibility. (3) Calculate the dimensionless well bottom-hole pseudo-pressure and dimensionless production rate of individual zone. (4) Apply Stehfest numerical inversion to calculate the above values in real space, according to Eqs.12 and 13, and the relationship of $m_w$ and $p_{wf}$, calculate $p_{wf}$ and $q_{gj}$, then go to step (1) and repeat above steps.

# 4 Results and discussion

## 4.1 Validation

### 4.1.1 Comparisons with the classical results

Lefkovits et al. firstly presented the analytical solutions to study the production and pressure behaviors in two-layer reservoir with uniform pressure[3]. They designed two cases that assumed that in the first case diffusivity ratio for the first and second layers are 0.91 and 0.09, for the second case the diffusivity ratio of the first and second layers are 0.55 and 0.45. The dimensionless radius for both cases are 2000. Additionally, they assumed that mobility ratio is identical with the diffusivity ratio for each layer. To verify the accuracy the proposed model in this paper, the presented model is simplified and the above parameters are used to calculate the fractional flow rate of the high permeable layer and the wellbore pressure. Fig. 3 shows the comparisons of the results obtained by Lefkovits et al.[3] and the presented model in this paper. It can be seen that there is a good agreement of production and pressure for two cases between the published results and this paper, which indicates that the proposed model for two-layer reservoir with uniform pressure is valid.

(a) Comparisons of dimensionless production

(b) Comparisons of dimensionless pressure

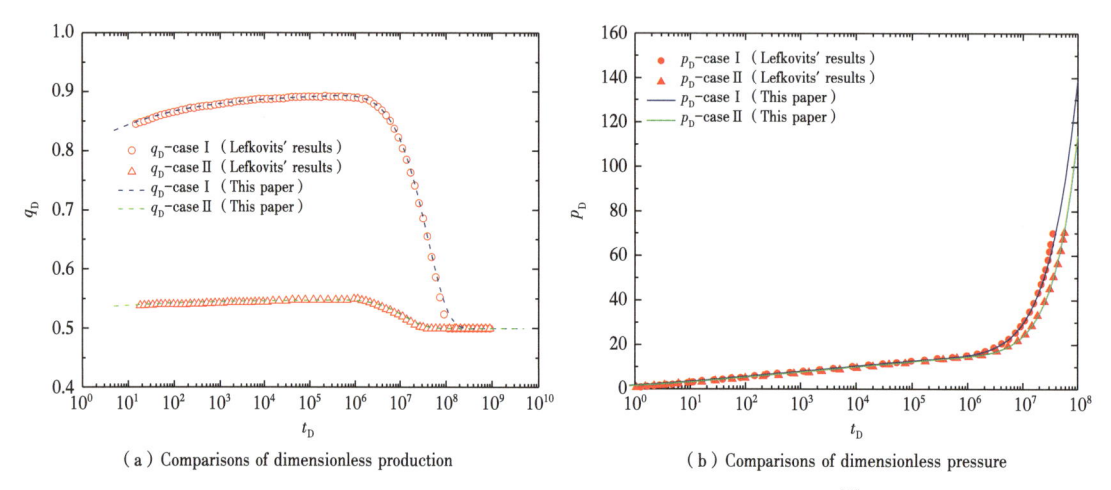

Fig. 3  Comparisons of production and pressure data taken from Lefkovits et al.[3] in Figs. 2 and 3

### 4.1.2  Field data matching

To verify the validity of the presented model further and demonstrate the application in practice, a filed case with two layers from Gaoshiti-Moxi carbonate gas reservoir is utilized to validate the accuracy of the proposed model. The relevant formation parameters for these two layers, such as formation thickness, radius, porosity, permeability and water saturation are determined through well logging and testing, and listed in Tab. 1. As the gas reservoir is in the initial stage of the exploitation, the well always produces with a constant flow rate, while it is difficult to estimate the production rate for each layer. Therefore, the well bottom-hole pressure date is monitored and collected to match the results calculated with the proposed model, as shown in Fig. 4. Seen from Fig. 4, the curve match between the field data and the presented model in this paper is excellent, which indicates the model can predict the well bottom-hole pressure in the future accurately. Moreover, the excellent match in Fig. 4 also demonstrates the validity of this model.

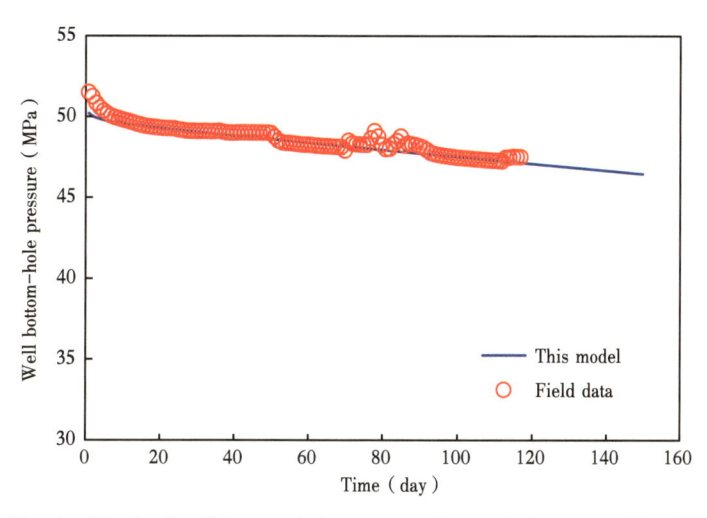

Fig. 4  Match of well bottom-hole pressure data with the proposed model

**Table 1  Formation, fluids and production parameters used in field and synthetic cases**

| Parameters | Field case Value(first layer) | Field case Value(second layer) | Synthetic case Value(first layer) | Synthetic case Value(second layer) |
|---|---|---|---|---|
| Formation thicknes(m) | 69.4 | 20.5 | 60 | 20 |
| Formation radius(m) | 860.8 | 615.7 | 800 | 600 |
| Natural fractures porosity(%) | — | 0.34 | — | 0.15 |
| Vugs porosity(%) | — | 2.5 | — | 1.5 |
| Matrix porosity(%) | 3.26 | 2.6 | 3 | 1.85 |
| Initial fractures permeability(mD) | — | 2.88 | — | 2.5 |
| Initial matrix permeability(mD) | 0.16 | — | 0.2 | — |
| Water saturation of matrix(%) | 15.5 | 14.3 | 15 | 15 |
| Interporosity flow coefficient between fractures and matrix | — | $1\times10^{-7}$ | — | $1\times10^{-7}$ |
| Interporosity flow coefficient between fractures and vugs | — | $1\times10^{-5}$ | — | $1\times10^{-5}$ |
| Stress-sensitive power exponent | 0.738 | 0.493 | 0.738 | 0.493 |
| Initial reservoir pressure(MPa) | 54.8 | 56.5 | 55 | 57.5 |
| Overburden stress(MPa) | 137.1 | 141.7 | 134.6 | 141.3 |
| Reservoir temperature(℃) | 152.4 | 156.9 | 150 | 156.5 |
| Wellbore radius(m) | 0.1 | | 0.1 | |
| Gas production rate(m³/d) | 220000 | | 200000 | |
| Production time(d) | 117 | | 1000 | |
| Specific gravity | 0.59 | | 0.59 | |
| Critical pressure(MPa) | 4.82 | | 4.82 | |
| Critical temperature(K) | 199.3 | | 199.3 | |

## 4.2 Sensitivity analysis

Formation properties and well conditions have significant impacts on production performance in multi-layer stress-sensitive carbonate gas reservoir, which include stress-sensitive power exponents, formation radius and permeability, and skin factors for each layer. Therefore, based on the data of field case, a synthetic case is set up in Tab.1 to analyze the above key factors on gas well production behaviors in multilayer commingled stress-sensitive carbonate gas reservoir.

### 4.2.1 Stress-sensitive power exponents

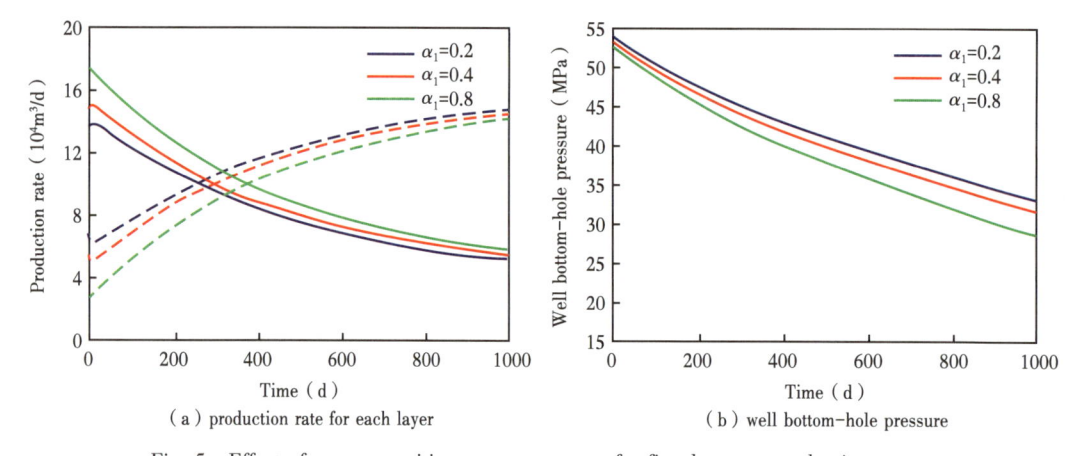

Figs. 5 and 6 show the effects of stress-sensitive power exponent for first and second layer on production performance of each layer and gas well. Note that in the figures of production rate for each layer, for instance, in Figs. 5a and 6a, the dashed and solid lines represent the production rate of the first and second layer, respectively. Since the formation permeability declines with the rise of stress-sensitive power exponent, consequently, seen from Figs. 5 and 6, with the increase of stress-sensitive power exponents, whether for first or second layer, it is obvious that the gas well bottom-hole pressure decreases, and the contribution of gas production rate declines. As the initial gas production mainly comes from second layer, in Fig. 6b the bottom-hole pressure decreases drastically and the differences of bottom-hole pressure are large for three cases at initial stage. While the first layer contributes mostly to the well production in the later period, with the increase of time the differences of bottom-hole pressure reduce in Fig. 6b and enlarge to some extent in Fig. 5b.

(a) production rate for each layer  (b) well bottom-hole pressure

Fig. 5  Effect of stress-sensitive power exponent for first layer on production curves

### 4.2.2 Formation radius

Figs 7 and 8 shows that the size of formation radius for first and second layer has diverse effect on production behaviors of each layer and gas well. When formation radius for certain layer enlarger, then the gas reserve of this layer will increase while other formation parameters keep constant. The first layer is tight and has lower permeability, and contributes to the gas well production at later

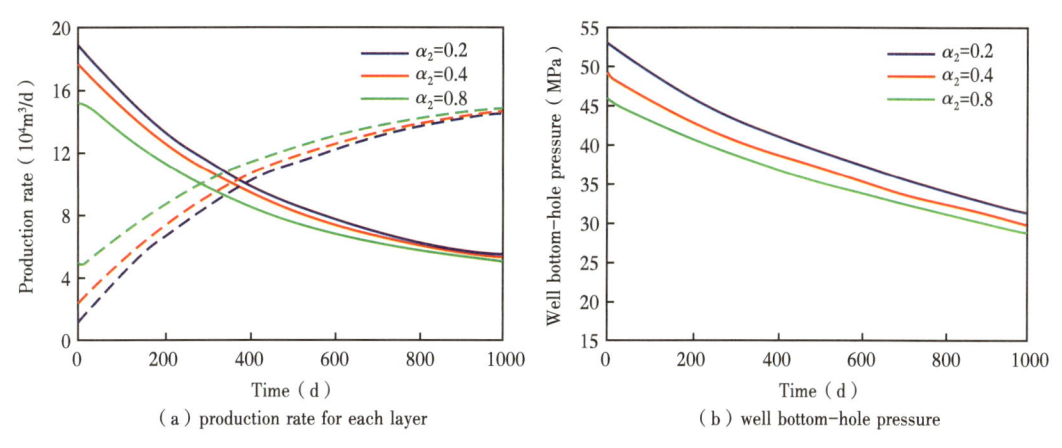

Fig. 6  Effect of stress-sensitive power exponent for second layer on production curves

stage. Hence, as seen in Fig. 7, initially, for cases with different formation radius of first layer, the production rate for each layer and the bottom-hole pressure are almost identical. When the time is longer than 400d, the first layer begin to contributes to the well production, which causes that the production of this layer and bottom-hole pressure increase with the formation radius. Whereas the second layer with larger permeability is full of fractures and vugs, and contributes to the well production initially, which leads to the rapid rise of production rate for second layer and bottom-hole pressure with the increase of formation radius, as shown in Fig. 8.

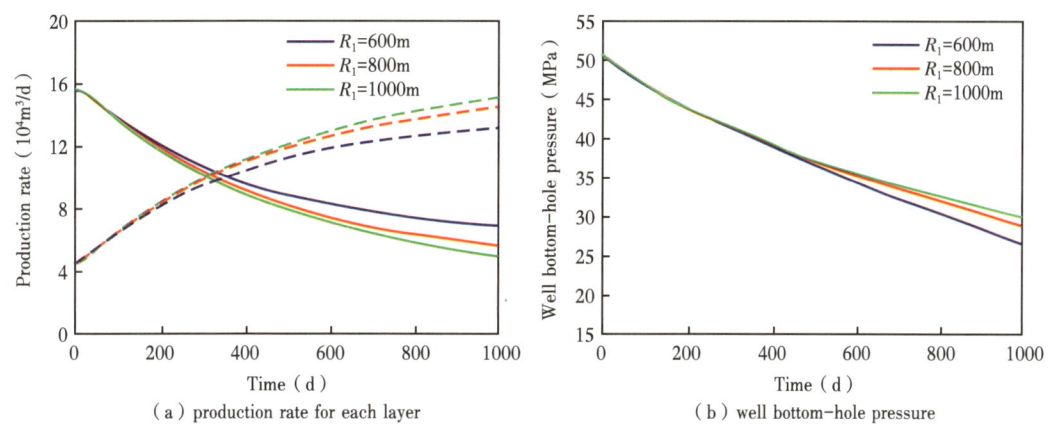

Fig. 7  Effect of formation radius for first layer on production curves

### 4.2.3  Formation permeability

The influences of formation permeability for each layer on production performances are shown in Figs. 9 and 10. Compared with Fig. 10, it is obvious that the variation of formation permeability for first layer has significant impact on production behaviors, as shown in Fig.9, which means that the flow capacity of tight layer determines the well production performance. The increase of formation permeability for first layer improves fluids flowing ability, which leads to the dramatic rise of

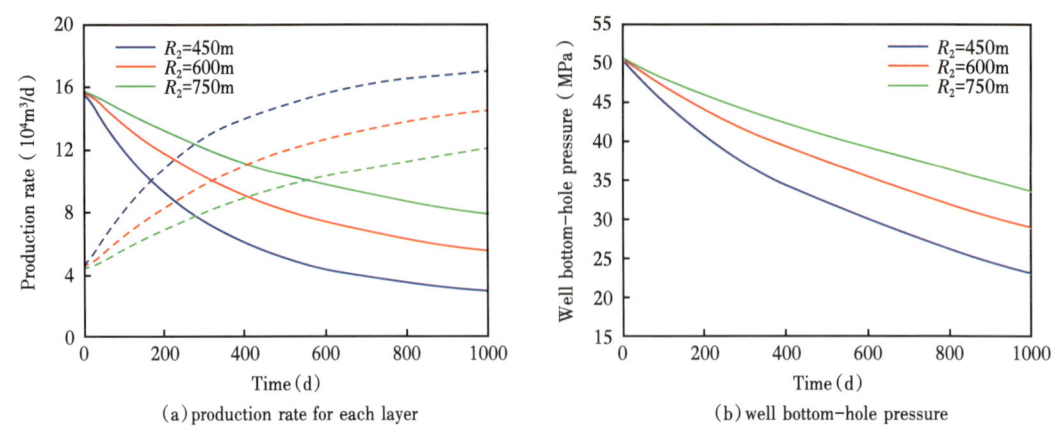

Fig. 8 Effect of formation radius for second layer on production curves

production rate for this layer. Influenced by the characteristics of successive production for multi-layer system, seen from Fig. 10, at initial stage the production rate of this layer and bottom-hole pressure increase with the formation permeability, while in Fig. 9 the discrepancies of bottom-hole pressure for different scenarios become larger in the later period. In addition, from Fig. 9, as the increase of permeability for first layer, it can be seen that the differences of bottom-hole pressure and production rate of first layer minimize, which demonstrates that there is optimal permeability to assure that the tight layer can be exploited efficiently.

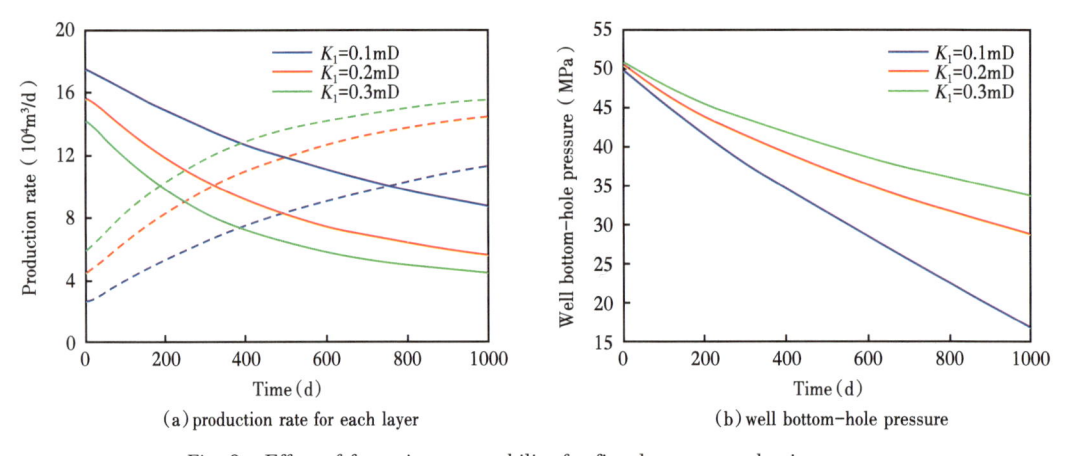

Fig. 9 Effect of formation permeability for first layer on production curves

#### 4.2.4 Skin factors

Figs. 11 and 12 show that the skin factors for each layer have important effects on the production proportion of individual layers and well bottom-hole pressure. The formation damage surrounding the wellbore increases with the skin factors, which results in the reduction of production rate and the decline of bottom-hole pressure. Thus seen from Figs. 11 and 12, the production rate for first or second layer and well bottom-hole pressure decrease with the rise of skin factors.

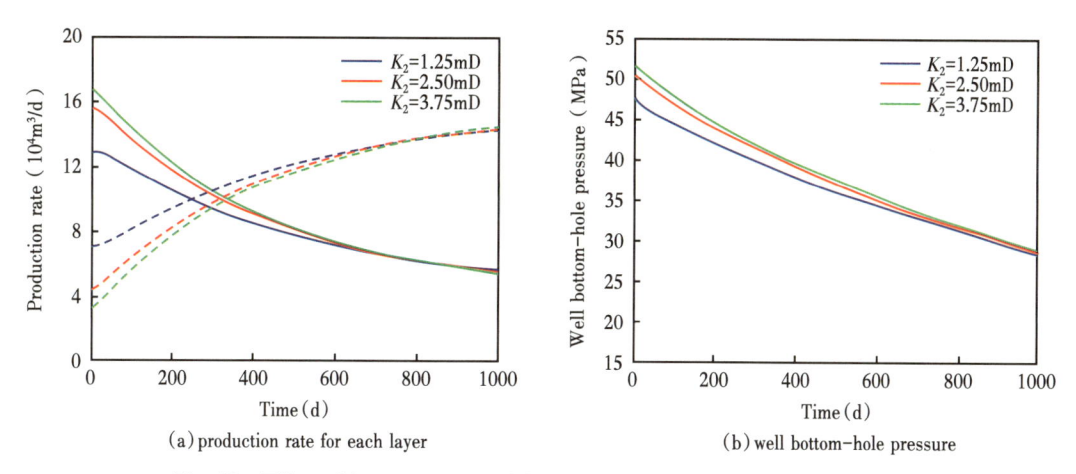

Fig. 10 Effect of formation permeability for second layer on production curves

However, there exist conspicuous differences in production rate for each layer and bottom-hole pressure in Figs. 11 and 12. The well condition of first layer on production behavior is more significant than second layer. The production rate of first layer and the bottom-hole pressure decrease greatly with the increase of skin factor of this layer, and with the extension of production time the gaps for different cases become larger. While for second layer in Fig. 12a, when the time is earlier than 400d, the production rate and the bottom-hole pressure decline moderately with the rise of skin factor. The above variation features for production behaviors also can be explained with the theory of successive production for multi-layer system. Additionally, through the comparisons of Figs. 11 and 12, it also can be concluded that mitigation of formation damage for tight layer could improve the gas well production performance.

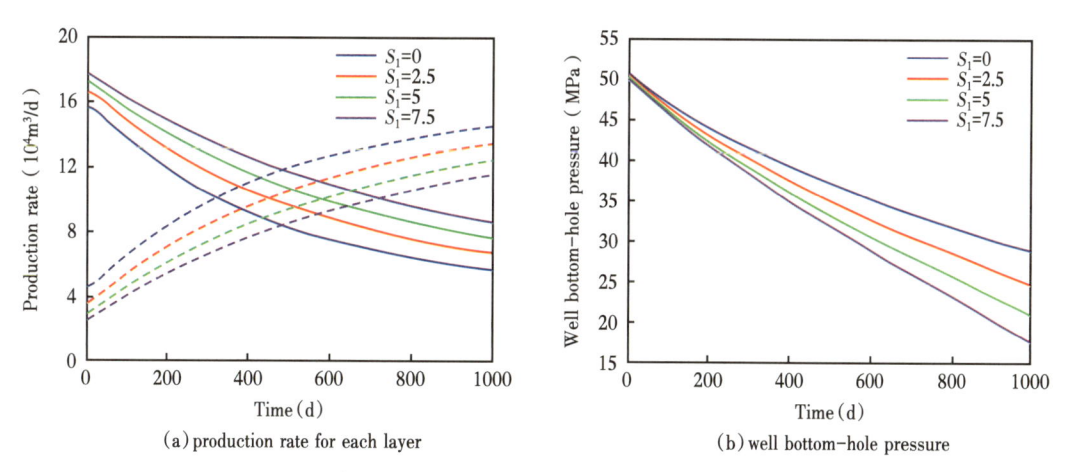

Fig. 11 Effect of skin factor for first layer on production curves

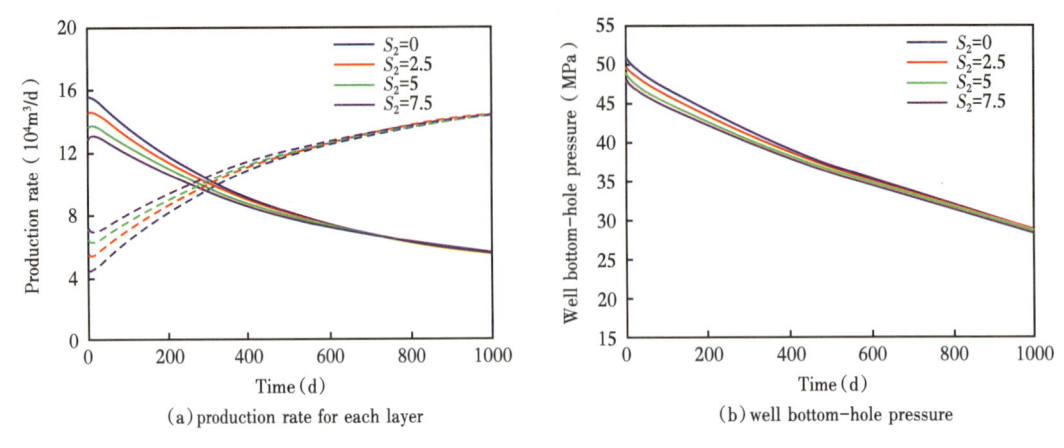

Fig. 12 Effect of skin factor for second layer on production curves

(a) production rate for each layer

(b) well bottom-hole pressure

## 5 Conclusions

A semi-analytical model is established to study the production performances in multi-layer stress-sensitive carbonate gas reservoir, in which the stress-sensitive permeability is determined with the laboratory experiments. The validity of the proposed model is verified with the Lefkovits's results, and a filed case in Gaoshiti-Moxi carbonate reservoir illustrates the applicability of this model. In addition, the effects of relevant factors on production behaviors are investigated. Several observations are highlighted:

(1) Stress sensitivity of permeability for carbonate gas reservoir can be described appropriately with power function rather than exponential function, and the stress-sensitive power exponents are estimated through the experiments. A large stress-sensitive power exponent for any of layers will lead to the reduction of production rate for this layer and the decline of well bottom-hole pressure.

(2) The characteristics of successive production for multi-layer system are demonstrated through the analysis on production behaviors of gas wells. Initially, the gas production primarily come from the layer with larger flow capacity, whereas the tight layer with lower permeability contributes mostly to the production in the later period.

(3) The formation permeability and skin effect for tight layer have more significant influences on the production performance than the layer with multiple fractures and cavities. The larger formation radius and permeability and the lower skin factor for layer with high permeability can improve the production rate of this layer and the well bottom-hole pressure at the initial stage.

## Appendix A. Derivation of dimensionless pseudo-pressure for layer $j$

According to the presented physical model and assumptions in chapter 3.1, and transport equation, equation of state and continuity equation are combined to derive the governing equations for layer $j$ in multilayer commingled stress-sensitive carbonate gas reservoir. To consider the permeability stress-sensitivity and the other pressure-dependent gas properties, such as viscosity and density, pseudo-time factor and pseudo-pressure are introduced. Thus the governing equations for natural fractures, matrix and vuggs systems in layer $j$ can be given by:

Natural fractures:

$$\frac{\partial^2 m_{fj}}{\partial r^2} + \frac{1}{r}\frac{\partial m_{fj}}{\partial r} = \frac{\phi_{fj}\mu_{gij}C_{tfj}}{a_t K_{fij}}\frac{\partial m_{fj}}{\partial(\beta_j t)} - \alpha_{mj}\frac{K_{mij}}{K_{fij}}(m_{mj} - m_{fj}) - \alpha_{cj}\frac{K_{cij}}{K_{fij}}(m_{cj} - m_{fj}) \quad (A.1)$$

Matrix:

$$\frac{\phi_{mj}\mu_{gij}C_{tmj}}{\alpha_t}\frac{\partial m_{mj}}{\partial(\beta_j t)} + \alpha_{mj}K_{mij}(m_{mj} - m_{fj}) = 0 \quad (A.2)$$

Vugs:

$$\frac{\phi_{cj}\mu_{gij}C_{tcj}}{a_t}\frac{\partial m_{cj}}{\partial(\beta_j t)} + \alpha_{cj}K_{cij}(m_{cj} - m_{fj}) = 0 \quad (A.3)$$

Where

$$C_{tfj} = C_{gij}(1 - S_{wfj}), C_{tmj} = C_{gij}(1 - S_{wmj}), C_{tcj} = C_{gij}(1 - S_{wcj}) \quad (A.4)$$

Initial conditions

Initially, the pressure prior to production for layer $j$ is uniform and equal to $p_{ij}$, so the initial conditions can be written as

$$m_{fj} = m_{mj} = m_{cj} = m_{ij} \quad (t = 0) \quad (A.5)$$

Boundary conditions

The outer boundary is assumed to be laterally closed, thus

$$\left.\frac{\partial m_{fj}}{\partial r}\right|_{r=R_{ej}} = 0 \quad (A.6)$$

In addition, the outer boundary condition also can be assumed to be infinite or constant-pressure. For different layers, the outer boundary condition may not be identical.

The damage skin effect for individual layer is considered, and then inner boundary condition with constant production rate can be given by

$$\left.\frac{K_{fij}h_j}{a_p q_{gj}\mu_{gij}B_{gij}}r\frac{\partial m_{fj}}{\partial r}\right|_{r=r_w} = 1, m_{fj} - S_j r\frac{\partial m_{fj}}{\partial r}\bigg|_{r=r_w} = m_{wfj} \quad (A.7)$$

In Eqs. (A.1) ~ (A.7), $m_{fj}$, $m_{mj}$, $m_{cj}$, $m_{wfj}$ and $m_{ij}$ are the pseudo-pressure of natural fractures, matrix, vugs, well bottom-hole and initial condition for layer $j$, respectively. $\beta_j$ is defined as pseudo-time factor of layer $j$, and can be computed approximately under the formation average pressure. The definition for these parameters can be expressed by

$$m_{fj} = \left(\frac{\mu_{gi}z_i}{p_i K_{fi}}\right)_j \int_{p_0}^{p_{fj}} \frac{K_{fj}p_j}{\mu_{gj}z_j} dp, \quad m_{mj} = \left(\frac{\mu_{gi}z_i}{p_i K_{fi}}\right)_j \int_{p_0}^{p_{mj}} \frac{K_{fj}p_j}{\mu_{gj}z_j} dp, \quad m_{cj} = \left(\frac{\mu_{gi}z_i}{p_i K_{fi}}\right)_j \int_{p_0}^{p_{cj}} \frac{K_{fj}p_j}{\mu_{gj}Z_j} dp$$

$$m_{wfj} = \left(\frac{\mu_{gi}z_i}{p_i K_{fi}}\right)_j \int_{p_0}^{p_{wfj}} \frac{K_{fj}p_j}{\mu_{gi}z_j} dp, \quad m_{ij} = \left(\frac{\mu_{gi}z_i}{p_i K_{fi}}\right)_j \int_{p_0}^{p_{ij}} \frac{K_{fj}p_j}{\mu_{gj}z_j} dp \tag{A.8}$$

$$\beta_j = \frac{1}{t}\left(\frac{\mu_{gi}C_{gi}}{K_{fi}}\right)_j \int_0^t \frac{K_{fj}}{\mu_{gj}C_{gj}} dt \approx \frac{1}{t}\left(\frac{\mu_{gi}C_{gi}}{K_{fi}}\right)_j \int_0^t \frac{K_{fj}(p_{avgj})}{\mu_{gj}(p_{avgj}) C_{gj}(p_{avgj})} dt$$

To simplify the presented mathematical model, we assume that $\beta_1 = \beta_2 = \cdots = \beta_j = \cdots = \beta_n$, and some specific parameters and dimensionless variables are defined, as shown in Eqs. (A.9) and (A.10):

$$\eta_j = \left[\frac{K_{fi}}{(\phi C_t)_{(f+m+c)}\mu_{gi}}\right]_j, (q_g B_{gi})_r = \sum_{j=1}^n (q_g B_{gi})_j \left(\frac{K_{fi}h}{\mu_{gi}}\right)_r = \sum_{j=1}^n \left(\frac{K_{fi}h}{\mu_{gi}}\right)_j \tag{A.9}$$

$$\eta_r = \left[\frac{K_{fi}}{(\phi C_t)_{(f+m+c)}\mu_{gi}}\right]_r = \sum_{j=1}^n \left[\frac{K_{fi}}{(\phi C_t)_{(f+m+c)}\mu_{gi}}\right]_j$$

$$m_{fjD} = \frac{(K_{fi}h/\mu_{gi})_r (m_{iJ} - m_{fj})}{a_p (q_g B_{gi})_r}, \quad m_{mjD} = \frac{(K_{fi}h/\mu_{gi})_r (m_{iJ} - m_{mj})}{a_p (q_g B_{gi})_r},$$

$$m_{cjD} = \frac{(K_{fi}h/\mu_{gi})_r (m_{iJ} - m_{cj})}{a_p (q_g B_{gi})_r}, \quad m_{ijD} = \frac{(K_{fi}h/\mu_{gi})_r (m_{iJ} - m_{ij})}{a_p (q_g B_{gi})_r},$$

$$m_{wfjD} = \frac{(K_{fi}h/\mu_{gi})_r (m_{iJ} - m_{wfj})}{a_p (q_g B_{gi})_r}, \quad t_D = a_t \left[\frac{K_{fi}}{(\phi C_t)_{(f+m+c)}\mu_{gi}}\right]_r \frac{\beta t}{r_w^2}$$

$$r_D = \frac{r}{r_w}, R_{ejD} = \frac{R_{ej}}{r_w}, \lambda_{mj} = \alpha_{mj}\frac{K_{mij}}{K_{fij}}r_w^2, \lambda_{cj} = \alpha_{cj}\frac{K_{cij}}{K_{fij}}r_w^2, \omega_{mj} = \frac{(\phi C_t)_{mj}}{(\phi C_t)_{(f+m+c)j}},$$

$$\omega_{cj} = \frac{(\phi C_t)_{cj}}{(\phi C_t)_{(f+m+c)j}}\omega_{fj} = \frac{(\phi C_t)_{fj}}{(\phi C_t)_{(f+m+c)j}}, \eta_{jD} = \frac{\eta_j}{\eta_r}, M_j = \frac{(K_{fi}h/\mu_{gi})}{(K_{fi}h/\mu_{gi})_r}, q_{jD} = \frac{q_{gj}B_{gij}}{(q_g B_{gi})_r} \tag{A.10}$$

In Eq. (A.10), $\eta_{jD}$, $M_j$ and $q_{jD}$ should be satisfied with the following equation:

$$\sum_{j=1}^n \eta_{jD} = 1, \sum_{j=1}^n M_j = 1, \sum_{j=1}^n q_{jD} = 1 \tag{A.11}$$

Through the application of dimensionless variables shown in Eq. (A.10) and the Laplace transformation, the governing equations and the initial and boundary conditions (Eqs. (A.1) ~ (A.7)) can be transformed into dimensionless terms in Laplace domain:

$$\frac{\partial^2 \overline{m}_{fjD}}{\partial r_D^2} + \frac{1}{r_D}\frac{\partial \overline{m}_{fjD}}{\partial r_D} = \frac{\omega_{fj}}{\eta_{jD}}(s\overline{m}_{fjD} - m_{ijD}) - \lambda_{mj}(\overline{m}_{mjD} - \overline{m}_{fjD}) - \lambda_{cj}(\overline{m}_{cjD} - \overline{m}_{fjD}) \tag{A.12}$$

$$\frac{\omega_{mj}}{\eta_{jD}}(s\overline{m}_{mjD} - m_{ijD}) + \lambda_{mj}(\overline{m}_{mjD} - \overline{m}_{fjD}) = 0 \tag{A.13}$$

$$\frac{\omega_{cj}}{\eta_{jD}}(s\overline{m}_{cjD} - m_{ijD}) + \lambda_{cj}(\overline{m}_{cjD} - \overline{m}_{fjD}) = 0 \tag{A.14}$$

$$\left.\frac{\partial \overline{m}_{fjD}}{\partial r_D}\right|_{r_D=R_{ejD}} = 0 \tag{A.15}$$

$$\left.M_j r_D \frac{\partial \overline{m}_{fjD}}{\partial r_D}\right|_{r_D=1} = -\overline{q}_{jD}, \left.\left(\overline{m}_{fjD} - S_j r_D \frac{\partial \overline{m}_{fjD}}{\partial r_D}\right)\right|_{r_D=1} = \overline{m}_{wfjD} \tag{A.16}$$

Substituting Eqs. (A.13) and (A.14) into Eq. (A.12) yields:

$$\frac{\partial^2 \overline{m}_{fjD}}{\partial r_D^2} + \frac{1}{r_D}\frac{\partial \overline{m}_{fjD}}{\partial r_D} = D_j(s)\overline{m}_{fjD} - \frac{D_j(s)}{s}m_{ijD} \tag{A.17}$$

Where

$$D_j(s) = s\left(\frac{\omega_{fj}}{\eta_{jD}} + \frac{\lambda_{mj}\omega_{mj}}{\omega_{mj}s + \lambda_{mj}\eta_{jD}} + \frac{\lambda_{cj}\omega_{cj}}{\omega_{cj}s + \lambda_{cj}\eta_{jD}}\right) \tag{A.18}$$

The general solution of Eq. (A.17) can be expressed by

$$\overline{m}_{fjD} = \frac{m_{ijD}}{s} + AI_0(\sqrt{D_j(s)}\,r_D) + BK_0(\sqrt{D_j(s)}\,r_D) \tag{A.19}$$

Furthermore, in Eq. (A.11), the term about the sum of dimensionless production can be transformed into Laplace space:

$$\sum_{j=1}^{n} \overline{q}_{jD} = \frac{1}{s} \tag{A.20}$$

## Appendix B. Calculation of pseudo-pressure and pseudo-time factor

To obtain well bottom-hole pressure through the method of interpolation, it is crucial to construct the relationship between the real pressure and pseudo-pressure. In Eq. (A.8), pseudo-pressure and pseudo-time factor are the function of fracture permeability, viscosity, deviation factor and compressibility. Fracture or matrix permeability of layer $j$, $k_{fj}$, is pressure-dependent, and can be calculated with Eq. (1) at different pressure. Note that the stress-sensitive exponent is varied with the types of formation. Gas deviation factor $z_j$, gas viscosity $\mu_{gj}$ and gas compressibility $C_{gj}$ in layer $j$ can be computed by the methods of Lee et al[36], Hall and Yarborough[37] and Dranchuk et al.[38], respectively. Since it is essential to estimate the average pressure of layer $j$ during the calculation of pseudo-time factor $\beta_j$, then MBE is used:

$$\frac{p_{avgj}}{z(p_{avgj})} = \frac{p_{ij}}{z(p_{ij})}\left(1 - \frac{G_{pj}}{G_{scj}}\right) \tag{B.1}$$

When production rate of layer $j$ is obtained with Eq. 13 at time step $k$, then the cumulative gas production can be expressed by

$$G_{pj}^{k} = G_{pj}^{k-1} + q_{gj}^{k} \tag{B.2}$$

The integral for pseudo-time factor can be calculated with following equation:

$$\beta_j = \frac{1}{t}\left(\frac{\mu_{gi}C_{gi}}{k_{fi}}\right)_j \int_0^t \frac{K_{fj}(p_{avgj})}{\mu_{gj}(p_{avgj})C_{gj}(p_{avgj})} dt$$
$$= \frac{1}{t}\left(\frac{\mu_{gi}C_{gi}}{K_{fi}}\right)_j \sum_{k=1}^n \frac{1}{2}\left[\frac{K_{fj}(p_{avgj}^{k-1})}{\mu_{gj}(p_{avgj}^{k-1})C_{gj}(p_{avgj}^{k-1})} + \frac{K_{fj}(p_{avgj}^k)}{\mu_{gj}(p_{avgj}^k)C_{gj}(p_{avgj}^k)}\right](t^k - t^{k-1}) \quad (B.3)$$

## Nomenclatures

| | | |
|---|---|---|
| $B_g$ | Gas formation volume factor, $m^3/Sm^3$ |
| $C_g$ | Gas compressibility, $MPa^{-1}$ |
| $C_t$ | Total compressibility, $MPa^{-1}$ |
| $G_p$ | Cumulative gas production, $10^8 m^3$ |
| $G_{sc}$ | Original gas in place, $10^8 m^3$ |
| $h$ | Formation thickness, m |
| $K$ | Permeability, mD |
| $m$ | Pseudo-pressure, MPa |
| $m_{wf}$ | Wellbore pseudo-pressure, MPa |
| $M$ | Mobility ratio between outer and inner regions, dimensionless |
| $p$ | Pressure, MPa |
| $p_{wf}$ | Wellbore pressure, MPa |
| $p_{avg}$ | Formation average pressure, MPa |
| $q_g$ | Gas production rate, $m^3/d$ |
| $r$ | Radial distance, m |
| $r_w$ | Wellbore radius, m |
| $R_e$ | Formation radius, m |
| $s$ | Laplace transform variable |
| $S$ | Skin factor |
| $S_w$ | Water saturation, fraction |
| $t$ | Time, day |
| $z$ | Gas deviation factor |
| $\alpha$ | Stress-sensitive power exponent, dimensionless |
| $\alpha_m, \alpha_c$ | Shape factors of matrix and vugs, $1/m^2$ |
| $\beta$ | Pseudo-time factor |
| $\lambda$ | Interporosity flow coefficient, dimensionless |
| $\omega$ | Storativity ratio, dimensionless |
| $\eta$ | Hydraulic diffusivity, dimensionless |
| $\mu_g$ | Gas viscosity, $mPa \cdot s$ |

| | | |
|---|---|---|
| $\varphi$ | | Porosity, fraction |
| $\sigma_s$ | | Overburden pressure, MPa |
| $a_t$, $a_p$ | | Constants, $a_t = 86.4$, $a_p = 1.842 \times 10^{-3}$ |

**Subscript**

| | |
|---|---|
| i | Initial condition |
| 0 | Standard condition |
| m | Matrix system |
| f | Natural fractures system |
| c | Vugs system |
| 1 | First layer |
| 2 | Second layer |
| j | jth layer |
| D | Dimensionless |

## References

[1] Zhou Z, Wang, X, Yin G, Yuan S, Zeng S. Characteristics and genesis of the (Sinian) Dengying formation reservoir in Central Sichuan, China. J. Nat. Gas Sci. Eng. 2016,29:311-321.

[2] Meng F, Lei Q, He D. et al. Production performance analysis for deviated wells in composite carbonate gas reservoirs. J. Nat. Gas Sci. Eng. 2018,56:333-343.

[3] Lefkovits H C, Hazebroek P, Allen E E, Matthews C S. A study of the behavior of bounded reservoirs composed of stratified layers. SPE J. 1961,March, 43-58.

[4] Cobb W M, Ramey H J, Jr Miller. Well-test analysis for wells producing commingled zones. J. Petrol. Technol. 1972,24(1):27-37.

[5] Raghavan R, Topaloglu H N, Cobb W M, Ramey H J. Well-test analysis for wells producing from two commingled zones of unequal thickness. J. Petrol. Technol. 1947,26(9):1035-1043.

[6] Tariq S M, Ramey H J. Drawdown behavior of a well with storage and skin effect communicating with layers of different radii and other characteristics. In: Paper Present at SPE Annual Fall Technical Conference and Exhibition, 1978,1-3 October, Houston, Texas, SPE-7453-MS.

[7] Spath J B, Ozkan E, Raghavan R. An efficient algorithm for computation of well responses in commingled reservoirs. In: Paper Present at Annual Technical Meeting, 1990,10-13 June, Calgary, Alberta, PETSOC-90-01.

[8] Sun H, Liu L, Lu Y, Zhou, F. Exact solution and typical curve for commingled reservoir. In: Paper Present at Middle East Oil Show, 2003,5-8 April Bahrain, SPE-81521-MS.

[9] Larsen L. Wells producing commingled zones with unequal initial pressures and reservoir properties. In: Paper Present at SPE Annual Technical Conference and Exhibition, 1981,2-5 October, San Antonio, Texas, SPE-10325-MS.

[10] Larsen L. Boundary effects in pressure-transient data from layered reservoirs. In: Paper Present at SPE Annual Technical Conference and Exhibition, 1989,8-11 October, San Antonio, Texas, SPE-19797-MS.

[11] Kuchuk F J, Wilkinson D J. Transient pressure behavior of commingled reservoirs. SPE Form. 1991,6(1):

111-120.

[12] Agarwal B, Chen H Y, Raghavan R. Buildup behaviors in commingled reservoir systems with unequal initial pressure distributions: Interpretation. In: Paper Present at SPE Annual Technical Conference and Exhibition, 1992,4-7 October, Washington, D.C., SPE-24680-MS.

[13] Gao C, Lee W J. Modeling commingled reservoirs with pressure–dependent properties and unequal initial pressure in different layers. In: Paper Present at SPE Annual Technical Conference and Exhibition, 1993,3-6 October, Houston, Texas, SPE-26665-MS.

[14] Shah P C, Spath J B. Transient wellbore pressure and flow rates in a commingled system with different layer pressures. In: Paper Present at SPE Production Operations Symposium, 1993,21-23 March, Oklahoma City, Oklahoma, SPE-25423-MS.

[15] Ahmed A,Lee, W J,Development of a new theoretical model for three-layered reservoirs with unequal. initial pressures. In: Paper Present at SPE Production Operations Symposium, 1995,2-4 April, Oklahoma City, Oklahoma, SPE-29463-MS.

[16] Anisur Rahman, N M Mattar. New analytical solution to pressure transient problems in commingled, layered zones with unequal initial pressures subject to step changes in production rates. J. Pet. Sci. Eng. 2007,56:283-295.

[17] Raghavan, R. Behavior of wells completed in multiple producing zones. SPE Form. 1989,4(2):219-230.

[18] Aly A, Chen H Y, Lee W J. Pre-production pressure analysis of commingled reservoir with unequal initial pressures. In: Paper Present at Permian Basin Oil and Gas Recovery Conference, 1994, 16-18 March, Midland, Texas, SPE-27661-MS.

[19] Aly A,Lee W J. A new pre-production well test for analysis of multilayered commingled reservoirs with unequal initial pressures. In: Paper Present at Permian Basin Oil and Gas Recovery Conference, 1994,16-18 March, Midland, Texas, SPE-27730-MS.

[20] Chao G, Jones J R, Raghavan R,Lee W J, Average reservoir pressure estimation of a layered commingled reservoir. SPE Form. 1994, 9(4):264-271.

[21] Aly A, Lee W J. Computational modeling of multi-layered reservoirs with unequal initial pressures: development of a new Pre-production well test. In: Paper Present at Low Permeability Reservoirs Symposium, 1995,20-22 March, Denver, Colorado, SPE-29586-MS.

[22] Chen H Y, Raghavan R,Poston S W. Average reservoir pressure estimation of a layered commingled reservoir. SPE J. 1997,2(1):3-15.

[23] Sui W, Zhu D. Determining multilayer formation properties from transient temperature and? pressure measurements in gas wells with commingled zones. J. Nat. Gas Sci. Eng. 2012,9:60-72.

[24] Shi W, Yao Y, Cheng S, Shi Z. Pressure transient analysis of acid fracturing stimulated well in multilayered fractured carbonate reservoirs: A field case in Western Sichuan Basin, China. J. Pet. Sci. Eng. 2020,184:1-17.

[25] Vieira Bela, R., Pesco S,Barreto A. Modeling falloff tests in multilayer reservoirs. J. Pet. Sci. Eng. 2019,174. 161-168.

[26] El-Banbi A H, Wattenbarger R A. Analysis of commingled tight gas reservoirs. In: Paper Present at SPE Annual Technical Conference and Exhibition, 1996,8-9 October, Denver, Colorado, SPE-36736-MS.

[27] El-Banbi A H,Wattenbarger, R A. Analysis of commingled gas reservoirs with variable bottom-hole flowing pressure and non-Darcy flow. In: Paper Present at SPE Annual Technical Conference and Exhibition, 1997, 5-8 October, San Antonio, Texas, SPE-38866-MS.

[28] Arevalo-Villagran J A, Wattenbarger R A,El-Banbi A.H. Production analysis of commingled gas reservoirs-

case histories. In: Paper Present at SPE International Petroleum Conference and Exhibition in Mexico, 2000, 1-3 February, Villahermosa, Mexico, SPE-58985-MS.

[29] Onwunyili C C, Onyekonwu M O. Coupled model for analysis of multilayer reservoir in commingled production. In: Paper Present at SPE Nigeria Annual International Conference and Exhibition, 2013, 30 July-1 Augst, Lagos, Nigeria, SPE-167588-MS.

[30] Ali T S S, Cheng Y, McVay D A, Lee W J. A practical approach for production data Analysis of multilayer commingled tight gas wells. In: Paper Present at SPE Eastern Regional Meeting, 2010, 12-14 October, Morgantown, West Virginia, SPE-138882-MS.

[31] Wang C W, Jia C S, Peng X L, et al. Effects of wellbore interference on concurrent gas production from multi-layered tight sands: A case study in eastern Ordos Basin, China. J. Pet. Sci. Eng. 2019, 179, 707-715.

[32] Zhao L, Chen Y, Ning Z, et al. Stress sensitive experiments for abnormal overpressure carbonate reservoirs: A case from the Kenkiyak fractured-porous oil field in the littoral Caspian Basin. Pet. Explor. Dev. 2013, 40(2): 208-215.

[33] Wang F, Li Y, Tang X, Chen J, Gao W. Petrophysical properties analysis of a carbonate reservoir with natural fractures and vugs using X-ray computed tomography. J. Nat. Gas Sci. Eng. 2016, 28: 215-225.

[34] Wang Y, Chen Y, Li D. Simulations and case studies for enhancing production in a stress-sensitive fractured carbonated reservoir. In: Paper Present at 50th U.S. Rock Mechanics/Geomechanics Symposium, 2016, 26-29 June, Houston, Texas, ARMA-2016-294.

[35] Seabra G S, Souza A L S, Fontoura S A B, et al. A coupled iterative hydromechanical analysis of a stress sensitive Brazilian carbonate reservoir. In: Paper Present at 51st U. S. Rock Mechanics/Geomechanics Symposium, 2017, 25-28 June, San Francisco, California, USA, ARMA-2017-0880.

[36] Lee A L, Gonzalez M H, Eakin B E. The viscosity of natural gases. J. Petrol. Technol. 1966, 18(8): 997-1000.

[37] Schmittroth L A. Numerical inversion of laplace transforms. Commun. ACM. 1960, 3(3): 171-173.

[38] Hall K.R, Yarborough L. A new equation of state for Z-factor calculations. Oil and Gas Journal. 1973, 71(7): 82-92.

[39] Dranchuk P M, Purvis R A, Robinson D B, Computer calculation of natural gas compressibility factors using the Standing and Katz correlation. Inst. of Petroleum Technical Institute Series. 1974, IP74-008, 1-13.

[40] Aly A. A new technique for analysis of wellbore pressure from multi-layered reservoirs with unequal initial pressures to determine individual layer properties. In: Paper Present at SPE Eastern Regional Meeting, 1994, 8-10 November, Charleston, West Virginia, SPE-29176-MS.

# 苏里格致密砂岩气藏大井组混合井网立体开发技术

张 吉 范倩倩 王 艳 侯科锋 王文胜 吴小宁

(中国石油长庆油田分公司)

**摘要**：苏里格气田属于典型的致密砂岩气藏，针对其纵向上多层系含气、横向上砂体连续性差的地质特征，形成多层系大井组混合井网立体开发技术。通过复合砂体垂向分期，单期河道砂体平面划界，明确有效砂体孤立型、切割叠置型、堆积叠置型和横向局部连通型四种叠置方式。利用现场干扰试验、气藏工程论证等方法优化井网，针对不同砂体叠置方式和储层特征，形成混合井网。考虑水平井、直井和定向井的开发优势，形成针对古生界多层系的大井组混合井网立体开发技术。该技术实现了山1段和盒8段河流相砂岩储层、马家沟组五段海相碳酸盐岩储层立体开发，整体储量一次动用，有效提高储量动用程度，降低开发成本，保障气田经济有效开发。

**关键词**：苏里格气田；致密砂岩气藏；砂体叠置方式；大丛式井组；混合井网；多井型开发

苏里格气田位于鄂尔多斯盆地伊陕斜坡西部，属于低渗低压低丰度致密砂岩气田，是中国最大的气田[1-3]。气井投产后，单井产量低，综合递减率达到23.8%，稳产难度大[4]。特别是2014年以来，稳产形势日趋复杂。首先，苏里格气田地质条件复杂，有效储层预测难度大，严重制约了井位部署和后期开发管理。其次，环保要求日益提高，产能建设困难加大。再次，由于征借地费用、钻前费用、搬迁费用等不断增加，使气田产能建设成本不断升高，降低开发成本迫在眉睫。

对于大井组混合井网立体开发技术，国内外学者的研究以工艺技术、地面配套等方面为主[5,6]。本文开展储层描述、井网优化、井型组合部署方式等关键技术研究，形成多层系大井组混合井网立体开发技术，有效缓解产能建设中遇到的难题，提高气田的开发效益，实现多套储层一套井网动用，有效提高气田最终采收率。

## 1 研究区概况

苏里格气田自2006年规模开发以来，大量井资料和动静态资料反映出苏里格气田地质特征的复杂性。主力气层下二叠统山西组1段(山1段)和中二叠统下石盒子组8段(盒8段)，主要为河流沉积的心滩、河道等砂体，沉积、成岩演化造成储层致密[7-9]，有效砂体分散，横向连续性差，且沉积的多期次造成储集层纵向上多层系含气(图1)。主力气层下奥陶统马家沟组五段碳酸盐岩，主要为碳酸盐潮坪沉积下的白云岩储层，受岩溶古地貌、白云岩成因等因素控制，有效储层局部发育[10]。

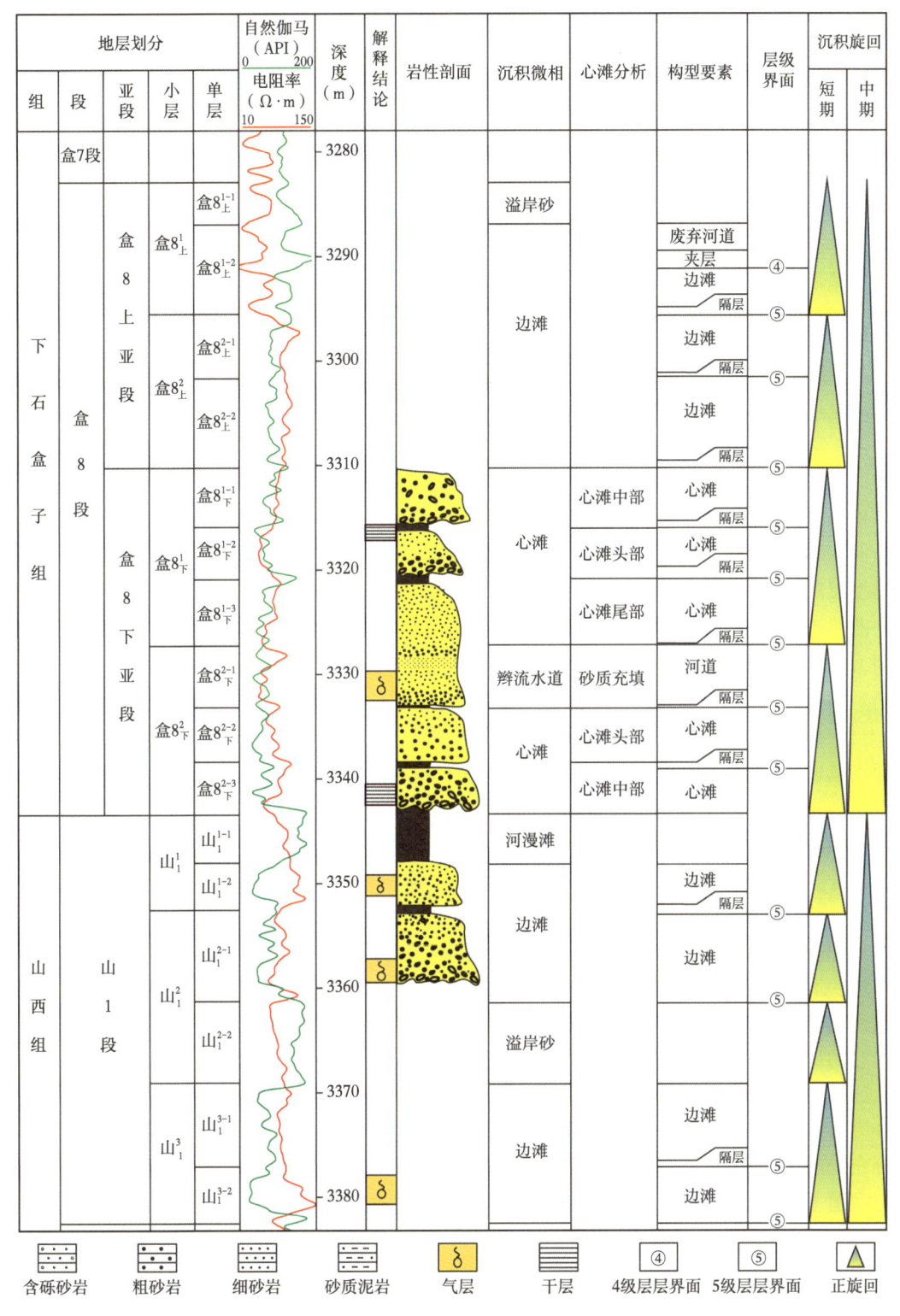

图 1 苏里格气田苏 3 加密区苏 3-J1 井综合柱状剖面

2009年开始部署大丛式井组,即4井丛及以上的井组。丛式井组可以减少井场、钻前道路土地占用,缩短钻井、试气搬家时间,便于后期井口设备维护等[11]。同时由于地质条件复杂,对井位部署的要求更高。细化对古生界储层的认识,提高有利区预测准确性;优化井网,提高气田最终采收率;进行大丛式井组混合井网立体开发,提高储量动用程度,成为提高气田开发效益的关键。

## 2 大井组混合井网立体开发技术

### 2.1 储层描述

#### 2.1.1 致密砂岩储层特征

苏里格气田主力气层的山1段和盒8段为致密砂岩储层,马家沟组五段为白云岩储层。为提高储层预测精度,高效准确地进行井位部署,前人开展了大量有利储层预测方法研究[12-14]。本文在小层划分基础上,对复合砂体进行垂向分期,单一河道砂体进行平面划界,深化对辫状河构型单元规模的认识,明确了有效砂体四种叠置方式。

根据测井、取心等资料,依据沉积旋回和岩心相特征,对苏里格气田苏3加密区苏3-J1井进行期次厘定,根据旋回特征,可将盒8下亚段划分为盒$8_下^1$小层和盒$8_下^2$小层。由于不同层理类型反映了水动力学机制的差异,根据此差异可寻找小层的分界线。其中,盒$8_下^1$小层电性测井曲线齿化严重,说明多期河道间彼此切叠程度低;而盒$8_下^2$小层电性测井曲线则齿化相对较弱,说明河道切叠程度高,细粒泥质成分被晚期河道冲刷改造彻底,造成多期河道间界限不清晰。而经连续取心资料标定表明,盒$8_下^1$小层与盒$8_下^2$小层之间,被一定厚度的平行层理分隔,上、下则分别以块状层理为主。盒$8_下^1$小层内部由于垂向叠置程度低,旋回具有三分性。借助连续取心资料识别精度高的优点,可以将测井曲线无法识别的泥岩隔层很好地识别出来,最下面隔层厚度约70cm,上部约60cm,为盒$8_下^1$小层划分为3期进一步提供了佐证。最终,结合自然伽马曲线旋回特征及岩心资料的高精度识别,可将盒8下亚段多期复合河道砂体识别出盒$8_下^{1-1}$、盒$8_下^{1-2}$、盒$8_下^{1-3}$、盒$8_下^{2-1}$、盒$8_下^{2-2}$和盒$8_下^{2-3}$共6个单层(图1),为后续单期河道砂体平面划界奠定了基础。

平面上对单一河道砂体进行划界,对刻画同一期河道的展布特征非常必要。单一河道边界界定一般有以下四个标志:河道间泥质细粒沉积;河道砂体顶面层位高程差异,不同河道砂体顶面出现明显差异;横向砂体规模的异常变化,河道沉积为中间厚两边薄,如果出现厚—薄—厚的砂体变化,可以判断属于不同期河道;韵律横向不协调(图2)。

通过复合砂体的垂向分期和单一河道平面划界,将砂体在纵横向的接触情况刻画出来,再结合大量测井数据分析,总结出苏里格气田有效砂体的四种叠置方式(图3)。孤立型砂体横向分布局限,厚度为2~5m,宽度为300~500m,长度为400~700m;切割叠置型砂体主要为心滩与河道下部粗岩相相连,复合砂体厚度为5~10m,薄层粗岩相延伸较远,在苏里格气田中部区域相对发育,但分布局限;堆积叠置型砂体为高能水道叠置带内多个有效砂体堆积叠置形成,但切割作用弱,砂体间有物性隔层,复合砂体规模与切割叠置型砂体基本一致;横向局部连通型砂体为低能水道下部粗砂岩,局部可形成分布范围较大的复合有效砂体。

对研究区盒8段沉积砂体进行系统分析可知,盒8下亚段属于辫状河沉积体系,叠置方式

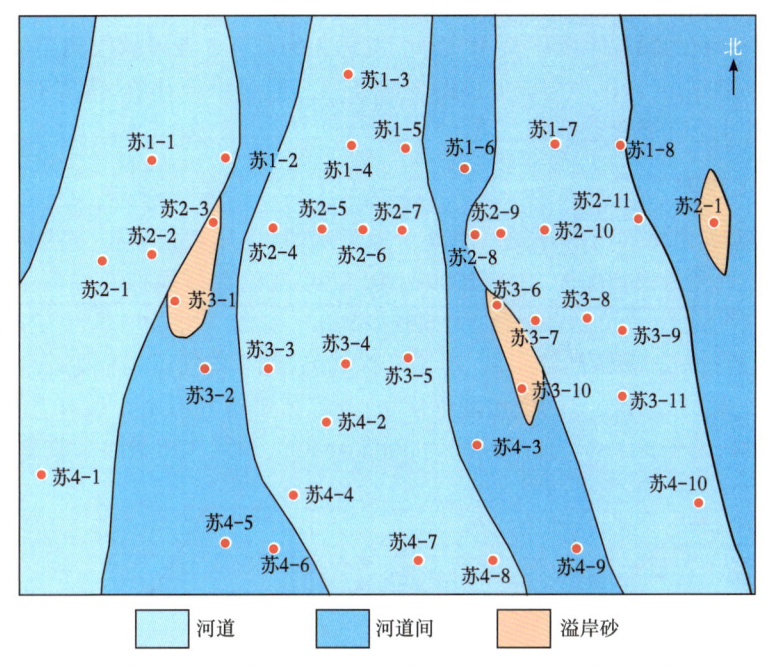

图 2　苏里格气田苏 3 加密区盒 $8_{下}^{2-1}$ 单层单一河道平面界限划分

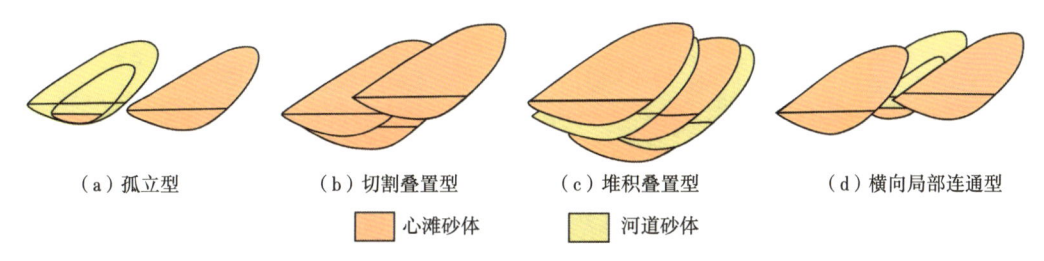

图 3　苏里格气田有效砂体叠置方式示意图

以切割叠置型和堆积叠置型为主,叠置带为多期河道砂体,局部砂体叠置宽度大,约为 2.4~3.6 km,过渡带发育单期河道或河道边部沉积;盒 8 上亚段为曲流河沉积体系,叠置方式以孤立型为主,砂体发育规模有限,叠置带与过渡带差异不明显。

#### 2.1.2　碳酸盐岩储层特征

苏里格气田下奥陶统马家沟组五段碳酸盐岩气藏主体位于鄂尔多斯盆地中央古隆起东侧,目的层为马五$_4^1$ 小层和马五$_5$ 亚段。与国外碳酸盐岩气藏多受构造控制不同,马五$_4^1$ 小层有效储层岩性为含膏粉—细晶白云岩,圈闭类型为岩性—古地貌圈闭[15],马五$_5$ 亚段储层岩性为粉晶及细—中晶白云岩,圈闭类型为岩性圈闭。

马五$_4^1$ 小层有效储层为风化壳储层,岩溶古地貌决定了储层受风化淋滤作用的强度,也就决定了储层的质量,因此岩溶古地貌为其主控因素。岩溶古地貌中残丘及洼地斜坡为储层发育区,地貌较低的古洼地和古沟槽为低产或无产区[11]。

马五$_5$ 亚段储层受风化淋滤作用较弱,储层的发育主要取决于成岩作用,通过深化白云岩形成机理研究,明确白云岩成因主要为埋藏白云岩化,而决定埋藏白云岩化的关键在于古沉积环境是否形成了有利的颗粒滩沉积微相,因此,沉积相为马五5亚段储层的主控因素。颗粒滩为最有利的沉积相,白云岩化是形成储层的先决条件。通过综合评价,落实有利目标区,为兼顾马家沟组五段井位部署奠定基础。

## 2.2 井网优化技术

2008年以前,苏里格气田的开发井网处于探索阶段,应用数值模拟方法优化开发井网为600m×1200m,预测气田采收率为22%;2009—2012年,通过数值模拟及油藏工程论证进行了开发井网优化,结合经济评价,井网优化为600m×800m,预测气田采收率为35%。

2013年后,通过对加密区砂体解剖、现场干扰试验(图4)、气藏工程论证、数值模拟、经济评价等多种方法,建立了开发井网优化理论模型,确定苏里格气田山1段和盒8段直井、定向井的合理控制面积为0.325km²,合理井网为500m×650m,预测气田最终采收率为50%。

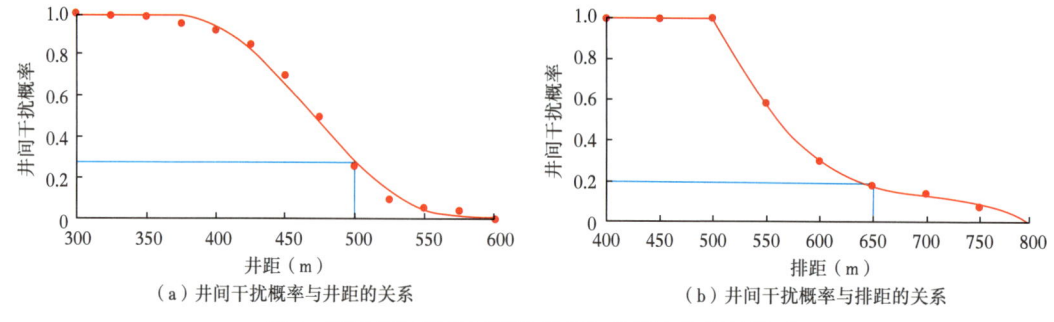

图4 苏里格气田井间干扰概率与井距和排距的关系

针对苏里格气田马家沟组五段气藏直井和定向井,综合密井网区地质解剖、不稳定分析控制范围预测结果,结合现场干扰试验及试井解释结果,推荐合理控制面积为2.000km²,合理井距为1200~1500m,合理排距为1300~1600m。

综合有效砂体解剖、现场干扰试验、产量不稳定分析和理论计算,确定苏里格气田山1段和盒8段水平井合理井距为500~600m,合理排距为1600~1800m。并根据古生界气藏的叠合程度,形成了满足多套储层开发的混合布井模式。

苏里格气田大丛式井组,常见井丛数为4~15,结合征地费用、井场建设、地面集输建设、试气等成本进行经济效果评价,认为8~10丛经济效益最佳,下面以9丛式为例,展示4种大丛式井组混合布井模式(图5)。模式①为直井和定向井组成的9丛式井组,适用于发育多个薄储层区域。模式②为直井、定向井和水平井组成的9丛式井组,为苏里格气田常见的3+6井网,即3口直井或定向井+6口水平井。直井和定向井可在前期进行储层评价,之后对可进行水平井部署的区域部署水平井。模式③为直井、定向井和水平井组成的9丛式井组,为7+2井网,即7口直井或定向井+2口水平井,可用于局部厚层砂体发育、适合部署水平井的区域;模式④与模式③相似,区别在于仅山1段和盒8段部署水平井,其他区域部署直井和定向井。具体部署哪种模式的井组,主要根据地质条件决定。

气藏应用篇

模式① 模式② 模式③ 模式④

| 马家沟组五段直井或定向井 | 山1段和盒8段直井或定向井 | 马家沟组五段水平井 | 山1段和盒8段水平井 |

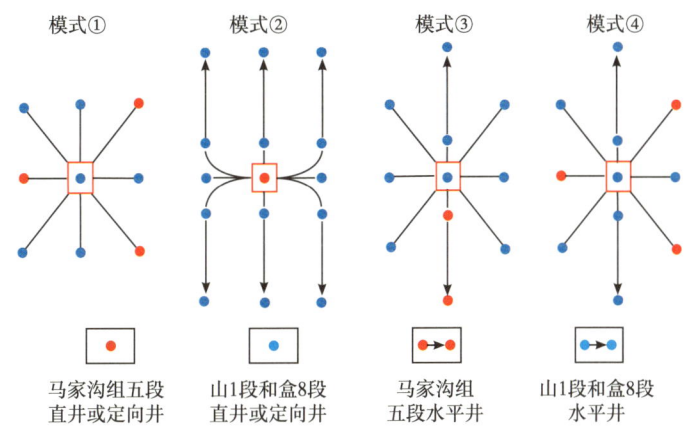

图 5 苏里格气田 9 丛式井组混合布井模式示意图

## 2.3 大丛式井组井型组合

苏里格气田目前常用井型有直井、定向井和水平井三种。垂向储量动用程度上,直井、定向井优于水平井;从平面储量动用范围来看,水平井优于直井。苏里格气田致密砂岩储层有效砂体厚度横向变化快、非均质性强,纵向上发育多套砂体,采用单一的水平井、直井或定向井开发,都不能满足垂向与平面两个方向上储量的充分动用,所以,采用合理组合井型,才能达到垂向和平面储量最大动用的目的,提高最终采收率。

针对苏里格气田四种不同叠置方式的砂体,最优化井型也有所不同。其中,堆积叠置型砂体根据夹层的不同,可分为具有渗透性夹层的堆积叠置型砂体和具有泥质夹层的堆积叠置型砂体。适合部署水平井的砂体类型包括:(1)大型孤立型砂体,气层厚度大于4m;(2)堆积叠置型砂体,砂地比大于70%,且泥质夹层厚度小于3m;(3)切割叠置型砂体,叠置砂体厚度大于5m;(4)横向局部连通型砂体,局部可形成规模较大砂体,具体井型需根据预测砂体规模确定,如预测砂体横向局部连通较远,砂体厚度大于3m,延伸长度大于1.5km,类似糖葫芦,可部署水平井,否则部署直井或定向井(图6)。适合部署直井或定向井的是孤立型砂体,砂体厚度小于4m;多层砂体发育,单个气层厚度小于4m。

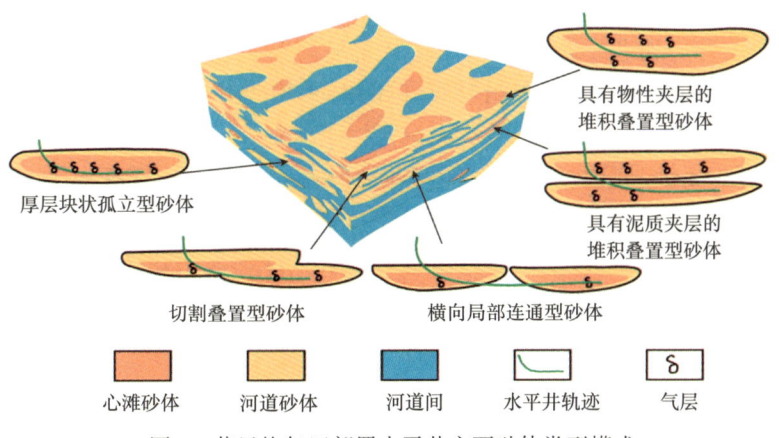

图 6 苏里格气田部署水平井主要砂体类型模式

— 375 —

苏里格气田中部和东部古生界均发育有利储层,对马家沟组五段气藏进行开发,可弥补山1段和盒8段致密砂岩气藏单井产量低的不足,因此古生界采用立体开发(图7)。形成了山1段和盒8段直井或定向井+马家沟组五段直井或定向井、山1段和盒8段水平井+马家沟组五段直井或定向井和山1段和盒8段直井或定向井+马家沟组五段水平井多层系立体开发技术,使储量动用程度大幅提高。

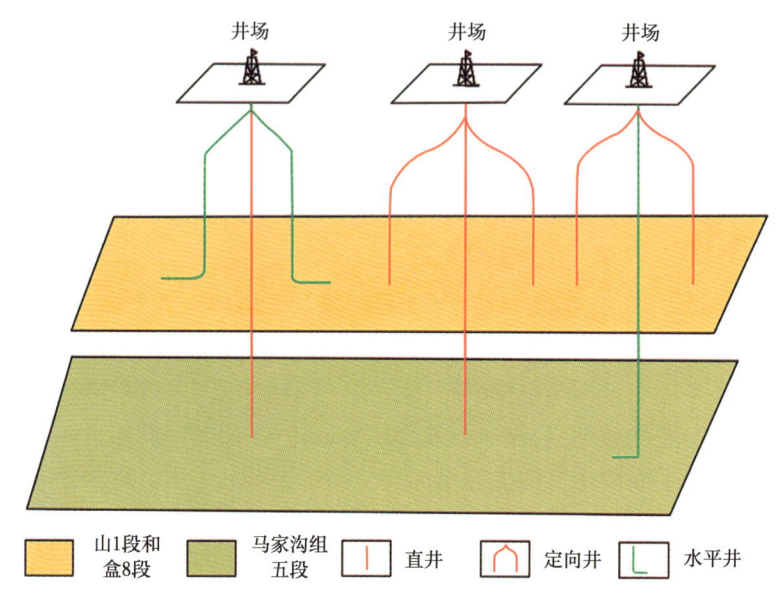

图7 苏里格气田中部和东部古生界立体开发模式

## 3 应用效果

对苏里格气田苏3加密区盒$8_下^1$小层沉积微相进行研究,刻画出心滩砂体和辫状河道砂体(图8a)。同时结合完钻井钻遇砂体情况,绘制砂体厚度分布(图8b)。在此基础上,在苏3加密区中部部署苏4-4大丛式井组,为9丛式井组,井距为500~550m,排距为650m。

大丛式井组工程效益主要为降低征地、钻前工程、钻井工程、试气工程和地面工程的费用。考虑到丛式井组单井增加进尺量,通过对丛式井组平均单井费用与单独建一个井场的单井费用进行对比,在苏3加密区,9丛式井组与直井相比,平均单井费用降低$53.2×10^4$元,综合降低$479.0×10^4$元。

同时,由于减少了钻机搬家、钻前修路等时间,气井平均建井周期缩短了8天。通过对砂体横向和垂向的刻画,水平井砂体钻遇率提高到85.3%,水平井单井产量达到直井的3~5倍。

大井组混合井网立体开发技术实现了山1段和盒8段河流相砂岩储层和马家沟组五段海相碳酸盐岩储层立体开发,整体储量一次动用。降低了气田开发成本,开发效果好,保障了气田的经济有效开发。

(a) 盒8下¹小层沉积相

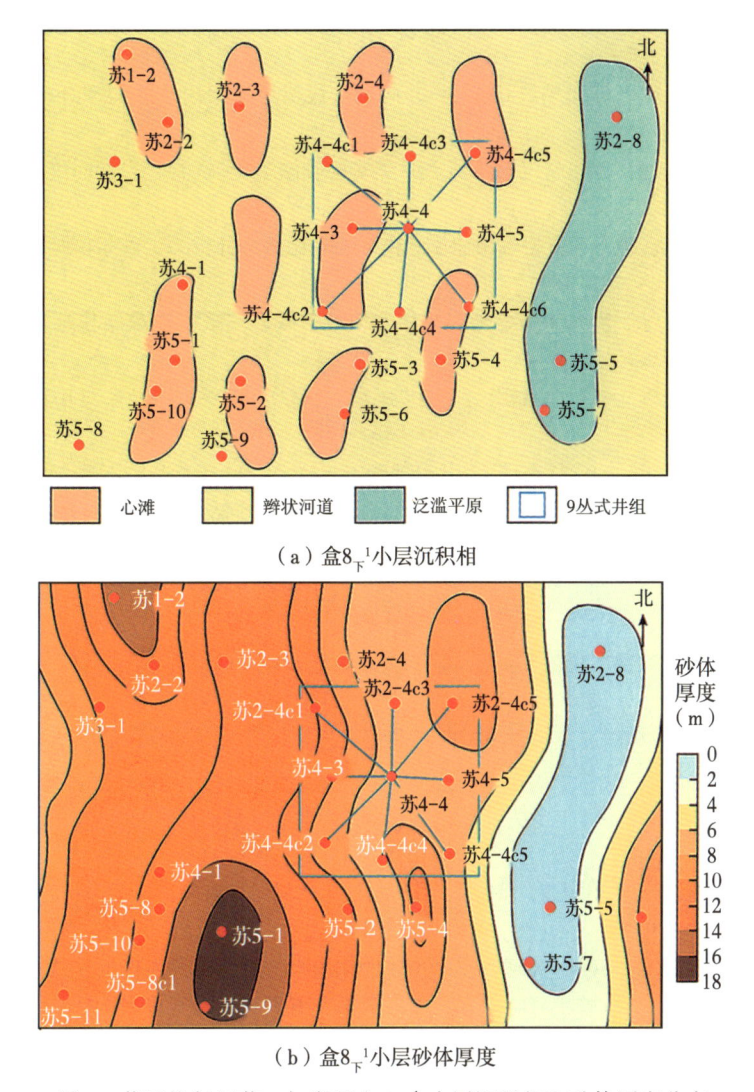

(b) 盒8下¹小层砂体厚度

图8 苏里格气田苏3加密区盒8下¹小层沉积相和砂体厚度分布

## 4 结论

(1)综合利用密井网岩心、测井、录井等多种资料,通过复合砂体垂向分期,单期河道砂体平面划界,明确了孤立型、切割叠置型、堆积叠置型和横向局部连通型四种有效砂体叠置方式。通过对马家沟组五段气藏储层主控因素的研究,形成其有利区优选方法。对储层的刻画,有效提高大井组混合井网立体开发效果。

(2)基于对储层的刻画,结合直井、定向井和水平井的优势,形成古生界多层系大井组混合井网立体开发技术,实现垂向多层系储量一次动用,有效提高了气田采收率。

(3)实践证明,大井组混合井网立体开发技术的应用,能有效提高苏里格气田采收率,增强气田稳产能力。一定程度上实现了苏里格气田降本增效,对于其他同类型多层系气藏的开发具有借鉴意义。

## 参 考 文 献

[1] 王涛,侯明才,王文楷,等. 苏里格气田召30井区盒8段层序格架内砂体构型分析[J]. 天然气工业,2014,34(7):27-33.

[2] 谢庆宾,谭欣雨,高霞,等. 苏里格气田西部主要含气层段储层特征[J]. 岩性油气藏,2014,26(4):57-65.

[3] 贾云超,尹楠鑫,李存贵,等. 苏里格气田储层建模方法对比优选分析——以苏6加密实验区块为例[J]. 断块油气田,2015,22(6):765-769.

[4] 卢涛,刘艳侠,武力超,等. 鄂尔多斯盆地苏里格气田致密砂岩气藏稳产难点与对策[J]. 天然气工业,2015,35(6):43-52.

[5] 何明舫,马旭,张燕明,等. 苏里格气田"工厂化"压裂作业方法[J]. 石油勘探与开发,2014,41(3):349-353.

[6] 王万庆,石仲元,杨光,等. G0-7"工厂化"井组钻井工艺技术[J]. 天然气与石油,2015,33(2):64-68.

[7] 赵忠军,李进步,马志欣,等. 苏36-11提高采收率试验区辫状河储层构型单元定量表征[J]. 新疆石油地质,2017,38(1):55-61.

[8] 付晓燕,杨勇,黄有根,等. 苏里格南区奥陶系岩溶古地貌恢复及对气藏分布的控制作用[J]. 天然气勘探与开发,2014,37(3):1-4.

[9] 杨华,付金华,李新社,等. 苏里格大型致密砂岩气藏形成条件及勘探技术[J]. 石油学报,2012,33(增刊1):27-36.

[10] 蒋传杰,杜孝华,张浩,等. 苏里格气田东区下奥陶统马五$_1$亚段白云岩成因[J]. 新疆石油地质,2017,38(1):41-48.

[11] 李进步,马志欣,张吉,等. 鄂尔多斯盆地苏里格气田降本增效系列技术[J]. 天然气工业,2018,38(2):51-58.

[12] 陈凤喜,王勇,张吉,等. 鄂尔多斯盆地苏里格气田盒8气藏开发有利区块优选研究[J]. 天然气地球科学,2009,20(1):94-99.

[13] 马志欣,张吉,薛雯,等. 一种辫状河心滩砂体构型解剖新方法[J]. 天然气工业,2018,38(7):16-24.

[14] 张吉,马志欣,王文胜,等. 辫状河储层构型单元解剖及有效砂体分布规律[J]. 特种油气藏,2017,24(2):1-5.

[15] 李进步,白建文,朱李安,等. 苏里格气田致密砂岩气藏体积压裂技术与实践[J]. 天然气工业,2013,33(9):65-69.

# 国内外大型碳酸盐岩气藏开发规律研究

孙玉平[1,2]　陆家亮[2]　刘　海[2]　万玉金[2]　唐红君[2]　张静平[2]

(1. 中国科学院大学；2. 中国石油勘探开发研究院)

**摘要**：以四川盆地元坝气田长兴组气藏和安岳气田磨溪区块龙王庙组气藏等一批大型碳酸盐岩气藏的勘探发现为标志,中国大型碳酸盐岩气藏勘探开发进入了新阶段。然而由于碳酸盐岩气藏特殊的成藏过程使得该类气藏地质条件复杂,开发面临的不确定性大,不同气田开发效果差异较大。为科学开发碳酸盐岩气藏,以国内外开发成熟的不同类型碳酸盐岩气藏为解剖对象,剖析了典型气藏开发效果之间的差异及影响因素,提出了影响碳酸盐岩气藏开发特征的主控因素,并在细分气藏类型的基础上,提出了不同类型气藏开发指标推荐取值。研究认为,影响碳酸盐岩气藏开发效果的主要因素为驱动类型、储层物性、裂缝发育程度及气藏管理措施,基于开发主控因素将碳酸盐岩气藏细分为弹性气驱、孔隙型水驱、裂缝型边水驱和裂缝型底水驱四个亚类;通过统计分析提出了可采储量采气速度、稳产期、递减率和采收率等开发指标规律,为国内类似气藏科学开发提供参考。

**关键词**：大型气田；碳酸盐岩；矿场统计；开发规律；驱动类型；开发指标；采收率

进入21世纪以来,中国大型碳酸盐岩气藏勘探开发进入了新一轮的黄金期,相继勘探发现了元坝气田长兴组气藏和安岳气田磨溪区块龙王庙组气藏等一批大型碳酸盐岩气藏,为大幅提高天然气国内生产能力、降低对外依存度和保障供气安全发挥了重要作用[1-3]。然而,从国内外碳酸盐岩开发历史看,大型气田开发效果差异明显。中国四川盆地卧龙河气田石炭系气藏为背斜构造控制的边水气藏,水体能量有限,储层有效孔隙度为4.3%~7.1%、渗透率为0.77mD,$H_2S$含量为0.2%~4.8%,属正常压力系统,针对该气藏特点确定稳产规模为$13×10^8m^3/a$,并以此速度连续稳定生产11年,目前气藏采出程度为80%,实现了高效开发;加拿大海狐狸气田为背斜构造控制的块状气藏,储层有效厚度为266m、孔隙度为2%~3%、渗透率为20~198mD,裂缝发育,边底水非常活跃,但是由于早期对气藏认识不足,该气田开发部署中仅部署了6口生产井,单井配产$93×10^4m^3/d$,约为无阻流量的1/3,探明地质储量采气速度为4.6%,结果气井过快见水,无法实现稳产,气藏废弃时采收率仅12%,气田开发效果差[4,5]。为此,本文以国内外开发中后期碳酸盐岩气藏为研究对象[6,7],深入剖析国内外大型气田开发特征、效果、影响因素和开发指标规律,以期为国内大型碳酸盐岩气田合理开发提供支撑。

## 1 全球大型碳酸盐岩气田分布

据美国石油地质学家协会(AAPG)资料,截至2002年底全球共发现95个大型碳酸盐岩气田,累计可采储量$73.8×10^{12}m^3$,占同期全球总可采储量的45%,主要分布在中东、俄罗斯西南部、欧洲大陆、北美和东南亚等地区。

## 2 大型碳酸盐岩气田开发影响因素分析

气田开发效果好的评价标准一般包括以下几个方面：最大限度地从气藏中采出天然气，也就是说最终采收率要高；在确保气藏不受伤害的条件下保持高且合理的开采速度；保证获得最高的经济效益。其中，采收率作为评价气藏开发效果的参数，应用最为广泛，因此以采收率作为首要的评价指标，以 CC Reservoir 和 IHS 两大数据库为数据来源，收集整理了 52 个气田的资料，剖析了影响气田开发效果的主要因素。通过对这 52 个已开发气田开发效果的综合分析，认为影响碳酸盐岩气田开发效果的主要因素为驱动类型、储层物性、裂缝发育程度及管理措施。

### 2.1 驱动类型

在影响天然气藏开发效果的各类因素中，驱动类型被认为是最重要的。图 1 为水驱和气驱碳酸盐岩气田采收率分布图，图中表明水驱气藏采收率范围为 11%~68%（平均 44%），气驱气藏采收率范围为 50%~93%（平均 74%），后者采收率明显高于前者。水驱气藏采收率较低并且变化范围较大，边底水不规则的侵入很容易造成气藏"水淹"。采收率低于 40% 的气田，其气藏特征是具有强烈的水驱和双孔介质系统。在气田开发早期阶段，配产过高、完钻层位接近气水界面，造成过快水窜，出水量较高。例如加拿大 Beaver River 气田和中国威远气田，采收率分别为 12% 和 31%。相反，当气藏储层具有较弱的含水层或者仅气藏局部与区域水层相连，这种情况下水侵会延迟出现，或者通过产量的精细控制和气井位置的控制，使产水量达到最小，采用这种方法的气田采收率超过 55%（法国 Meillon 气田、美国 Carthage 气田、马来西亚 Luconia-F6 气田和巴基斯坦 Sui 气田）[8]。

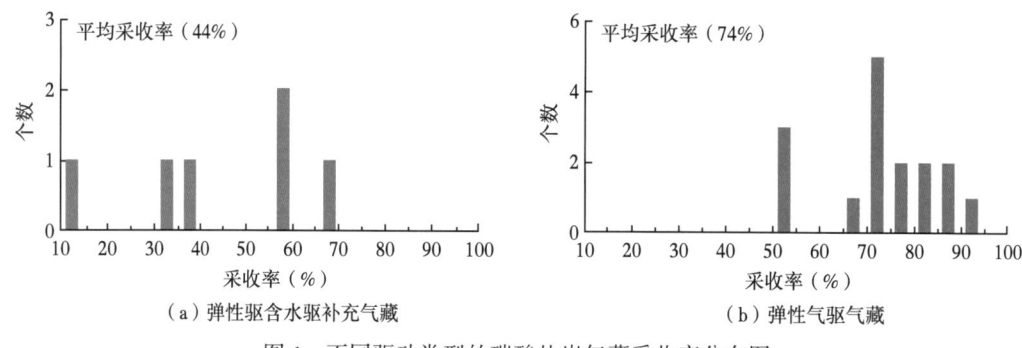

图 1 不同驱动类型的碳酸盐岩气藏采收率分布图
(a) 弹性驱含水驱补充气藏　(b) 弹性气驱气藏

矿场统计分析表明，天然气生产过程中引起水侵主要与以下因素有关：含水层展布、储层类型、采气速度及完井层段。通常，水侵容易发生在具有双孔隙系统或者活跃水层的气藏中；过高的采速或生产层段太接近气水界面是引起早期水侵的两个重要因素。

### 2.2 储层物性

储层物性也是影响气藏采收率的重要因素。统计表明，裂缝欠发育气藏采收率与基质渗透率呈正相关（图 2），中高渗透弹性气驱气藏采收率普遍高于 70%，开发效果好，采收率高受益于储层物性好、气藏连通性高的优势。低渗透致密裂缝欠发育储层采收率小于 70%,与中高渗透储层形成鲜明对比。

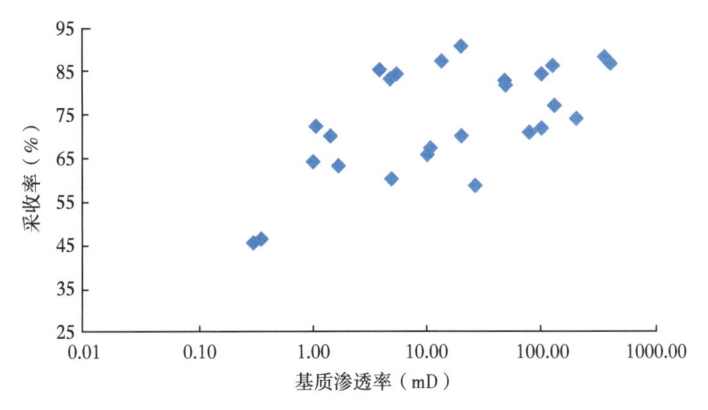

图 2　裂缝不发育的弹性气驱为主气藏的采收率与渗透率关系图

但是当低渗透致密储层中存在大量天然裂缝时,裂缝能够极大地提高储层有效渗流能力,从而提升气井产量和气田开发效果。法国 Lacq 气田基质渗透率低( <1mD),但裂缝非常发育,有效渗透率大于 400mD,产能系数值介于 1000~20000mD·m,极大地提高了气井生产能力,发现井初产量 $980×10^4m^3/d$,气藏最终采收率达到 80%。裂缝型碳酸盐岩储层性质变化大,并且很难预测,大多数这类储层具有相对较低的基质孔隙度和渗透率,单井产量主要与该井钻遇的裂缝数量有关,高产和稳产不仅需要较高的基质孔隙度,更需要较高的裂缝渗透率。

## 2.3　裂缝发育程度

相对碎屑岩,碳酸盐岩脆性大,更容易产生裂缝。为了量化评价裂缝影响,应用数值模拟方法并设置了三类储层:第一类储层为低渗透孔隙型储层,没有裂缝,基质渗透率为 0.1mD;第二类储层为低渗透裂缝型储层,基质渗透率为 0.1mD、裂缝渗透率为 10mD,有效渗透率为 8.6mD;第三类储层为中高渗透孔隙型储层,基质渗透率为 10mD。模拟结果表明:(1)裂缝能极大提高气田开发效果。基质渗透率为 0.1mD 的孔隙型储层,气藏采收率为 41.8%;基质渗透率为 0.1mD、裂缝渗透率为 10mD 的储层,气藏采收率可达 61%。可见当发育裂缝时,裂缝增产贡献明显。(2)有效渗透率相当时,裂缝孔隙型储层气藏开发效果不如孔隙型储层气藏。基质渗透率为 10mD 的孔隙型储层气藏,采收率为 98%,高于有效渗透率为 8.6mD 的裂缝型储层气藏。因而,裂缝孔隙型储层气藏开发效果的预测不能简单类比相同有效渗透率的孔隙型储层气藏。

## 2.4　管理措施

对比分析采收率较高的 24 个气田(采收率大于 75%)的地质特点表明,造成它们采收率较高的原因多样。其中 8 个储层基质渗透率大于 10mD,8 个气田发育连通性很好的天然裂缝或者喀斯特岩溶体系。剩下气藏成功开发的关键是依赖于有效的开发管理策略和适应的开发技术,而不是依赖于储层特征因素。

### 2.4.1 压裂酸化

一般来讲,并没有简单的标准来预测哪个气藏是否适合水力压裂。一些开发效果最差的气藏和一些开发效果最好的气藏都实施过压裂。对于裂缝不发育的储层,水力压裂的作用是依靠人工造缝,使得储层中发育裂缝的独立部位相互连通。对于裂缝发育的储层,水力压裂主要的作用是避免井筒损伤,且打开之前闭合的裂缝。依靠采用大规模水力压裂,好几个储层裂缝不发育的气藏都获得了不错的采收率。在美国科罗拉多 Beecher Island 气藏,最终采收率与压裂次数及支撑剂的体积密切相关。美国俄克拉何马的 Berlin 气藏是另一个成功应用水力压裂来提高产量的典型实例,其储层基质渗透率低(0.1mD),自然裂缝不发育,但该气藏构造简单,具备地质上和经济上对整个气藏实施水力压裂改造的可能,通过大范围大规模的水力压裂措施使单井产量普遍提高 3~4 倍,最终采收率 80%。此外,酸化压裂是获取较高采收率的关键手段。在阿联酋 Sajaa 气藏碳酸盐岩储层中,酸化压裂可以溶解堵塞裂缝限制气流的方解石胶结物,压开裂缝,使得天然气产量提高 400%~600%。在俄罗斯 Korobkov 气藏,酸化压裂很好地提高了钻井中受到钻井液伤害井的产量。

### 2.4.2 井网加密

对于渗流能力较好的常规气藏,稀疏的井网即可以动用全部储量,增加井数只能提高气藏采气速度,对提高气藏最终采收率作用不大。而当气藏物性较差、非均质性强和连通性差时,随着井网加密,气藏采收率会得到显著提升。中国四川盆地老气田开发效果统计表明,当储层渗透率大于 13mD 时,不同类型气藏采收率与井距的关系趋势大体一致,而当储层渗透率低于 13mD 时,气藏采收率随井网密度增加而增加,从低于 30% 增加到 75% 以上,通常井距小于 2km 时才可能保证采收率较高。加密井措施可以调节井距来获得最有效的单井泄流面积,该措施也使得裂缝不发育的气藏提高了采收率。例如,英国 Leman 气田,加密井的数量已经远大于一个正常北部海海上气藏应该具有的钻井数,钻加密井是为了更有效地泄流那些离断层较远且不发育裂缝的分割部分。

### 2.4.3 防水治水

有 4 个开采效果好的气藏都经历了水侵的影响,但它们还是获得了较高的采收率,主要是因为采取了有效积极的治水措施。控制水侵最有效的方法是谨慎小心的选择完井井段。在俄罗斯 Vuktyl 气藏以及墨西哥 Catedral and Muspac 气藏,开发井都部署在构造较高部位,远远高于气水界面,目的就是控制水锥。在 Catedral and Muspac 气藏,目前已部署的大斜度井都与北西—南东向主断层平行,目的是避免与裂缝区域(主断层)相交,避免给水锥提供高渗流通道。在 Vuktyl 气藏,产量完全依赖于降低早期水窜的措施。在俄罗斯 Korobkov 气藏,井位部署都很小心地避开了高产水的区域。提高裂缝不发育低产井的压降,同时降低裂缝发育的高产井压降,这可以使得裂缝发育区水侵伤害最小化。尽管 Korobkov 气藏遭受了局部水窜影响,但是它还是获得了十分满意的采收率(92%)。上述的治理措施延长了裂缝气藏的生产周期,使得最终采收率较高。多年谨慎积极的开发管理工作最终延长了气藏的稳产期,取得很好的生产效果,与此相对的是,缺乏开发管理策略的气藏,都是达到产量顶峰后产量很快递减,稳产期短,生产效果差,例如 Beaver River 气藏等。

## 3 大型碳酸盐岩气田开发指标

驱动类型、储层物性和裂缝发育程度是影响碳酸盐岩气藏开发效果最主要的客观因素,据此将碳酸盐岩气田分为气驱、孔隙型水驱、裂缝型边水驱和裂缝型底水驱四种类型。四类气藏开发指标规律见表1。

表1 不同类型碳酸盐岩气藏开发指标取值表(括号内为平均值)

| 气藏类型 | 气田数(个) | 可采储量采气速度 | 稳产期(a) | 稳产期末采出程度 | 递减率 | 采收率 |
|---|---|---|---|---|---|---|
| 弹性气驱 | 28 | 2.0%~13.6% (5.1%) | 1.0~20.0 (10.1) | 19.0%~70.0% (54.8%) | 3.8%~22.0% (10.2%) | 孔隙型:<br>A. 中高渗:70%~90%,平均82%;<br>B. 低渗,miskar(多层)40%,berlin(整装)80%<br>裂缝型:<br>A. 低孔高渗 80%~90%,平均87%;<br>B. 低孔低渗 60%~80%,平均69%;<br>C. 高孔低渗,50%~60%,平均55% |
| 孔隙型水驱气藏 | 7 | 2.8%~4.9% (3.8%) | 5.0~18.0 (9.3) | 28.0%~31.0% (29.3%) | 5.0%~19.0% (11.3%) | 中高渗:50%~80%,平均71% |
| 裂缝型边水驱气藏 | 7 | 3.2%~12.2% (7.7%) | 1.0~17.0 (7.8) | 31.4%~72.5% (54.4%) | 14.0%~29.0% (21.4%) | 中活跃 55%~75%,平均64% |
| 裂缝型底水驱气藏 | 10 | 2.3%~14.2% (8.8%) | 1.0~14.0 (3.7) | 18.4%~56.6% (40.1%) | 5.1%~20.2% (14.5%) | A:措施有效,40%~50%,平均45%;<br>B:开发失败,10%~30%,平均20% |

### 3.1 弹性气驱气藏

弹性气驱气藏不受水侵的影响,开采方式采用衰竭式开发,开发相对容易,采气速度范围大,可采储量采气速度为2%~14%,平均为5%,通常会有较长的稳产期,平均10年,稳产期末采出程度高达55%,最终采收率由储集空间类型和储层物性决定。对于中高渗透储层,具有非常好的压力传导性,开发相对容易,该类气藏最终采收率通常在80%以上,其中孔隙型中高渗气藏平均为82%,裂缝型中高渗气藏平均87%。对裂缝型低渗气藏,有效渗透率往往能达到基质渗透率的2~4个数量级,气井产能高,气藏仍能取得较好开发效果,平均采收率为69%。对于发育少量裂缝的低渗透率储层,特别是白云质灰岩气藏,孔隙度很高,但是裂缝不发育,基质渗流能力差,气田整体开发效果一般,平均采收率约为55%。

### 3.2 孔隙型水驱气藏

主要为礁滩相储层,基质渗透率较高,开发指标与中高渗透砂岩水驱气藏类似,可采储量采气速度平均为3.8%,稳产期平均9.3年,最终采收率为50%~80%,平均为71%。

## 3.3 裂缝型边水驱气藏

相对底水驱气藏,边水驱气藏开发风险略低,特别是对具有中等—弱边水驱的裂缝性气藏而言,采收率与气田管理紧密相关。单井产量和完钻井段的精确控制对于提高最终天然气采收率非常重要。法国 Meillon 气田是具有中等采收率(55%~65%)气田的典型代表,该气田 1968 年投产,稳产至 1979 年,1978 年出现局部的水窜,天然气产量从 1982 年到 1990 年快速降低。通过动静态结合、多方法对比,深化了裂缝发育规律认识,为 Meillon 气田开发调整提供了科学依据,并取得较好调整效果。Meillon 气田 B1 井 1968 年投产,1978 年水侵,1988 年在对气井水侵和裂缝发育规律认识的基础上认为可以恢复生产,测试获日产气 $10\times10^4 m^3$,1989 年开始排水,1990 年 5 月重新投产,初始日产气 $22\times10^4 m^3$,日产水 $90 m^3$,至 1991 年 5 月,累计增加产气量 $1300\times10^4 m^3$,累计产水 $10\times10^4 m^3$,使该井天然气产量得到显著增加,且使相邻天然气井又重新恢复生产[8]。

## 3.4 裂缝型底水驱气藏

以底水驱动为主的气藏,开采特征是具有较短的稳产期和较低的采收率,调研分析发现所有快速水淹气田全部为裂缝型底水驱气藏。对于底水驱气藏,加强气田管理、气井产量控制和射开层位可以起到一定的延缓底水锥进和暴性水淹的风险,防水比治水更关键。加拿大 Beaver river、Kotaneelee 和 Pointed Mountain 三个气田属于同一构造带,彼此邻近且同为裂缝型底水驱气藏。Beaver river 最早开发,对水体活跃程度和裂缝发育程度判断不准,对由此带来的风险准备不知,配产偏高,气井快速水淹,排水采气效果不理想,气藏采收率极低。Kotaneelee 和 Pointed Mountain 开发相对较晚,开发部署时充分吸取了 Beaver River 的经验教训,采取了防止水锥的措施,采收率虽不及弹性气驱等类型气藏,但相对 BeaverRiver 已经提高不少(Kotaneelee 采收率为 50%,Pointed Mountain 采收率为 39%)。

# 4 结论

(1)影响碳酸盐岩气田开发效果的最主要因素为驱动类型,其次为储层物性、裂缝发育程度及气藏管理措施。

(2)基于开发主控因素将碳酸盐岩气田细分为弹性气驱、孔隙型水驱、裂缝型边水驱和裂缝型底水驱四个亚类,通过统计分析提出了可采储量采气速度、稳产期、递减率和采收率等开发指标规律,不同类型气藏应选取有针对性的开发指标体系组合。

### 参 考 文 献

[1] 贾爱林,闫海军,郭建林,等. 全球不同类型大型气藏的开发特征及经验[J]. 天然气工业,2014,34(10):33-46.
[2] 马新华. 创新驱动助推磨溪区块龙王庙组大型含硫气藏高效开发[J]. 天然气工业,2016,36(2):1-8.
[3] 郭旭升,郭彤楼,黄仁春,等. 中国海相油气田勘探实例之十六四川盆地元坝大气田的发现与勘探[J]. 海相油气地质,2014,19(4):57-64.
[4] 唐泽尧,杨天泉. 卧龙河气田地质特征[J]. 天然气勘探与开发,1994,16(2):1-12.

[5] Morrow, D W, Miles, W C, The Beaver River structure: a cross-strike discontinuity of possible crustal dimensions in the Southern Mackenzie Fold Belt, Yukon and Northwest territories, Canada[J]. Bulletin of Canadian Petroleum Geology, 2001,48(1):19-29.

[6] Halbouty MT. Giant oil and gas fields of the decade, 1990—1999[M]. Tulsa: AAPG, 2003.

[7] 孙玉平,韩永新,张满郎. 全球碳酸盐岩气田开发调研报告[R]. 廊坊：中国石油勘探开发研究院廊坊分院,2013.

[8] 孙玉平,陆家亮,万玉金,等. 法国拉克、麦隆气田对安岳气田龙王庙组气藏开发的启示[J]. 天然气工业,2016, 36(11): 37-45.

[9] C&CReservoirs. Gas/Condensate Production, Carbonate Reservoirs[R]. USA: C&CReservoirs,1998.

[10] C&CReservoirs. Fractured, Tight and Unconventional Reservoirs[R]. USA: C&CReservoirs,2005.

# 黄骅坳陷千米桥潜山凝析气藏开发经验与启示

初广震[1]　韩永新[1]　周宗良[2]　周兆华[1]　郑国强[1]

(1. 中国石油勘探开发研究院;2. 中国石油大港油田公司勘探开发研究院)

**摘要**:大港油田"十三五"规划中提出到 2020 年天然气产量达到 $10×10^8 m^3$,根据目前大港油田天然气探明地质储量和开发现状,千米桥潜山气藏是大港油田天然气上产的关键和现实领域。千米桥潜山气藏以地质条件复杂,开发难度大著称,1999 年发现至今,试采评价 18 年,目前仍未上报开发动用。千米桥气田探明天然气地质储量 $266.09×10^8 m^3$,截至目前累计产气 $8.91×10^8 m^3$,采气速度为 0.19%,采出程度为 3.35%。千米桥潜山气藏的开发一直困扰着科研工作者,2016 年中国石油勘探开发研究院针对大港油田"十三五"规划 $10×10^8 m^3$ 产量的目标,提出重新评价千米桥和建产再上千米桥的研究对策,对千米桥潜山进行新一轮的研究和评价,具体开展了地震精细解释落实千米桥潜山构造,明确优质储量区;裂缝预测与储层反演评价,优选出板深 4、板深 7 和板深 8 区块为今后重点开发的"甜点"区块;提出了分断块—井一策分步实施的二次开发策略;部署开发井位,力争准确可靠、高产稳产,实现千米桥潜山气藏成功开发。

**关键词**:千米桥潜山;凝析气藏;奥陶纪;气田开发;经验与启示

## 1　气田地质特征

千米桥潜山位于黄骅坳陷北大港构造带大张坨断层与向北倾斜的港西断层的上升盘,是一个深埋在古近系构造层之下的低位序潜山构造,被千米桥西断层和港 8 井断层分成西、中、东三排潜山,即西潜山、主潜山和东潜山,本次研究区域主要为主潜山,面积达 $56 km^2$。

### 1.1　构造特征

千米桥潜山储层为下古生界奥陶系,为一套碳酸盐岩储层。从奥陶系沉积后,本区经历了 3 次主要构造运动,使奥陶系 2 次暴露地表遭受溶蚀。第 1 次是晚加里东期—早海西期的整体抬升运动,奥陶系碳酸盐岩遭受了长达 130Ma 的风化侵蚀。第 2 次是印支期—早燕山期的北西—南东挤压应力场使千米桥地区逆冲成山,石炭—二叠系遭剥蚀后,潜山大部分地区再次暴露在地表并遭受风化淋滤。第 3 次是燕山期中、晚期—喜马拉雅期,区域应力场由南东—北西向挤压转变为南东—北西向拉张,潜山随盆地一起急剧下陷,接受了巨厚的古近系和新近系沉积,致使潜山深埋,形成了目前的构造格局[1,2]。

潜山主体由下古生界构成,是一个在中生代逆冲褶皱背景上,经白垩纪晚期较强烈剥蚀改造的深埋潜山构造[3]。奥陶系顶面构造总体表现为南高北低,由北向南划分为板深 4、板深 7、板深 8、板深 703 和板深 6 等 5 个局部圈闭。板深 4 圈闭独立于潜山主体东北部,呈现中部凹陷的构造形态。其他 4 个构造圈闭形态呈现"F"形,由东侧轴向近东西的板深 7 背斜和西侧近南北向断裂背斜组成,根据地震资料解释结果,中生代期间形成的北东向古逆冲断层系已被部分北东向的正断层所切割改造[3]。

本轮解释的潜山构造采用了大港油田 3150 地震数据体,与前人构造解释整体形态一致,但在局部存在较大差异,主要表现在:(1)板深 16-17 高点两侧断层与港 8 井断层相连,板深 4 区块为独立断块;(2)板深 7 北部洼陷解释构造变陡,致使局部含气面积减少;(3)中央地区小断层及断层组合有明显的变化(图 1)。沿用相同的气水界面,新解释构造含气面积较之前减少 13.58km$^2$。

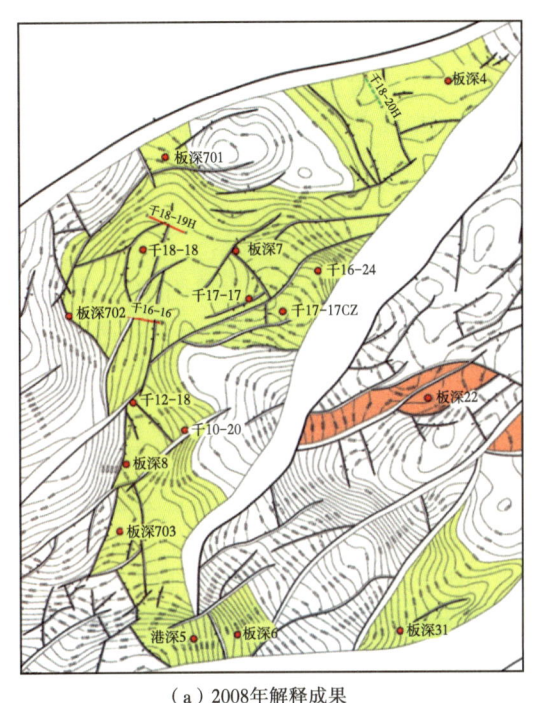

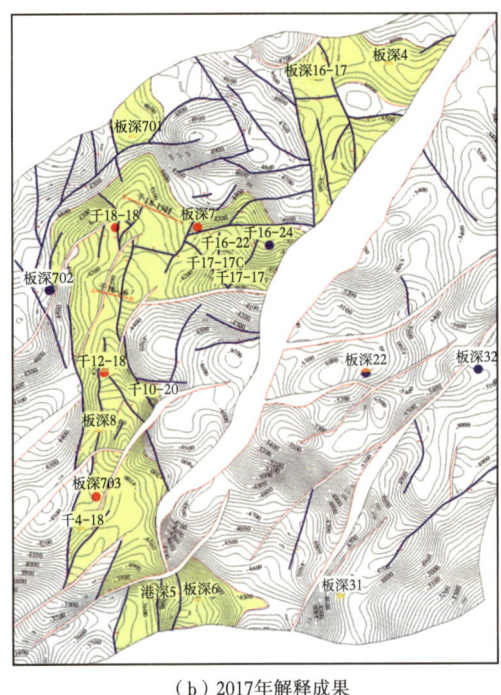

(a)2008年解释成果　　　　　　　　(b)2017年解释成果

图 1　两期构造解释成果对比图

## 1.2　油气藏成因与类型

前人对千米桥潜山凝析油气藏的成因归纳起来大致有四种观点:一种观点认为,千米桥潜山是多次运聚成藏,早期低成熟油气与晚期高成熟油气混合,属于油藏气侵型凝析气藏,即早期进入潜山的以油为主,后期烃源岩达到高成熟阶段以气为主[4-6];第二种观点的学者通过碳同位素、包裹体分析认为,千米桥潜山油气来自板桥、歧口 2 个凹陷,板深 4 井、板深 701 井天然气来自西侧的板桥凹陷,板深 8 井、板深 6 井天然气来自东侧的歧口凹陷,而板深 4 井、板深 7 井、板深 8 井的凝析油都来自板桥凹陷,并且指出千米桥潜山曾发生了 3 期油气运移、4 幕油气充注[7,8];第三种观点的研究人员通过地球化学分析认为千米桥潜山的高蜡凝析油来自东侧的歧口凹陷,并指出充注方向是从北东向南西,即从板深 4 井逐次向板深 7 井、板深 8 井充注[9];第四种观点认为千米桥潜山油气来源于板桥凹陷沙三段,充注方向从北向南,存在两条充注路径,一条是从板深 701 井注入向板深 7 井、板深 8 井方向运聚,另一条是注入板深 4 井区,由于断层遮挡不再向南运移[10]。

关于千米桥潜山油藏类型,于学敏等[11]认为,千米桥潜山为受风化壳控制的、带较大油环的高饱和凝析气藏。何炳振等[10]指出千米桥气藏为高饱和中高凝析油含量层状凝析气藏,无油环及活跃边底水。张亚光等[12]认为千米桥潜山主体为以中高凝析油含量和饱和型凝析气藏为主体的复合气藏,而潜山边部的板深4井和板深701井为特低凝析油含量的凝析气藏,没有活跃的边底水。陶自强[13]则认为千米桥潜山为常温常压中等凝析油含量的凝析气藏。

## 1.3 储层地质特点

区域上华北地台在早古生代是一个巨大的陆表海,千米桥地区位于这个巨大的陆表海的东北部,以碳酸盐岩台地为主,同一时期沉积环境相对稳定、均一,随海平面的周期性的升降,沉积演化上呈现开阔台地和局限台地交互出现。因此,沉积相对储层的控制主要体现在垂向的分布上。

千米桥潜山自上至下揭开地层依次为新生界(第四系、新近系、古近系)、中生界(白垩系、侏罗系)、上古生界(二叠系、石炭系)和下古生界(奥陶系)。其中上古生界因印支、燕山早期运动的挤压隆起,而遭受风化剥蚀,潜山顶部已剥蚀殆尽,使得下古生界奥陶系亦遭受长期的风化淋滤形成良好储层。

地层厚度分布稳定,一般厚100～700m,发育有下奥陶统冶里组、亮甲山组,中奥陶统下马家沟组、上马家沟组和峰峰组。其中亮甲山组主要发育深棕褐色、浅灰色厚层块状中—细白云岩为特点;下马家沟组下部、上马家沟组顶界主要发育云灰坪微相深灰、灰色白云岩,其他主要为泥晶灰岩;峰峰组主要发育褐灰、浅灰、灰色泥晶灰岩、砾屑灰岩、砂屑灰岩夹泥岩、泥云岩等[14]。千米桥潜山油气主要储集在奥陶系上马家沟组上段和峰峰组。钻探证实,区内峰峰组主要分布于板深8井区、千16-24井附近和板深703井以南的区域,其他地区全部剥蚀殆尽。马家沟组分布广泛,仅潜山东北部板深4井区剥蚀程度较大[15]。

储层具有以下四个特点:

(1)天然气主要分布在顶面厚度200m以内的风化壳储层内。

对区域内19口钻井的统计分析,钻遇到下马家沟组的钻井共3口井,这3口井都没有钻遇生产气层;10口获工业油气流钻井中所有的生产层都位于风化壳储层顶面至200m厚度的储层内,其中主要产气层位为峰峰组和上马一段。

(2)各个构造高点为一个非均质性强的缝洞型储集体。

非均质性强是千米桥潜山储层的一个重要的特点。潜山天然气分布在风化壳储层内,宏观上层状分布,微观上呈孤立"蜂窝状"独立成藏分布,藏内多有残水,连通性差。具体来说,储层纵向连通性好,横向连通性差。板深8井生产层位为上马一段,生产后期对上部峰峰组未动用的优质储层进行射孔,测试产量低,测得压力与生产层位压力相同,从而证实储层纵向穿层连通。以板深8区域3口井为研究目标,对储层横向连通性进行了分析。三口井井距在1200～1300m之间,首先井间干扰测试未发现板深8、千12-18和千10-20井之间存在连通关系;其次2001年6月板深8井关井实测压力恢复值32.785MPa,而千12-18井实测恢复压力折算至气层中部的地层压力不足20MPa,证明两口井之间不存在连通关系;最后,多种方法综合分析单井最大控制半径在500m左右,平均单井控制半径只有300m,反映了潜山储层在平面上连通性较差(图2,表1)

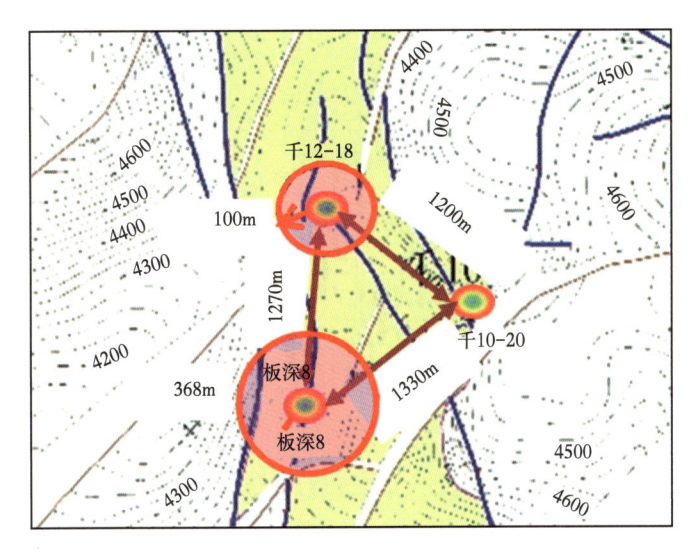

图2 板深8区块井位井距图

表1 多种方法计算单井控制半径表

| 井号 | 压降法 | AIWI解释 | 试井解释 | RTA解释 |
| --- | --- | --- | --- | --- |
| 板深7 | 188.5 | 342.4 | 235.6 | 157 |
| 板深8 | 458.6 | 505.7 | 530.9 | 368 |
| 千18-18 | 207.3 | 289.0 | 213.6 | 263 |
| 千12-18 | 91.1 | 100.5 | | |
| 板深703 | 292.1 | 449.2 | | |

(3)存在多期裂缝并以纵向缝为主、孔洞缝发育。

千米桥潜山储层总体上可以分为两类：一类是以裂缝为主，储层具有较高的渗透率，但是孔隙度却较低；另一类以晶间孔、溶孔、小型溶洞为主，岩性主要为泥（粉）晶白云岩、角砾状白云岩，孔隙度相对较高，渗透率却并不太高。本次研究在前人研究的基础上，对上马组大尺度裂缝进行了预测，以叠后地震资料为基础，运用倾角导向滤波方法预测裂缝发育。预测结果从剖面上显示高角度裂缝发育，纵向连通，有利于形成"甜点区"（图3）。平面显示潜山高部位与断裂系统耦合控制裂缝发育，主要潜力区域也是裂缝密度高值区，主要分布于板深8、板深7和板深4区块，这也是今后继续评价和开发的重点区域（图4）。

(4)气藏具有边底水，不同构造单元具有不同气水界面。

各气藏之间具有不完全相同的油气水界面。不同构造区之间气藏的气水界面高度相差较大，相隔最远的构造相差达600多米；同一构造区带内的不同断块之间，气水界面也不相同，相邻的构造相差100多米（板深7井和千18-18井），使得不同的构造具有各自不同的水动力系统，但不同的构造之间的气水界面高度变化有一定的规律性，整体表现为南部气水界面高，越向北，气水界面越低。

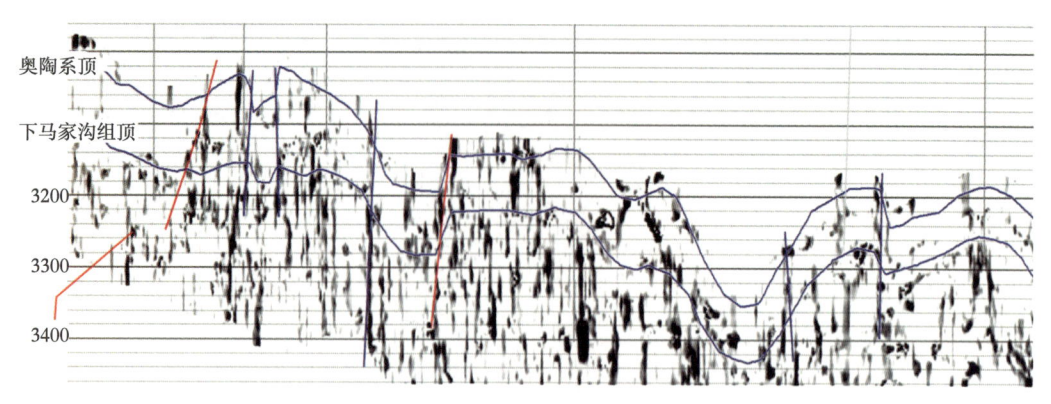

图 3　千米桥潜山裂缝发育剖面预测图

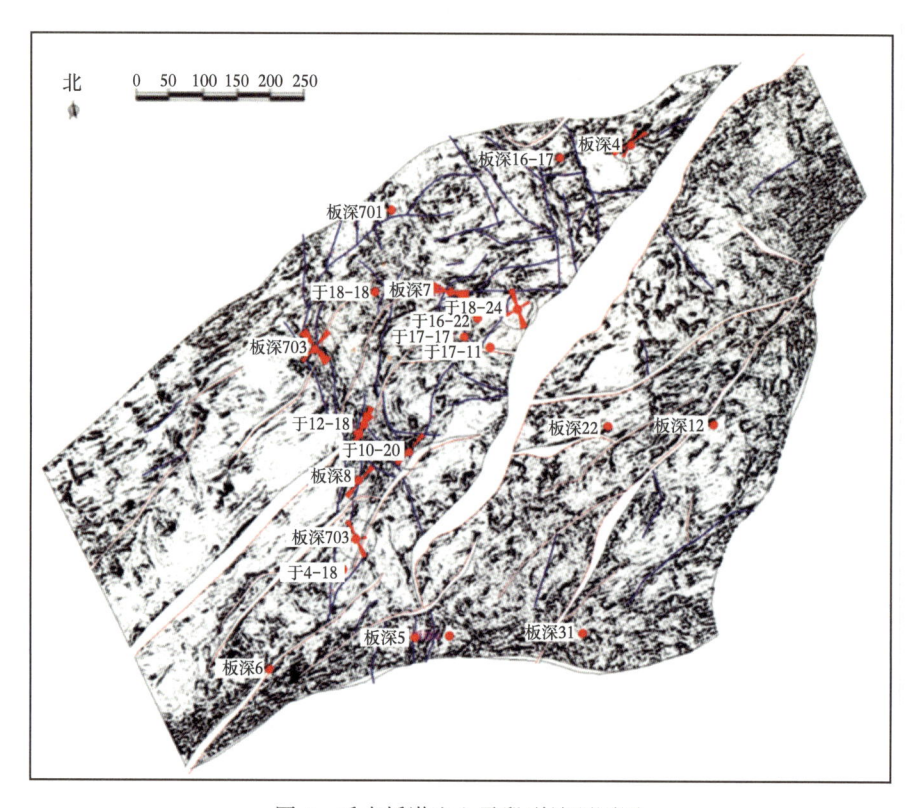

图 4　千米桥潜山上马段裂缝预测图

## 2　开发历程

千米桥潜山油气田发现于 1999 年,发现井为板深 7 井,于 1998 年 4 月开钻,完钻井深 5191.96m。对主力产层进行了酸化压裂,15.88mm 油嘴试油,折日产油 609.3m³,天然气 45.4×10⁴m³,随后又在板深 8 井、千 12-18 井、千 18-18 井获得高产油气流,宣告了千米桥潜山气田的发现。吴永平等[16]对千米桥潜山勘探发现历程有着详细的阐述。截至目前,千米桥潜山气田共完钻 19 口井,其中 10 口获工业油气流,3 口井开采衰竭,3 口井正常开采。

千米桥潜山气藏的开发至今历经18年,总体上可以划分为三个阶段:(1)气田建产期(1999—2002年)。这段时间是千米桥潜山开发的高潮,分批部署实施钻井16口,其中板深7、板深8、板深4、板深701等井发现高产油气流,2000年上报探明天然气地质储量305.1×$10^8m^3$。但在2000年以后陆续进行的9口开发评价井中,钻探成效不理想,千米桥潜山油气开发随即陷入低谷。(2)开发停滞期(2003—2007年)。在这5年期间,千米桥气藏没有实施一口新钻井,仅仅依靠板深8、千18-18、板深7等井维持产量。2006年大港油田进行了探明储量套改,套改后的千米桥潜山天然气探明地质储量为266.09×$10^8m^3$。(3)开发上升期(2008—2015年)。这个时期共完成5口钻井,其中包括2口水平井。3口井正常生产,千16-16井2008年6月投产,初期日产气11×$10^4m^3$,日产油40.86t,目前已生产9年,日产气2.25×$10^4m^3$,日产油10.29t,累计产气0.86×$10^8m^3$,累计产油4.64×$10^4$t。由于气井产水量大,从2011年至今依靠排水采气维持了较长期稳产,表现了水平井稳产年限长的特征。这口井的成功也给后续出水量大气井进行排水采气提供了可靠的借鉴。3口井平均日产气3.3×$10^4m^3$,累计产气1.85×$10^8m^3$。2口井由于工程问题导致无法生产,千18-19H井2008年完钻,在817米的水平段潜山地层钻进过程中共点燃火把36次,经统计本井至少钻遇了7组裂缝带,并且在5204~5347m井段存在大溶洞。试油后井口撞坏,因险情压井污染储层,一直未投、关井;千16-22井2014年实施的直井,中途测试压力高,六次压井无效,漏失$2487m^3$钻井液,无法生产。两口井在奥陶系目的层都显示出了较大的生产潜力。

## 3 气田开发的认识与启示

### 3.1 碳酸盐岩潜山储层非均质性强,平面连通性差,垂向沟通程度较好,如何寻找甜点区是后续开发的关键问题

千米桥潜山碳酸盐岩储层非均质性极强,在勘探开发过程中表现出气水界面复杂多变、油气藏压力系统多变、储层中油气混合程度差、井流物成分差别较大等特点,说明发育在储层中的缝洞系统横向连通性较差,存在渗流屏障。千米桥潜山油气藏早期被认为是受奥陶系顶面古风化壳岩溶控制的块状油气藏,高孔渗储层主要发育在距潜山顶面50~250m范围内的潜流带上[11,17-19],后期研究认为潜山油气充注受不连通缝洞系统以及优势充注网络的共同控制,潜山油气藏属于裂缝性油气藏,而非岩溶块状油气藏[3]。

气藏不同区块气水界面高程相差较大,已经证实的板深8井气水界面埋深4380m,而板深7井气水界面埋深达到4560m,板深703气水界面埋深4300m,北部山头的板深4井气水界面约为4600m。不同断块具有不同的气水界面,油气藏压力系统和井流物,表明了各井储层之间连通性较差。

千米桥潜山油气藏不同部位产液性和气油比也存在变化。如处于同一山头高部位的千17-17井和千16-24井产水,不含油气;而位于北翼的板深7井却获得高产。板深7井奥陶系产层气油比为$2200m^3/t$,而南侧的板深703井则为$4300m^3/t$,北翼的板深701井只产气。即便位于同一局部构造、层位相同、埋深相近的板深8井与千12-18井,气油比亦表现为较明显的差异性,前者气油比为$4000m^3/t$,后者则为$20500m^3/t$。

## 3.2 气井差异大,压力恢复速度不同,单井控制有效半径小

从生产特征看,各井产气能力差异大,高效井累产高,板深 8 井累计产量高达 $3.68 \times 10^8 m^3$,年递减约 10%,井控储量 $6.89 \times 10^8 m^3$;而低效井递减达 30%~50%,生产时间短,累产量低,井控储量 $(0.5 \sim 1) \times 10^8 m^3$。

多种方法综合分析单井最大控制半径在 500m 左右,平均单井控制半径只有 300m,反映了潜山储层在平面上连通性较差(表2)。

表2 多种方法计算单井控制半径

| 井号 | 压降法 | AIWI 解释 | 试井解释 | RTA 解释 |
| --- | --- | --- | --- | --- |
| 板深 7 | 188.482m | 342.408m | 235.6m | 157m |
| 板深 8 | 458.639m | 505.759m | 530.9m | 368m |
| 千 18-18 | 207.33m | 289.005m | 213.6m | 263m |
| 千 12-18 | 91.0995m | 100.524m | | |
| 板深 703 | 292.147m | 449.215m | | |

## 3.3 树立千米桥潜山气藏能够实现高效开发的信心,搞清潜山构造、寻找高效井、提高工程质量是千米桥潜山气藏二次开发关键

千米桥发现于 1999 年,由于技术的原因当时解释的构造成果与实际存在一定误差,多口井因为解释误差的问题导致了钻井失利,例如千 10-20 井,按当时资料解释此井正常钻入潜山储层内,而根据 2008 年的新资料解释此井只在距潜山顶界约 20m 深度的地层钻进,未进入潜山主体,最后钻出断层,进入邻块的中生界,因而井底出现紫红色泥岩。

寻找高效井是解决千米桥开发难题的钥匙。2008 年以前共完钻 17 口井,仅有 3 口高效井,累计产量 $6.31 \times 10^8 m^3$,占总产气量的 90%。2008 后,由于地质认识和技术水平的提高,开发出了板深 16-17 和千 16-16 等高效井,提高了千米桥天然气产量,但是目前开发的水平离预期还有一段距离,这需要在地震、地质和气藏工程等多方面继续提高认识,搞清楚天然气主控因素和富集规律,为发现高效井提供可靠的地质保障。

工程问题是千米桥开发面临的另外一个重要的问题,也是一个直接决定成败的关键因素。22 口井中有 7 口井出现工程问题,例如卡钻、钻头脱落、测试仪器落井、压井污染储层等一系列的问题,这些问题导致了一些原本可以高效生产的井低效生产甚至地质报废,同时也给地质和气藏认识带来了相当大的干扰。

## 4 结论

千米桥潜山气藏经历了从勘探到试采评价 18 年的漫长历程,这种情况在国内气藏开发史上极其少见,期间出现的诸多问题与其说是开发失败的原因不如说是宝贵的开发经验,这些问题对后续气田的开发以及相对与碳酸盐岩潜山气藏来说都具有重要的借鉴意义。千米桥潜山气藏成功开发需要重新对潜山构造形态、储层富集规律和主控因素进行深入分析和研究,同时要注重钻完井、后期改造等一系列配套工作。鉴于潜山储层非均质性强的特点,建议今后开发

过程中分断块寻找甜点,一井一策分步实施,部署开发井位,力争准确可靠、高产稳产,实现千米桥潜山气藏成功开发。

## 参 考 文 献

[1] 郑亚斌,肖毓祥,龚幸林,等. 千米桥潜山油气储集成藏模式探讨[J]. 天然气地球科学,2007,18(6):848-853.

[2] 齐振琴,程昌茹,孙秀会,等. 千米桥古潜山岩溶地貌演化及古岩溶洞穴发育特征[J]. 海相油气地质,2008,13(4):37-43.

[3] 付立新,杨池银,肖敦清. 大港千米桥潜山储层形成对油气分布的控制[J]. 海相油气地质,2007,12(2):33-38.

[4] 姜平. 千米桥潜山构造油气藏成藏分析[J]. 石油勘探与开发,2000,27(3):14-16.

[5] 卢鸿,王铁冠,王春江,等. 黄骅坳陷千米桥古潜山构造凝析油气藏的油源研究[J]. 石油勘探与开发,2001,28(4):17-21.

[6] 罗霞,胡国艺,张福东,等. 千米桥奥陶系潜山天然气气源对比[J]. 石油勘探与开发,2002,29(4):41-43.

[7] 杨池银. 千米桥潜山凝析气藏成藏期次研究[J]. 天然气地球科学,2003,14(3):181-185.

[8] 杨池银. 千米桥潜山凝析气藏流体非均质性控制因素[J]. 天然气工业,2004,24(11):34-37.

[9] 姜平,王建华. 大港地区千米桥潜山奥陶系古岩溶研究[J]. 成都理工大学学报(自然科学版),2005,32(1):50-53.

[10] 何炳振,王振升,苏俊青. 千米桥古潜山凝析气藏成因探索[J]. 特种油气藏,2003,10(4):1-3.

[11] 于学敏,苏俊青,王振升. 千米桥潜山油气藏基本地质特征[J]. 石油勘探与开发,1999,26(6):7-9.

[12] 张亚光,苏俊青,朱银霞,等. 千米桥潜山凝析气藏地质特征[J]. 天然气地球科学,2003,14(4):264-266.

[13] 陶自强. 千米桥潜山凝析气藏生产井出水原因分析[J]. 天然气地球科学,2003,14(4):295-297.

[14] 陈昭年. 黄骅坳陷千米桥潜山形成演化与油气成藏史[D]. 北京:中国地质大学博士论文,2003.

[15] 杨树合,王树红,王连敏,等. 裂缝性潜山凝析气藏评价与开发——以千米桥潜山凝析气藏为例[J]. 天然气地球科学,2006,17(6):857-861.

[16] 吴永平,杨池银,付立新,等. 中国海相油气田勘探实例之九——渤海湾盆地千米桥凝析油气田的勘探与发现[J]. 海相油气地质,2007,11(3):44-52.

[17] 李建英,卢刚臣,孔凡东,等. 千米桥潜山奥陶系储层特征及孔隙演化[J]. 石油与天然气地质,2001,22(4):367-371.

[18] 陈恭洋,何鲜,刘树明,等. 千米桥碳酸盐岩古潜山裂缝预测[J]. 中国石油勘探,2003,8(3):92-94.

[19] 陈恭洋,何鲜,陶自强,等. 千米桥潜山碳酸盐岩古岩溶特征及储层评价[J]. 天然气地球科学,2003,14(5):375-379.

# 苏6区块气藏剩余储量评价及提高采收率对策

董 硕 郭建林 郭 智 孟德伟 冀 光 程立华

(中国石油勘探开发研究院)

**摘要**：针对致密气剩余储量分布及提高采收率对策不明晰的问题,以苏6区块为例,通过气藏精细描述,明确了区块有效砂体展布特征及剩余气分布类型,估算各类型剩余储量占比;基于剩余储量分布特点,应用经济技术指标评价法,确定合理的直井井网密度,提出直井井网加密提高采收率技术对策。研究结果显示,苏6区块剩余气分布类型主要为井网未控制型、水平井遗留型、直井遗留型等三种类型,直井遗留型又可细分为射孔不完善型、复合砂体内部阻流带型两个亚类,其中,井网未控制型为主力剩余气,占总剩余地质储量的67.7%,为主要挖潜对象;合理的直井井网密度为4口/km$^2$,加密后区块采收率可提高至48%左右。

**关键词**：致密砂岩气藏；剩余储量；采收率；苏6区块

苏里格气田为典型的致密砂岩气藏,非均质性较强,气井单井产量低,综合递减率高,稳产难度大。明确气田剩余储量类型及分布,并提出相应的提高采收率措施,对于气田稳产具有重要意义。针对致密砂岩气剩余储量分布描述,根据成因可细分为多种类型,并可通过地层压力等指标进行表征,预测剩余气的动用潜力[1-4]。对于致密气藏提高采收率,主要在于井网井距的优化,通常以储层空间描述为基础,针对不同类型储层或不同开发方式气藏,综合利用气藏工程、数值模拟、矿场试井分析、现场加密实验等多种方法,构建井网密度与采收率关系,最终确定合理的井网密度及井网井距[5-8]。

综合上述研究,可发现对于剩余气储量分布的研究局限于前期地质评价方面,未能对气藏开发后的剩余气赋存类型、占比进行精确描述。为此,以下将苏里格气田苏6区块作为研究对象,在充分分析气藏地质特征的基础上,对剩余储量类型进行划分,并针对性提出提高采收率的举措。

## 1 地质背景

苏6区块位于苏里格气田中部,总面积约484km$^2$,探明地质储量1038.82×10$^8$m$^3$,开发井网完善,是具有代表性的生产试验区。苏6区块气源来自于本溪—山西组煤系烃源岩,具有"广覆式"生烃的特征。有效砂体在盒5—山2段都有发育,主要开发层位为下二叠统山西组山1段和中二叠统下石盒子组盒8段。上石盒子组发育的河漫湖相泥岩是气藏的区域盖层,储层上覆的泥岩构成了气藏的直接盖层,而储集砂体侧向延伸尖灭,侧方发育的泥岩构成了气藏侧面的封堵条件[9]。大面积分布的烃源岩、良好的盖层条件,使得砂体的发育成为成藏的主控因素[10-12]。

## 2 剩余储量分类与描述

从地质及开发的角度,根据剩余储量的成因,苏6区块剩余储量类型划分为井网未控制型、直井遗留型、水平井遗留型三种类型,其中直井遗留型又可细分为射孔不完善型、复合砂体内部阻流带型。

## 2.1 井网未控制型

苏6区砂体可细分为河道、心滩等[13],据井网资料统计,单期河道的厚度介于2~15m之间,宽度介于1000~2500m之间,单个心滩砂体厚度为2~10m,长度为500~1200m,宽度为200~600m。作为典型的辫状河沉积控制下的低渗透—致密气田,其有效储层主要分布于河道沉积的底部与心滩位置[12]。经密井网解剖,苏6区有效砂体通常呈孤立状,厚度范围介于2~10m,宽度范围主要为100~500m,长度范围主要为300~600m(图1)。

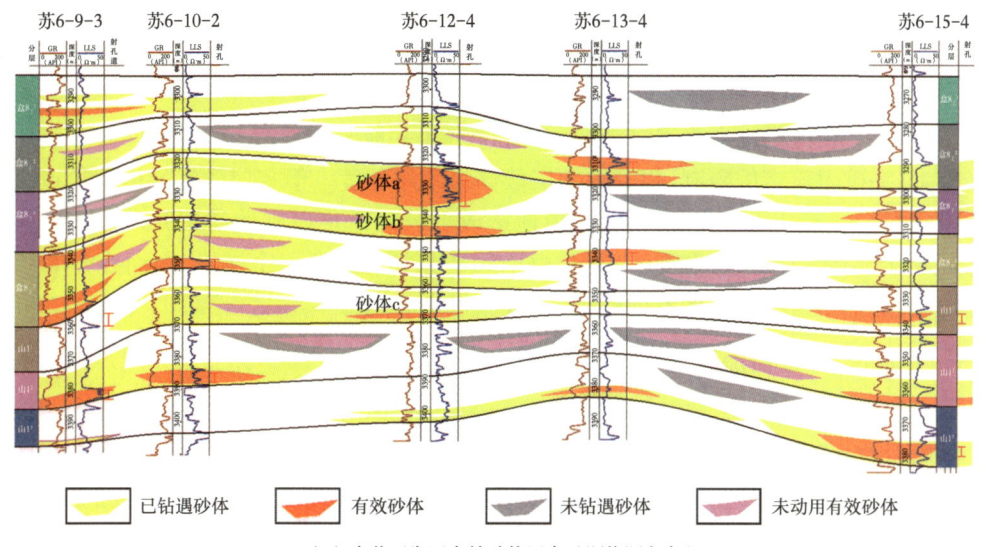

(a)直井开发区有效砂体展布(顺物源方向)

(b)水平井开发区有效砂体展布

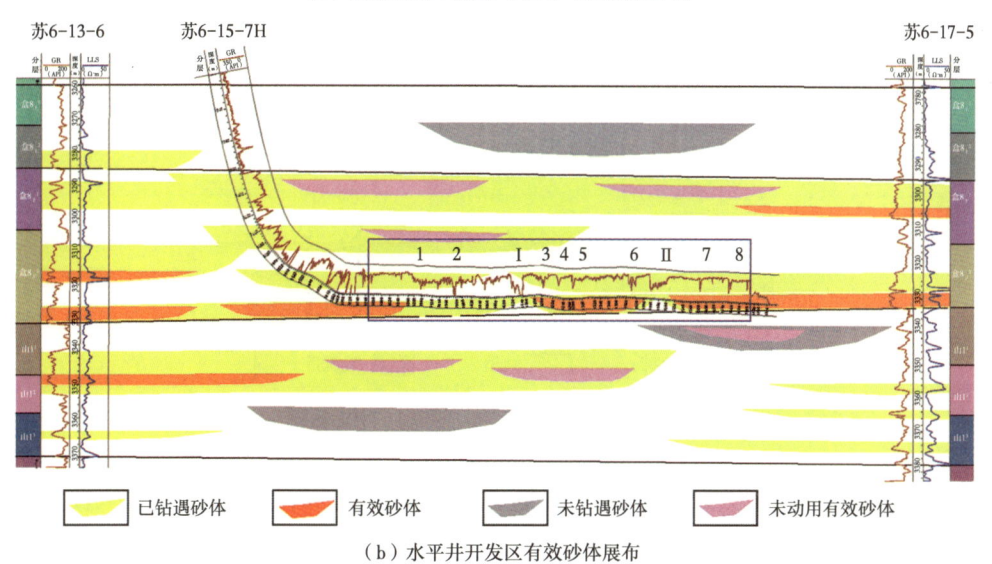

图1 苏6区块有效砂体展布图

由于有效砂体的展布规模较小,长度多小于600m,而在井距较大的区域,如图1a中的井距可达1000~2000m,稀疏的井网无法全部控制井区内的有效砂体,不完善的井网遗留了大量的井间砂体,造成了大量的未动用储量。目前区内主体开发井网为600m×800m,井网对有效砂体控制程度不足,形成了井网未控制型剩余储量。

## 2.2 射孔不完善型

研究区纵向上发育多个供气层,部分气井为单层主力供气,主力层厚度大,连续性好,产气贡献率在70%以上,其他小层产气能力较小。气井投产时,优先开发主力层和供气能力较好的层位,对于部分井段个别质量较差的小层,常不射孔生产,遗留有部分储量未动用。图1a中,苏6-12-4气井在山1—盒8层段钻遇3套有效砂体a、b、c,其中砂体a厚度大,含气性好,已射孔生产,而b、c两套砂体厚度小,含气性差,未进行射孔,形成未动用储量。

## 2.3 水平井遗留型

除了直井以外,区内钻有水平井,水平井通过增加井筒与储层接触面积、采取多段压裂改造等措施,可突破阻流带的限制,提升主力层段储量动用程度。但多层含气的地质特点决定了水平井纵向多层储量动用不充分,不可避免地造成部分储量难以有效动用开发。如图1b中,水平井苏6-15-7H目的层为盒$8_{下}^2$小层,由于水平井只能钻遇一个层段,因而在盒$8_{下}^2$小层上下层位,还遗留有未动用的有效砂体,造成储量未动用。

## 2.4 复合砂体内部阻流带型

在单一辫状河河道内部,发育心滩和辫状河河道沉积,在单个心滩砂体内部发育落淤层,在单砂体间发育由废弃河道泥岩与沟道泥岩组成的夹层[14-16]。在图1b中,苏6-15-7H水平段长度为1095m,钻遇砂体长度为1081m,有效储层长度为1020m,通过GR曲线的解剖,在水平段中识别出了两个较厚的夹层Ⅰ、Ⅱ,以及8个较薄的夹层1—8。水平井开发可有效突破夹层的遮挡,但在直井开发中,由于这些夹层的遮挡,形成了"阻流带"(图2),阻碍了气体在砂体内的流动,试气资料表明[17],直井在砂体范围内存在流动边界,证实"阻流带"可影响复合砂体渗透能力和直井储量动用程度,形成剩余储量。

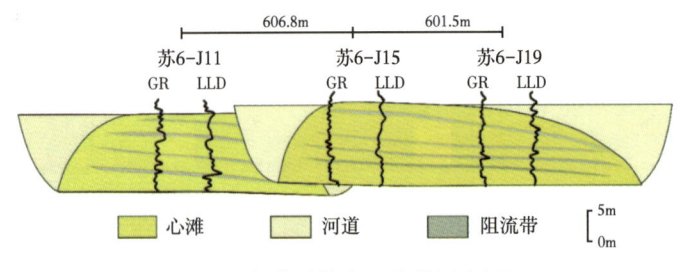

图2 复合砂体内阻流带展布图

在以上剩余储量分布类型中,根据剩余储量的地质成因,可将射孔不完善型、复合砂体内部阻流带型统称为直井遗留型。因此,苏6区块剩余储量分布类型主要为井网未控制型、直井遗留型、水平井遗留型三种类型。

## 3 剩余储量评价

### 3.1 区块总剩余储量计算

苏里格型砂质辫状河沉积体系非均质性强,由规模不等的小透镜状砂体多期切割叠置而成。对砂体剖面图(图1a)进行分析,苏6区600m×800m主力开发技术井网仅能控制部分含气砂体,井网钻遇区以外的砂体很难得到控制,在井间和层间形成剩余储量(图1)。因此,可运用井区地质储量与总井控动态储量之差估算总剩余储量。在此,利用丰度分布平面图计算全区静态地质储量,应用产量递减分析方法计算各单井动态控制储量,各单井动态控制储量之和即为苏6区块动态控制储量。

根据储量丰度大小(图3),将研究区划分为$<0.5\times10^8m^3/km^2$、$(0.5\sim1)\times10^8m^3/km^2$、$(1\sim1.5)\times10^8m^3/km^2$、$(1.5\sim2)\times10^8m^3/km^2$、$(2\sim2.5)\times10^8m^3/km^2$、$>2.5\times10^8m^3/km^2$共6个丰度等级,设定其分布面积分别为$A_1$、$A_2$、$A_3$、$A_4$、$A_5$和$A_6$。

$$R_s = \sum_{i=1}^{6} f_i A_i \tag{1}$$

式中　$R_s$——静态地质储量,$10^8m^3$;
　　　$f_i$——第$i$等级平均储量丰度,$10^8m^3/km^2$;
　　　$A_i$——第$i$等级分布面积,$km^2$。

根据总剩余储量估算原则,其计算式为:

$$R_r = R_s - R_d \tag{2}$$

式中　$R_r$——总剩余储量,$10^8m^3$;
　　　$R_d$——总井控动态储量,$10^8m^3$。

经统计计算(表1),静态地质储量为$819.42\times10^8m^3$,总井控动态储量$83.79\times10^8m^3$,由此根据式(2)可得总剩余储量$735.63\times10^8m^3$。

表1　苏6井区静态地质储量计算表

| 储量丰度<br>($10^8m^3/km^2$) | 井区面积<br>($km^2$) | 平均储量丰度<br>($10^8m^3/km^2$) | 地质储量<br>($10^8m^3$) |
| --- | --- | --- | --- |
| <0.5 | 13.53 | 0.31 | 4.19 |
| 0.5~1.0 | 95.80 | 0.74 | 70.89 |
| 1.0~1.5 | 195.96 | 1.27 | 248.87 |
| 1.5~2.0 | 131.93 | 1.76 | 232.20 |
| 2.0~2.5 | 74.54 | 2.25 | 167.72 |
| >2.5 | 30.24 | 3.16 | 95.56 |
| 总计 |  |  | 819.43 |

## 3.2 不同类型剩余储量估算

### 3.2.1 井网未控制型

井网未控制储量为井网未控制区面积与储量丰度的乘积。对于井网控制区面积，忽略井间干扰的影响，可视为各单井控制面积之和，而单井控制面积，可综合单井控制储量、射开储层厚度等参数，通过容积法计算求得。将 6 个丰度等级对应的井网控制区面积设为 $B_1$、$B_2$、$B_3$、$B_4$、$B_5$ 和 $B_6$，进而可由式(3)计算出井网未控制型剩余储量。

$$R_1 = \sum_{i=1}^{6} f_i(A_i - B_i) \tag{3}$$

式中　$R_1$——井网未控制型储量，$10^8 m^3$；

　　　$B_i$——井网控制区总面积，$km^2$。

对图 3 进行面积和储量丰度数据统计，并将得到的结果代入式(3)，求得直井井网未控制型剩余储量为 $498.22 \times 10^8 m^3$，占总剩余储量的 67.7%（表 2）。

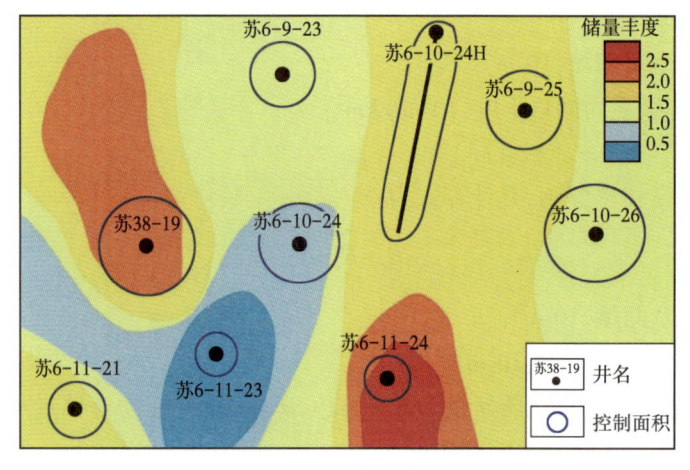

图 3　苏 6 区块储量丰度—控制面积分布图

表 2　井网未控制型剩余储量计算表

| 储量丰度<br>($10^8 m^3/km^2$) | 井区面积<br>($km^2$) | 平均储量丰度<br>($10^8 m^3/km^2$) | 地质储量<br>($10^8 m^3$) |
| --- | --- | --- | --- |
| <0.5 | 9.36 | 0.30 | 2.81 |
| 0.5~1.0 | 68.56 | 0.74 | 42.25 |
| 1.0~1.5 | 155.94 | 1.30 | 202.72 |
| 1.5~2.0 | 68.44 | 1.84 | 125.93 |
| 2.0~2.5 | 37.18 | 2.17 | 80.68 |
| >2.5 | 12.68 | 3.22 | 40.83 |
| 总计 | | | 498.22 |

## 3.2.2 射孔不完善型

由于该类剩余储量在直井井网控制区范围内,为准确求取剩余储量大小,定义单井静态控制储量:

$$R_{so} = A_o f_o \tag{4}$$

式中 $R_{so}$——单井静态控制储量,$10^8 \text{m}^3$;
$A_o$——单井控制面积,$\text{km}^2$;
$f_o$——单井所在区域平均储量丰度,$10^8 \text{m}^3/\text{km}^2$。

要求得未射孔层位(砂体)储量,即射孔不完善型剩余储量,则首先需明确未射孔砂体展布规律。根据砂体连井剖面图可知,砂体厚度与其长宽规模有一定联系,Kelly[18]给出砂体展布参数关系式如下:

$$W_b = 11.413 h_d^{1.4182} \tag{5}$$

$$L_b = 4.9517 W_b^{0.9676} \tag{6}$$

式中 $W_b$——未射孔砂体宽度,m;
$L_b$——未射孔砂体长度,m;
$h_d$——未射孔砂体厚度,m。

将式(5)、式(6)作为未射孔有效砂体的长宽厚关系式,并将砂体近似为半椭球体,依据测井解释数据获取砂体的孔隙度与含水饱和度,由此可计算砂体储量为:

$$R_{2o} = \frac{1}{6}\pi h_d W_b L_b \phi_b (1 - S_{wb}) \tag{7}$$

式中 $R_{2o}$——单井射孔不完善型储量,$10^8 \text{m}^3$;
$\phi_b$——砂体孔隙度,%;
$S_{wb}$——砂体含水饱和度,%。

因钻遇砂体厚度 $h_d$ 未必是该砂体最厚之处(即非椭球体的轴 $c$),导致该计算结果偏小,因而设其钻遇至 $0.5c$ 处,给定修正系数 $\alpha = 13.8$。

由于苏6区井数过多,难以实现对每口井未射孔砂体的储量精确计算。为此,定义未射孔储量比 $\beta$(射孔不完善型剩余储量与单井静态控制储量之比),选取15口具代表性的直井统计计算,发现井均射孔不完善型剩余储量占单井静态控制储量的0.56%左右,据此可推算苏6区射孔不完善型剩余储量约为 $4.59 \times 10^8 \text{m}^3$,占总剩余储量的0.6%(表3)。

表3 射孔不完善型剩余储量统计表

| 井名 | 未射层位 | $h_d$ (m) | $\phi_b$ (%) | $S_{wb}$ (%) | $\alpha^* R_{2o}$ ($10^4 \text{m}^3$) | $A_o$ ($\text{km}^2$) | $R_{so}$ ($10^8 \text{m}^3$) | $\beta$ (%) |
|---|---|---|---|---|---|---|---|---|
| 苏38-16-1 | 1 | 5.4 | 8.7 | 48.4 | 11.54 | 0.16 | 0.16 | 0.69 |
| 苏6 | 1 | 2.3 | 8.5 | 51.1 | 0.42 | 0.23 | 0.57 | 0.07 |
|  | 2 | 3.9 | 7.3 | 46.8 | 2.91 |  |  |  |
|  | 3 | 2.7 | 7.9 | 49.2 | 0.75 |  |  |  |

续表

| 井名 | 未射层位 | $h_d$<br>(m) | $\phi_b$<br>(%) | $S_{wb}$<br>(%) | $\alpha^* R_{2o}$<br>($10^4 m^3$) | $A_o$<br>($km^2$) | $R_{so}$<br>($10^8 m^3$) | $\beta$<br>(%) |
|---|---|---|---|---|---|---|---|---|
| 苏6-11-10 | 1 | 7.8 | 9.0 | 59.6 | 37.67 | 0.2 | 0.3 | 1.52 |
| | 2 | 2.5 | 6.6 | 56.6 | 0.40 | | | |
| | 3 | 4.8 | 7.2 | 38.8 | 7.25 | | | |
| 苏6-11-8 | 1 | 5.7 | 10.6 | 54.2 | 15.32 | 0.05 | 0.07 | 2.48 |
| | 2 | 3.4 | 8.2 | 42.4 | 2.10 | | | |
| | 3 | 2.6 | 8.7 | 46.3 | 0.75 | | | |
| 苏6-J15 | 1 | 5.3 | 11.9 | 55.9 | 12.57 | 0.05 | 0.08 | 1.66 |
| | 2 | 1.8 | 7.3 | 42.8 | 0.17 | | | |
| 苏40-14 | 1 | 4.4 | 9.4 | 57.2 | 4.76 | 0.1 | 0.2 | 0.41 |
| | 2 | 3.8 | 8.3 | 52.3 | 2.69 | | | |
| 苏6-8-14 | 1 | 1.8 | 6.4 | 54.5 | 0.12 | 0.18 | 0.09 | 0.14 |
| | 2 | 3.0 | 7.8 | 34.2 | 1.42 | | | |
| 苏6-0-11 | 1 | 6.1 | 8.9 | 60.5 | 14.34 | 0.28 | 0.75 | 0.14 |
| 苏6-4-7 | 1 | 2.1 | 9.3 | 72 | 0.19 | 0.21 | 0.06 | 0.06 |
| | 2 | 1.9 | 7.7 | 64.6 | 0.13 | | | |
| 苏6-17-11 | 1 | 2.1 | 9.3 | 70.1 | 0.20 | 0.2 | 0.16 | 0.03 |
| | 2 | 1.8 | 5.0 | 53.8 | 0.09 | | | |
| | 3 | 1.7 | 7.8 | 64.5 | 0.09 | | | |
| 苏6-4-18 | 1 | 2.4 | 7.4 | 33.6 | 0.58 | 0.18 | 0.22 | 0.04 |
| | 2 | 2.3 | 8.1 | 47.3 | 0.43 | | | |
| 苏6-12-3 | 1 | 4.1 | 10.9 | 45.2 | 5.41 | 0.18 | 0.38 | 0.14 |
| | 2 | 2.8 | 10.1 | 24.4 | 1.63 | | | |
| 苏6-23-14 | 1 | 1.7 | 7.2 | 34.7 | 0.15 | 0.15 | 0.27 | 0.14 |
| | 2 | 4.0 | 7.0 | 35.3 | 3.73 | | | |
| | 3 | 2.3 | 8.2 | 24.9 | 0.62 | | | |
| 苏6-3-10 | 1 | 6.1 | 11.3 | 61.9 | 17.57 | 0.19 | 0.43 | 0.41 |
| | 2 | 2.5 | 7.8 | 46.1 | 0.58 | | | |
| 苏6-4-2 | 1 | 5.1 | 11.6 | 53.9 | 11.07 | 0.17 | 0.17 | 0.55 |
| 平均 | | | | | | | | 0.56 |

### 3.2.3 水平井遗留型

与直井类似，水平井动态控制储量及控制面积可通过Blasingame、FMB等产量递减分析方法求得，由于水平井可突破阻流带的限制，因而在计算此类剩余储量时，阻流带型剩余储量可忽略不计，该类剩余储量即为单井静态控制储量与动态控制储量之差，其表达式为：

$$R_{3o} = R_{so} - R_{do} \tag{8}$$

式中 $R_{3o}$——水平井单井遗留型剩余储量,$10^8 \mathrm{m}^3$;

$R_{do}$——单井动态控制储量,$10^8 \mathrm{m}^3$。

根据图 3 进行数据统计分析,发现水平井仅能控制井控范围内地质储量的 60%~70%,会形成 30%~40% 的剩余储量(表4),苏 6 区内水平井遗留型剩余储量约为 $85.37 \times 10^8 \mathrm{m}^3$,占剩余总储量的 11.6%。

表 4 典型水平井遗留型剩余储量统计表

| 井名 | $A_o$($\mathrm{km}^2$) | $R_{so}$($10^8 \mathrm{m}^3$) | $R_{do}$($10^8 \mathrm{m}^3$) | $R_{3o}$($10^8 \mathrm{m}^3$) | $R_{3o}/R_{so}$ |
| --- | --- | --- | --- | --- | --- |
| 苏 6-10-24H | 0.41 | 0.82 | 0.44 | 0.38 | 0.46 |
| 苏 6-13-5H | 0.23 | 0.55 | 0.31 | 0.24 | 0.44 |
| 苏 6-16-1H | 0.65 | 1.56 | 1.10 | 0.46 | 0.29 |
| 苏 6-2-10H | 0.54 | 1.30 | 0.91 | 0.39 | 0.30 |
| 苏 6-2-10H1 | 0.40 | 0.88 | 0.67 | 0.21 | 0.24 |
| 苏 6-21-12H | 0.62 | 1.43 | 0.99 | 0.44 | 0.31 |
| 苏 6-4-10H1 | 0.84 | 1.93 | 1.13 | 0.80 | 0.42 |
| 苏 6-4-10H2 | 0.38 | 0.99 | 0.59 | 0.40 | 0.40 |
| 苏 6-22-23H1 | 0.19 | 0.36 | 0.21 | 0.15 | 0.42 |
| 苏 6-4-21H | 0.25 | 0.45 | 0.30 | 0.15 | 0.34 |

#### 3.2.4 复合砂体内部阻流带型

由于阻流带的规模、展布及其对气体的阻碍能力无法评价,故该类剩余储量无法准确计算,因此,该类剩余储量可由总剩余储量与其余三种剩余储量相减计算得出。

$$R_4 = R_r - R_1 - R_2 - R_3 \tag{9}$$

式中 $R_4$——复合砂体内部阻流带型及井控范围内未钻遇砂体剩余储量,$10^8 \mathrm{m}^3$;

$R_2$——射孔不完善型储量,$10^8 \mathrm{m}^3$;

$R_3$——水平井遗留型剩余储量,$10^8 \mathrm{m}^3$。

根据总剩余储量及已计算得到的剩余储量(井网未控制、射孔不完善及水平井遗留),计算得到该类剩余储量为 $147.45 \times 10^8 \mathrm{m}^3$,占总剩余储量的 20.1%。

### 3.3 剩余储量占比及动用方式

通过对五种剩余储量进行统计对比分析可发现:井网未控制型剩余储量占总剩余地质储量的 67.7%,为主力剩余气;其余剩余储量类型占比为 32.3%,相对较小,为非主力剩余气(表5)。

表 5  苏里格气田富集区剩余储量分类占比统计表

| 剩余储量类型 | | 赋存方式 | 比例(%) |
|---|---|---|---|
| 直井遗留型 | 射孔不完善型 | 直井未射孔遗留的薄层或含气层 | 0.6 |
| | 复合砂体内部阻流带型 | 心滩内部阻流带控制的滞留气 | 20.1 |
| 水平井遗留型 | | 水平井控制区内遗留的非主力层 | 11.6 |
| 井网未控制型 | | 开发井网未控制的孤立砂体气 | 67.7 |

对于直井垂向射孔不完善型,主要改善措施为查层补孔,此类剩余储量以差气层为主,由于规模较小,且产气能力较差,因此提高采收率效果有限;复合砂体内部阻流带型剩余储量处于现有直井控制范围内,考虑经济成本,无法通过井网加密进行动用;对于水平井遗留型,侧钻水平井风险大,作业成本高,当前缺乏有效的动用方式。

## 4 提高采收率对策

井网未控制型储量作为剩余储量的主体,其成因主要为井网不完善,为此,应对现有井网进行加密,提高井网完善程度。因而,以下提出直井井网加密挖潜技术。

井网密度超过一定值后,井网密度增大,井间干扰愈发严重,虽采收率不断增加,但增加幅度越来越小。因此,需选用合理的方法,确定最佳的井网密度,提高气藏采收率,同时获得较好的经济效益。前人曾综合运用定量地质模型法、动态泄气范围法等多种方法确定井网加密合理井网密度[19],其中经济技术指标评价法可综合考虑成本与技术问题,应用效果较好,为此,本文采用该方法确定合理井网密度。

经济技术指标评价法是结合当前经济技术条件,在气井产能指标评价的基础上,采用数值模拟手段,建立"井网密度—单井最终累计产气量—采收率"关系模型,明确井间开始产生干扰时对应的最优技术井网密度、最小经济极限产量对应的最小经济极限井网密度,两者之间为井网可调整加密的区间范围,综合采收率最终确定合理的井网密度(图4)。

利用财务净现值计算经济极限累计产气量[20],其表达式为:

$$FNPV = \sum_{t=1}^{n}(C_1 - C_0)_t(1 + i_\varepsilon)^{-t} \tag{10}$$

式中　$t$——时间,a;

　　　$i_\varepsilon$——收益率,%;

　　　$C_1$——当年现金流入,万元;

　　　$C_0$——当年现金流出,万元;

　　　$n$——计算期,a;

　　　$FNPV$——第 $n$ 年财务净现值。

根据苏6区的生产情况,单井综合成本取 $740×10^4$ 元/井,生产成本 $0.13$ 元/m³,税费 $0.021$ 元/m³,天然气价格 $1.1$ 元/m³,收益率12%。取单井稳产日产气量 $1.1×10^4$ m³/d,稳产3年,之后以指数形式递减,共生产15年计算,当FNPV为0的时候,达到经济极限累计产气量,为 $0.17×10^8$ m³。根据图7可得出,对应的井网密度(经济极限井网密度)为 $4.5$ 口/km²。

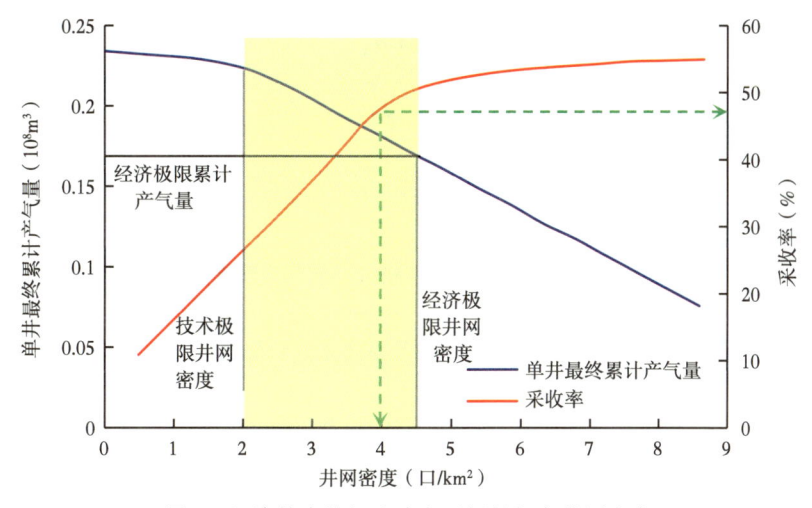

图 4 经济技术指标法确定可调整加密井网密度

基于经济技术指标评价法,分析认为苏 6 区致密气藏最优技术井网密度约为 2 口/km²,最小经济极限井网密度约为 4.5 口/km²,即可调整加密的区间为 2~4.5 口/km²,且根据图 7 中采收率曲线的增长趋势,在井网密度达到 4 口/km² 时,采收率增长幅度减缓,因此,综合经济与技术指标,认为该区块最优井网密度为 4 口/km²,此时采收率可提高到 48% 左右。

## 5 结论

(1) 气藏精细描述结果表明,剩余储量可分为井网未控制型、水平井遗留型、直井遗留型等三种类型,直井遗留型又可细分为射孔不完善型、复合砂体内部阻流带型;

(2) 井网未控制型剩余储量占总剩余地质储量的 67.7%,为主力剩余气,水平井遗留型、直井遗留型占比分别为 11.6% 和 20.7%,为非主力剩余气,井网未控制型剩余气应作为主要挖潜对象;

(3) 针对苏 6 区块主力剩余气分布特点,可采用直井井网加密提高采收率技术,经济技术指标评价法计算结果显示,在目前的经济技术条件下,直井井网可加密至 4 口/km²,采收率可提高到 48% 左右。

### 参 考 文 献

[1] 王昔彬,刘传喜,郑祥克,等. 低渗特低渗气藏剩余气分布的描述[J]. 石油与天然气地质,2003,24(4):401-403,416.

[2] 赵正具,王顺玉. 低渗砂岩气藏提高采收率措施研究[J]. 天然气勘探与开发,2012,35(3):44-48,56,84.

[3] 张云鹏,徐霜,李东,等. 文 13 西气藏储层特征及剩余气开发潜力[J]. 石油与天然气地质,2001,22(3):261-263.

[4] 卜淘,李忠平,詹国卫,等. 川西坳陷低渗砂岩气藏剩余气类型及分布研究[J]. 天然气工业,2003,23(S1):13-15.

[5] 贾爱林,王国亭,孟德伟,等. 大型低渗—致密气田井网加密提高采收率对策——以鄂尔多斯盆地苏里

格气田为例[J]. 石油学报, 2018, 39(7):802-813.

[6] 李跃刚, 徐文, 肖峰, 等. 基于动态特征的开发井网优化——以苏里格致密强非均质砂岩气田为例[J]. 天然气工业, 2014, 34(11):56-61.

[7] 王东, 梁倚维, 马力, 等. 致密气井有效井距数值试井模拟分析[J]. 石油化工应用, 2016, 35(12):94-97.

[8] 吕志凯, 唐海发, 刘群明, 等. 苏里格大型致密砂岩气田储层结构与水平井提高采收率对策[J]. 现代地质, 2018, 32(4):832-841.

[9] 刘圣志, 李景明, 孙粉锦, 等. 鄂尔多斯盆地苏里格气田成藏机理研究[J]. 天然气工业, 2005, 25(3):4-6,191.

[10] 卢涛, 张吉, 李跃刚, 等. 苏里格气田致密砂岩气藏水平井开发技术及展望[J]. 天然气工业, 2013, 33(8):38-43.

[11] 王国勇. 致密砂岩气藏水平井整体开发实践与认识:以苏里格气田苏53区块为例[J]. 石油天然气学报, 2012, 34(5):153-157.

[12] 郭智, 贾爱林, 薄亚杰, 等. 致密砂岩气藏有效砂体分布及主控因素——以苏里格气田南区为例[J]. 石油实验地质, 2014, 36(6):684-691.

[13] 林志鹏, 单敬福, 陈乐, 等. 苏里格气田苏6区块盒8段古河道砂体演化规律[J]. 油气地质与采收率, 2018, 25(05):1-9,23.

[14] 牛博, 高兴军, 赵应成, 等. 古辫状河心滩坝内部构型表征与建模——以大庆油田萨中密井网区为例[J]. 石油学报, 2015, 36(1):89-100.

[15] 孙天建, 穆龙新, 赵国良. 砂质辫状河储层隔夹层类型及其表征方法——以苏丹穆格莱特盆地Hegli油田为例[J]. 石油勘探与开发, 2014, 41(1):112-120.

[16] 卢志远, 马世忠, 何宇, 等. 鄂尔多斯盆地砂质辫状河夹层特征——以苏东27-36密井网区为例[J]. 断块油气田, 2018, 25(6):704-708,714.

[17] 罗瑞兰, 雷群, 范继武, 等. 低渗透致密气藏压裂气井动态储量预测新方法——以苏里格气田为例[J]. 天然气工业, 2010, 30(7):28-31,128.16.

[18] Kelly S. Scaling and hierarchy in braided rivers and their deposits: Examples and implications for reservoir modeling[M]// Sambrook Smith G H, Best J L, Bristow C S, et al. Braided rivers:Process, deposits, ecology and management. Oxford, UK: Blackwell Publishing, 2006:75-106.

[19] 何东博, 王丽娟, 冀光, 等. 苏里格致密砂岩气田开发井距优化[J]. 石油勘探与开发, 2012, 39(4):458-464.

[20] 高嘉祺, 陈明强. 鄂尔多斯盆地低渗透气藏水平井经济开采预测模型[J]. 西安石油大学学报(自然科学版), 2017, 32(2):81-85.

# Distribution Characteristics of the Mudstone Interlayer and Their Effects on Water Invasion in Kela 2 Gas Field

Yongzhong Zhang[1]　Yong Sun[2]　Zhaolong Liu[1]　Hualin Lin[1]

(1. PetroChina Research Institute of Petroleum Exploration & Development, Beijing, China
2. Research Institute of Exploration and Development, Tarim Oilfield Company,
PetroChina, Korla, Xinjiang, China)

**Abstract**: The reservoirs of Kela 2 gas field are mainly thick-very thick-bedded sandstone inter-bedded with mudstone. In the past 15 years of high-speed development, Kela 2 gas field has been facing the challenge of inhomogeneous water invasion. In order to understand the water invasion laws, based on seismic, logging, core, and production dynamic data, the characteristics of the non-pay interlayers and their sealing ability are analyzed. Then the effects of interlayer, faults, fractures as well as the high permeability formations on the migration of edge and bottom water were discussed. The research results show that: (1) Mudstone interlayers can be divided into two types, one is continuously distributed in the whole area with a thickness of about 10~50m, the other is discontinuously distributed in the field with an individual thickness generally less than 2m. (2) The interlayer sealing ability is closely related to the thickness of interlayer and fault throw. The interlayer at top of the Baxigai Formation, together with the interlayers at bottom of the Bashijiqike Formation has a stronger sealing ability than the others. (3) Controlled by the geological characteristics of the gas field, water invasion laws are varied in the different regions. In the west and east regions, Edge water can induce the seriously inhomogeneous water invasion through the relatively high permeability layer; while in the south-central region, the bottom water invasion is relatively moderate because the thick-bedded interlayer weakens the energy of water from deeper bed; in addition, in the SW region, bottom water breakthroughs along faults vertically and then invades along the natural fractures and the relatively high permeability layer, which led to water flooding of gas wells in the area. This study is of great significance to water control and stable production of the gas field.

**Keywords**: Kela 2 gas field; Mudstone interlayer; Fault; Water invasion.

# 1 Introduction

There are many types of impermeable or very low permeable interlayers in clastic reservoir. The interlayers are formed due to the change of river hydrodynamic conditions or the difference of sediment lithology caused by the diagenesis after sedimentation[1]. The shape, scale and spatial combination of the interlayers are quite different[2-5]. It is very important to study the interlayer for further understanding the law of migration of edge and bottom water and the distribution of remaining oil or gas[6-8].

The reservoirs of the Kela-2 gas field in Kuqa depression are mainly braided-river-delta and fan delta sandstones inter-bedded with siltstone[9], which have a thickness of more than 460m in total and average net-to-gross ratio of 75%. In the past 15 years of high-speed development, Kela 2 gas field is now facing problems of rapid water breakthrough in gas wells and seriously inhomogeneous water invasion in the gas field[10]. The mudstone interlayers and the widely developed faults[10], together with the locally developed fractures[11] play an important role in water invasion. Thus, it is necessary to study the spatial distribution relationship between interlayer, fault and fracture in the gas field. There are abundant data of core, well logs, and production performance in Kela 2 gas field, which provide good conditions for the study of interlayer. In this paper, based on the well data, combined with outcrops and seismic interpretation results, the distribution characteristics of the interlayers are analyzed and the water invasion laws are discussed in Kela 2 Gas Field. This study can provide a basis for optimization of development program of Kela 2 gas field.

## 2  Geological setting

Kuqa Depression in Tarim Basin where Kela-2 gas field is located belongs to a Meso-Cenozoic sedimentary foreland basin[12], which can be divided into 3 secondary structural units, including Kelasu thrust belt, Qiulitage thrust belt and Baicheng sag[13]. The Kelasu thrust belt is a thrust fold belt, where folds and fault block structures controlled by a series of imbricated thrusts occur in the under-salt structural formation. Kela-2 gas field is in the east section of Kelasu structural belt of Kuqa foreland basin on the north of the Tarim Basin, between Kela-1 structure and Kela-3 structure (Fig. 1).

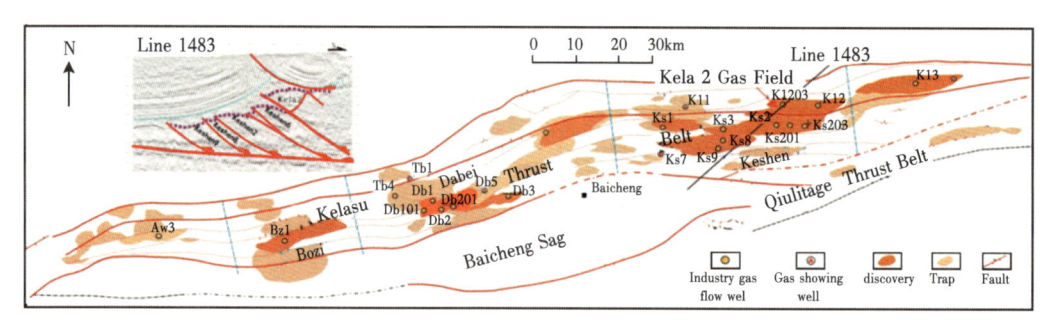

Fig. 1  Location of the Kela-2 gas field and subdivision of the Kuqa Depression

Neogene, Paleogene and Cretaceous are developed from top downwards in the drilled strata of Kela 2 gas field. The Neogene is divided from top downwards into the Kuqa ($N_2k$), Kangcun ($N_{1-2}k$) and the Jidike ($N_1j$) Formations; while the Paleogene is divided into Suweiyi ($E_{2-3}s$) and the Kumugeliemu ($E_{1-2}km$) Formations; the Lower Cretaceous Bashijiqike ($K_1bs$), Baxigai ($K_1bx$), and Shushanhe Formations ($K_1s$) are drilled in the field. The Upper Cretaceous Series is absent in Kela-2 due to erosion. The upper boundary of the Bashijiqike corresponds to a tectonic regional unconformity. The Lower Palaeogene rests unconformably on the red beds of the lower Cretaceous System. The gas bearing rocks are the Cretaceous Bashijiqike ($K_1bs$), Baxigai ($K_1bx$) Formations and the Dolomite Member ($E_{1-2}km_3$) and the Glutenite unit in the Paleogene Kumugeliemu Formation (Table 1).

Table 1  The concise table of stratum in Kela 2 gas field

| Series | Strata Fm. | Member | Code | Thickness (m) | Lithostratigraphy | NTG |
|---|---|---|---|---|---|---|
| Paleocene-Eocene | Kumugeliemu | Mudstone | $E_{1-2}km_1$ | 164~215 | large set of thick mudstone, and the content of gypsum in the lower part is obviously increased | |
| | | Gypsum and salt rock | $E_{1-2}km_2$ | 431~757 | alternating beds with different thickness of white thick-very thick bedded gypsiferous salt rock and brown mudstone | |
| | | Dolomite | $E_{1-2}km_3$ | 4~9 | mainly grey dolomitic micrites, bioclastic rocks, and sparry dol-arenite | 0.93 |
| | | Gypsum mudstone | $E_{1-2}km_4$ | 14~44 | grey brown, medium thick and thin-bedded mudstone with gypsum, gyp mudstone | |
| | | Glutenite | $E_{1-2}km_5$ | 0~21 | light grey, brown, medium and hick-bedded, pebbled fine sandstone inter-bedded with fine sandstone, interbeds of siltstone and brown thin-bedded mudstone | 0.53 |
| Lower Cretaceous | Bashijiqike | I | $K_1bs_1$ | 57~143 | brown, thick-very thick-bedded sandstone inter-bedded with siltstone, dark brown mudstone | 0.83 |
| | | II | $K_1bs_2$ | 143~165 | brownish red, light brown thick-very thick-bedded sandstones interbedded with siltstone and thin-bedded mudstone with GR values often in excess of 130API | 0.83 |
| | | III | $K_1bs_3$ | 92~111 | thick-very thick brown, gray brown lithic fine conglomerate followed by brownish red hugely thick-bedded sandstone | 0.53 |
| | Baxigai | I | $K_1bx_1$ | 11~22 | brown, thin-medium bedded mudstone, with minor thin (less than one meter thick) silty mudstone | |
| | | II | $K_1bx_2$ | 87~106 | brown, thick-bedded, massive, silty fine and very fine grained sandstone, inter-bedded with dark brown mudstone and silty mudstone | 0.84 |

# 3  Characteristics of the interlayer

The reservoir of Kela 2 gas field is a set of dolomite and clastic rock composite deposition with a total thickness of more than 460m. According to the well correlation within the Kela-2 gas field and outcrop observation within the Kuqa Basin of the Palaeogene and Cretaceous systems, there are two types of interlayers in Kela 2 gas field: one type is continuously distributed in the whole area, which has a thickness of about 10~50m, the other is discontinuously distributed in the field, which has an individual thickness generally less than 2m.

## 3.1 Continuously distributed interlayer

According to the well correlation within the Kela-2 gas field, there are three sets of continuously distributed interlayer from top downwards in the drilled strata: the Gypsum mudstone member ($E_{1-2}km_4$) of the Kumugeliemu Formation, the first Sub-member ($K_1bs_3^1$) of the Member III within the Bashijiqike Formation and the Member I ($K_1bx_1$) of the Baxigai Formation (Fig. 2).

The $E_{1-2}km_4$ interlayer. The Gypsum mudstone member ($E_{1-2}km_4$) is a thick-interlayer between the Dolomite ($E_{1-2}km_3$) and the Glutenite ($E_{1-2}km_5$) Member within the Paleogene Kumugeliemu Formation ($E_{1-2}km$). In the initial stage of the Kumugeliemu Formation deposited, the Kuqa basin was an inter-tidal lagoon bay area of a shallow marine sea and later it became a sand starved basin with strong evaporation, forming large sets of gypsum salt and gypsum mudstone. From top to bottom, the Kumugeliemu can be further divided into 5 lithologic Members (Table 1): Mudstone ($E_{1-2}km_1$), Gypsum and salt rock ($E_{1-2}km_2$), Dolomite ($E_{1-2}km_3$), Gypsum mudstone ($E_{1-2}km_4$), and Glutenite ($E_{1-2}km_5$). In the Kumugeliemu Formation, the gas bearing rocks are confined to the Dolomite Member ($E_{1-2}km_3$) and to the Glutenite unit ($E_{1-2}km_5$). The non-pay Gypsum mudstone member ($E_{1-2}km_4$) interbed within the Dolomite ($E_{1-2}km_3$) and the Glutenite ($E_{1-2}km_5$) Member. The rocks of the $E_{1-2}km_4$ interlayer are grey brown, medium thick and thin-bedded mudstone with gypsum, gyp mudstone, inter-bedded with thin-bedded muddy gypsiferous rocks. It is capped by a 2 to 3m thick anhydrite bed, and is easily identifiable by high natural gamma ray value and low resistivity. All the wells in Kela 2 gas field meet the formation. The drilling thickness of the interlayer is 14~44m.

The $K_1bs_3^1$ interlayer. The Bashijiqike Formation is main gas bearing formation within Kela 2 gas field. Based upon correlation between wells, it can be divided into three members named from top downwards $K_1bs_1$, $K_1bs_2$ and $K_1bs_3$. The non-pay interlayers are classified with reference to the logs. The Bashijiqike Formation interbeds total more than 30 layers. Among these interlayers, there is a set of thick-interlayers at the uppermost 15~20m of $K_1bs_3$ (named $K_1bs_3^1$). In the middle part there is sandstone, while in the lower and upper part mainly are mudstone inter-bedded with shaly siltstone. The natural gamma curve is characterized by "W" type. The thicknesses of the interlayers vary between 4 and 9m. The two interlayers represent a lacustrine deposit and as such are interpreted to extend over the whole field.

The $K_1bx_1$ interlayer. It is the top part of Baxigai Formation. The Baxigai Formation is divided into two sub-units ($K_1bx_1$ and $K_1bx_2$) and correspond to a mudstone and a sandstone member respectively. The Upper Mudstone Member ($K_1bx_1$) has a drilled thickness of 11~22m. The rocks consist of brown, thin-medium bedded mudstone, with minor thin (less than one meter thick) silty

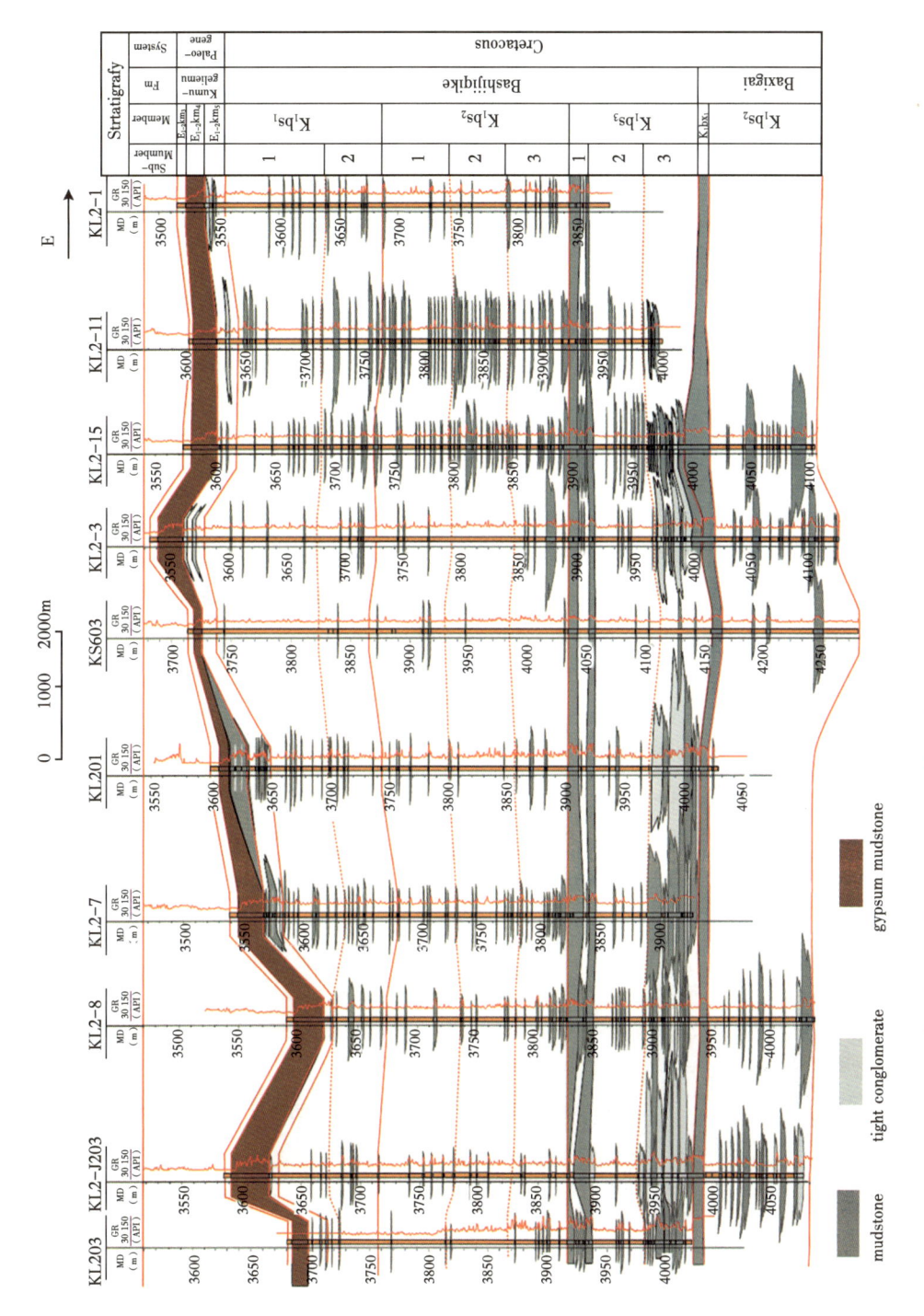

Fig. 2　Well correlation profile with interlayer distribution in the Kela-2 gas field

mudstone. The natural gamma curve has a very distinct box shape, with values from 95 to 150API. The average deep resistivity value is $3 \sim 9\Omega \cdot m$. The mudstones, together with the overlying mudstone and tight conglomerate interlayers at the bottom of Bashijiqike Formation are predicted to reduce the effect of bottom water production.

## 3.2 Discontinuously distributed interlayer

The clastic reservoirs ($E_{1-2}km_5-K_1bx$) dominate the Kela-2 gas field, which have a total thickness of more than 400m. By comparing the sedimentary facies of the drilling wells with that of the outcrop profile, the reservoirs ($E_{1-2}km_5-K_1bx$) are characterized by sedimentary facies of fan delta front ($E_{1-2}km_5$ and $K_1bs_3$) and of braided delta front ($K_1bs_1$, $K_1bs_2$ and $K_1bx_2$)[9]. In the Early Cretaceous, there developed many fan-shaped depositional systems along the north frontier and formed a series of braided-channel delta or with multiple material source areas by mutual connection and superposition, accordingly formed the sand bodies with steady extensive distributions interbedded with thin and discontinuous muddy layers.

The discontinuous interlayers mainly consist of mudstone deposited in distributary inter-channel, followed by channel lag sediments at the bottom of underwater channels and mud gravel strips deposited inside channels. They are classified into five lithological types: mudstone, silty-mudstone, muddy-siltstone and conglomeratic mudstone and tight conglomerate (Fig. 3). Tight conglomerate interlayer is rare, which commonly developed in the bottom of the $K_1bx_3$, while mudstone interlayer is common.

According to the outcrop statistics, single layer thickness is $0.1 \sim 4.4m$, and the average frequency of interlayer is 0.16 per-metre, while the average density is 0.1m/m. Mudstone and silty mudstone, which are located at the top of single channel sand body or deposited in distributary inter-channel, have a thin single layer thickness of $0.5 \sim 2m$ in general, and a horizontal extension of less than 200m in general, showing discontinuous distribution. Generally, channel lag sediments at the bottom of underwater channels and mudgravel strips deposited inside channels have a thin single

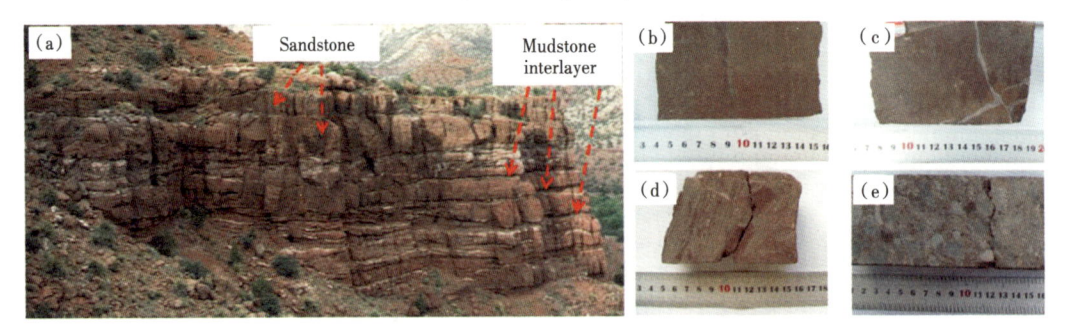

Fig. 3 Outcrop and core pictures showing the interlayer characteristics (a): Outcrop of the $K_1bs$ showing mudstone interlayers within a sandy braided fluvial system; (b), (c), (d), and (e) are core pictures from wells within the Kela-2 gas field. (b): the left part is silty-mudstone and the right part is mudstone; (c): muddy-siltstone; (d): conglomeratic mudstone; (e): tight conglomerate)

layer thickness of 0.1~0.3m, and its extension length measured in the field is less than 50m, mostly between 10~30m.

According to 21 well logging interpretation statistics, interlayer thicknesses are generally low. Around 80% of the total interbeds are less than 1m thick and <2% are greater than 5m thick which are mainly at the bottom of the $K_1bs_3$. The frequency of interlayer is 0.08~0.24m/m, with the average of 0.14/m, while the density is 0.07~0.36m/m, with the average of 0.18m/m, showing the heterogeneity of sedimentation within the field.

## 3.3 Interlayer sealing potentiality

Interlayer sealing performance is closely related to its own property such as the thickness and the lateral extension, and related to the intersecting relationship between faults/fractures and interlayers. Geological analysis and production performance show that both the discontinuous interlayers and the continuously distributed interlayers do not have complete sealing property due to developed faults in the field.

Geological analysis. The analysis of the relationship between faults/fractures and interlayers showed that it is mainly the faults that break out the continuously distributed interlayers leading to the connectivity of gas reservoirs instead of the fractures.

When the interlayer is cut by a fault, the sealing ability of the interlayer depends on the sealing ability of the fault. The relationship between the thickness of interlayer and fault throw is often used to evaluate the fault sealing ability[14]. The greater the thickness of mudstone joint at the two sides of the fault is, the stronger the ability to prevent the vertical migration of fluid along the fault is. There are numerous faults developed in the field. According to the structure map of top $E_{1-2}km$, $K_1bs_3$ and $K_1bx_1$ of KeLa 2 Structure, the faults identified in Kela-2 have throws of several meters to several hundred meters, and majority of faults within gas bearing boundary have throws of less than 50 m except for a few faults in the SW flank which have throws of 60~100m. The statics show that there are 22 and 18 faults with the fault throws greater than the thickness of the $E_{1-2}km_4$, $K_1bs_3^1$ interlayer respectively; while only 1~2 faults in the SW flank have throws greater than the set of superimposed interlayer which is composed of the $K_1bx_1$ mudstone and overlying mudstone and tight conglomerate interlayers at the bottom of $K_1bs_3^3$. The set of superimposed interlayer are predicted to have stronger sealing potentiality than the others.

Analysis of the core and FMS logs reveals that there are natural fractures with different development degree in reservoirs of the field; fracture density is generally lower than 0.1/m except the KL203, KL2-14, KL2-J203 and KL2-12 which have a fracture density of about 0.4/m; otherwise fractures within the muddy interlayers are rare, small and are mainly filled with calcareous material. Further study of the relationship between fractural parameter and stress variation suggests that the maximum thickness for the fractures extending in the mudstone is 4 meters[15]; therefore, the fracture has little effect on the sealing performance of the three sets of continuously distributed interlayers described above, but the fractures developed with a relatively higher density in the sandstone are predicted to increase the effect of bottom water production.

The pressure test analysis. The pressure test results show the gas reservoirs characteristic of good internal connectivity during development. KL2-H1 is a horizontal well completed in 2008, and the target is Dolomite Member ($E_{1-2}km_3$) in which other gas wells are not perforated. The pressure test results show that the formation pressure drop are more or less the same as the other production formation in other gas wells, which indicates that the gas sealing ability of the $E_{1-2}km_4$ interlayer is poor. The pressure data from KL2-J203, Ks603 and Ks604 show that the pressure differences between the lower and upper reservoirs of the $K_1bs_3^1$ interlayer is only 0.2~0.5MPa, therefore the fluid sealing ability of the $K_1bs_3^1$ interlayer is also very small; while the pressure difference of 13.5MPa between the upper and lower reservoirs of the $K_1bx_1^1$ interlayer from Ks603 illustrates that this set of interlayer weakens the energy of the bottom water and limits the bottom water invasion.

# 4 Effects of interlayer on water invasion

## 4.1 Characteristics of water invasion

Kela 2 gas field structure belongs to an anticlinal deep net gas reservoir with edge and bottom water. Since it was put into development in December 2004, there have been 9 gas wells producing water (KL2-14, KL203, KL2-13, KL2-8, KL205, KL2-12, KL2-1, KL2-10, KL204). Production tests such as fluid production profile and various saturation tests were carried out in the 9 water produced wells and in an non-water produced well (KL2-11), as well in a inspection wells (KL2-J203 drilled in 2015), which can be used to interpret the water invasion; Besides, there are logging data from 3 wells including a inspection wells (KL2-J3) drilled in 2019 and 2 wells of which the drilling target is adjacent Keshen 6 block at footwall of the Kela fault, and therefore drilled through the target formation of Kela 2 gas field (Ks 603 and Ks 604, drilled in 2015) which can be used to interpret the change of gas water interface. These data indicate that water invasion presents seriously inhomogeneous in the gas field. The rise height of edge and bottom water varies greatly in different regions. The rise height in the SW flank and in the east flank is significantly higher than that in other areas. The gas water interface rose about 300m in the southwest and the east flank, 142m in the North flank and about 90m in the south central area. According to the wells in the west and east regions (e.g. KL2-14, KL2-12 and KL2-10), water intrusion is seriously inhomogeneous along various formations (Fig. 4). Such performance are largely determined by the geological conditions of the reservoir, such as the distribution of interlayer, fault, fracture and relatively high permeable layer, and their crosscutting relationship. Interlayer play a important role in the water invasion.

## 4.2 Effect of interplayers on water invasion

Of the geologic factors which influence the water invasion in Kela 2 gas field, the interlayer has a considerable effect on reducing the bottom water intrusion; while the main function of fault is to break the barrier interlayer leading to the bottom water intrudes vertically along the fault; and some

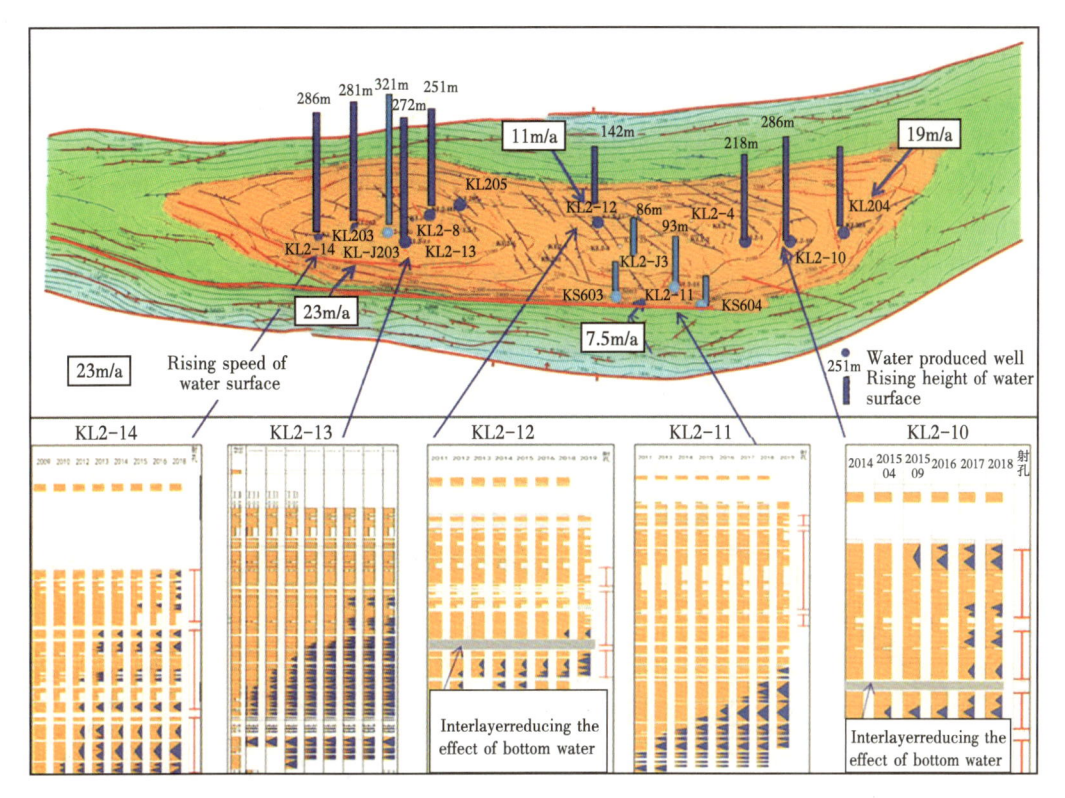

Fig. 4  Figure showing the water invasion performance in Kela 2 gas field (The figure above shows the lifting height and speed of gas water interface, and the figure below shows the production test profile of production wells in the recent years)

faults developed in the South of the gas reservoir also have the ability of sealing edge water; in addition, the locally developed fractures can strengthen the invasion of bottom water, and the relatively high permeability formation is conducive to the laterally inhomogeneous water invasion.

The SW area of Kela 2 gas field is the extremely deformed and thinned forelimb of the fold with steep southerly dip. Interpretation of seismic data show that the area is characterized by having a high density of seismic and sub-seismic faults. The $K_1bs_3^1$ and $K_1bx_1$ interlayers are both penetrated by faults, leading to the sealing ability of the interlayers are greatly weakened (Fig. 5b). the core and FMS log data reveal that natural fracture density is high in this area, which increase the bottom water production. Besides, the edge water in the west, the volume of which is about twice of the gas r reservoir, can provide large elastic energy in the early-middle stage of development and elastic energy will reduce owing to the energy release with the gas reservoir development. Thus, bottom water breakthrough the interlayers and ascends quickly as a whole along faults, fractures, in the early stage of development, and water intrusion gradually slows down with the production of gas. No Obviously fracture heterogeneous water channeling appears in the area owing to the numerous of interlayers restraining the coning of bottom water; while as there occurs inhomogeneous water intrusion along the high permeability zone in the edge area such as in the KL2-14 (Fig. 4).

In the east area, the reservoir properties are much better than that in the west area. The mean porosity of the sandstones in the west area is 12%, and the mean permeability is about 7 mD; while those in the east area are individually about 15% and about 35 mD. The $K_1bs_3^1$ and $K_1bx_1$ interlayers are both emerged under the gas water interface (Fig. 5a). The pressure test shows the Bashijiqike Formation in the Kela 2 gas field is connected with that in the Kela 3 gas field 20km away from Kela 2 gas field, therefore there is a large edge water volume which is about 6 times volume of gas reservoir. Thus, inhomogeneous water intrusion along the high permeability zone dominate the gas well in the east area. Take KL2-10 for an example, edge water fingering is very obvious. The formations at the top and bottom of the well are medium high water intrusion, while those at the middle are only three layer water flooded, and are medium low water flooded. In addition, it appears the interlayer restraining the bottom water vertical invasion in the KL2-10 well (Fig. 5a).

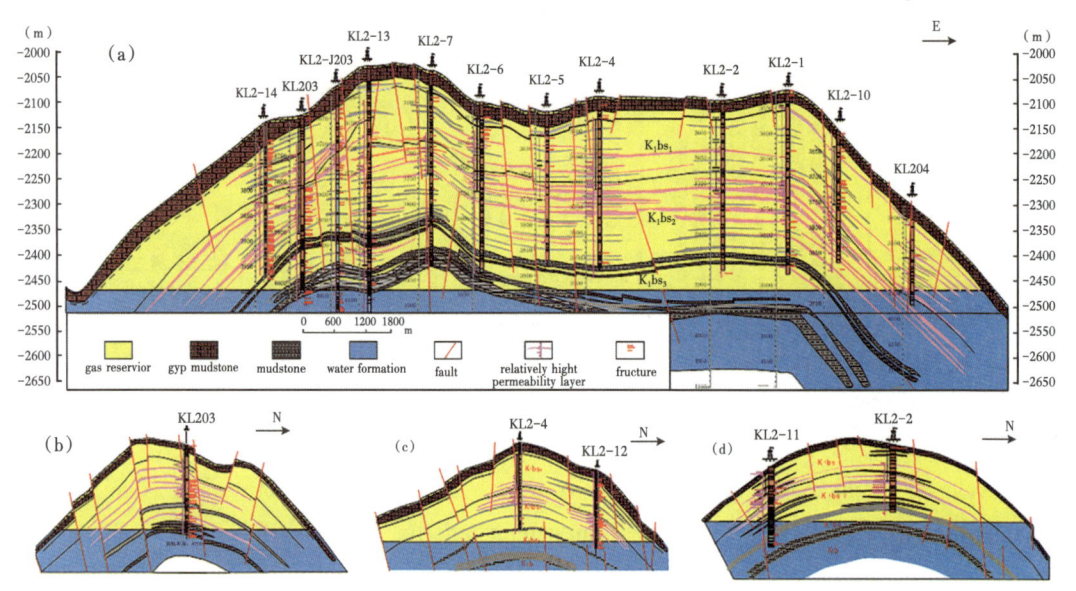

Fig. 5 Figure showing EW and SN gas reservoir profiles with characteristics of interlayers, faults, fractures, as well as relatively high permeable formations

Affected by the interlayer, the characteristics of lateral water intrusion in the northern region are obvious. Take KL2-12 as an example, faults are developed in small scale and fractures are developed with a density of about 0.47/m, which lead to the water rapidly rising nearly vertically. $K_1bs_3^1$ and $K_1bx_1$ interlayers are both emerged under the gas water interface and distance between perforation zone and edge water is 290 meter (Fig. 5c), which can illustrate the two interlayers have little effects on the bottom water intrusion vertically; where as there is a interlayer with a thickness of about 7m at interval 3802~3807m with no fracture, which strongly inhibit the invasion of bottom water. The bottom water hasn't broken the barrier in four years (Fig. 4), therefore the lateral water intrusion dominates the gas well in the recent years. In the northern region, the volume of edge water is much smaller than that of the East and West flanks, and the energy of edge and bottom water is weak. The well in the area can produce with low water gas ratio for a long time.

In the south central area, the edge water is blocked by a sealing fault (Fig. 4 and Fig. 5d). The pressure data shows that the pressure difference between the upper and lower walls of the fault is about 20MPa. meanwile, energy of bottom water is largely weaken by the $K_1bx_1$ interlayer. Furthermore, the density of fracture in this area is low. Thus the bottom water rises slowly. It can be considered that the bottom water in this area is not terrible, but the horizontal invasion of formation water from adjacent areas brings risks to the gas field development. Measures such as adjusting individual well production and deploying horizontal well wells with relatively low permeability formations as target should be taken to weaken the lateral invasion of formation water in the adjacent area.

# 5 Conclusions

There are 3 set of interlayers extend over the whole field, which are the Gypsum mudstone member ($E_{1-2}km_4$) of the Kumugeliemu Formation with a thickness of 14~44m., the first Sub-member ($K_1bs_3^1$) of the Member III within the Bashijiqike Formation with a thickness of 15~20m, and the mud stone Member ($K_1bx_1$) of the Baxigai Formation with a thickness of 11~22m. They divide the reservoir of Kela 2 gas field into a dolomite units and 3 Clastic units. The 3 Clastic units are composed of thick-very thick-bedded sandstone inter-bedded with discontinuously distributed mudstone interlayer with an individual thickness generally less than 2m, and <2% of the total interbeds are greater than 5m thick which are mainly at the bottom of the $K_1bs_3$ Overlying the $K_1bx_1$. Both the discontinuous interlayers and the continuously distributed interlayers do not have complete sealing property due to the penetration of fault to interlayer. The greater the thickness of mudstone joint at the two sides of the fault is, the stronger the ability to prevent the vertical migration of fluid along the fault is. The $K_1bx_1$ interlayer, together with the interlayers at bottom of the Bashijiqike Formation has a stronger sealing ability than the others.

The interlayer, fault, fracture and relatively high permeable layer, and their crosscutting relationship are important geological factors affecting the water invasion. Controlled by the geological characteristics of the gas field, water invasion laws are varied in the different regions. In the west and east regions, Edge water can induce the seriously inhomogeneous water invasion through the relatively high permeability layer; in the north area, the lateral water intrusion dominates the gas well due to a thickness interlayer inhibition to bottom water verticaly asends; while in the south-central region, the bottom water invasion is relatively moderate because the thick-bedded interlayer weakens the energy of water from deeper bed; in addition, in the SW region, bottom water breakthrough the interlayers and ascends quickly as a whole along faults, fractures, and the relatively high permeability layer, which led to water flooding of gas wells in the area.

## References

[1] Zhang Changmin, Yin Taiju, Zhang Shangfeng, et al. Hierarchy analysis of mudstone barriers in Shuanghe Oilfield. Acta Petrolei Sinica. 2004,25(3):48-52.

[2] Yu Xinghe, Ma Xingxiang, Mu Longxin, et al. Reservoir geology model and analysis of hierarchy surface. Petroleum Industry Press,2004.

[3] Lynds R, Hajek E.Conceptual model for predicting mudstone dimensions in sandy braided river reservoirs. AAPG Bulletin. 2006,90(8):1273-1288.

[4] Yin Senlin, Wu Shenghe, Feng Wenjie, et al.Muddy interlayer style characterization of alluvial fan reservoir: A case study on lower Karamay Formation, Yizhong Area, Karamay Oilfield. Petroleum Exploration and Development. 2013,40(6):757-763.

[5] Sun Tianjian, Mu Longxin, Zhao Guoliang. Classification and characterization of barrier-intercalation in sandy braided river reservoirs: Taking Hegli Oilfield of Muglad Basin in Sudan as an example. Petroleum Exploration and Development. 2014,41(1):112-120.

[6] Zou Zhiwen, Si Chunsong, Yang Mengyun. Origin and distribution of interbeds and the influence on oil-water layer:An example from Mosuowan area in the hinterland of Junggar Basin. Lithologic Reservoirs. 2010,22(3): 66-70.

[7] Cui Jian, Li Haidong, Feng Jiansong, et al.Barrier-beds and inter-beds characteristics and their effects on remaining oil distribution in braided river reservoirs: A case study of the Ng IV oil unit in shallow north Gaoshangpu oilfield. Special Oil & Gas Reservoirs. 2013,20(4):26-31.

[8] Wang Min, Zhao Guoliang, Feng Min, et al.Distribution pattern of intercalations and its impact on migration of edge and bottom water in sandy braided-river reservoirs-A case study of Fal structure in P Oilfield, South Sudan. Petroleum Geology and Recovery Efficiency. 2017,24(2):8-21.

[9] Jia Jinhua, Gu Jiayu, Guo Qingyin, et al.Sedimentary facies of cretaceous reservoir in Kela 2 gas field of tarim basin. Journal of Palaeogeography. 2011,3(3):67-75.

[10] Jiang Tongwen, Zhang Hui, Wang Haiying, et al.Effects of faults geomechanical activity on water invasion in Kela 2 gas field, Tarim Basin. Natural Gas Geoscience. 2017,28(11):1735-1744.

[11] Feng Zhendong, Dai Junsheng, Deng Hang, et al.Quantitative evaluation of fractures with fractal geometry in Kela -2 gas field. Oil & Gas Geology. 2011,32(54):928-939.

[12] Tian Zuoji, Song Jianguo. Tertiary structure characteristics and evolution of Kuqa foreland basin. Acta Petrolei sinica. 1999,20(4):7-13.

[13] Xu Zhenping, Li Yong, Ma Yu-jie, et al.Future gas exploration orientation based on a new scheme for the division of structural units in the central Kuqa Depression, Tarim Basin. Geology and exploration. 2011,31(3), 31-36.

[14] Fu Guang, Li Feng-jun, Bai Ming-xuan. Analysis of the relationship between lateral sealing and vertical sealing of faults.Petroleum Geology & Oilfield Development in Daqing. 1998. 17(2), 9-12.

[15] Zhang Hongguo, Dai Junsheng, Feng Zhendong, et al. Study of shale interlayer sealing in Kela-2 gas field. Xinjiang Petroleum Geology. 2011,32(4), 363-365.

# 基于数值试井法的神木气田多层压裂气井产能评价

刘姣姣[1,2],刘志军[1,2],刘 倩[1,2],左海龙[1,2]

(1. 中国石油长庆油田分公司勘探开发研究院;2. 低渗透油气田勘探开发国家工程实验室)

**摘 要**:为准确、快速评价多层压裂气井产能,针对目前常规评价方法的局限性,利用数值试井法建立数值模型,模拟修正等时试井过程,评价气井产能。研究表明:(1)通过模拟,等时测试阶段的合理生产间隔受储层渗透率影响较大,当气井储层有效渗透率小于0.2mD时,测试时间间隔定在36~48小时较合理;有效渗透率大于0.2mD时,测试时间间隔定在24小时较为合理,产量序列依次按照试气无阻流量$q_{AOF}$的10%、20%、40%、60%确定;在延续测试阶段,合理延续时间以30天为宜,以$q_{AOF}$的30%为测试产量,根据该原则确定试井工作制度,以获得准确的产能方程。(2)数值试井法与矿场修正等时测试结果相比,相对误差低于10%,评价结果可靠,为多层压裂气井产能评价提供了一种实用的技术方法。

**关键字**:多层压裂;致密气藏;历史拟合;产能评价

神木气田是典型的多层系致密砂岩气藏,纵向发育多套含气层系,但优势层不突出,储层致密、非均质性强,单层产量低[1,2]。以单层增产、多层动用为目的,引入分层压裂多层合采技术[3],能有效提高气井产能,改善开发效果。那么,针对多层压裂直井的产能评价就成为开发人员的首要工作,其准确性直接关系到气田的产能部署、气井生产制度制定等。

前人针对多层压裂气井的产能评价研究,较多是对分层产量动态变化规律开展研究[4-7],而如何快速评价气井合采产能是现场最关心的问题。一般来说,气井产能主要通过两种方法确定:解析法和矿场试验法[8]。气井经过多层压裂改造后,近井地带气体渗流符合高速非达西渗流规律,产能方程满足二项式[9]。由于多层系气藏层间干扰现象以及压裂裂缝的复杂性,并不能准确推导出方程系数$A$、$B$值的表达式[10-12],所以解析法评价多层压裂气井产能存在一定的局限性。一点法测试是现场产能测试中最常用的方法,式中的$\alpha$值表征储层非均质性,属于气田经验值[13],对于多层系气藏,砂体连通性差,气井射孔层位、层数的不确定性,很难准确确定出反映气田特征的$\alpha$值,所以一点法对于多层系气藏误差较大。而实施修正等时试井可以得到较准确的产能公式,但是存在耗时长、费用高等缺点[14,15]。

针对目前常规方法评价多层压裂气井产能的局限性,本文提出数值试井方法,即利用气井日常生产数据($q$、$p$)进行拟合分析,获取较准确的储层及井筒参数,在此基础上利用数值试井设计,模拟修正等时试井过程,最终评价气井产能,既节省了实际测试时间,又节约了开发成本。

## 1 理论基础

经压裂措施改造后,多层系致密砂岩气藏气井获得工业气流,压裂裂缝为气体主要渗流通道[16,17]。气体的流速非常高,紊流造成的附加压降不可忽略,此时达西定律已经失效,渗流更符合"Forchheimer"定律,即

$$\frac{\mathrm{d}p}{\mathrm{d}r} = \frac{\mu}{K}v + \beta'\rho v^2 \tag{1}$$

式中　$p$——压力,MPa;

　　　$r$——距离,m;

　　　$\mu$——流体黏度,mPa·s;

　　　$K$——地层有效渗透率,mD;

　　　$v$——流体的渗流速度,$m^3/d$;

　　　$\beta$——湍流系数,$m^{-1}$;

　　　$\rho$——流体密度,$kg/m^3$。

因此,多层系致密砂岩气藏直井的产能公式满足二项式形式:

$$\psi(p_e) - \psi(p_{wf}) = Aq + Bq^2 \tag{2}$$

式中　$\psi(p_e)$,$\psi(p_{wf})$——边界拟压力、井底流压拟压力,$MPa^2/(mPa·s)$;

　　　$q$——气井标准状况下的产气量,$10^4 m^3/d$;

　　　$A$——层流项系数;

　　　$B$——紊流项系数。

由式(2)可知,为求取气井绝对无阻流量$q_{AOF}$,需要确定准确的$A$、$B$值。本文提出数值试井方法评价多层压裂气井产能,其具体流程如图1所示。借助Ecrin软件平台,通过对气井生产数据的拟合,得到气井储层及井筒相关参数,建立气井数值模型;在此基础上模拟修正等时试井过程,确定产能方程。

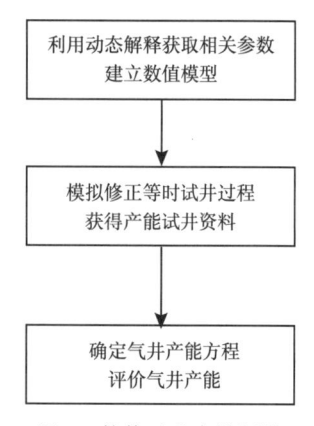

图1　数值试井产能评价方法流程图

## 2　模型参数的确定

应用数值试井法评价多层系气藏直井产能,首先要建立可靠的数值模型。由于实际有效渗透率与生产测井解释得到的参数有较大差别,而且压裂施工后并没有对裂缝参数(如裂缝半长、裂缝导流能力)进行监测,因此,快速有效地确定模型基础参数是确定气井产能方程的第一步。目前对有效参数值的获取主要通过不稳定压力恢复试井解释得到,但是对于神木气田这种典型致密气藏,压力传播慢,试井方法存在耗时长、费用高的局限性。因此,本文依托Ecrin动态流动分析平台下的Topaze生产分析软件,在考虑气井多层地质模型的基础上,采用产量不稳定分析法对气井生产数据(产量和压力历史)进行分析拟合,实现不关井条件下获取气井储层及井筒相关参数。Topaze生产分析软件中的解释分析成果可发送至同平台下的Saphir试井解释软件中,直接进行产能试井设计,方便快捷。

### 2.1　单井数值模型的建立

神木气田主力层段为石盒子组盒8段、山西组山1段和山2段以及太原组,主要发育三角

洲平原分流河道,砂体相互叠置呈条带状[18],近南北向展布,开发方式采用衰竭式开采,无能量补充,因此所建立的模型采用封闭边界,控制区域为长条形地带,裂缝设为有限导流裂缝。

以实际井为例进行分析,双 A 井所射穿层位为 4 层,各生产层之间有良好的非渗透隔层。在 Topaze 生产分析软件的多层模块中,给定每个小层物性参数"初值",建立多层气井数值模型。由于储层致密不能自然建产,各生产层均实施压裂改造,初步设计裂缝半长为 90m,裂缝导流能力为 20mD·m。考虑神木气田砂体规模,初选取 800m×600m 长条形封闭区域为单井控制区域,建立 PEBI 网格模型,气藏内单元格尺寸为 30m,裂缝周围逐渐加密(其中裂缝改造参数、控制面积大小最终在拟合结果中获得)。图 2 为双 A 井多层数值模型示意图。

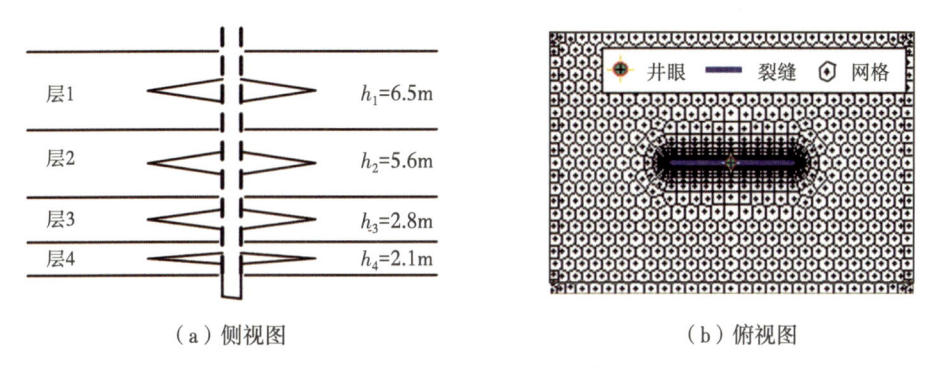

（a）侧视图　　　　　　　　　　（b）俯视图

图 2　气井多层数值模型示意图

## 2.2　开发动态历史拟合

气井生产动态历史拟合的过程实际上是验证模型的过程,不断修正地质模型和物性参数,从而得到更符合实际情况的数学模型。Topaze 从渗流理论出发,模型中考虑流量变表皮的影响,建立产量和压力之间的关系。如图 3 所示,日产气量、累计产气量拟合曲线与实际生产曲线相符度较高,井底流压变化趋势与气井实际情况一致。在生产过程中,气井井底流压不断下降,这是由于气井有效控制范围内的平均地层压力不断下降,其下降速率与控制半径有关。

另外,Topaze 生产分析软件的一个优势是可以应用 Log-Log 和 Blasingame 典型曲线为诊

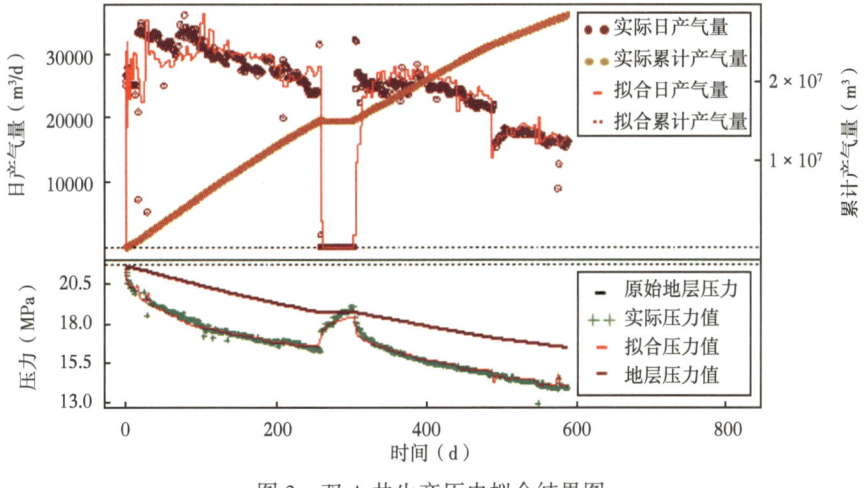

图 3　双 A 井生产历史拟合结果图

断手段,辅助识别致密储层流动阶段及其特征,描述井的状态。其原理是将变化的压力、流量进行处理,转化成等效的稳定压力和稳定产量,对重整流量和压力进行积分和积分求导,得到规整化的产量和压力,在双对数坐标曲线上会出现类似于试井解释中的拟稳态状态的特征线段。因此,对比双 A 井的 Log-Log 和 Blasingame 典型拟合曲线,可以验证生产历史拟合得到的模型合理性。如图 4 所示,根据双 A 井规整化压力积分与积分导数曲线后半段的 45°线,其表征气井进入拟稳定流动阶段,可反演评价气井控制面积,调整模型中的控制面积参数直至拟合曲线与实际曲线相符。如图 5 所示,规整化产量、累计产量积分与积分导数曲线在不稳定流动阶段的斜率受模型渗透率、控制半径等参数的影响,通过调整参数直到三条拟合曲线的变化趋势与实际曲线一致。在 Topaze 生产动态分析软件中,调整气井模型参数,气井生产历史曲线、双对数曲线和 Blasingame 典型曲线同时拟合,当三种曲线拟合符合率达到 90%,即可有效降低单一生产历史拟合带来的多解性,提高了模型准确度。最终确定双 A 井原始地层压力为 24.61MPa,

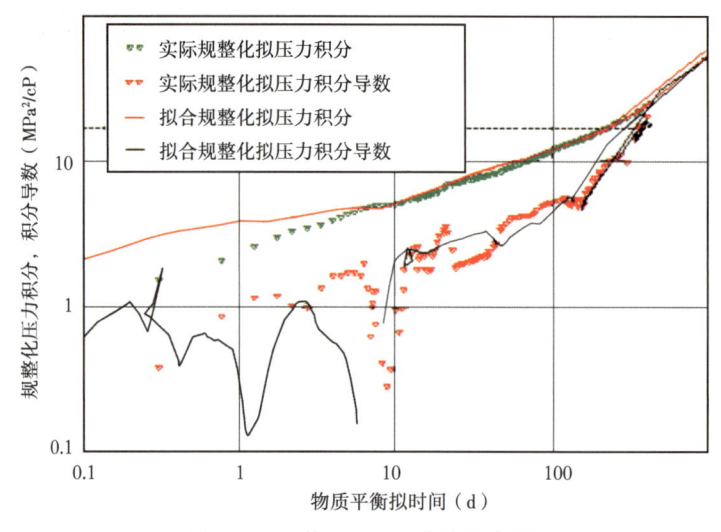

图 4 双 A 井 Log-Log 曲线拟合图

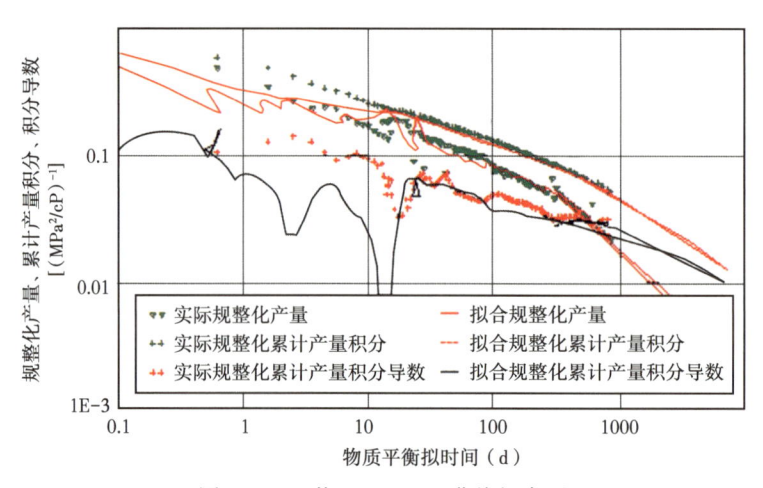

图 5 双 A 井 Blasingame 曲线拟合图

控制面积为 0.05km², 裂缝半长为 70m, 导流能力为有限导流(20mD·m), 其他相关参数具体数值见表 1。

表 1　双 A 井拟合参数解释表

| 原始地层压力 (MPa) | 控制面积 (km²) | 各层物性参数 层位 | 有效厚度 (m) | 渗透率 (mD) | 孔隙度 (%) | 井筒压裂改造参数 裂缝半长 (m) | 裂缝导流能力 (mD·m) |
|---|---|---|---|---|---|---|---|
| 24.61 | 0.05 | Layer 1 | 6.5 | 0.21 | 6.25 | 70 | 20 |
|  |  | Layer 2 | 5.6 | 0.05 | 7.39 |  |  |
|  |  | Layer 3 | 2.8 | 0.15 | 6.00 |  |  |
|  |  | Layer 4 | 2.1 | 0.40 | 6.93 |  |  |

## 3　数值试井法产能评价

通过上述气井开发动态历史拟合, 得到可信的相关地层参数及压裂改造参数值。将该模型同步到同平台下的 Saphir 试井分析软件中, 优化设计修正等时试井测试参数, 模拟产能测试过程, 确定产能方程 $A$、$B$ 值, 进而评价气井产能。

### 3.1　模拟产能测试制度的设计与制定

#### 3.1.1　产量序列的确定

在进行气井修正等时试井测试过程中时, 首先要确定产能测试的产量序列。一般来说, 修正等时试井测试产量序列是有一定要求的:(1)最低产气量大约是无阻流量 $q_{AOF}$ 的 10%;(2)最高产气量不应大于无阻流量 $q_{AOF}$ 的 75%;(3)等时测试阶段产量序列采取递增序列。

对于气田的实际开发, 气井投产前会进行一点法试气测试, 利用陈元千提出的一点法无阻流量公式[19]初步估算气井无阻流量, 即

$$q_{AOF} = \frac{2(1-\alpha)q_{sc}}{\alpha\left(\sqrt{1+4\left(\frac{1-\alpha}{\alpha^2}\right)p_D} - 1\right)} \tag{3}$$

$$p_D = \frac{p_R^2 - p_{wf}^2}{p_R^2}$$

式中　$q_{sc}$——气井试气日产气量, $10^4 m^3$;
　　　$\alpha$——气田经验参数, 神木气田 $\alpha$ 值为 0.86。

双 A 井投产前进行一点法测试求产, 原始地层压力为 24.6MPa, 在井底流压为 21.5MPa 条件下, 井口产量为 $2.93 \times 10^4 m^3/d$, 计算试气无阻流量为 $11.14 \times 10^4 m^3/d$。根据产量序列确定原则, 依次按照试气无阻流量 $q_{AOF}$ 的 10%、20%、40%、60% 确定等时测试阶段的产量序列;按照 $q_{AOF}$ 的 30% 确定延续流动阶段的产量。

#### 3.1.2　等时生产时间与延续生产时间的确定

神木气田储层致密, 要求模拟产能测试时的等时生产时间与延续生产时间达到一定数值,

以获得稳定的产能方程系数 $A$、$B$ 值。对于等时生产时间,开井流动影响应该超过井筒储集和措施改造区的范围,达到地层径向流的影响范围。随着测试时间延长,产能方程系数 $B$ 值不断增大,到一定时间时为一恒定常数。如图 6 所示,根据产能方程 $B$ 值与等时生产时间的关系曲线,确定双 $A$ 井合理等时生产时间为 24h。

在气井延续生产段,只有压力波传播到边界后,$A$ 值才为准确数值。若延续生产时间过短,将导致 $A$ 值过小,造成无阻流量偏大。如图 7 所示,根据产能方程 $A$ 值与延续生产时间的关系曲线,确定合理延续生产时间为 30 天。

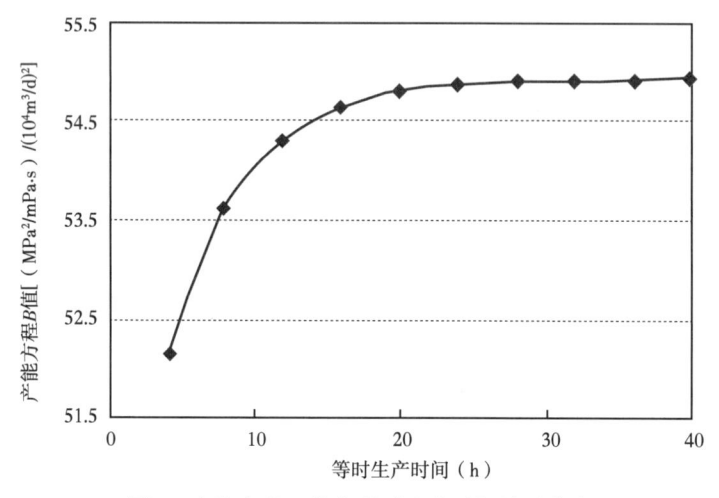

图 6　产能方程 $B$ 值与等时生产时间关系曲线

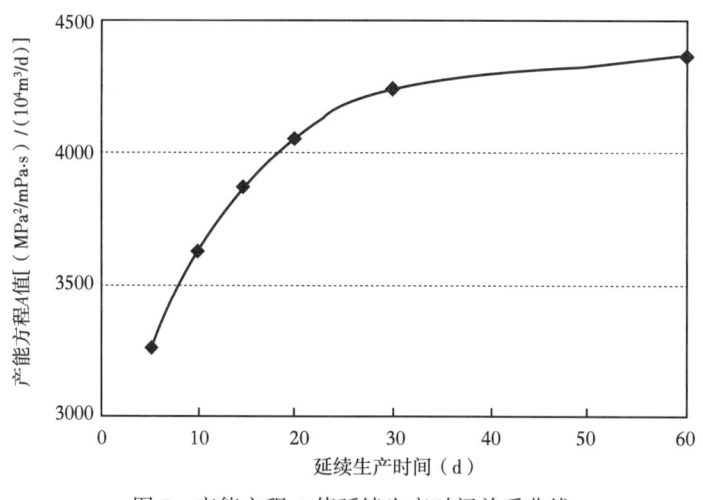

图 7　产能方程 $A$ 值延续生产时间关系曲线

通过模拟,统计 16 口典型井的合理等时生产时间在 18~48 小时,合理延续时间均为 30 天。地层渗透率对测试时间影响较大,如图 8 所示,绘制合理等时生产时间与气井有效渗透率（由开发动态历史拟合确定）关系曲线图。从图中可以看出:气井有效渗透率越高时,合理等时生产时间越短;当气井有效渗透率大于 0.2mD 时,测试时间间隔定在 24 小时较为合理;当气井有效渗透率在 0.2mD 以下时,测试时间间隔定在 36~48 小时较合理。

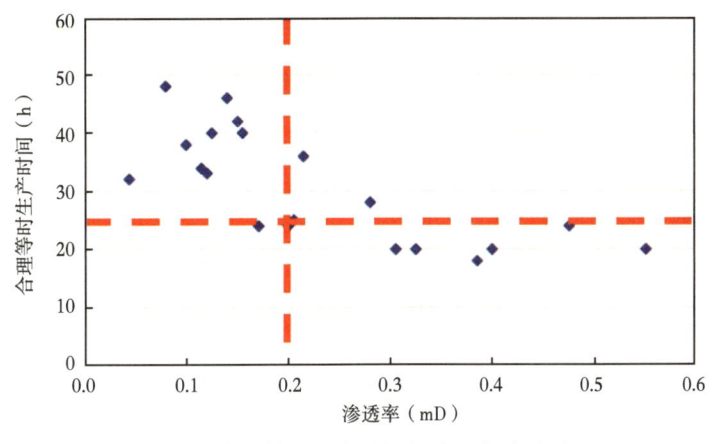

图 8　合理等时生产时间与渗透率关系图

## 3.2　二项式产能方程的确定

在双 A 井多层数值模型建立的基础上,应用 Saphir 试井分析软件,最终确定试井工作制度(表 2),模拟修正等时试井测试过程。图 9 为产能试井曲线,可获得产能试井资料。

表 2　双 A 井数值法模拟修正等时试井工作制度设计表

| 井号 | 等时阶段 ||||| 延续阶段 ||
|---|---|---|---|---|---|---|---|
| | 等时间隔 (h) | $q_1$ ($10^4 m^3/d$) | $q_2$ ($10^4 m^3/d$) | $q_3$ ($10^4 m^3/d$) | $q_4$ ($10^4 m^3/d$) | $q_g$ ($10^4 m^3/d$) | 生产时间 (d) |
| 双 A 井 | 24 | 1.0 | 2.0 | 4.0 | 6.0 | 3.0 | 30 |

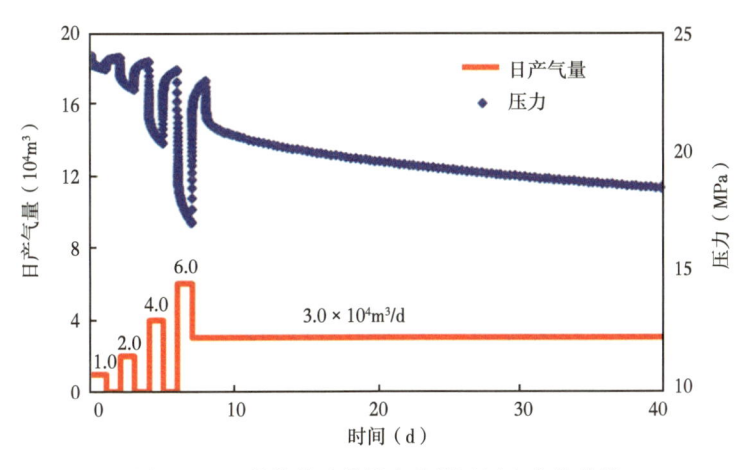

图 9　双 A 井数值法模拟产能测试压力变化曲线

通过模拟修正等时试井测试资料分析,得到其二项式产能曲线,如图10所示。

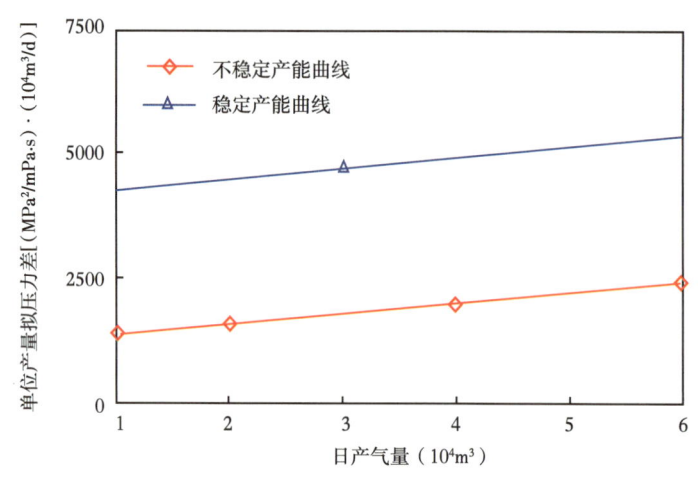

图 10 双 A 井数值法模拟二项式产能曲线

应用二项式方程回归得到双 A 井产能方程为

$$\psi_R - \psi_{wf} = 4233.71q + 54.8813q^2 \tag{4}$$

由式(3),当井底流压为 1 个大气压时,计算得到双 A 井的绝对无阻流量为 $9.09×10^4m^3/d$。2017 年 9 月,该井以 $1.0×10^4m^3/d$、$2.0×10^4m^3/d$、$4.0×10^4m^3/d$、$6.0×10^4m^3/d$ 四个工作制度进行矿场修正等时试井测试,延续阶段以 $3.06×10^4m^3/d$ 进行生产,试采 33 天,其原始地层压力为 24.61MPa,各工作制度下开井井底流压及关井井底压力数据见表3。

表 3 双 A 井实际矿场产能试井压力数据表

| 项目 | 日产气量 ($10^4m^3$) | 实际测试 井底流压(MPa) | 关井井底压力(MPa) |
|---|---|---|---|
| 第 1 等时阶段 | 1.0007 | 23.410 | 24.600 |
| 第 2 等时阶段 | 2.0143 | 21.660 | 24.364 |
| 第 3 等时阶段 | 4.0122 | 19.366 | 24.015 |
| 第 4 等时阶段 | 6.0028 | 16.887 | 23.595 |
| 延续流动阶段 | 3.0600 | 21.135 | 24.000 |

根据表 3 数据,确定双 A 井矿场测试的产能方程为 $\psi_R-\psi_{wf}=4101.56q+61.3548q^2$,所得无阻流量为 $9.54×10^4m^3/d$。对比发现,该井模拟测试得到的绝对无阻流量与实际测试结果相差仅为 4.9%,证明了应用数值试井法所得结果相对准确。

## 4 应用效果分析

神木气田历年修正等时产能测试共 16 井次,应用数值试井法分别建立每口井的数值模型、优化合理测试时间、制定相应的试井制度,最终得到气井的绝对无阻流量。如图 11 所示,数值试井法与矿场测试的结果对比,其相对误差均低于10%,说明该方法所得气井产能相对准确,能够满足工程计算的需要,可广泛应用于气井产能评价,有效降低多层系气藏经验公式评价产能的误差。

图11 神木气田数值试井法评价产能结果与矿场测试结果对比柱状图

数值试井方法为多层压裂气井产能评价提供了新的思路,因此,针对神木气田生产时间超过三年的气井,应用该方法评价产能480余口。图12为评价气井无阻流量柱状图,从图中可以看出,神木气田气井产能主要分布在$(5.0\sim15.0)\times10^4m^3/d$,平均为$10.4\times10^4m^3/d$,明确了气井的开发潜力,为气田气井生产制度的制定提供了技术支撑。

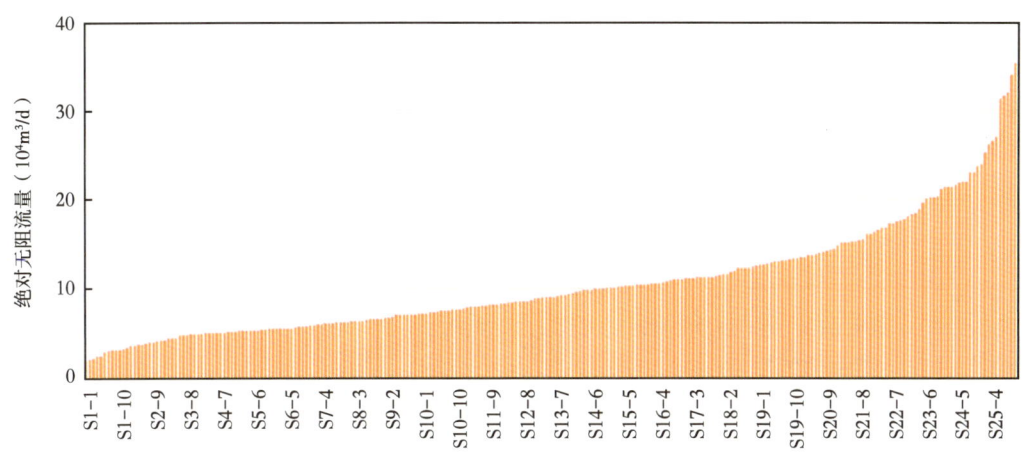

图12 神木气田数值试井法评价气井无阻流量柱状图

## 5 结论

(1)模拟修正等时试井过程时,等时测试阶段的合理生产时间间隔受储层渗透率影响较大,当气井储层有效渗透率小于0.2mD时,测试时间间隔定在36~48小时较合理;有效渗透率大于0.2mD时,测试时间间隔定在24小时较为合理,以获得稳定的气井产能方程。

(2)数值试井产能评价方法步骤简便易操作,相对于矿场试验,两者结果相对误差低于10%,准确度高且不必关井、耗时少,可快速有效评价多层压裂气井产能,指导气井生产。

## 参 考 文 献

[1] 王东旭. 多层系砂岩气藏大丛式井组开发布井技术研究[D]. 成都:西南石油大学,2017:1-10.

[2] 杨华,刘新社,闫小雄,等. 鄂尔多斯盆地神木气田的发现与天然气成藏地质特征[J]. 天然气工业,2015,35(6):1-13.

[3] 王亚娟,张燕明,何明舫,等. 神木气田多层系致密砂岩气层增产技术研究[J]. 油气井测试,2017,26(3):66-68.

[4] 王晓东,刘慈群. 分层合采油井产能分析[J]. 石油钻采工艺,1999,21(2):56-61.

[5] 赵广民. 多层合采气井产能优化分析[D]. 北京:中国地质大学,2005:17-30.

[6] 冯毅,魏攀峰,段长江,等. 室内定量试验评价临兴地区致密砂岩气两层合采产量变化[J]. 非常规油气,2017,4(6):40-44.

[7] 王文举,潘少杰,李寿军,等. 致密气藏高低压多层合采物理模拟研究[J]. 非常规油气,2016,3(2):59-64.

[8] 庄惠农. 气藏动态描述和试井[M]. 北京:石油工业出版社,2004:56-102.

[9] 刘方玉,马华丽,蒋凯军,等. 压裂后气井的产能评价方法分析[J]. 油气井测试,2010,19(5):35-38,47.

[10] Liu C Q, Wang X D. Transient 2D flow in layered reservoirs with crossflow[A]. SPE-25086-PA,1993.

[11] Pascal H, Kingston J D. Analysis of vertical fracture length and Non-Darcy flow coefficient using variable rate test[A]. SPE9348,1980.

[12] Noman R, Archer S J. The effect of pore structure on Non-Darcy gas flow in some low-permeability reservoir rocks[A]. SPE16400,1987.

[13] 徐云林,辛翠平,施里宇,等. 利用不稳定试井资料对延安气田进行初期产能评价[J]. 非常规油气,2018,5(6):62-69.

[14] 何逸凡,廖新维,徐梦雅,等. 低渗致密气藏压裂水平井产能公式确定新方法[C]. 油气藏监测与管理国际会议论文集,2011.

[15] 杨波,唐海,周科,等. 多层合采气井合理配产简易新方法[J]. 油气井测试,2010,19(1):66-68.

[16] 刘秉谦,张遂安,李宗田,等. 压裂新技术在非常规油气开发中的应用[J]. 非常规油气,2015,2(2):78-86.

[17] 吴则鑫. 苏里格气田致密气井产能主控因素分析[J]. 非常规油气,2018,5(5):62-67.

[18] 孟德伟,贾爱林,郭智,等. 致密砂岩气藏有效砂体规模及气井开发指标评价[J]. 中国矿业大学学报,2018,47(5):1046-1054.

[19] 黄炳光,李小平. 气藏工程分析方法[M]. 北京:石油工业出版社,2004:5-128.